Behavioural Responses to a Changing World

Behavioural Responses to a Changing World

Challenges and Applications

EDITED BY

Bob B. M. Wong

and

Ulrika Candolin

OXFORD
UNIVERSITY PRESS

Great Clarendon Street, Oxford, OX2 6DP,
United Kingdom

Oxford University Press is a department of the University of Oxford.
It furthers the University's objective of excellence in research, scholarship,
and education by publishing worldwide. Oxford is a registered trade mark of
Oxford University Press in the UK and in certain other countries

Published in the United States of America by Oxford University Press
198 Madison Avenue, New York, NY 10016, United States of America

British Library Cataloguing in Publication Data
Data available

Library of Congress Control Number: 2023948860

ISBN 9780192858979
ISBN 9780192858986 (pbk.)

DOI: 10.1093/oso/9780192858979.001.0001

Printed and bound by
CPI Group (UK) Ltd, Croydon, CR0 4YY

Links to third party websites are provided by Oxford in good faith and
for information only. Oxford disclaims any responsibility for the materials
contained in any third party website referenced in this work.

The manufacturer's authorised representative in the EU for product safety is
Oxford University Press España S.A. of el Parque Empresarial San Fernando de
Henares, Avenida de Castilla, 2 – 28830 Madrid (www.oup.es/en).

For Olivia, Lauren, Connor, Harper, Lily, Alice, Linnéa, and Gustav.

Foreword

Why does behaviour matter? It matters for understanding how animals respond to their environment, and for understanding the variation among taxa in crucial life history characteristics, such as parental care, foraging, or predator avoidance. Until recently, however, behaviour has not been a focus for scientists studying the effects of anthropogenic change on organisms. Ecologists and conservation biologists were more concerned about population-level alterations due to habitat fragmentation, or the way that communities changed in response to climate variation.

Yet for animals, behaviour is the first interface with the environment, a link between physiology and the outside world, and hence an essential tool in coping with environmental change. This timely and important book explores what happens to animal behaviour when humans change the world, whether by creating forest fragments exposed to traffic, lighting the night sky, or introducing new pathogens. Because of its inherent plasticity, behaviour can take the lead in adaptation to such altered circumstances. The companion volume to this one, *Behavioural Responses to a Changing World: Mechanisms and Consequences*, laid the groundwork for understanding what those behavioural responses are and how they affect animal populations. This book takes those findings and expands upon them, taking advantage of modern genomic and technological tools.

It can sometimes seem as if works on the effects of anthropogenic change on biodiversity are simply cataloguing disaster, modern-day doomsday books full of hopelessness. This volume is different. In addition to documenting behavioural change, the chapters go further and make concrete recommendations for ways to ameliorate harm. For example, as temperatures rise, shade canopies can help birds and other small animals drink at the water's edge without experiencing deadly heat. Noise pollution can be mitigated by using bubble curtains around the monopiles in windfarm construction, decreasing the temporary loss of habitat for harbour porpoises. Other suggested actions include education programmes discouraging the public from feeding feral pigeons, simple changes that can be remarkably effective.

The effects of human-caused change on behaviour can be subtle, and are sometimes non-intuitive, offering a good lesson in not making too many assumptions about which animals will be harmed by human activities and how that harm might happen. The chapter on harvesting (hunting and fishing) is a case in point: although it seems intuitive that greater activity levels in targeted animals would increase their encounter rate with harvesters and hence their risk, that isn't necessarily the case. Brown bears that moved less, for instance, were more vulnerable to hunters, perhaps because more active bears were better able to avoid people. Counterexamples also exist, of course, but the message is that because such behavioural responses are so variable and harvesting activities—and probably other anthropogenic environmental changes—so complex, fixes may be harder than we might have thought.

That behavioural complexity also leads many of the authors in the book to question other assumptions, such as the ability to generalize from experiments done in captivity to an animal's situation in the wild. For instance, we often examine heritability in the lab, and it can be hard to know how such a genetic component of behaviour translates to more natural conditions. What does that mean for the operation of natural or sexual selection in the field, and the implications for evolutionary adaptation of behaviour to environmental change? Several of the chapters point both to places where our understanding of the genetic mechanisms behind behavioural responses has improved our ability to predict further change, and to large gaps in our ability to connect genetic changes to behaviour. Improvements in genomic techniques, for example, have made it possible to look at detailed differences in adaptation among populations subject to different anthropogenic pressures.

The broad approach in the chapters also shows how animal behaviour influences far more than an individual animal's actions. Even plants are affected by anthropogenic changes to animals, perhaps most notably in the case of pollinators. The effect of changes to plant traits on pollinators has long been known, but changes in pollinator behaviour can similarly alter plant characteristics. Movement of animals because of anthropogenic change can also move their pathogens and parasites, further influencing the ecosystems they encounter.

The book also draws new connections between behaviour and topics usually reserved for discussions of community and population ecology, helping us see that what individual animals do reflects changes in habitats at a much greater scale. Habitat loss and fragmentation affect how animals interact, from dominance behaviour to movement patterns, and if behaviour is sufficiently flexible, those new interactions can still allow populations to persist. Individual differences in exploratory propensity,

for example, might mean that at least some members of a species colonize new fragments, ensuring population continuity, though this solution is by no means guaranteed. Habitat restoration is sometimes seen as a panacea to fragmentation or habitat loss, but if the restored sites contain appropriate food and shelter but do not provide for a species' behavioural proclivities, the animals may not persist. Solutions may include, for instance, bridges between fragments that allow a species to maintain its characteristic movement patterns.

Finally, the volume helps us gain insights into broader principles of animal behaviour. Animals in cities often behave differently from their less urban counterparts, and of course some species— raccoons, coyotes, mynas—are more likely to settle in urban areas to begin with. Why is that? Careful comparisons of urban and non-urban taxa can help us understand how to ameliorate human-animal impacts (a term preferred to 'conflicts' because it does not imply antagonism on either part). Furthermore, studying those animals that do succeed in cities sheds light on broader issues such as the evolution of cognition and the flexibility of circadian rhythms. We can hope that the work highlighted in this book will give us tools to both understand the behaviour of animals in a changing world and give those animals a chance to persist.

Marlene Zuk
Department of Ecology, Evolution and Behavior,
University of Minnesota, Minneapolis, USA

Preface

Human activities pose existential challenges to wildlife and ecosystems. From large-scale agricultural expansion to the discharge of pollution into our rivers and seas, humans have modified the global environment in unprecedented ways. In many species, behavioural adjustments are often the first response to rapid environmental change, enabling wildlife to quickly improve their survival and reproductive prospects when their habitats are altered. In fact, some species are so adept at taking advantage of anthropogenic change that they are now prospering under human-modified conditions.

But not all behavioural adjustments are beneficial. Wildlife, for example, can easily be led astray by 'ecological traps', when altered conditions cause animals to make poor behavioural decisions. Environmental disturbances, such as noise from busy city roads or exposure to hormone-mimicking pollutants, can also cause sensory and physiological impairments that undermine the capacity of animals to mount appropriate behavioural responses. And even when behavioural adjustments are possible, they may not always be enough to counter the scale and magnitude of the disturbances that are taking place.

More broadly, by affecting who lives and who dies, wildlife behaviour can have wide-ranging repercussions for the persistence of populations and, ultimately, the survival of species. And because species are connected through complex ecological networks, such changes have the potential to cascade well beyond the initial disturbance to affect the health and stability of entire ecosystems. Behavioural responses (or lack thereof) can also have important evolutionary implications by affecting the ability of organisms to adapt to changes in the longer term, and by altering the very

mechanisms that underpin the process of speciation itself. Such profound consequences highlight the critical role that behaviour plays in determining why some species are able to thrive in a changing world while others spiral into oblivion.

Clearly, behaviour matters. But how can an understanding of wildlife behavioural responses to anthropogenic change—and knowledge of animal behaviour more generally—contribute to solving real-world problems? This, in many respects, is the 'million dollar' question. Some commentators argue that, as a discipline, behavioural ecology has been far too preoccupied with pursuing theoretical and intellectual advances into our understanding of animal behaviour, while making very few practical contributions to conserving and managing vulnerable wildlife populations. While progress has been historically slow, with the emergence of new active fields of research such as conservation behaviour, the tide is surely turning. And the timing could not be more pressing, especially given novel and incipient threats, from rising concerns over worldwide pollinator declines to the ecological and human health impacts of emerging infectious diseases.

This book focuses on the many challenges facing wildlife and their habitats, and how a deeper understanding of animal behavioural responses can be used to predict and mitigate the anthropogenic impacts. The book is a companion to our earlier volume, *Behavioural Responses to a Changing World: Mechanisms and Consequences*. Since the publication of that first book more than a decade ago, there has been a massive surge in interest and research on the topic. The current volume builds on the foundations of our first—but with a translational perspective that focuses on how different forms of anthropogenic change can impact behaviour, and how we can better harness this understanding to deliver practical and effective solutions to palpable risks and challenges of global environmental concern.

The book is organized into two parts. Part 1 focuses on the myriad environmental challenges that wildlife are being subjected to, including the impacts of rising CO_2 levels and climate change, pollution, biological invasions, wildlife harvesting, habitat loss and fragmentation, and urbanization. We consider the behavioural consequences of these challenges, the ways in which behavioural knowledge can be used to contribute to our understanding of the wildlife and environmental impacts, as well as potential solutions and interventions for addressing them. Part 2 extends upon these themes by examining how application of behavioural knowledge can contribute to better practical outcomes for animal conservation, the management of human-wildlife interactions, animal welfare, the control of pests and diseases, and even improvements to the health, security, and well-being of human society. The book concludes with a synthesis and overview of the key themes and lessons for shaping the discipline so that it is better equipped to address the many challenges that the world is facing, both now and into the future. A common element of all chapters is the use of case studies, including chapter 'boxes', to illustrate the diverse ways that animal behaviour knowledge and behavioural interventions can be effectively deployed. As we continue to strengthen the link between basic and applied behavioural research, we hope the book will inspire readers to build upon the existing knowledge base and elevate the application of this knowledge to the next level of efficacy for real-world benefits.

Acknowledgements

This book took shape during a pandemic. In an era of lockdowns, home quarantine, and social distancing, the project connected us through a common purpose, and gifted us the opportunity to work with amazing colleagues from around the globe.

We begin by expressing our immense gratitude to our chapter contributors for their dedication, perseverance, and hard work.

We thank the reviewers for sharing their expertise, insights, and constructive feedback on the chapter drafts.

We thank the teams at OUP and Integra Software Services, especially Ian Sherman for his energy and guidance, as well as Katie Lakina, Adam Brevik and Rajeswari Azayecoche for their support

Lastly, a special thanks to our families for their love and understanding, and the immense joy they bring into our lives.

Bob B. M. Wong
Melbourne, Australia

Ulrika Candolin
Helsinki, Finland

Contents

List of contributors

Marlene Ågerstrand Department of Environmental Science, Stockholm University, Stockholm, Sweden

Iain Barber Department of Life Sciences, Aberystwyth University, Wales, UK

Oded Berger-Tal Mitrani Department of Desert Ecology, The Jacob Blaustein Institutes for Desert Research, Ben-Gurion University of the Negev, Midreshet Ben-Gurion, Israel

Michael G. Bertram Department of Wildlife, Fish, and Environmental Studies, Swedish University of Agricultural Sciences, Umeå, Sweden
Department of Zoology, Stockholm University, Stockholm, Sweden
School of Biological Sciences, Monash University, Melbourne, Australia

Daniel T. Blumstein Department of Ecology & Evolutionary Biology, Institute of the Environment & Sustainability, University of California Los Angeles, Los Angeles, USA

Tomas Brodin Department of Wildlife, Fish, and Environmental Studies, Swedish University of Agricultural Sciences, Umeå, Sweden

Robert C. Brooks Evolution & Ecology Research Centre, and School of Biological, Earth and Environmental Sciences, University of New South Wales, Sydney, Australia

Culum Brown School of Natural Sciences, Macquarie University, University of New South Wales, Sydney, Australia

Rachel Y. Chock Conservation Science and Wildlife Health, San Diego Zoo Wildlife Alliance, Escondido, USA

Timothy D. Clark School of Life and Environmental Sciences, Deakin University, Geelong, Australia

Jeff C. Clements Fisheries and Oceans Canada, Moncton, Canada

Daphne Cortese School of Biodiversity, One Health, and Veterinary Medicine, University of Glasgow, Glasgow, UK
MARBEC, University of Montpellier, CNRS, Ifremer, IRD, Sète, France

Susan J. Cunningham FitzPatrick Institute of African Ornithology, DSI-NRF Centre of Excellence, University of Cape Town, Cape Town, South Africa

Jeremy J. Cusack Centro de Modelación y Monitoreo de Ecosistemas, Universidad Mayor, Santiago, Chile
Okala, Stirling, UK

Amy-Charlotte Devitz Department of Ecology, Evolution and Behavior, University of Minnesota, Twin Cities, St Paul, USA

Beatriz Díaz Pauli Department of Biological Sciences, University of Bergen, Bergen, Norway

Marco Festa-Bianchet Département de biologie, Université de Sherbrooke, Sherbrooke, Québec, Canada

Lisieux Fuzessy Center for Ecological Research and Forestry Applications, Bellaterra, Catalonia, Spain

César González-Lagos Facultad de Artes Liberales, Universidad Adolfo Ibáñez, Santiago de Chile, Chile; Center of Applied Ecology and Sustainability (CAPES), Santiago, Chile

Alison L. Greggor Conservation Science and Wildlife Health, San Diego Zoo Wildlife Alliance, Escondido, USA

Robin Hale Arthur Rylah Institute for Environmental Research, Department of Energy, Environment and Climate Change, Melbourne, Australia

Scarlett R. Howard School of Biological Sciences, Monash University, Melbourne, Australia

Therésa M. Jones School of BioSciences, The University of Melbourne, Melbourne, Australia

Fredrik Jutfelt Department of Biology, Norwegian University of Science and Technology, Trondheim, Norway

Shaun S. Killen School of Biodiversity, One Health, and Veterinary Medicine, University of Glasgow, Glasgow, UK

Kathryn B. McNamara School of BioSciences, The University of Melbourne, Melbourne, Australia

Yolanda Melero Center for Ecological Research and Forestry Applications, Bellaterra, Catalonia, Spain
Department of Evolutionary Biology, Ecology and Environmental Sciences & Biodiversity Research Institute (IRBio), Universitat de Barcelona, Barcelona, Spain

Melissa J. Merrick Conservation Science and Wildlife Health, San Diego Zoo Wildlife Alliance, Escondido, USA

Amelia Munson School of Biodiversity, One Health, and Veterinary Medicine, University of Glasgow, Glasgow, UK

Lauren M. Pintor The Ohio State University, School of Environment and Natural Resources, Columbus, USA

Rocío A. Pozo Pen & Paper Science Studio, Curacautín, Chile

Andrew N. Radford School of Biological Sciences, University of Bristol, Bristol, UK

Lindsey S. Reisinger University of Florida, School of Forest, Fisheries & Geomatics Sciences, Gainesville, USA

Zong-Xin Ren Kunming Institute of Botany, Chinese Academy of Sciences, Kunming, China

Debra M. Shier Conservation Science and Wildlife Health, San Diego Zoo Wildlife Alliance, Escondido, USA
Department of Ecology and Evolutionary Biology, University of California Los Angeles, Los Angeles, USA

Michael Sievers Coastal and Marine Research Centre, Australian Rivers Institute, School of Environment and Science, Griffith University, Gold Coast, Australia

Andrew Sih Department of Environmental Science & Policy, University of California at Davis, USA

Emilie Snell-Rood Department of Ecology, Evolution and Behavior, University of Minnesota, Twin Cities, St Paul, USA

Daniel Sol Institute of Evolutionary Biology (CSIC-Universitat Pompeu Fabra), Barcelona, Catalonia, Spain
Center for Ecological Research and Forestry Applications, Bellaterra, Catalonia, Spain

Josefin Sundin Department of Aquatic Resources, Swedish University of Agricultural Sciences, Drottningholm, Sweden

Stephen E. Swearer National Centre for Coasts and Climate, School of BioSciences, The University of Melbourne, Melbourne, Australia

Sarah Wahltinez Nautilus Collaboration, Lucaston, Australia

Colleen L. Wisinski Conservation Science and Wildlife Health, San Diego Zoo Wildlife Alliance, Escondido, USA

Bob B. M. Wong School of Biological Sciences, Monash University, Melbourne, Australia

Marlene Zuk Department of Ecology, Evolution and Behavior, University of Minnesota, Minneapolis, USA

Challenges

CHAPTER 1

Climate change

Susan J. Cunningham

Overview

Anthropogenic climate change is a pervasive driver of global change, affecting 'pristine' and modified habitats alike. Wildlife behavioural responses to climate change are underpinned by their physiological tolerances. Impacts on behaviour can occur across a wide range of spatial and temporal scales—from adjustments in microhabitat use and diel timing of behaviour, to broad-scale shifts in phenology and geographic ranges. Different species respond in different ways and at different rates, resulting in changed species interactions, with implications for population persistence and ecosystem function. Changing selection pressures may also alter patterns of animal behaviour in unpredictable ways. Conservation actions range from hands-on management 'fixes', such as provision of resources and thermally buffered refuges within habitats, to landscape-scale planning for protected areas likely to include future climate refugia. All such actions are ultimately 'patches': if we want to conserve current biodiversity, we must make fundamental changes to reduce emissions and restabilize the global climate.

1.1 Introduction

Climate change is among the most pervasive of global change drivers, affecting remote wilderness and heavily human-modified landscapes alike. Climate change is driven by elevated levels of greenhouse gases in the Earth's atmosphere, associated with fossil fuel burning and modern agricultural practices. Average global surface temperature has already increased by 1.1°C compared to historic baselines, and the last decade was warmer than any period in the last 125,000 years (IPCC 2023). This warming is associated with large-scale changes in weather patterns, including increased mean surface temperatures, changes in precipitation, and increased frequency of extreme weather events. In the oceans, climate change is associated with warming sea surface temperatures and marine heatwaves, sea ice loss, sea level rise, and changes in ocean currents, salinity, and pH (see also Chapter 5). Climate change therefore alters the abiotic conditions to which animals are exposed in multiple and dramatic ways. Wildlife behavioural responses, in turn, are underpinned by their physiological tolerances of these changing abiotic conditions.

One way animals respond behaviourally to climate change is by shifting activity in space and time. Animal responses manifest on multiple scales—from changes in diurnal cycles of activity and microscale habitat use (e.g. Buckley et al. 2013), to seasonal shifts in the timing of behaviour and global changes in range distributions (e.g. Parmesan and Yohe 2003). These behavioural changes can carry fitness benefits (compared to a lack of response). By shifting activity in space and time, animals can reduce exposure to extremes of temperature or desiccation, minimizing physiological costs of maintaining homeostasis in increasingly harsh environments (Buckley et al. 2015). However, behavioural responses may also involve costly trade-offs when they conflict with an animal's capacity to perform other important behaviours (Cunningham et al. 2021). For example, changes in timing and location

Susan J. Cunningham, *Climate change*. In: *Behavioural Responses to a Changing World*. Edited by: Bob B. M. Wong and Ulrika Candolin, Oxford University Press.
© Oxford University Press (2024). DOI: 10.1093/oso/9780192858979.003.0001

of activity may impact activities such as foraging, territory defence, mate attraction, or parental care. Furthermore, sympatric species may respond to changing abiotic conditions differently, altering interactions between them. Behavioural changes in response to changing climate conditions therefore have consequences across multiple scales, from individual fitness to population persistence and ecosystem function.

This chapter examines the influence of climate change on animal behaviour, with a focus on vertebrates in terrestrial ecosystems. I begin with the mechanisms underlying relationships between thermal physiology and behaviour (Section 1.2), then discuss risks and opportunity costs associated with behavioural changes to mitigate physiological demands and the role of resource availability in these (Section 1.3). From here, I assess impacts on fitness mediated through changes in breeding and parental care behaviours and the potential for climate change to alter selection pressures on breeding systems (Section 1.4), then broaden out to describe large-scale patterns that emerge from these underlying processes, including phenological and range shifts and the implications of these (Section 1.5). I briefly address consequences of climate change for frequency and intensity of wildfires, and the impact of these on animal behaviour (Section 1.6), and finally touch on interspecific variation in responses and implications for individuals, populations, and species interactions (Section 1.7). In Section 1.8, I examine how animal behaviour can inform conservation in a changing climate. Throughout, I attempt to draw on examples from a variety of latitudes and habitats—not just the well-studied Northern Hemisphere temperate zones—to highlight the truly global impact of this global change driver on animal behaviour.

1.2 Relationships between behaviour and thermal physiology

Animals' physiological tolerances of changing climate conditions—and consequently, their behavioural responses to these—can be context dependent, based on adaptation to historic climates. For example, hot weather constrains animal behaviour and survival in polar, temperate, tropical, and subtropical latitudes alike; although the definition of a dangerous 'heat event' is very different in these different parts of the world. For example, heat-related mortality of incubating Brünnich's guillemots *Uria lomvia* has been recorded on days when the air temperature exceeded ~16°C in arctic Hudson Bay (Gaston et al. 2002). By contrast, similar mortality events involving birds and bats occur in subtropical southern Africa (McKechnie et al. 2021) and Australia (Welbergen et al. 2008) only when air temperatures exceed 40°C. Animals often use behaviour to moderate the impact of environmental conditions on physiology, but changes in behaviour can carry costs that animals must balance against the physiological demands of exposure to unfavourable abiotic conditions (Cunningham et al. 2021). Animal physiology is therefore fundamental to understanding and predicting behavioural responses to climate change.

All animals have a range of body temperatures across which their physiological performance is maximized. The exact shape of the relationship between body temperature and performance is species-specific, and parameters such as optimum body temperature, peak physiological performance, and the breadth of the tolerable range of body temperatures are all under selection pressure (Boyles et al. 2011; Figure 1.1). However, performance is compromised if body temperature falls too low, or rises too high. This is because changes in enzyme function and cell membrane structure occur at sub-optimally low or high body temperatures, affecting vital functions (Angilletta et al. 2010). Experimental research with ectothermic animals (e.g. lizards), in which body temperature can be modified directly by changing the ambient temperature in the laboratory, shows that performance (measured using sprint speed, bite strength, etc. as proxies) declines more rapidly at supra-optimal (too hot) than sub-optimal (too cold) body temperatures (Angilletta et al. 2010; Boyles et al. 2011). Zero performance and death lie at each extreme. Although more difficult to demonstrate empirically, similar asymmetric 'thermal performance curves' likely also exist in

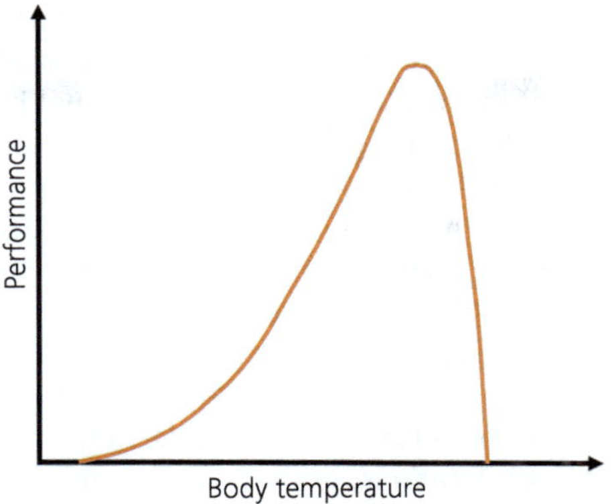

Figure 1.1 Physiological performance is related to body temperature in a highly non-linear manner. The breadth of body temperatures across which performance is maximized, height of peak performance, and the range of tolerable body temperatures are all highly species-specific and under selection pressure. Performance declines more quickly at body temperatures that are too high than body temperatures that are too low (Boyles et al. 2011).

endothermic animals such as birds and mammals (Angilletta et al. 2010; Boyles et al. 2011; Levesque and Marshall 2021).

Animals therefore face the challenge of maintaining body temperature within a range that facilitates adequate physiological performance. This they must do in the face of environmental temperatures that can be extremely variable in space and time. The thermal challenge an animal experiences at any given moment depends on heat exchange through a variety of processes, including solar and reflected radiation, convection (e.g. wind), and conduction (with the air or water and any other substrates with which the animal is in contact). In terrestrial environments, available operative environmental temperatures (sensu Bakken et al. 1985) can vary to a large degree over very small spatial scales. For example, at an air temperature of 37°C, we measured operative environmental temperatures ranging from 41°C inside a tree canopy, to 65°C on the ground in the sun, in the Kalahari Desert, southern Africa (Cunningham et al. 2021). Carroll et al. (2016) recorded operative environmental temperatures ranging from 27°C to 79°C at air temperatures above 25°C in shrubland in western Oklahoma.

1.3 'Behavioural buffering': risks and opportunity costs

Ectothermic animals, like reptiles, amphibians, invertebrates, and most fishes, manage body temperature, and therefore physiological performance, directly through behavioural use of thermal environments ('behavioural thermoregulation'). For example, lizards in terrestrial environments sun-bask during cool periods and shuttle between sun and shade as environmental temperatures rise (Huey and Tewksbury 2009). By contrast, endothermic animals (e.g. birds and mammals) are defined by a greater capacity to maintain body temperatures at variance with the environment, via metabolic heat production and evaporative cooling. The energy and water requirements of this process depend upon environmental temperature in a highly non-linear manner and increase rapidly at environmental temperatures below or above thermoneutrality (McNab 2012; Figure 1.2). Behavioural thermoregulation is therefore important for endothermic animals, too, as by selecting thermally favourable locations within landscapes they can substantially reduce energy and water costs.

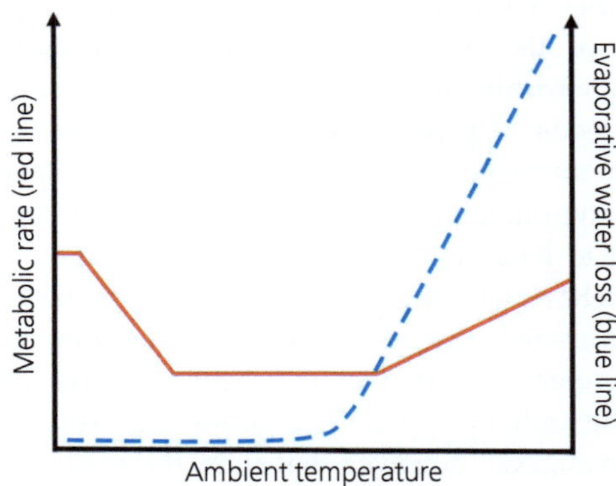

Figure 1.2 Endothermic animals maintain body temperatures at odds with the temperature of their environment by altering rates of metabolic heat production (red solid line) and evaporative water loss (blue dashed line). The range of ambient or environmental temperatures across which metabolic rate is minimized is known as the 'thermoneutral zone' (TNZ). Costs of keeping warm, or cool, increase rapidly at temperatures above and below thermoneutrality (Scholander et al. 1950; Calder and King 1974).

Under climate change, animals could take advantage of these benefits of behavioural thermoregulation to buffer costs of increasingly warm weather conditions, potentially reducing their vulnerability—described as 'behavioural buffering' (Huey et al. 2012; Beever et al. 2017). Animals at all latitudes have the potential to do this: for example, during periods of high air temperature, southern yellow-billed hornbills *Tockus leucomelas* in the subtropical Kalahari seek shady locations inside the canopies of acacia trees (at air temperatures > 35°C; van de Ven et al. 2019), while white-winged snow finches *Montifringilla nivalis* select cooler, shadier, and snowier foraging locations above the treeline in the European Alps (at air temperatures > 10°C; Alessandrini et al. 2022). Warming can also result in dramatic changes in animals' diel activity patterns: for example, during hot weather, Arabian oryx *Oryx leucoryx* increase shade-seeking during the day, and shift activity patterns from diurnal to crepuscular to nocturnal (Hetem et al. 2012; Figure 1.3a). The energy and water savings so made can be instrumental in animals' ability to maintain energy and water balance (Carroll et al. 2015), and could help individuals avoid lethal body temperatures under increasingly challenging thermal conditions.

However, behavioural buffering carries risks and costs of its own. For example, in Oklahoma, male bobwhite quails *Colinus virginianus* caring for broods of precocial chicks select locations in grassy and wooded landscapes that buffer operative environmental temperatures by > 10°C compared to landscape averages (Carroll et al. 2015). This behavioural microhabitat selection ensures that chicks are never exposed to operative environmental temperatures that exceed heat tolerance limits. The authors' climate change projections for the region suggest, however, that such 'thermally safe' locations in the landscape will become increasingly hard to find in future. In the meantime, this behavioural buffering may shield the quails from directional selection for greater thermal tolerances, reducing their capacity for adaptation to rising temperatures (Carroll et al. 2015). Such a process has already been observed in *Scleroporus* lizards in Texas, with biophysical models suggesting that the lizards' currently effective behavioural buffering of

heat stress may result in abrupt population declines in future, as thermal micro-refuges in the lizards' habitats become too hot, but lizards have not been exposed to directional selection for greater thermal tolerance (Buckley et al. 2015).

Behavioural thermoregulation also carries insidious costs in the form of missed opportunities. Opportunity costs result from partial or complete incompatibility between thermoregulatory behaviours (e.g. shade-seeking) and other components of fitness. For example, retreat into shaded/cool locations during hot weather may force animals to forgo exploitation of profitable foraging patches (Clark 1987; van de Ven et al. 2019), or alter their capacity to attract or retain mates (Santee and Bakken 1987), or defend territory or offspring (Campagna and Le Boeuf 1988; Oswald et al. 2008; Cook et al. 2020). Larger-scale shifts in habitat use (e.g. moving upslope in alpine environments or seeking damper, shadier forest types in mixed habitats) may come at the cost of reduced forage availability and quality (Street et al. 2016; Mason et al. 2017). The consequences of opportunity costs are often non-trivial, associated with reduced survival and breeding success. For example, retreat into rock crevices to shelter from heat reduces the capacity of Mexican *Sceloporus* lizards to forage, compromising their body condition and capacity to breed during hot weather (Sinervo et al. 2010). Cumulative impacts of such opportunity costs caused population declines and local extinctions of these lizards under warming trends between the 1960s and early 2000s (Sinervo et al. 2010).

Costs associated with behavioural thermoregulation are often mediated by loss of foraging intake. These costs may therefore be buffered by resource conditions that allow animals to increase foraging efficiency. For example, research on relationships between air temperature, foraging behaviour, and body mass maintenance in subtropical birds commonly shows reduced foraging effort during hot weather, associated with loss of body condition (e.g. in southern pied babblers *Turdoides bicolor* and southern yellow-billed hornbills (du Plessis et al. 2012; van de Ven et al. 2019)). Work on birds living at similar latitudes in suburban and urban environments shows similar relationships between air temperature and foraging behaviour

(Edwards et al. 2015; Stofberg et al. 2022). However, in both urban red-winged starlings *Onychognathus morio* (in Cape Town) and suburban Australian magpies *Gymnorhyna tibicen* (in Perth), declines in foraging at air temperatures > 35°C were not associated with body mass loss, suggesting abundant, calorie-dense anthropogenic food and artificial water resources may allow urban individuals to compensate for reduced foraging hours during heatwaves (Edwards et al. 2015; Stofberg et al. 2022).

Extreme temperatures and resource scarcity make arid environments useful test cases for examining interactions between resource availability and temperature under natural conditions. Experimental provision of supplementary food and water changed diel activity patterns of hoopoe larks *Alaemon alaudipes* in the Arabian Desert, highlighting the state-dependent nature of animals' behavioural responses to changing abiotic conditions (Tieleman and Williams 2002). Cycles of drought and good rainfall years in the southern Kalahari Desert also highlight the importance of rain and associated influx of primary productivity for thermoregulation and survival of arid-zone insectivores. For example, drought and associated decline in availability of termite prey for aardvarks *Octeropus afer* resulted in changed activity patterns as the animals attempted to reduce energetic costs of thermoregulation, severe dips and spikes in body temperature (likely associated with significant physiological performance costs), and in many cases, death (Weyer et al. 2020). Conversely, good rainfall and associated increases in primary productivity and invertebrate abundance can buffer costs of high temperatures. Southern pied babblers breed during the hot summer months in the Kalahari Desert, in association with the 'greening up' of the dominant camelthorn *Vachellia erioloba* trees in this environment (Hunt et al. 2023). Especially hot summers are associated with dramatic reductions in juvenile and adult survival of these insectivorous birds over the following winter (Bourne et al. 2020a). However, these losses are buffered in high rainfall years (for adults, overwinter survival after hot summers with good rainfall was ~90%, but ~25% after hot summers combined with drought; juvenile overwinter survival was ~75% and ~10%, respectively). High resource availability in good rainfall years likely reduces

both the physiological and behavioural costs of thermoregulation, allowing birds to enter winter in better body condition, boosting survival.

The availability and configuration of shade within habitats is another resource with the potential to buffer opportunity costs. Sears et al. (2016) used computer simulations and experimental data collection on real lizards in outdoor pens to show that lizards maintain body temperatures closer to their preferred optimum at lower energetic cost in environments in which shade is found in scattered patches, rather than large clumps. Similarly, white-browed sparrow-weavers *Plocepasser mahali* (~45 g subtropical African passerine birds) with greater access to shade in their home ranges panted less and maintained higher foraging intensity at high air temperatures, than conspecifics in sunnier home ranges, suggesting shade availability buffers foraging-thermoregulation trade-offs in this species (J. Whyte unpub. data).

1.4 Breeding and parental care

Behavioural responses to changing climate conditions can include alterations in parental care behaviour, with direct implications for breeding success, offspring quality, and lifetime fitness of individuals. The strength of climate change impacts on breeding outcomes may differ between populations with different breeding systems. For example, thermoregulatory trade-offs experienced by individual parents may depress net parental care provided to offspring in animals with uniparental care. However, declines in provision of care may be buffered in pair- or cooperative breeders where more than one individual cares for the same offspring. Changes in the costs of parental care associated with climate-driven changes in temperature, precipitation, and resource availability therefore have the potential to shift selection pressures associated with different breeding systems.

1.4.1 Provisioning patterns and offspring quality and survival

Parental care is common in mammals and birds globally, but costs and benefits of parental care behaviours may be altered by climate change.

Figure 1.3 Wildlife respond behaviourally to climate change in numerous ways. (a) Over short time scales, animals may change behaviour patterns to escape increasingly high temperatures. During heatwaves, Arabian oryx in the hot deserts of the Arabian Peninsula seek shade during the day, and shift diel timing of activity, becoming increasingly crepuscular or even nocturnal (Hetem et al. 2012). White-winged snow finches in the European Alps seek shadier and snowier foraging locations when air temperature exceeds 10°C (Alessandrini et al. 2022). (b) Reductions in provisioning rates to offspring are a ubiquitous response of birds to hot weather conditions, affecting species from ring ouzels at temperatures above 7.5°C in the Swiss Bernese Alps (Barras et al. 2021), to southern pied babblers at temperatures above 35°C in the Kalahari Desert in southern Africa (Wiley and Ridley 2016). (c) Species interactions are strongly affected by climate change, via, for example, changes in phenology and range shifts that draw species apart or push them together in space and time. This can result in mismatches between predators and prey, exemplified by changing relationships between European woodland birds such as pied flycatchers and the caterpillars they rely on to raise their offspring (e.g. Nicolau et al. 2021). Alternatively, changes in habitat and climate suitability can cause species to shift ranges, bringing them into contact with novel ecological partners. For example, climate change-driven loss of sea ice is forcing polar bears into coastal terrestrial habitats where they now prey on seabird colonies (Prop et al. 2015).

Photos: (a, left) Alamy.com/Inaki Relanzon; (a, right) Shutterstock.com; (b, left) Alamy/Carl Day; (b, right) Nicholas Pattinson; (c, left) Alamy/Andy Sands; (c, right) Alamy/Alex Holder

Increasing costs associated with parental care and reductions in offspring provisioning have been documented during periods of high temperatures and/or drought in a range of environments. In subtropical Africa, for example, reduced hunting and food provision during hot weather results in lower pup survival in wild dogs *Lyceon pictus* (Woodroffe et al. 2017). Similar reductions in breeding output during hot weather have been documented in meerkats *Suricatta suricatta* in arid savanna environments, where they are exacerbated by drought (van de Ven et al. 2020). Lactation capacity of female mammals may be limited by their ability to dissipate body heat (Speakman and Król

2010), an issue that could become increasingly problematic under global temperature rise. The propensity of female fur seals *Arctocephalus philipii* to abandon lactation and seek thermal refuge in the sea during hot periods was already documented in Chile in the early 1990s (Francis and Boness 1991), and more recent work in small mammals (Simons et al. 2011; Zhao et al. 2020) suggests heat dissipation limitations on lactation could be widespread (although not ubiquitous, e.g. Sadowska et al. 2019).

In birds, declines in nest provisioning rates at high air temperatures are a pervasive part of avian responses to hot weather conditions in all environments (Figure 1.3b). For example, provisioning rates decline at air temperatures above ~35°C in several species in subtropical Africa and the Mediterranean; above ~20°C in cape rockjumpers *Chaetops frenatus* inhabiting the Cape Fold mountains of South Africa; and above ~7.5°C in alpine ring ouzels *Turdus torquatus alpestris* in the Swizz Bernese Alps (Wiley and Ridley 2016; van de Ven et al. 2020; Barras et al. 2021; Oswald et al. 2021). Reduced provisioning is generally associated with reduced offspring growth, delayed fledging, and reduced fledging probability—outcomes that have been documented in habitats from arid subtropical deserts to temperate, coastal, and alpine zones (Cunningham et al. 2013; Catry et al. 2015; Wiley and Ridley 2016; van de Ven et al. 2020; Oswald et al. 2021). These poor nest outcomes may be driven by lower provisioning, but high nest temperatures likely also directly suppress nestling growth via physiological means. Studies employing path analyses on two African species—a cup-nesting passerine (southern pied babbler) and a cavity-nesting Buceroti-forme (southern yellow-billed hornbill)—both suggest ~30% of the temperature-related decline in offspring growth can be explained by reduced provisioning rates and the remaining ~70% by direct effects of high temperatures (Bourne et al. 2020b; van de Ven et al. 2020).

1.4.2 Changing selection pressures on breeding systems

'What drives variation in breeding systems of animals?' is a classic question in behavioural ecology. For example, parental care can be provided by a single parent (uniparental), a breeding pair (biparental), or a group of individuals (cooperative breeding and eusociality)—or not provided at all. Cooperative breeding is intuitively puzzling, because 'helpers' forgo breeding to raise the offspring of others. Core hypotheses about the evolution of cooperation rely on ideas about relatedness between group members, degree of kinship with the offspring helped, and the impact of environmental conditions (Hamilton 1964; Brown 1987; Jetz and Rubenstein 2011). The 'ecological constraints' hypothesis suggests cooperation evolved in climatically stable, saturated environments where offspring find it difficult to obtain a breeding position and instead help their parents raise their siblings (Brown 1987; Hatchwell and Komdeur 2000). However, cooperative breeding is also common in unstable, harsh environments where rainfall and resource availability is unpredictable within and between years (e.g. the arid savannas of Australia and southern Africa) (Jetz and Rubenstein 2011). This pattern has led to the development of the alternate 'hard life' (Koenig and Mumme 1987) and related 'bet-hedging' hypotheses (Rubenstein 2011), which suggest cooperation is favoured because helpers buffer the impacts of environmental variation on breeding output. The 'dual benefits framework' proposed by Shen et al. (2017) integrates these hypotheses, proposing that a shift from stable to unpredictable climate conditions may favour cooperative breeding in species that already live in social groups.

This likely role of climate in the evolution of breeding systems highlights the potential for climate change to alter selection pressures and cause shifts in parental care behaviour. Biparental and cooperative care may increase while uniparental care declines, as the cost of parental care increases under increasingly harsh and unpredictable conditions. For example, Kentish plovers *Charadrius alexandrinus* switch from uniparental to biparental incubation dependent on local climate: both parents share the thermal costs of incubation when temperature conditions are extreme (AlRashidi et al. 2011). In Australia, jacky winters *Microeca fascinans* show similar flexibility within single breeding attempts. Only females of this small arid-zone bird incubate, but males will contribute by shading eggs in extreme heat, and the likelihood of this behaviour occurring increases with temperature (Sharpe et al.

2021). In both cases, this parental flexibility reduces risk of nest failure due to abandonment, if thermal conditions overwhelm the capacity of a single parent to cope.

It would seem logical that cooperative breeding animals, where helpers as well as parents provide care, could benefit even more from sharing the load between individuals when environmental conditions are harsh (Rubenstein 2011). However, declines in breeding success during hot and dry weather were not buffered in larger groups in detailed studies of southern pied babblers (Bourne et al. 2020b), superb starlings *Lamprotornis superbus* (Guindre-Parker and Rubenstein 2020), meerkats (van de Ven et al. 2020), or sociable weavers (D'amelio et al. 2022); but see Capilla-Lasheras et al. (2021) on white-browed sparrow-weavers. These surprising results may be explained by several factors. Social status may influence animal responses, such that decline in offspring care by subordinate helpers under increasingly challenging environmental conditions is steeper than that of dominant breeders, as seen in southern pied babblers (Wiley and Ridley 2016; Bourne et al. 2023). This could be driven by lower relatedness between helpers and offspring than between breeders and offspring. Alternatively, subordinate individuals in social groups may have poorer thermoregulatory capacity than dominants and may therefore suffer greater performance costs during hot weather, limiting their ability to invest in offspring care (Cunningham et al. 2017). Most empirical work on this question has been carried out in environments that are already hot and arid. However, arid zones are experiencing rapid rates of warming and it is possible that the capacity of sociality to buffer harsh conditions has already been exceeded by recent climate change. Alternatively, or in addition, increased resource constraints associated with heatwaves and drought may outweigh benefits of load-lightening for larger groups.

1.5 Phenological and spatial changes in animal behaviour

In addition to the in-situ behavioural changes described above, animals may also respond to climate change by shifting activity in time and space at extremely large spatial and temporal scales.

For example, climate change-related changes in the timing of behaviour include full-scale phenological shifts across whole ecosystems (especially well documented in the context of advancing spring phenology in the temperate zones of the Northern Hemisphere). Spatially, notable shifts in range distributions are occurring across taxa globally as climate conditions become more favourable for species in new geographical areas, but less favourable in parts of their historic range (Parmesan and Yohe 2003).

1.5.1 Phenology

Changes in timing are among the best documented animal behavioural responses to climate change. Phenology (the timing of events in the natural world) is incredibly important for fitness in a seasonal world, as animals must match essential life cycle components (breeding, migration, etc.) with favourable environmental conditions on annual (or longer) timescales. They must also synchronize these events with the life cycles of their ecological partners, to ensure sufficient access to food and minimize predation and parasite pressure (Stenseth and Mysterud 2002). Climate change impacts phenology because important life history events such as breeding and migration are often cued by climate variables, including temperature (typically at mid to high latitudes) and rainfall (typically at low latitudes) (Cohen et al. 2018).

Most data on phenological shifts come from study systems in the temperate zones of the Northern Hemisphere (Brown et al. 2016). In temperate regions, advances in spring phenology associated with warmer temperatures have emerged as a dominant response to climate change across taxa (Cohen et al. 2018). A large literature exists on changes in timing of spring breeding behaviour in the small woodland birds of Europe (e.g. tits [Paridae] and flycatchers [Muscicapidae]; Visser et al. 2006; Burger et al. 2012; Nicolau et al. 2021; and many others). Spring migration dates of temperate zone bird species have also received much attention (e.g. Charmantier and Gienapp 2013; Gill et al. 2013; Helm et al. 2019), with the majority of species advancing their arrival dates in warmer springs. In rarer cases, spring phenology may be delayed in association with climate warming (e.g. Antarctic seabirds are nesting later due to reduction in sea ice cover in East Antarctica; Barbraud and Weimerskirch 2006).

At tropical and subtropical latitudes, phenology is often driven by precipitation (Cohen et al. 2018) and changes in rainfall patterns have potential to cause consequential phenological change. Large pulses of primary productivity follow rainfall in subtropical arid zones (Dickman et al. 1999; Palmer 2010). Correlated increases in abundance of invertebrates and small vertebrates are relied on by migrant and breeding birds, mammals, and reptiles in these environments (Dickman et al. 1999; Lloyd 1999; Palmer 2010). Animals time their breeding to match these increases in resource availability (Lloyd 1999; Hidalgo Aranzamendi et al. 2019). In deserts in the summer-rainfall regions of southern Africa, delayed or absent pulses of prey availability may explain delayed or absent breeding by birds during drought years, for example (Pattinson et al. 2022). However, relationships between climate conditions and breeding in arid-zone birds can be complex, with some studies suggesting a role for temperature as well as rainfall in current and previous breeding seasons in the probability of breeding, laying dates, and length of the breeding season (Mares et al. 2017).

Climate change has the potential to result in phenological mismatches between ecological partners, as different taxa may respond in different ways. For example, species may respond to change in the same climate cue (e.g. temperature or rainfall) at different rates, or respond to different cues (e.g. temperature versus daylength) (Cohen et al. 2018). Animal species with larger body mass tend to shift their phenology more slowly in response to climate change than smaller species (Cohen et al. 2018), meaning trophic and parasite-host interactions are at particular risk of being decoupled as climate change advances. In a similar vein, ectotherms show stronger phenological advances in response to climate warming than endotherms, perhaps due to underlying differences in thermal physiology; and herbivores respond more strongly than carnivores, likely due to closer coupling to changes in primary productivity (Cohen et al. 2018).

In tightly seasonal high temperate latitudes, small differences in species' phenological responses to warming temperatures have the potential to drive large phenological mismatches. These have been documented in detail in European woodland bird assemblages provisioning caterpillars to their offspring in early spring (e.g. Both et al. 2006, 2009; Bauer et al. 2010; Figure 1.3c). In these systems, resident and migrant passerine bird species rely on timing the laying of their eggs to tightly match peak nestling demand to peak availability of caterpillars. The caterpillars (larvae of a range of moth species), in turn, rely on the budburst of woodland trees to provide their food. Each trophic level in this tripartite partnership (trees, moths, and birds) responds independently to warming temperatures in spring (Burgess et al. 2018). The rate of spring phenological advancement consequently varies between plant, moth, and bird taxa, and also varies with latitude and location. This means the degree of phenological mismatch between ecological partners varies in space and time in a complex manner. The evolutionary and population-level consequences of mismatches remain unclear, likely because they are intricately linked to other ecological factors in ways that are still poorly understood (e.g. density-dependent recruitment and other complex selection pressures, which also vary in time and space). Despite the depth of literature on this topic, this therefore remains an active field of investigation (Visser and Gienapp 2019; Zhemchuzhnikov et al. 2021; Kharouba and Wolkovich 2023).

1.5.2 Range shifts

Alongside phenological shifts, shifts in geographical ranges are among the most notable impacts of climate change on animal behaviour (Pecl et al. 2017). Early meta-analyses showed range shifts poleward and uphill (i.e. towards cooler areas) were occurring across taxa, leading to claims of a globally coherent 'fingerprint' for biological responses to climate change (e.g. Parmesan and Yohe 2003). However, datasets upon which these syntheses drew were heavily biased towards the temperate zones of the Northern Hemisphere, due to the paucity of information from other regions. Newer data (e.g. Tingley et al. 2012; Van der Wal et al. 2013) show that shifts in other directions (e.g. equatorward or longitudinal; downhill) can also be consistent with species tracking favoured climate envelopes as they shift in space and time due to climate change. This is because precipitation, as well as temperature, can limit species ranges (e.g. Tingley et al. 2012), and 'precipitation envelopes' are less likely to shift predictably towards higher latitudes and altitudes as the world warms. Furthermore, the direction of shifts in climate envelopes is less predictable at lower latitudes than it is in the temperate zones. For example, in Australia, climate envelopes suitable for a range of different bird taxa are indeed shifting southward (poleward) overall in temperate grasslands, savannas, shrublands, and forests. However, in tropical, arid, and Mediterranean climate regions of the continent, the climate is shifting idiosyncratically, such that bird species may need to shift their ranges to the west or northeast, for example, to keep up with climate change (Van der Wal et al. 2013).

Keeping up with climate envelopes as they shift through space and time is important for several reasons. The first is that natural selection may have left animals with behavioural and physiological adaptations that suit them to certain suites of climate conditions. Tracking the climate envelope to which they are adapted therefore allows them to reduce, for example, the behavioural and physiological costs of thermoregulation, improving performance and maximizing fitness. Second, tracking climate envelopes may be critical for keeping up with ecological partners in space and time. Indeed, there is evidence that species that successfully shift both their phenology and their geographic range in line with climate change have better conservation trajectories than those that do not (e.g. Niven et al. 2009). This raises the question of what promotes capacity for species to move in space and time in response to changing climates, and what constrains this?

Species that fail to track climate envelopes might be constrained by factors inherent in their ecology. Species with low mobility or slow dispersal strategies may be unable to keep up with the rapid pace at which climate envelopes are shifting, for instance. Obligately social animals may be slow to colonize new areas because dispersing individuals must quickly find pre-existing social groups to join. Southern pied babblers, for example, progressively lose body condition the longer they spend without a social group (Ridley et al. 2008). Specialist ecological relationships may also hinder animals' ability to colonize new ranges if their partner species

or habitat does not disperse at the same rate. For example, animals that rely on old growth forests may find their climate envelope shifting faster than their forest habitat can mature. This issue of 'no receiving habitat' may also apply in cases where climate envelopes shift into entirely unsuitable areas: for example, the boreal forests of Eurasia and North America are invading the northern tundras from the south, tracking the warming climate. However, the northern limit of tundra is constrained by the presence of the Arctic Ocean, meaning this habitat is squeezed between the advancing forests and the sea, and there is no capacity for tundra plants or their associated fauna to move north (Zöckler and Lysenko 2000). In mountain areas, the capacity of animals to follow climate envelopes upslope may be limited by the ever-reducing area available as mountain peaks are reached (Dirnböck et al. 2011) or by oxygen limitations (Jacobsen 2020).

Other aspects of anthropogenic global change may also shape the capacity of animal species to shift their ranges in response to climate change. For example, habitat modification (e.g. the draining of wetlands, urbanization) may create barriers to movement or lack of suitable receiving habitat as climate envelopes shift into human-modified areas (Robillard et al. 2015). On the other hand, habitat modification can also facilitate climate-driven range shifts. One example of this comes from the treeless landscapes of South Africa's Karoo Desert. This region has experienced significant climate warming since the 1990s, making the climate more suitable for a native corvid species, the pied crow *Corvus albus*. However, crows are tree nesters, and the Karoo landscape is lacking in trees. Despite this, pied crow numbers have increased dramatically in the Karoo in line with climate change. Citizen science data show that crow numbers have increased specifically in areas where electricity pylons provide nest sites. The crows appear to have used the pre-existing electrical infrastructure as a stepping-stone to invade new habitat as the climate became suitable (Cunningham et al. 2016).

1.6 Fire

In many ecosystems, fire is an important ecological driver, and wildlife behavioural responses to fire have been honed by natural selection (Pausas and

Parr 2018). Fire provides opportunity, and animals that live in fire-prone landscapes may be adapted to take advantage of this. Raptors and other predators, for example, may follow fire fronts to capture fleeing invertebrate and small vertebrate prey (Hovick et al. 2017). Other species may seek out post-burn environments to access newly accessible food resources (e.g. McGregor et al. 2016). Some birds, for example bronze-winged coursers *Rhinoptilus chalcopterus* in Africa, make use of post-burn landscapes for breeding and will travel large distances to access these (Maclean and Kirwan 2020). Other species may experience higher breeding success in years following fire because reduced vegetation improves their ability to detect predators or creates habitat structure in which they can more easily find prey (Corbett et al. 2003). Pulses of nutrient input into soils following fire, together with newly open landscapes, can result in rapid plant growth, thus providing nutrient-rich food resources for herbivores (Parrini and Owen-Smith 2010) and nectarivores (Pausas and Parr 2018). Fires may also drive changes in distribution of mobile animals that rely on older vegetation for food and shelter and must move to new areas following burns. Fire therefore structures ecosystems in important ways, and in fire-prone ecosystems, fire frequencies and the existence of mosaics of burned and unburned habitat are extremely important to support biodiversity (Bowman et al. 2016).

However, climate change is driving an increase in frequency of wildfires, and in the occurrence of fires in systems in which they were previously rare. Hotter weather, especially when combined with landscape management policies including fire suppression (which cause increases in fuel load through time), is also causing fires to burn hotter, longer, and over larger areas (Jolly et al. 2022). These changes put additional stress on wildlife already dealing with other components of climate change (Pausas and Parr 2018; Kay et al. 2021). In addition to the increased risk of direct mortality in more severe fires (Jolly et al. 2022), multiple components of animal behaviour may be affected. For example, complex changes in movement ecology associated with changed fire frequency have been documented in caribou *Rangifer tarandus groenlandicus* in North American boreal forests (Rickbeil et al. 2017), and mountain lions *Puma concolor*

in urbanized landscapes may increase risk-taking behaviour in the aftermath of wildfires (Blakey et al. 2022). Fires may alter species interactions in-situ by changing the spatial structure of environments and the hormonal and nutritional status of animals, with implications for behaviour and population dynamics (Verdolin 2006; Doherty et al. 2022). For example, thatch ants *Formica obscuripes* in Idaho experience changes in stress hormone levels post-fire that alter their responses to threats: individual workers were more likely to flee disturbance than attempt to defend their nest immediately post-fire, but the opposite was true in unburned landscapes (Kay et al. 2021). Understanding and predicting the behavioural responses of animals to fire is an emerging field, but one of great importance if we are to manage the impacts of increased frequency and severity of wildfires on our warming planet (Michel et al. 2023).

1.7 Species interactions

Changes in animal behaviour in response to climate change have the potential to alter species interactions, in turn affecting how each species responds to further change (Gilman et al. 2010). Virtually every type of species interaction can be affected by climate change (Tylianakis et al. 2008). Changes in interactions occur because species respond to climate change in different ways (as described above), and these responses may also depend on the presence or absence of other species (Davis et al. 1998a, b). Interspecific variation in climate change responses across all scales can disturb existing relationships between species, and establish new relationships, with the potential to profoundly affect individual fitness, population persistence, and ecosystem function. Three major processes could be particularly important: (1) changes in the relative abundances of species within ecosystems; (2) changes in species presence in time (phenology) and space (range shifts); and (3) changes in behaviour in-situ that alter the way animals encounter and interact with one another, shifting the strength and types of interactions between them.

Changes in relative abundance of species within ecosystems can occur because of different climate change tolerances between interacting species—driven, for example, by different physiological tolerances to temperature rise (Tylianakis et al. 2008). Changes in relative abundance of interacting species can have cascading effects through ecosystems. For example, changes in relative abundance of available prey could cause prey-switching in generalist predators in line with classic ecological theory (Solomon 1949; Holling 1959). Phenological mismatches between interacting species (Section 1.5.1), can result in trophic cascades (Both et al. 2009). Range shifts can cause predator release or new predation pressures, changes in competitive interactions and parasite-host relationships, and novel interactions between species that did not encounter each other before, with varied and unpredictable outcomes (Tylianakis et al. 2008). One particularly compelling example of this is the climate change-related incursions of polar bears *Ursus maritimus* into eider duck *Somateria mollissima* colonies in northern Canada. Reductions in sea ice cause polar bears to shift foraging ranges into terrestrial systems and switch prey from seals hunted on the ice to eiders nesting on the shore, with important implications for eider breeding success (Iverson et al. 2014). Similar patterns of increased polar bear incursions into coastal bird breeding colonies, associated with reduced sea ice, have also been documented in Svalbard and Greenland (Prop et al. 2015; Figure 1.3c).

Animals may also change behaviour in subtle ways in-situ in response to climate warming, changing the timing of their active periods, or location, or amount of activity, to avoid the heat. These fine-scale changes in behaviour can fundamentally alter species interactions even if relative abundances of species are unchanged. For example, changes in diel activity patterns can temporally separate sympatric species, or bring them together in novel ways, for example, by bringing a diurnal species into new contact with nocturnal predators (Veldhuis et al. 2019). Changes in foraging locations or parental care behaviours can likewise bring together or draw apart sympatric species in habitat space, promoting changes in interactions and altering fitness outcomes (e.g. Jones et al. 2001; Oswald et al. 2008). The subtle behavioural changes animals use to manage exposure to increased environmental temperatures may therefore have broad-scale ecosystem-level effects. The now well-documented population- and ecosystem-level effects of animals'

behavioural responses to changing predation risk (i.e. the 'landscape of fear'; Laundré et al. 2010) provides a framework we can use to understand the likely impacts of these subtle changes in animal behaviour as they navigate the 'landscape of heat' in a warming world (Cunningham et al. 2021).

1.8 Animal behaviour and conservation in a changing climate

Climate change affects population trajectories of different species in different ways, creating both 'winners' and 'losers' within and across taxon groups (e.g. Clucas et al. 2014; Tayleur et al. 2016; Kafash et al. 2018; Jackson et al. 2022; Fekete et al. 2023). A range of species traits underlie these differences in vulnerability and resilience, including morphological (e.g. body mass), ecological (e.g. habitat or diet specialism), physiological (e.g. broad versus narrow thermal tolerance ranges), and behavioural (e.g. sociality) characters. Behavioural flexibility is a hallmark of species that cope well with environmental change (e.g. Tuomainen and Candolin 2011), and climate change 'winners' include adaptive species that can take advantage of novel opportunities (as seen in the example of pied crows and the use of powerlines discussed in Section 1.5.2; Cunningham et al. 2016).

However, although animals that display behavioural flexibility may succeed in a changing world, alterations in their behaviour that change their interactions with humans can also exacerbate conservation problems (Abrahms 2021). Climate change is exacerbating negative human-wildlife interactions globally because it alters resource availability in natural systems and may result in large changes in habitat suitability (see also Chapter 13). Some wildlife species respond behaviourally by changing their habitat use as they seek new sources of food and shelter (Abrahms et al. 2023). For example, in Africa and North America, drought has increased visitation rates of elephants *Loxodonta africana* and tapirs *Tapirus bairdii* to human settlements in search of food, causing crop damage and retaliatory killings (Mukeka et al. 2020; Pérez-Flores et al. 2021). Loss of sea ice has driven polar bears into terrestrial habitats, causing increasing numbers of direct encounters with, and attacks on, humans

(Wilder et al. 2017). Changes in the movement ecology of animals associated with climate change can bring individuals into greater contact with roads and other infrastructure, increasing collision mortality (e.g. Popp et al. 2018).

Climate change is also causing immediate conservation problems through direct effects on species. For example, increased temperatures and reduced rainfall in the Mojave Desert over the last five decades have caused collapse of the bird community, with surveyed locations losing on average 43% of their species over that time (Iknayan and Beissinger 2018); and opportunity costs associated with behavioural thermoregulation have caused the extinction of *Scleroporus* lizard populations in Mexico under climate warming trends (Sinervo et al. 2010). Detailed data on the behavioural mechanisms underlying these population-level changes can provide guidance for effective conservation management (Box 1.1).

Box 1.1 Using behaviour data to inform practical conservation interventions under climate change

Detailed data on behavioural and physiological responses of animals to changing climate conditions can aid the design of targeted, practical conservation interventions. For example, work on physiological correlates of a climate change-driven avian community collapse in the Mojave Desert (Iknayan and Beissinger 2018) revealed that species' declines were largely driven by evaporative cooling requirements (Riddell et al. 2019). In hot and arid environments, visiting surface water sources during the day can be dangerous for small animals because of high temperatures (e.g. > 60°C) at the water's edge (Abdu et al. 2018). Installing shade canopies at waterholes in arid landscapes may therefore keep water available during heatwaves for birds and other small animals, by reducing environmental temperatures (Figure 1.4).

Figure 1.4 Providing artificial shade at desert waterholes can dramatically reduce temperatures at the water's edge, reducing the thermal risk experienced by small animals drinking during the heat of the day. By keeping water accessible when they need it most, this simple intervention could help animals survive heatwaves.

Photo: S. Cunningham

Box 1.1 *Continued*

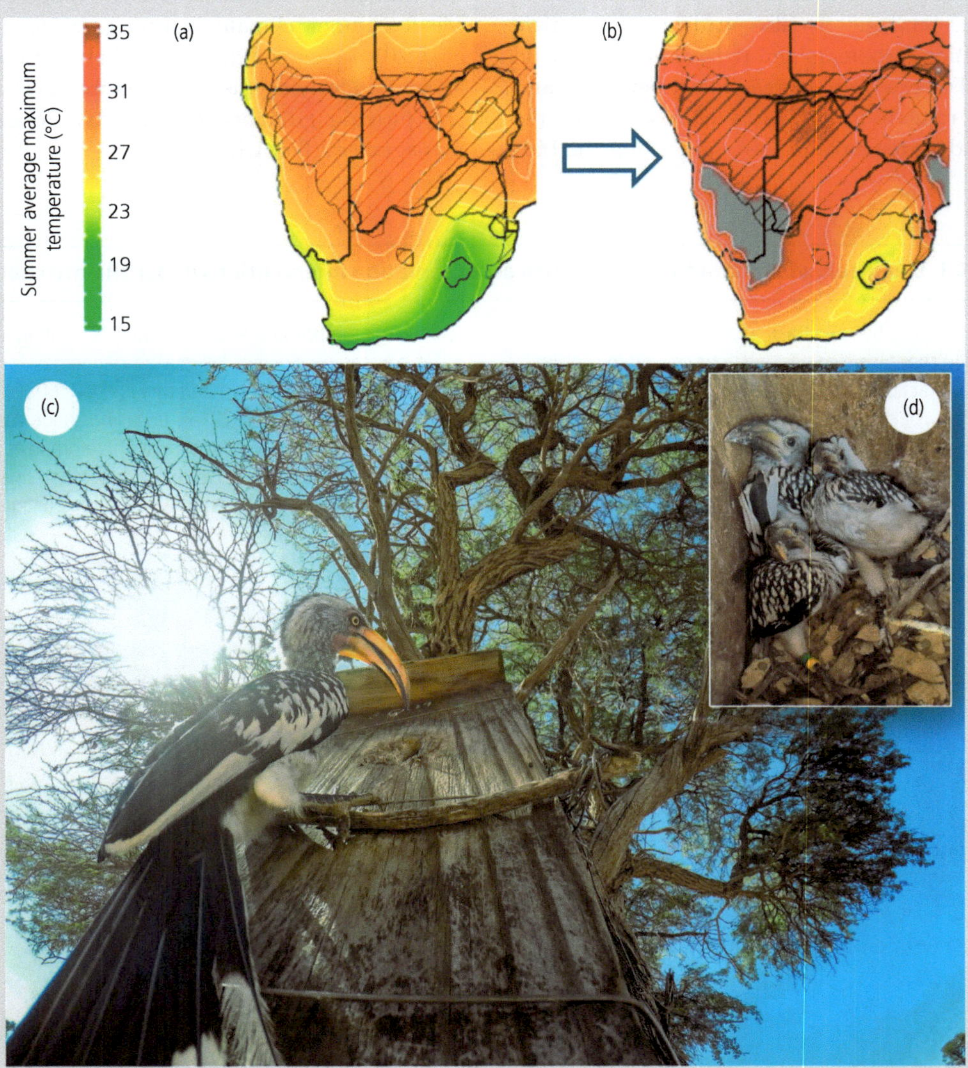

Figure 1.5 Southern yellow-billed hornbills *Tockus leucomelas* are summer-breeding cavity-nesting birds from southern Africa. Their nest success drops below 50% when average maximum daily air temperatures during the nesting period exceed 35°C (van de Ven et al. 2020). The maps of southern Africa show summer average maximum air temperatures (a) now and (b) in 2080. The hashed area is the hornbills' geographic range; the grey portions in (b) show areas where hornbill breeding success will be < 50% by the end of the century. Hornbills readily breed in nest boxes (c, a male hornbill provisioning his nest; d, inset, chicks inside the nest box). Providing thermally buffered nest boxes for these birds might help improve nest success during hot weather. Mapping hornbill vulnerability in space and time using behaviour data highlights areas of greatest climate vulnerability within their current range for conservation planning.

Maps from Conradie et al. (2019); photo: Lisa Nupen

Box 1.1 *Continued*

When trialled in the Kalahari Desert, this simple intervention allowed more birds to drink at hotter times of day and improved water access for several small-bodied species, though the presence of shade structures reduced visitation rates of some larger birds, perhaps due to increased real or perceived predation risk, highlighting the importance of careful design and monitoring of species responses (Abdu et al. 2018). Detailed behavioural data on parental care, breeding outcomes, and survival in cavity-nesting and roosting species further highlight the importance of careful design when providing nest boxes as conservation tools. These can either buffer animals from extreme weather events, including heatwaves (when suitably designed), or become ecological traps (e.g. Larson et al. 2018; Figure 1.5). Fine-scale data on physiological heat tolerance limits and heat stress-related behaviours of different flying fox species have enabled development of best-practice guidelines for management and prevention of heat-related mass mortality events of these large fruit bats in Queensland, Australia (Bishop 2020). Finally, data from urban wildlife with access to abundant anthropogenic food and water supplies suggest easy access to resources might buffer mass losses associated with reduced foraging during hot weather (e.g. Edwards et al. 2015; Stofberg et al. 2022). This suggests interesting avenues for conservation interventions involving resource provision.

Behavioural responses to changing abiotic conditions are often non-linear, corresponding to non-linear patterns in animals' underlying physiology (e.g. Figures 1.1 and 1.2). Multiple studies identify tipping points in air temperatures, above which fitness costs associated with heat exposure become rapidly more severe (e.g. du Plessis et al. 2012; van de Ven et al. 2019; Oswald et al. 2021). By mapping thresholds in space and time based on climate change projections, areas of high risk and potential thermal refugia can be identified within species ranges (e.g. Conradie et al. 2019; Figure 1.5). These maps can aid spatial conservation planning: identifying areas suitable for protection and allowing assessment of the future suitability of existing protected areas. More controversially, behaviour data like these could be used to identify areas in which the climate is becoming suitable for species outside their current ranges, into which they could be introduced ('assisted migration'; Ste-Marie et al. 2011).

1.9 Conclusion

Global temperatures continue to rise as the concentration of greenhouse gases in our atmosphere climbs (IPCC 2023). Animals are responding behaviourally at multiple spatial and temporal scales, from micro-adjustments in habitat use and daily activity patterns, to shifts in geographic ranges and phenology of life-history events. Many behavioural responses stem from phenotypic plasticity; however, climate change is also driving changes in the costs and benefits of different behaviours, altering selection pressures and potentially resulting in evolutionary change. Interspecific variation in animals' responses to climate change is re-ordering biological communities, changing species interactions, and altering ecosystem function, with wide-reaching and unpredictable effects. The remarkable behavioural flexibility displayed by animals may provide some buffer against the negative impacts of climate change, and some species are thriving under new climate regimes. But climate change 'losers' outnumber the 'winners', and climate change is a major contributor to global biodiversity loss (Dirzo et al. 2014). Aided by detailed data on animal behaviour and physiology (e.g. Box 1.1), we may be able to mitigate some climate change impacts on wildlife in the short term. If we are to conserve biodiversity as we know it in the long term, we must embrace new technologies for power and food production and make lifestyle changes to reduce greenhouse gas emissions. Climate change is happening everywhere, not just where people live. When it comes to this defining environmental issue of our age, human attitudes at all levels of society matter to wildlife both on our doorsteps and in remote locations.

References

Abdu, S., McKechnie, A.E., Lee, A.T.K., et al. (2018). Can providing shade at water points help Kalahari birds beat the heat? *Journal of Arid Environments*, 152, 21–27.

Abrahms, B. (2021). Human-wildlife conflict under climate change. *Science*, 373, 484–485.

Abrahms, B., Carter, N.H., Clark-Wolf, T.J., et al. (2023). Climate change as a global amplifier of human–wildlife conflict. *Nature Climate Change*, 13, 224–234.

Alessandrini, C., Scridel, D., Boitani, L., et al. (2022). Remotely sensed variables explain microhabitat selection and reveal buffering behaviours against warming in a climate-sensitive bird species. *Remote Sensing in Ecology and Conservation*, 8, 615–628.

AlRashidi, M., Kosztolányi, A., Shobrak, M., et al. (2011). Parental cooperation in an extreme hot environment: natural behaviour and experimental evidence. *Animal Behaviour*, 82, 235–243.

Angilletta, M.J., Cooper, B.S., Schuler, M.S., et al. (2010). The evolution of thermal physiology in endotherms. *Frontiers in Bioscience*, 2, 861–881.

Bakken, G.S., Santee, W.R., and Erskine, D.J. (1985). Operative and standard operative temperature: tools for thermal energetics studies. *American Zoologist*, 25, 933–943.

Barbraud, C., and Weimerskirch, H. (2006). Antarctic birds breed later in response to climate change. *Proceedings of the National Academy of Sciences*, 103, 6248–6251.

Barras, A.G., Niffenegger, C.A., Candolfi, I., et al. (2021). Nestling diet and parental food provisioning in a declining mountain passerine reveal high sensitivity to climate change. *Journal of Avian Biology*, 52, jav.02649.

Bauer, Z., Trnka, M., Bauerová, J., et al. (2010). Changing climate and the phenological response of great tit and collared flycatcher populations in floodplain forest ecosystems in Central Europe. *International Journal of Biometeorology*, 54, 99–111.

Beever, E.A., Hall, L.E., Varner, J., et al. (2017). Behavioral flexibility as a mechanism for coping with climate change. *Frontiers in Ecology and the Environment*, 15, 299–308.

Bishop, T. (2020). *Interim Flying-Fox Heat Stress Guideline*. Queensland Government.

Blakey, R.V., Sikich, J.A., Blumstein, D.T., et al. (2022). Mountain lions avoid burned areas and increase risky behavior after wildfire in a fragmented urban landscape. *Current Biology*, 32, 4762–4768.e5.

Both, C., Bouwhuis, S., Lessells, C.M., et al. (2006). Climate change and population declines in a long-distance migratory bird. *Nature*, 441, 81–83.

Both, C., Van Asch, M., Bijlsma, R.G., et al. (2009). Climate change and unequal phenological changes across four trophic levels: constraints or adaptations? *Journal of Animal Ecology*, 78, 73–83.

Bourne, A.R., Cunningham, S.J., Spottiswoode, C.N., et al. (2020a). Hot droughts compromise interannual survival across all group sizes in a cooperatively breeding bird. *Ecology Letters*, 23, 1776–1788.

Bourne, A.R., Cunningham, S.J., Spottiswoode, C.N., et al. (2020b). High temperatures drive offspring mortality in a cooperatively breeding bird. *Proceedings of the Royal Society B: Biological Sciences*, 287, 20201140.

Bourne, A.R., Ridley, A.R., and Cunningham, S.J. (2023). Helpers don't help when it's hot in a cooperatively breeding bird, the Southern Pied Babbler. *Behavioral Ecology*, arad023.

Bowman, D.M.J.S., Perry, G.L.W., Higgins, S.I., et al. (2016). Pyrodiversity is the coupling of biodiversity and fire regimes in food webs. *Philosophical Transactions of the Royal Society B: Biological Sciences*, 371.

Boyles, J.G., Seebacher, F., Smit, B., et al. (2011). Adaptive thermoregulation in endotherms may alter responses to climate change. *Integrative and Comparative Biology*, 51, 676–690.

Brown, C.J., O'Connor, M.I., Poloczanska, E.S., et al. (2016). Ecological and methodological drivers of species' distribution and phenology responses to climate change. *Global Change Biology*, 22, 1548–1560.

Brown, J.L. (1987). *Helping Communal Breeding in Birds: Ecology and Evolution*. Princeton University Press, Princeton, New Jersey.

Buckley, L.B., Ehrenberger, J.C., and Angilletta, M.J. (2015). Thermoregulatory behaviour limits local adaptation of thermal niches and confers sensitivity to climate change. *Functional Ecology*, 29, 1038–1047.

Buckley, L.B., Tewksbury, J.J., and Deutsch, C.A. (2013). Can terrestrial ectotherms escape the heat of climate change by moving? *Proceedings of the Royal Society B: Biological Sciences*, 280, 20131149.

Burger, C., Belskii, E., Eeva, T., et al. (2012). Climate change, breeding date and nestling diet: how temperature differentially affects seasonal changes in pied flycatcher diet depending on habitat variation. *Journal of Animal Ecology*, 81, 926–936.

Burgess, M.D., Smith, K.W., Evans, K.L., et al. (2018). Tritrophic phenological match–mismatch in space and time. *Nature Ecology & Evolution*, 2, 970–975.

Calder, W.A., and King, J.R. (1974). Thermal and caloric relations of birds. In: D.S. Farner and J.R. King (eds), *Avian Biology, Volume IV*, pp. 259–413. Academic Press, New York.

Campagna, C., and Le Boeuf, B.J. (1988). Thermoregulatory behaviour of southern sea lions and its effect on mating strategies. *Behaviour*, 107, 72–90.

Capilla-Lasheras, P., Harrison, X., Wood, E.M., et al. (2021). Altruistic bet-hedging and the evolution of cooperation in a Kalahari bird. *Science Advances*, 7, 1–11.

Carroll, J.M., Davis, C.A., Elmore, R.D., et al. (2015). Thermal patterns constrain diurnal behavior of a ground-dwelling bird. *Ecosphere*, 6, 1–15.

Carroll, J.M., Davis, C.A., Fuhlendorf, S.D., et al. (2016). Landscape pattern is critical for the moderation of thermal extremes. *Ecosphere*, 7, 1–16.

Catry, I., Catry, T., Patto, P., et al. (2015). Differential heat tolerance in nestlings suggests sympatric species may face different climate change risks. *Climate Research*, 66, 13–24.

Charmantier, A., and Gienapp, P. (2013). Climate change and timing of avian breeding and migration: evolutionary versus plastic changes. *Evolutionary Applications*, 7, 15–28.

Clark, L. (1987). Thermal constraints on foraging in adult European starlings. *Oecologia*, 71, 233–238.

Clucas, G.V., Dunn, M.J., Dyke, G., et al. (2014). A reversal of fortunes: climate change 'winners' and 'losers' in Antarctic Peninsula penguins. *Scientific Reports*, 4, 1–7.

Cohen, J.M., Lajeunesse, M.J., and Rohr, J.R. (2018). A global synthesis of animal phenological responses to climate change. *Nature Climate Change*, 8, 224–228.

Conradie, S.R., Woodborne, S.M., Cunningham, S.J., et al. (2019). Chronic, sublethal effects of high temperatures will cause severe declines in southern African arid-zone birds during the 21st century. *Proceedings of the National Academy of Sciences*, 116, 14065–14070.

Cook, T.R., Martin, R., Roberts, J., et al. (2020). Parenting in a warming world: thermoregulatory responses to heat stress in an endangered seabird. *Conservation Physiology*, 8, coz109.

Corbett, L.K., Anderson, A.N., and Muller, W.J. (2003). Terrestrial vertebrates. In: A.N. Anderson, G.D. Cook, and R.J. Williams (eds), *Fire in Tropical Savannas*, pp. 126–152. Springer, New York.

Cunningham, S.J., Gardner, J.L., and Martin, R.O. (2021). Opportunity costs and the response of birds and mammals to climate warming. *Frontiers in Ecology and the Environment*, 19, 300–307.

Cunningham, S.J., Madden, C.F., Barnard, P., et al. (2016). Electric crows: powerlines, climate change and the emergence of a native invader. *Diversity and Distributions*, 22, 17–29.

Cunningham, S.J., Martin, R.O., Hojem, C.L., et al. (2013). Temperatures in excess of critical thresholds threaten nestling growth and survival in a rapidly-warming arid savanna: a study of common fiscals. *PLOS ONE*, 8, e74613.

Cunningham, S.J., Thompson, M.L., and McKechnie, A.E. (2017). It's cool to be dominant: social status alters short-term risks of heat stress. *Journal of Experimental Biology*, 220, 1558–1562.

D'amelio, P.B., Ferreira, A.C., Fortuna, R., et al. (2022). Disentangling climatic and nest predator impact on reproductive output reveals adverse high-temperature

effects regardless of helper number in an arid-region cooperative bird. *Ecology Letters*, 25, 151–162.

Davis, A.J., Jenkinson, L.S., and Lawton, J.H. (1998a). Making mistakes when predicting shifts in species range in response to global warming. *Nature*, 391, 783–786.

Davis, A.J., Lawton, J.H., Shorrocks, B., et al. (1998b). Individualistic species responses invalidate simple physiological models of community dynamics under global environmental change. *Journal of Animal Ecology*, 67, 600–612.

Dickman, C.R., Mahon, P.S., Masters, P., et al. (1999). Long-term dynamics of rodent populations in arid Australia: the influence of rainfall. *Wildlife Research*, 26, 389–403.

Dirnböck, T., Essl, F., and Rabitsch, W. (2011). Disproportional risk for habitat loss of high-altitude endemic species under climate change. *Global Change Biology*, 17, 990–996.

Dirzo, R., Young, H.S., Galetti, M., et al. (2014). Defaunation in the Anthropocene. *Science*, 345, 401–406.

Doherty, T.S., Geary, W.L., Jolly, C.J., et al. (2022). Fire as a driver and mediator of predator–prey interactions. *Biological Reviews*, 97, 1539–1558.

du Plessis, K.L., Martin, R.O., Hockey, P.A.R., et al. (2012). The costs of keeping cool in a warming world: implications of high temperatures for foraging, thermoregulation and body condition of an arid-zone bird. *Global Change Biology*, 18, 3063–3070.

Edwards, E.K., Mitchell, N.J., and Ridley, A.R. (2015). The impact of high temperatures on foraging behaviour and body condition in the Western Australian Magpie *Cracticus tibicen dorsalis*. *Ostrich*, 86, 137–144.

Fekete, J., De Knijf, G., Dinis, M., et al. (2023). Winners and losers: cordulegaster species under the pressure of climate change. *Insects*, 14, 348.

Francis, J.M., and Boness, D.J. (1991). The effect of thermoregulatory behaviour on the mating system of the Juan Fernandez fur seal, *Arctocephalus philippii*. *Behaviour*, 119, 104–126.

Gaston, A.J., Hipfner, J.M., and Campbell, D. (2002). Heat and mosquitoes cause breeding failures and adult mortality in an arctic-nesting seabird. *Ibis*, 144, 185–191.

Gill, J.A., Alves, J.A., Sutherland, W.J., et al. (2013). Why is timing of bird migration advancing when individuals are not? *Proceedings of the Royal Society B: Biological Sciences*, 281, 20132161.

Gilman, S.E., Urban, M.C., Tewksbury, J., et al. (2010). A framework for community interactions under climate change. *Trends in Ecology & Evolution*, 25, 325–331.

Guindre-Parker, S., and Rubenstein, D.R. (2020). Survival benefits of group living in a fluctuating environment. *The American Naturalist*, 195, 1027–1036.

Hamilton, W.D. (1964). The genetical evolution of social behaviour. II. *Journal of Theoretical Biology*, 7, 17–52.

Hatchwell, B.J., and Komdeur, J. (2000). Ecological constraints, life history traits and the evolution of cooperative breeding. *Animal Behaviour*, 59, 1079–1086.

Helm, B., Van Doren, B.M., Hoffmann, D., et al. (2019). Evolutionary response to climate change in migratory pied flycatchers. *Current Biology*, 29, 3714–3719.e4.

Hetem, R.S., Strauss, W.M., Fick, L.G., et al. (2012). Activity re-assignment and microclimate selection of free-living Arabian oryx: responses that could minimise the effects of climate change on homeostasis? *Zoology*, 115, 411–416.

Hidalgo Aranzamendi, N., Hall, M.L., Kingma, S.A., et al. (2019). Rapid plastic breeding response to rain matches peak prey abundance in a tropical savanna bird. *Journal of Animal Ecology*, 88, 1799–1811.

Holling, C.S. (1959). The components of predation as revealed by a study of small-mammal predation of the European pine sawfly. *The Canadian Entomologist*, 91, 293–320.

Hovick, T.J., McGranahan, D.A., Elmore, R.D., et al. (2017). Pyric-carnivory: raptor use of prescribed fires. *Ecology and Evolution*, 7, 9144–9150.

Huey, R.B., Kearney, M.R., Krockenberger, A., et al. (2012). Predicting organismal vulnerability to climate warming: roles of behaviour, physiology and adaptation. *Philosophical Transactions of the Royal Society B: Biological Sciences*, 367, 1665–1679.

Huey, R.B., and Tewksbury, J.J. (2009). Can behavior douse the fire of climate warming? *Proceedings of the National Academy of Sciences*, 106, 3647–3648.

Hunt, K., Marais, L., Cunningham, S.J., et al. (2023). Camelthorn and blackthorn trees provide important resources for Southern Pied Babblers (*Turdoides bicolor*) in the Kalahari. *Ibis*, 13, 82–94.

Iknayan, K.J., and Beissinger, S.R. (2018). Collapse of a desert bird community over the past century driven by climate change. *Proceedings of the National Academy of Sciences*, 115, 8597–8602.

IPCC. (2023). *Climate Change 2023: Synthesis Report*. C.W. Team, H. Lee, J. Romero (eds). IPCC, Geneva, Switzerland.

Iverson, S.A., Gilchrist, H.G., Smith, P.A., et al. (2014). Longer ice-free seasons increase the risk of nest depredation by polar bears for colonial breeding birds in the Canadian Arctic. *Proceedings of the Royal Society B: Biological Sciences*, 281, 20133128.

Jackson, H.M., Johnson, S.A., Morandin, L.A., et al. (2022). Climate change winners and losers among North American bumblebees. *Biology Letters*, 18, 20210551.

Jacobsen, D. (2020). The dilemma of altitudinal shifts: caught between high temperature and low oxygen. *Frontiers in Ecology and the Environment*, 18, 211–218.

Jetz, W., and Rubenstein, D.R. (2011). Environmental uncertainty and the global biogeography of cooperative breeding in birds. *Current Biology*, 21, 72–78.

Jolly, C.J., Dickman, C.R., Doherty, T.S., et al. (2022). Animal mortality during fire. *Global Change Biology*, 28, 2053–2065.

Jones, M., Mandelik, Y., and Dayan, T. (2001). Coexistence of temporally partitioned spiny mice: roles of habitat structure and foraging behavior. *Ecology*, 82, 2164–2176.

Kafash, A., Ashrafi, S., Ohler, A., et al. (2018). Climate change produces winners and losers: differential responses of amphibians in mountain forests of the Near East. *Global Ecology and Conservation*, 16, e00471.

Kay, C.B., Delehanty, D.J., Pradhan, D.S., et al. (2021). Climate change and wildfire-induced alteration of fight-or-flight behavior. *Climate Change Ecology*, 1, 100012.

Kharouba, H.M., and Wolkovich, E.M. (2023). Lack of evidence for the match-mismatch hypothesis across terrestrial trophic interactions. *Ecology Letters*, 26, 955–964.

Koenig, W.D., and Mumme, R.L. (1987). *Population Ecology of the Cooperatively Breeding Acorn Woodpecker*. Princeton University Press, Princeton, New Jersey.

Larson, E.R., Eastwood, J.R., Buchanan, K.L., et al. (2018). Nest box design for a changing climate: the value of improved insulation. *Ecological Management & Restoration*, 19, 39–48.

Laundré, J.W., Hernández, L., and Ripple, W.J. (2010). The landscape of fear: ecological implications of being afraid. *The Open Ecology Journal*, 3, 1–7.

Levesque, D.L., and Marshall, K.E. (2021). Do endotherms have thermal performance curves? *Journal of Experimental Biology*, 224, jeb141309.

Lloyd, P. (1999). Rainfall as a breeding stimulus and clutch size determinant in South African arid-zone birds. *Ibis*, 141, 637–643.

Maclean, G.L., and Kirwan, G.M. (2020). Bronze-winged courser (*Rhinoptilus chalcopterus*), version 1.0. In: J. del Hoyo, A. Elliott, J. Sargatal, et al. (eds), *Birds of the World*. Cornell Lab of Ornithology, Ithaca, New York.

Mares, R., Doutrelant, C., Paquet, M., et al. (2017). Breeding decisions and output are correlated with both temperature and rainfall in an arid-region passerine, the sociable weaver. *Royal Society Open Science*, 4, 170835.

Mason, T.H.E., Brivio, F., Stephens, P.A., et al. (2017). The behavioral trade-off between thermoregulation and foraging in a heat-sensitive species. *Behavioral Ecology*, 28, 908–918.

McGregor, H.W., Legge, S., Jones, M.E., et al. (2016). Extraterritorial hunting expeditions to intense fire scars by feral cats. *Scientific Reports*, 6, 1–7.

McKechnie, A.E., Rushworth, I.A., Myburgh, F., et al. (2021). Mortality among birds and bats during an extreme heat event in eastern South Africa. *Austral Ecology*, 46, 687–691.

McNab, B.K. (2012). *Extreme Measures: The Ecological Energetics of Birds and Mammals*. University of Chicago Press, Chicago, Illinois.

Michel, A., Johnson, J.R., Szeligowski, R., et al. (2023). Integrating sensory ecology and predator-prey theory to understand animal responses to fire. *Ecology Letters*, 26, 1050–1070.

Mukeka, J.M., Ogutu, J.O., Kanga, E., et al. (2020). Spatial and temporal dynamics of human–wildlife conflicts in the Kenya Greater Tsavo Ecosystem. *Human-Wildlife Interactions*, 14, 255–272.

Nicolau, P.G., Burgess, M.D., Marques, T.A., et al. (2021). Latitudinal variation in arrival and breeding phenology of the pied flycatcher *Ficedula hypoleuca* using large-scale citizen science data. *Journal of Avian Biology*, 52, e02646.

Niven, D.K., Butcher, G.S., and Bancroft, G.T. (2009). Christmas bird counts and climate change: northward shifts in early winter abundance. *American Birds*, 109, 10–15.

Oswald, K.N., Smit, B., Lee, A.T.K., et al. (2021). Higher temperatures are associated with reduced nestling body condition in a range-restricted mountain bird. *Journal of Avian Biology*, 52, 1–10.

Oswald, S.A., Bearhop, S., Furness, R.W., et al. (2008). Heat stress in a high-latitude seabird: effects of temperature and food supply on bathing and nest attendance of great skuas *Catharacta skua*. *Journal of Avian Biology*, 39, 163–169.

Palmer, C.M. (2010). Chronological changes in terrestrial insect assemblages in the arid zone of Australia. *Environmental Entomology*, 39, 1775–1787.

Parmesan, C., and Yohe, G. (2003). A globally coherent fingerprint of climate change impacts across natural systems. *Nature*, 421, 37–42.

Parrini, F., and Owen-Smith, N. (2010). The importance of post-fire regrowth for sable antelope in a Southern African savanna. *African Journal of Ecology*, 48, 526–534.

Pattinson, N.B., van de Ven, T.M.F.N., Finnie, M.J., et al. (2022). Collapse of breeding success in desert-dwelling hornbills evident within a single decade. *Frontiers in Ecology and Evolution*, 10, 842264.

Pausas, J.G., and Parr, C.L. (2018). Towards an understanding of the evolutionary role of fire in animals. *Evolutionary Ecology*, 32, 113–125.

Pecl, G.T., Araújo, M.B., Bell, J.D., et al. (2017). Biodiversity redistribution under climate change: impacts on ecosystems and human well-being. *Science*, 355, eaai9214.

Pérez-Flores, J., Mardero, S., López-Cen, A., et al. (2021). Human-wildlife conflicts and drought in the greater Calakmul Region, Mexico: implications for tapir conservation. *Neotropical Biology and Conservation*, 16, 539–563.

Popp, J.N., Hamr, J., Chan, C., et al. (2018). Elk (*Cervus elaphus*) railway mortality in Ontario. *Canadian Journal of Zoology*, 96, 1066–1070.

Prop, J., Aars, J., Bårdsen, B.J., et al. (2015). Climate change and the increasing impact of polar bears on bird populations. *Frontiers in Ecology and Evolution*, 3, 1–12.

Rickbeil, G.J.M., Hermosilla, T., Coops, N.C., et al. (2017). Barren-ground caribou (*Rangifer tarandus groenlandicus*) behaviour after recent fire events; integrating caribou telemetry data with Landsat fire detection techniques. *Global Change Biology*, 23, 1036–1047.

Riddell, E.A., Iknayan, K.J., Wolf, B.O., et al. (2019). Cooling requirements fueled the collapse of a desert bird community from climate change. *Proceedings of the National Academy of Sciences*, 116, 21609–21615.

Ridley, A.R., Raihani, N.J., and Nelson-flower, M.J. (2008). The cost of being alone: the fate of floaters in a population of cooperatively breeding pied babblers *Turdoides bicolor*. *Journal of Avian Biology*, 39, 389–392.

Robillard, C.M., Coristine, L.E., Soares, R.N., et al. (2015). Facilitating climate-change-induced range shifts across continental land-use barriers. *Conservation Biology*, 29, 1586–1595.

Rubenstein, D.R. (2011). Spatiotemporal environmental variation, risk aversion, and the evolution of cooperative breeding as a bet-hedging strategy. *Proceedings of the National Academy of Sciences*, 108 (supplement 2), 10816–10822.

Sadowska, J., Gębczynski, A.K., Lewoc, M., et al. (2019). Not that hot after all: no limits to heat dissipation in lactating mice selected for high or low BMR. *Journal of Experimental Biology*, 222, jeb204669.

Santee, W.R., and Bakken, G.S. (1987). Social displays in red-winged blackbirds (*Agelaius phoeniceus*): sensitivity to thermoregulatory costs. *Auk*, 104, 413–420.

Scholander, P.F., Hock, R., Walters, V., et al. (1950). Heat regulation in some arctic and tropical mammals and birds. *Biology Bulletin*, 99, 237–258.

Sears, M.W., Angilletta, M.J., Schuler, M.S., et al. (2016). Configuration of the thermal landscape determines thermoregulatory performance of ectotherms. *Proceedings of the National Academy of Sciences*, 113, 10,595–10,600.

Sharpe, L.L., Bayter, C., and Gardner, J.L. (2021). Too hot to handle? Behavioural plasticity during incubation in a small, Australian passerine. *Journal of Thermal Biology*, 98, 102921.

Shen, S.F., Emlen, S.T., Koenig, W.D., et al. (2017). The ecology of cooperative breeding behaviour. *Ecology Letters*, 20, 708–720.

Simons, M.J.P., Reimert, I., Van Der Vinne, V., et al. (2011). Ambient temperature shapes reproductive output during pregnancy and lactation in the common vole (*Microtus arvalis*): a test of the heat dissipation limit theory. *Journal of Experimental Biology*, 214, 38–49.

Sinervo, B., Méndez-de-la-cruz, F., Miles, D.B., et al. (2010). Erosion of lizard diversity by climate change and altered thermal niches. *Science*, 328, 894–899.

Solomon, B.Y.M.E. (1949). The natural control of animal populations. *Journal of Animal Ecology*, 18, 1–35.

Speakman, J.R., and Król, E. (2010). Maximal heat dissipation capacity and hyperthermia risk: neglected key factors in the ecology of endotherms. *Journal of Animal Ecology*, 79, 726–746.

Ste-Marie, C., Nelson, E.A., Dabros, A., et al. (2011). Assisted migration: introduction to a multifaceted concept. *The Forestry Chronicle*, 87, 724–730.

Stenseth, N.C., and Mysterud, A. (2002). Climate, changing phenology, and other life history traits: nonlinearity and match-mismatch to the environment. *Proceedings of the National Academy of Sciences*, 99, 13,379–13,381.

Stofberg, M., Amar, A., Sumasgutner, P., et al. (2022). Staying cool and eating junk: influence of heat dissipation and anthropogenic food on foraging and body condition in an urban passerine. *Landscape and Urban Planning*, 226, 104465.

Street, G.M., Fieberg, J., Rodgers, A.R., et al. (2016). Habitat functional response mitigates reduced foraging opportunity: implications for animal fitness and space use. *Landscape Ecology*, 31, 1939–1953.

Tayleur, C.M., Devictor, V., Gaüzère, P., et al. (2016). Regional variation in climate change winners and losers highlights the rapid loss of cold-dwelling species. *Diversity and Distributions*, 22, 468–480.

Tieleman, B., and Williams, J.B. (2002). Effects of food supplementation on behavioural decisions of hoopoe-larks in the Arabian Desert: balancing water, energy and thermoregulation. *Animal Behaviour*, 63, 519–529.

Tingley, M.W., Koo, M.S., Moritz, C., et al. (2012). The push and pull of climate change causes heterogeneous shifts in avian elevational ranges. *Global Change Biology*, 18, 3279–3290.

Tuomainen, U., and Candolin, U. (2011). Behavioural responses to human-induced environmental change. *Biological Reviews*, 86, 640–657.

Tylianakis, J.M., Didham, R.K., Bascompte, J., et al. (2008). Global change and species interactions in terrestrial ecosystems. *Ecology Letters*, 11, 1351–1363.

Van der Wal, J., Murphy, H.T., Kutt, A.S., et al. (2013). Focus on poleward shifts in species' distribution underestimates the fingerprint of climate change. *Nature Climate Change*, 3, 239–243.

Van de Ven, T.M.F.N., Fuller, A., and Clutton-Brock, T.H. (2020). Effects of climate change on pup growth and survival in a cooperative mammal, the meerkat. *Functional Ecology*, 34, 194–202.

van de Ven, T.M.F.N., McKechnie, A.E., and Cunningham, S.J. (2019). The costs of keeping cool: behavioural trade-offs between foraging and thermoregulation are associated with significant mass losses in an arid-zone bird. *Oecologia*, 191, 205–215.

van de Ven, T.M.F.N., McKechnie, A.E., Er, S., et al. (2020). High temperatures are associated with substantial reductions in breeding success and offspring quality in an arid-zone bird. *Oecologia*, 193, 225–235.

Veldhuis, M.P., Kihwele, E.S., Cromsigt, J.P.G.M., et al. (2019). Large herbivore assemblages in a changing climate: incorporating water dependence and thermoregulation. *Ecology Letters*, 22, 1536–1546.

Verdolin, J.L. (2006). Meta-analysis of foraging and predation risk trade-offs in terrestrial systems. *Behavioral Ecology and Sociobiology*, 60, 457–464.

Visser, M.E., and Gienapp, P. (2019). Evolutionary and demographic consequences of phenological mismatches. *Nature Ecology & Evolution*, 3, 879–885.

Visser, M.E., Holleman, L.J.M., and Gienapp, P. (2006). Shifts in caterpillar biomass phenology due to climate change and its impact on the breeding biology of an insectivorous bird. *Oecologia*, 147, 164–172.

Welbergen, J.A., Klose, S.M., Markus, N., et al. (2008). Climate change and the effects of temperature extremes on Australian flying-foxes. *Proceedings of the Royal Society B: Biological Sciences*, 275, 419–425.

Weyer, N.M., Fuller, A., Haw, A.J., et al. (2020). Increased diurnal activity is indicative of energy deficit in a nocturnal mammal, the aardvark. *Frontiers in Physiology*, 11, 1–15.

Wilder, J.M., Vongraven, D., Atwood, T., et al. (2017). Polar bear attacks on humans: implications of a changing climate. *Wildlife Society Bulletin*, 41, 537–547.

Wiley, E.M., and Ridley, A.R. (2016). The effects of temperature on offspring provisioning in a cooperative breeder. *Animal Behaviour*, 117, 187–195.

Woodroffe, R., Groom, R., and Mcnutt, J.W. (2017). Hot dogs: high ambient temperatures impact reproductive success in a tropical carnivore. *Journal of Animal Ecology*, 86, 1329–1338.

Zhao, Z.J., Hambly, C., Shi, L.L., et al. (2020). Late lactation in small mammals is a critically sensitive window of vulnerability to elevated ambient temperature. *Proceedings of the National Academy of Sciences*, 117, 24,352–24,358.

Zhemchuzhnikov, M.K., Versluijs, T.S.L., Lameris, T.K., et al. (2021). Exploring the drivers of variation in trophic mismatches: a systematic review of long-term avian studies. *Ecology and Evolution*, 11, 3710–3725.

Zöckler, C., and Lysenko, I. (2000). Water birds on the edge: first circumpolar assessment of climate change impact on arctic breeding water birds. *WCMC Biodiversity Series No. 11: Water*.

CHAPTER 2

Noise pollution

Andrew N. Radford

Overview

Anthropogenic (human-made) noise pervades all ecosystems across the globe. Human activities both increase the amount of noise and generate sounds with different characteristics to those occurring naturally, meaning that wildlife now faces novel and unprecedented acoustic challenges compared to pre-industrial times. Behavioural changes are the most widespread response to noise pollution because such changes can result from even moderate noise levels occurring far from a source; they are also the most well-studied. Here, I begin with an overview of the different behavioural responses generated by anthropogenic noise. I then describe reasons why we see variation in these responses—within and between species, and across time—and the consequences for individual fitness, populations, communities, and ecosystems. Throughout, I provide examples illustrating the diverse range of taxa affected. Finally, I discuss what we still need to learn about noise impacts and the importance of implementing and testing mitigation and management strategies.

2.1 Anthropogenic noise: sources and problems

Terrestrial and aquatic environments across the globe have become substantially noisier since the Industrial Revolution because of sounds produced by human activities (Barber et al. 2010; Buxton et al. 2019; Duarte et al. 2021). Major sources of anthropogenic (human-made) noise on land include transportation networks, resource extraction, and urban development, whilst commercial shipping, recreational boating, offshore construction, habitat exploration, and energy production all generate noise in aquatic habitats. In some cases, such as the use of seismic surveys and sonar arrays, humans produce sounds deliberately. Many other human activities, such as vehicle use, pile-driving, and infrastructure operation, produce noise as an unintended by-product. Anthropogenic noises range on a continuum from intermittent sounds to those of a more continuous nature, but they also vary in many other characteristics, including frequency, rise time, duty cycle, impulsiveness, and source level (Francis and Barber 2013; Duarte et al. 2021). Crucially, though, all human sources not only increase the amount of noise in an area but also produce sounds with different acoustic features to those generated by the physical environment (abiotic sounds) and by organisms (biotic sounds) (Hildebrand 2009). For instance, much anthropogenic noise is a predominantly chronic disturbance with the sound energy concentrated at low frequencies (typically below 2 kHz); such sounds can readily permeate the environment across time and space. Animals in the natural world are therefore experiencing a very different soundscape, and thus a novel acoustic challenge, to that which existed in pre-industrial times.

In the last two decades, there has been a huge increase in research investigating the impacts of anthropogenic noise on wildlife (reviewed in Kight and Swaddle 2011; Francis and Barber 2013; Morley et al. 2014; Shannon et al. 2016; Duarte et al. 2021). High-intensity sound can cause death, physical damage, and hearing loss, whilst noises from a variety of anthropogenic sources have been shown

Andrew N. Radford, *Noise pollution*. In: *Behavioural Responses to a Changing World*. Edited by: Bob B. M. Wong and Ulrika Candolin, Oxford University Press. © Oxford University Press (2024). DOI: 10.1093/oso/9780192858979.003.0002

Table 2.1 Four reasons why noise can cause behavioural changes; these are not mutually exclusive

Mechanism	Explanation
Acoustic masking	Noise can increase the threshold for detection or discrimination of important sounds of similar frequencies (Moore 2012). Masking can be complete, when an acoustic signal or cue is not detected at all, or partial, when the acoustic signal or cue is detectable by the listener but the content is hard to understand.
Perception as a threat	Noise may be perceived as a threat and thus generate responses (e.g. fleeing, hiding, additional vigilance) similar to those seen in a predation context (reviewed in Francis and Barber 2013).
Distraction	Noise can divert an individual's finite attention away from their primary task or goal, interfering with biologically important decision-making (Chan and Blumstein 2011). Stimuli from one modality (in this case, sound) can therefore interfere with the processing of information obtained in other modalities (e.g. visual or olfactory cues and signals).
Generation of stress	Noise can induce physiological stress as evidenced by, for instance, increases in the production of the steroid hormones cortisol, corticosterone, and aldosterone (Kight and Swaddle 2011). Elevated stress can, in turn, lead to behavioural changes; short-term stress may be beneficial (e.g. priming an animal to avoid a dangerous area) but chronic stress can have detrimental consequences.

to affect animal physiology, development, and behaviour. Behavioural impacts are likely the most temporally and spatially widespread because they can result from even moderate noise levels occurring at large distances from a noise source. Moreover, they are the most well-studied of responses and are particularly important because behavioural alterations represent the first line of defence for organisms in a changing world (Candolin and Wong 2012). Anthropogenic noise can affect animal behaviour in four main ways, which are not mutually exclusive: it can mask other sounds, be perceived as a threat, act as a distraction, and cause stress (Table 2.1). Thus, anthropogenic noise can hinder the processing of acoustic information (so-called unimodal effects), affect the detection and use of information in other sensory modalities (cross-modal interference), and generate a wide range of other behavioural changes.

In this chapter, I begin with an overview of the different behavioural responses generated by anthropogenic noise. Our understanding of noise impacts has come from a variety of complementary research approaches—including studies conducted in captivity and the wild, studies comparing areas with different noise levels, and studies where noise levels have been manipulated—which vary in their behavioural and acoustic validity, as well as the level of experimental control, but all add valuable knowledge. I then discuss reasons why we see variation in behavioural responses—both within and between species, as well as across

time—and what the consequences are for individual fitness, populations, communities, and ecosystems. Throughout, I provide specific examples illustrating the diverse range of taxa known to be affected (Figure 2.1). I also provide some case studies of species that have each been used as model organisms to investigate several different questions relating to noise pollution. Finally, I look to the future—both in terms of what we still need to learn about noise impacts, including their mechanistic underpinnings and their effects in a multi-stressor context (see also Chapter 10), and the importance of implementing and testing mitigation and management strategies such that we can reduce the impact of this globally pervasive pollutant.

2.2 Behavioural responses to anthropogenic noise

Many behavioural studies investigating the impacts of anthropogenic noise have focused on the responses of individual animals, as this is logistically the simplest scenario to consider. But there is also increasing evidence that social interactions, both between members of the same species and between different species, can be affected. It is important to note, though, that noise does not always have a discernible effect—for instance, there was no detected adjustment in the vocalizations of harbour seals *Phoca vitulina* or Pacific chorus frogs *Pseudacris regilla* to compensate for noise-induced masking (Nelson et al. 2017; Matthews et al. 2020).

Figure 2.1 Animals from all major taxa have been shown to change their behaviour in response to anthropogenic noise. For example: (a) orcas *Orcinus orca*, (b) European robins *Erithacus rubecula*, (c) gray treefrogs *Hyla versicolor*, (d) ambon damselfish *Pomacentrus amboinensis*, (e) greater mouse-eared bats *Myotis myotis*, (f) field crickets *Gryllus campestris*, (g) prairie dogs *Cynomys ludovicianus*, and (h) hermit crabs *Pagurus bernhardus*.

Photos: (a) Shutterstock.com/Alessandro De Maddalena; (b) Shutterstock.com/Claire Norman; (c) Shutterstock.com/Deatonphotos; (d) Shutterstock.com/scubaluna; (e) Shutterstock.com/Agami Photo Agency; (f) Shutterstock.com/Sanjay M Dalvi; (g) Shutterstock.com/Marie Dirgova; (h) Shutterstock.com/Myroslav Orshak

Such examples are likely under-represented in the literature due to the general publication bias towards research with statistically significant findings.

Animals may respond to increased noise in an area by moving away, but those remaining may exhibit important changes to individual behaviours, such as vigilance, foraging, movement, orientation, navigation, and settlement. Several studies have indicated noise-avoidance behaviour by different taxa: for example, seismic survey activity led to an 88% decrease in sightings of baleen whales and a 53% decrease in sightings of toothed whales (Kavanagh et al. 2019), whilst the creation of a 'phantom road' with an array of loudspeakers playing traffic noise resulted in a 31% decrease in migrating birds in a roadless area of the USA (Ware et al. 2015). For many animals, though, avoidance of noisier areas is not a possibility— a species might not be capable of relocating, the relevant area may hold critical resources, and/or there may be no viable alternative to move to— and there is strong evidence that those remaining can be negatively affected. For instance, anthropogenic noise can cause individuals to become more vigilant, as seen in white-crowned sparrows *Zonotrichia leucophrys*, prairie dogs *Cynomys ludovicianus*, and dwarf mongooses *Helogale parvula* (Shannon et al. 2014; Ware et al. 2015; Figure 2.2). Increased vigilance can reduce time available for foraging (Shannon et al. 2014; Ware et al. 2015), but foraging can also be affected directly: for example, traffic noise decreases the foraging efficiency of greater mouse-eared bats *Myotis myotis* and three-spined sticklebacks *Gasterosteus aculeatus* (Siemers and Schaub 2011; Voellmy et al. 2014). In terms

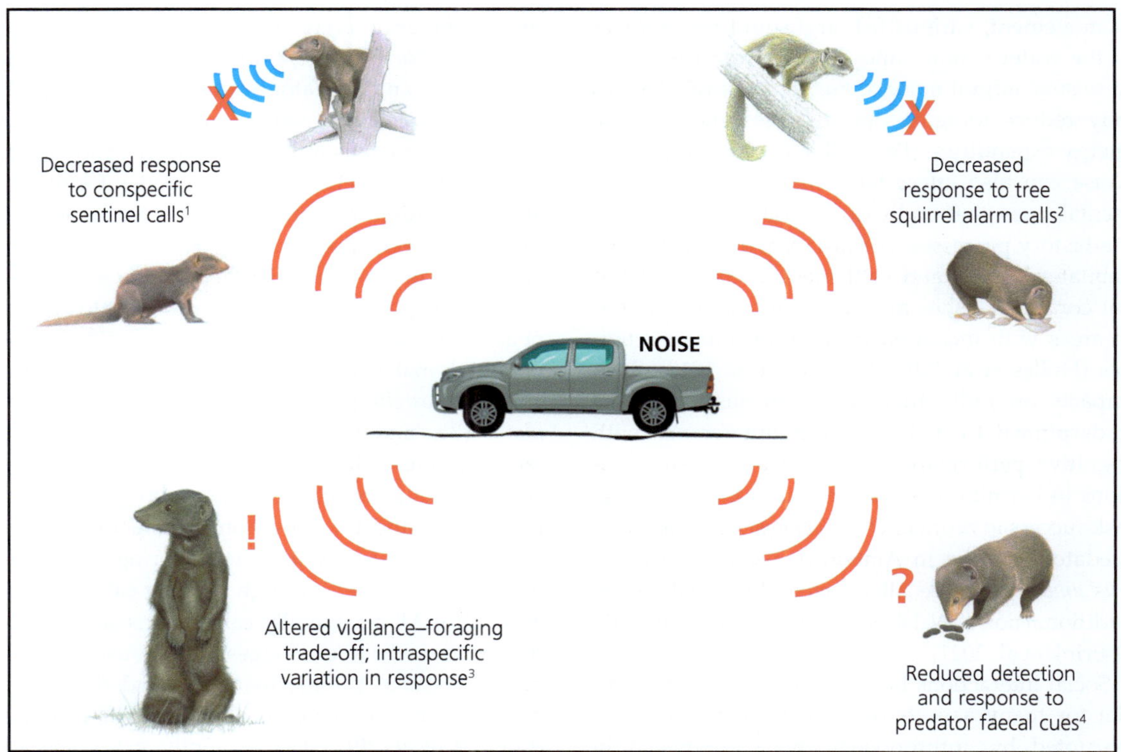

Figure 2.2 Impacts of traffic noise on the behaviour of dwarf mongooses *Helogale parvula*. [1]Kern and Radford (2016); [2]Morris-Drake et al. (2017); [3]Eastcott et al. (2020); [4]Morris-Drake et al. (2016).

Original artwork: Martin Aveling (@AvelingArtworks)

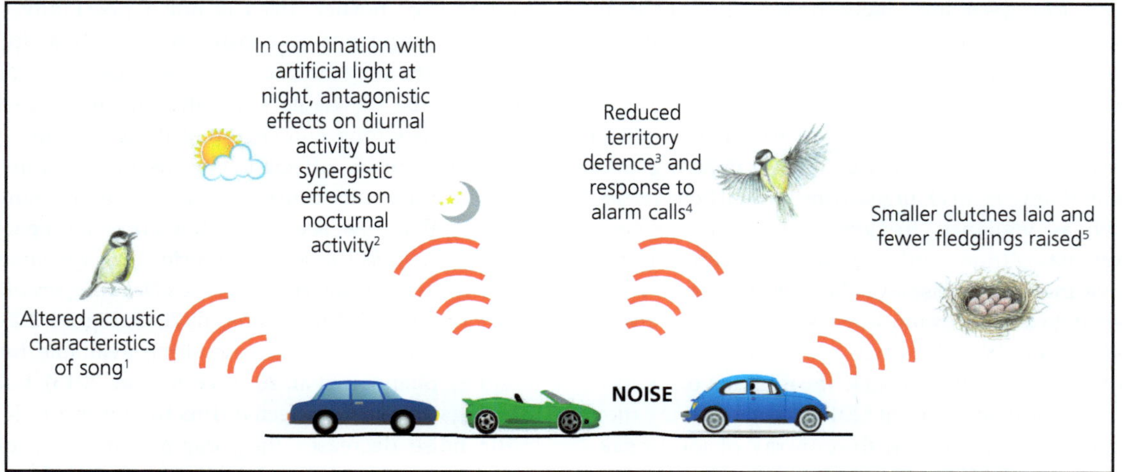

Figure 2.3 Impacts of traffic noise on the behaviour of great tits *Parus major*. [1]Slabbekoorn and Peet (2003); [2]Dominoni et al. (2020); [3]Mockford and Marshall (2009); [4]Templeton et al. (2016); [5]Halfwerk et al. (2011).

Original artwork: Martin Aveling (@AvelingArtwork)

of movement, various fish and squid move lower in the water column and swim faster in response to seismic airgun noise; these behavioural changes may reduce foraging opportunities and increase energy expenditure (Fewtrell and McCauley 2012). Noise can also affect the perception of environmental cues that individuals use to inform key life-history processes. For instance, the orientation, habitat-selection, and settlement decisions of larval coral reef fishes and invertebrates may suffer in areas with increased noise from human activities (Holles et al. 2013; Lecchini et al. 2018). Such impacts on individual behaviour may often be underpinned by noise-induced interference with cognitive performance; resource-assessment decisions in hermit crabs *Coenobita clypeatus*, foraging-task success in zebra finches *Taeniopygia guttata*, and predator learning in Ambon damselfish *Pomacentrus amboinensis* are all affected detrimentally by additional noise (Walsh et al. 2017; Ferrari et al. 2018; Osbrink et al. 2021).

Social interactions between conspecifics are crucial for many aspects of animal life and may be disrupted by anthropogenic noise (reviewed in Fisher et al. 2021). Perhaps most obviously, acoustic communication can be affected; such communication is often vital for survival and reproductive success, with numerous studies demonstrating that the behaviour of receivers is compromised by noise pollution. For instance, additional-noise playback reduced the anti-predator responses of wild superb fairy-wrens *Malurus cyaneus* and great tits *Parus major* to alarm calls indicating danger (Zhou et al. 2019; Figure 2.3). Use of other vocal information about predation risk can also be affected, as seen in the responses of dwarf mongooses to the surveillance calls of sentinels (Figure 2.2). From a reproductive perspective, noise has been shown to impact negatively the acoustic attraction or detection of potential mates by, for example, gray treefrogs *Hyla chrysoscelis* and field crickets *Gryllus bimaculatus* (Bee and Swanson 2007; Bent et al. 2018), territory defence in red-mouthed gobies *Gobius cruentatus* and great tits (Sebastianutto et al. 2011; Figure 2.3), and the provision of parental care by blue tits *Cyanistes caeruleus* to offspring that communicate their need through begging calls (Lucass et al. 2016). More generally, calculations of acoustic-transmission distances suggest that acoustic signals of, for instance, marine mammals and fish likely have a reduced range of effect in noisier areas (Putland et al. 2017). Consequently, many species exhibit changes in their acoustic behaviour to maintain detection and information flow. For example, a wide variety of invertebrates, anurans, birds, and both terrestrial and marine mammals alter the amplitude, frequency, rate, and duration of vocalizations, as well as where and when they vocalize, when competing with anthropogenic noise (reviewed in Kunc and Schmidt 2020; Duquette et al. 2021).

In addition to effects on communication, or as a consequence of them, noise pollution might impact group formation, social decision-making, and interactions between members of the same and different groups. For instance, both tufted titmice *Baeolophus bicolor* and Carolina chickadees *Poecile carolinensis* increased their grouping (reduced nearest-neighbour distances) and rates of social interactions in traffic noise, potentially to enhance communication and information transfer (Owens et al. 2012). In the group-living cichlid *Neolamprologus pulcher*, additional-noise playback not only affected individual behaviours (e.g. reduced nest digging and defence against egg predators) but caused increases in the amount of aggression and submission exhibited among group members (Bruintjes and Radford 2014). Alterations in social interactions between conspecifics can, in turn, affect social organization with respect to dominance hierarchies, collective behaviour, and mating systems (reviewed in Fisher et al. 2021). For instance, since noise can impact communication and social recognition, there are likely knock-on consequences for collective action and social coordination. As one example, groups of juvenile seabass *Dicentrarchus labrax* exposed to pile-driving playback became less cohesive, less directionally ordered, and were less correlated in speed and directional changes (Herbert-Read et al. 2017). Understanding how noise affects social interactions is crucial because many animals rely on such behaviours for their survival and reproductive success, and any potential impacts could have fundamental ecological and evolutionary implications.

Anthropogenic noise can also affect interspecific interactions, be that use of heterospecific cues or direct encounters between species. Since many animals eavesdrop on heterospecific acoustic signals (e.g. alarm calls), noise may mask this information, as demonstrated in northern cardinals *Cardinalis cardinalis* and dwarf mongooses (Grade and Sieving 2016; Figure 2.2). Similarly, there may be masking of acoustic cues indicating the presence of predators or prey. For instance, because greater

mouse-eared bats find ground-running arthropods by listening for the faint rustling sounds created as these species move, prey detection is compromised by anthropogenic noise (Siemers and Schaub 2011). Moreover, there may be disruption to the use of information in other sensory modalities: for example, dwarf mongooses are less likely to respond appropriately when encountering predator faeces if they are exposed to traffic-noise playback (Figure 2.2). In terms of direct interactions, acoustic communication between different species (e.g. prey signalling their quality to predators) may be negatively affected by anthropogenic noise (Brumm and Slabbekoorn 2005). Beyond communication, there can be changes in both mutualistic interactions (e.g. between cleanerfish and clients) and those of an antagonistic nature (e.g. between predators and prey). For instance, bluestreak cleaner wrasses *Labroides dimidiatus* were less cooperative towards their client fish species when exposed to motorboat noise (Nedelec et al. 2017), whilst several species, including hermit crabs and Ambon damselfish, respond less often and less rapidly to simulated predators when experiencing additional noise (Chan et al. 2010; Simpson et al. 2016). Thus, in conflict scenarios and under certain circumstances, one party in an interspecific interaction may actually benefit from the additional noise. Indeed, anthropogenic environments could even provide new opportunities for species (reviewed in Fleming and Bateman 2018). But the 'winners' and 'losers' in any dyadic interaction influenced by noise will depend on the relative hearing sensitivities and noise tolerances of the species involved (see below).

2.3 Variation in behavioural responses

As research on the impacts of anthropogenic noise has expanded, it has become increasingly clear that there is extensive variation in behavioural responses. This variation occurs both between and within species, as well as over different time frames (i.e. initial responses can change over hours, days, or weeks of repeated exposure, and there can be adaptation across generations). Understanding such variation is crucial for improving the management of captive animals, the monitoring of wild populations, the modelling of species responses, and the mitigation of noise-pollution effects on wildlife.

Differences are expected between species in their behavioural responses to anthropogenic noise for a variety of reasons. Some studies have tested explicitly how different species respond to the same noise source. For instance, Voellmy et al. (2014) showed that whilst both European minnows *Phoxinus phoxinus* and three-spined sticklebacks consumed fewer prey when there was additional noise, the reason differed: minnows shifted their activity away from foraging behaviour (exhibiting greater inactivity and more social behaviour), whereas sticklebacks maintained foraging effort but made more mistakes. Studies comparing terrestrial areas with natural variation in noise levels, and those that experimentally applied traffic noise to roadless areas, have both demonstrated that bird species differ in their avoidance of noise and in their behavioural responses if they remain in noisier sites: some species were adversely impacted while others were not affected at all (Senzaki et al. 2020a, b). Interspecific differences might arise because of variation in, for example, hearing ability, physiological stress response, existing vocal repertoire, anti-predator strategies, and diet. For instance, Pieniazek et al. (2020) found a general trend for freshwater fish with more sensitive hearing to exhibit a greater decrease in foraging during boat-noise exposure. Senzaki et al. (2020a, b) found that those bird species most likely to abandon noisy areas have low-frequency acoustic signals; relative effects of noise on reproductive timing and hatching success were dependent on interspecific differences in vocalization frequency, nesting location, and diet. Such differences will likely affect community composition and structure both directly, through differences in the relative success of each species when disturbed, and indirectly, through altered interactions between species (see later).

Individuals of the same species also differ in myriad ways, so variation in their behavioural responses to the same noise exposure is expected (reviewed in Harding et al. 2019). One source of intraspecific variation is differences in intrinsic characteristics such as age, sex, dominance status, personality, body size, and condition. For example, younger birds of various species showed greater avoidance than older individuals when exposed to traffic-noise playback (McClure et al. 2017), whilst European eels *Anguilla anguilla* in poorer condition exhibited a noise-induced reduction in responses to a simulated predatory strike that was not apparent in better-quality individuals (Purser et al. 2016). Intraspecific variation in responses to noise might also be the consequence of differences in extrinsic factors, including environmental or social context, prior experience of noise, and the presence and magnitude of additional stressors. For instance, traffic-noise playback did not affect the song-bout duration of tree frogs *Hyla arborea* singing alone but did lead to an alteration by those singing in a chorus (Lengagne 2008), and there was a synergistic effect of artificial light at night and noise on great tit nocturnal activity but an antagonistic effect of the two stressors on diurnal activity (Figure 2.3). A systematic review of the literature revealed that only 10% of papers examining noise impacts tested experimentally for intraspecific response variation, but that 75% of those considering intraspecific variation reported significant effects (Harding et al. 2019). Varied responses among conspecifics may affect relative mortality risk and the ability to emigrate or react flexibly, will determine the evolutionary potential of post-disturbance populations, and could have far-reaching consequences for communities and ecosystem functioning (see later).

Behavioural responses to noise can also vary in time frame, from initial individual plasticity through to evolutionary change across generations (see Tuomainen and Candolin 2011). The first response of individual animals to any environmental change is often a plastic alteration in behaviour. This flexibility has been demonstrated by numerous experimental studies where animals in a variety of taxa changed their behaviour in response to anthropogenic-noise playback. As just two examples, Amazonian treefrogs *Dendropsophus triangulum* altered their call rate, whilst silvereyes *Zosterops lateralis* adjusted their call frequency, amplitude, and duration, when experiencing noise playbacks (Kaiser and Hammers 2009; Potvin and Mulder 2013). However, there are differences in the capacity for such behavioural plasticity: some species may have limited vocal flexibility, for instance, if their songs or calls have little existing variation in frequency (Brumm and Slabbekoorn 2005). The initial reaction of individuals to noise may change through learning (LaZerte et al. 2016) or because repeated or chronic exposure leads to either increased (e.g. through sensitization or reduced tolerance) or decreased (e.g. through desensitization, increased tolerance, or habituation) responses (Nedelec et al. 2016; Neo et al. 2018). For instance, whilst juvenile threespot dascyllus *Dascyllus trimaculatus* initially hid more in the coral reef when exposed to motorboat-noise playback, they no longer did so after one or two weeks of exposure (Nedelec et al. 2016). Behavioural responses can also change over time through innovations and cultural inheritance, via social transmission of new behavioural patterns within and across generations (Tuomainen and Candolin 2011). Ultimately, there may be evolution of behavioural responses (i.e. genetic change over generations). For instance, the minimum frequency of white-crowned sparrow songs in a particular population increased over a 35-year period as urban noise increased (Luther and Derryberry 2012). Even if initial behavioural plasticity does not buffer a species fully against the challenges of anthropogenic noise, it might give the population additional time to adapt genetically to the new acoustic environment.

2.4 Consequences of behavioural changes

Behavioural changes in response to noise ultimately matter only if there are fitness consequences for individuals. Noise can directly affect fitness (e.g. causing mortality, or through physiological and developmental impacts), but the focus in this chapter is on how behavioural responses can lead to fitness consequences. Any such impacts on survival and reproductive success can have a profound influence on the persistence and evolution of populations, which, in turn, can alter community composition and structure, as well as the functioning of whole ecosystems.

Behavioural responses to noise can either be maladaptive or adaptive (reviewed in Tuomainen and Candolin 2011). Maladaptive responses are those that lead to a decrease in an individual's fitness, either because previous behavioural patterns continue but these are now detrimental in the changed environment, or because the behavioural alterations exhibited have negative consequences for survival or reproductive success. For example, if animals continue to produce vocalizations that are now masked by anthropogenic noise (Brumm and Slabbekoorn 2005), there could be a decreased likelihood of attracting a mate or of hearing the needs of offspring. If acoustic cues indicating viable or healthy habitat are masked (Holles et al. 2013; Lecchini et al. 2018), animals may end up in poorer-quality areas with potential reductions in fitness. In terms of behavioural changes, if noise leads to unnecessary increases in vigilance (Shannon et al. 2014; Ware et al. 2015), for instance, the resulting reduction in time available for activities such as foraging or parental care could, in principle, increase the risk of starvation for adults or young. By contrast, some noise-induced changes are viewed as adaptive; that is, they (likely) increase the survival and reproductive success of individuals. For example, vocal adjustments in the amplitude, temporal structure, frequency, and complexity of vocalizations are argued to improve signal detection and discrimination in noisy areas (Brumm and Slabbekoorn 2005), and are thus suggested to have benefits in terms of territory defence, mate attraction, anti-predator behaviour, and parental care. However, such vocal adjustments might also result in many direct or indirect fitness costs due to, for instance, reduced transmission distances, increased risk of predation or parasitism, altered energy budgets, and loss of vital information (reviewed in Read et al. 2014). As with all behavioural ecology, there are likely trade-offs between costs and benefits.

Many behavioural studies speculate about potential fitness consequences, but caution is needed, especially when extrapolating from demonstrations that acute noise exposure causes short-term responses. That is because there may be compensation in quieter periods or reduced noise impacts with repeated or chronic exposure, and even long-standing behavioural changes may not actually affect survival or reproductive success. Ideally, what is needed are studies that explicitly examine behaviour and quantify the resulting fitness consequences (Kunc et al. 2016); such work is relatively rare and mostly focused on reproductive success. For example, traffic noise resulted in female great tits laying smaller clutches and pairs raising fewer fledglings, likely due to masking of acoustic communication (Figure 2.3). Similarly, ash-throated flycatchers *Myiarchus cinerascens* had lower reproductive success if their nests were exposed to noise, in this case due to higher rates of abandonment at the incubation stage (Mulholland et al. 2018). Spiny chromis *Acanthochromis polyacanthus* broodguarding males exposed to motorboat-noise playback reduced feeding and offspring interactions, resulting in a lower likelihood of offspring survival (Nedelec et al. 2017). In a study that combined laboratory and field experiments, a reduced response to predatory threats by Ambon damselfish translated into a higher mortality rate when faced with predatory dusky dottybacks *Pseudochromis fuscus* (Simpson et al. 2016). There is also considerable correlational evidence that the use of military sonar likely causes behavioural changes that lead to fatal mass strandings in a variety of cetaceans (Nowacek et al. 2007).

Anthropogenic noise can change populations both in terms of the abundance of a species and also its composition. In general, if individuals avoid noisier areas, or if they remain but respond maladaptively and suffer reductions in survival or reproductive success (see above), local population declines can result. Conversely, some species may increase in abundance if they are released from competitive or predation pressure due to noise-induced declines in others (Francis et al. 2009; Senzaki et al. 2020b). But intraspecific variation in responses to noise may also lead to changes in population structure. For instance, McClure et al. (2017) found age differences in noise-avoidance responses to a phantom road—first-year birds responded more strongly than adults, potentially because of different foraging–predation trade-offs and site-selection decisions—resulting in changed population demographics. More generally, whilst initial behavioural responses to human-induced environmental change may help to maintain a viable population and facilitate adaptation to new conditions, they also mean that there will likely be a greater proportion of individuals capable of rapid change moving forward. The genetic make-up of the population will have changed, with noise altering the evolutionary potential of a post-disturbance population (Tuomainen and Candolin 2011). It is also possible that avoidance behaviour could result in reduced genetic connectivity of the population, if it becomes fragmented, increasing the risk of inbreeding and thus lost genetic variation (see Brook et al. 2002). Long-term studies of population consequences of noise are rare, so predictive modelling is especially important. Such modelling can include determining how noise propagates across landscapes and its likely effect on populations (Barber et al. 2011; Mortensen et al. 2021).

Changes in populations can have direct and indirect consequences for communities (Kok et al. 2023). For example, studies of arthropods, anurans, and birds have found noise-induced changes in community composition (Bunkley et al. 2017; Grace and Noss 2018; Senzaki et al. 2020a). Most obviously, if a species declines in number, or even goes locally extinct, due either to avoidance behaviour or strong negative fitness impacts of noise, then it will be less represented in the community (Ware et al. 2015; Senzaki et al. 2020a, b). At the same time, there may be increases in the prevalence of some species if they gain through behavioural interactions with another species that is more susceptible to noise. For instance, this can happen where a predator benefits at the expense of a prey species whose antipredator responses are compromised by noise, as found in the dottyback–damselfish predator–prey relationship (Simpson et al. 2016). Likewise, hosts may benefit if their parasites are compromised by noise, as is the case with frog-biting midges *Corethrella* spp. being unable to detect their túngara frog *Engystomops pustulosus* hosts (McMahon et al. 2017). There can also be indirect, knock-on consequences arising from a noise-induced change in the local abundance of a species. For instance, if avian pollinators or seed-dispersers move away from noisy areas, then there can be negative consequences for the plant and tree species that rely on them (Phillips et al. 2021). Conversely, if nest-predators are less common in noisy areas, then reproductive success of their prey can increase (Francis et al. 2009). Similarly, noise may change the balance of interspecific competitive interactions in favour of one species: for example, shore crabs *Carcinus maenas* are less likely to aggregate at a food source if it is noisy, resulting in reduced competition for sympatric common shrimps *Crangon crangon* (Hubert et al. 2018). Moreover, there can be carryover effects into other areas: for example, grasshoppers and odonates were less common at sites where their avian predators had moved when avoiding noise elsewhere (Senzaki et al. 2020b). Thus, there can be 'winners' and 'losers' within communities exposed to anthropogenic noise, and noise pollution can reverberate through wider communities by disrupting or enhancing interspecific interactions and ecological services.

2.5 The future

2.5.1 Greater understanding of noise impacts

Compared to 20 years ago, we now have a much greater understanding of the impacts of anthropogenic noise on wildlife. But there are still important gaps in knowledge where additional work is crucial. I highlight three general ones here, which are not mutually exclusive: expansion in scope, mechanistic underpinnings, and multi-stressor effects.

Behavioural studies of the impacts of anthropogenic noise would benefit from an expansion in various respects. First, there is still a taxonomic bias towards vertebrates in general (Morley et al. 2014) and to birds and marine mammals in particular (Shannon et al. 2016; Jerem and Mathews 2020). Second, most field studies have been conducted in Europe or North America, and there is a relative dearth of work both in nations with developing or emerging economies, and in rural areas likely to experience imminent major increases in urbanization (Jerem and Mathews 2020). Ideally, there would therefore be both a taxonomic and geographic expansion in knowledge, especially in biodiversity hotspots that likely have little prior exposure to anthropogenic noise and thus where new human activities might have the greatest initial effect. Third, as indicated earlier, additional work would usefully explore variation in responses to noise—for instance, how members of the same species are affected by different noise sources, and how the same noise source can have different impacts on individuals of the same and different species—and what causes this variation

(see Harding et al. 2019). Using existing studies to extrapolate likely effects of noise will always be difficult but is made especially challenging at present because our understanding of intraspecific and interspecific variation is so limited. Fourth, there is a need for more research that measures the longer-term effects of noise exposure, including quantifying survival and reproductive success, because there can be changes in behaviour with chronic exposure and, ultimately, individual fitness consequences underpin population viability. Finally, further investigation of how interspecific interactions are affected by noise would be beneficial. Assessment is needed of the benefits and costs to both parties (e.g. predator and prey, host and parasite, two competitor species), not only because this is important in its own right—all animals interact with others throughout their lives—but because it is a stepping-stone to understanding the community-level effects. Whilst early studies of impacts of anthropogenic noise on wildlife understandably focused on short-term effects of acute exposures, the goal now is a more holistic knowledge from individuals to ecosystems.

For a fuller understanding of the behavioural impacts of anthropogenic noise, we need more detailed studies of the mechanistic underpinnings. At one level, 'mechanism' can refer to whether a behavioural response results from masking, distraction, threat, and/or stress (Table 2.1). Studies explicitly teasing apart which of these is the underlying reason for a noise-induced change in behaviour are currently rare (Zhou et al. 2019). Another mechanistic level concerns physiological, neurological, developmental, cellular, immunological, and genetic processes (Kight and Swaddle 2011). Whilst an extensive literature exists on how these might be affected by different environmental noises, a more integrated, interdisciplinary approach marrying such research with behavioural work would be of benefit. There is also a need to establish more fully the interplay between the sound-detection capabilities of a species and the acoustic characteristics of noise sources that cause (detrimental) impacts (Kunc et al. 2016); establishing the link between hearing mechanisms and vulnerability to noise. Understanding the mechanisms of noise impacts is important because it can help not only to predict effects but also, potentially, to propose successful mitigation methods and noise-management plans (Kight and Swaddle 2011; Francis and Barber 2013).

Anthropogenic noise is rarely, if ever, isolated from other forms of human disturbance, but there is a dearth of studies investigating multi-stressor effects on behaviour (reviewed in Halfwerk and Slabbekoorn 2015; Chapter 10). Major sources of human-made noise often generate other stressors too: for instance, urbanization concurrently modifies temperature, light, and noise levels, whilst roads and their associated traffic don't just add noise to the environment but also artificial light, chemical pollution, and the risk of collisions. More generally, animals do not experience anthropogenic noise in isolation, but must simultaneously cope with other major disturbances (e.g. climate change, habitat destruction, ocean acidification). The effects of multiple stressors could be additive, multiplicative, synergistic, or antagonistic, or one stressor could dominate another (Halfwerk and Slabbekoorn 2015; Harding et al. 2019). Attempting to extrapolate likely combined responses from single-stressor studies is difficult because, whilst populations might show no adverse effects from individual pollutants, the addition of another stressor may cause a markedly different response or take individuals beyond their physiological limit (Côté et al. 2016). Furthermore, other anthropogenic changes can affect noise-pollution levels and thus their effects: for instance, ocean acidification results in decreased water pH, which reduces sound absorption, whilst increasing temperatures lead to a reduction in the speed at which sound travels through the ocean (Kunc et al. 2016). Studies investigating how other stressors modify the impacts of anthropogenic noise on wildlife are therefore crucial but are currently rare (for exceptions, see McMahon et al. 2017; McCormick et al. 2018a; Senzaki et al. 2020a).

As much as possible, any such future research needs to record and to make available accurate and relevant measures of sound sources; without this information, comparisons between studies are challenging (see McKenna et al. 2016). In terms of relevance, those recordings need to be of sound pressure, particle motion, and/or vibrations depending on the species in question. Full reporting of acoustic metrics includes, for instance, information on natural and playback power spectra or frequency (e.g. range and peak levels), equipment specifications, and reference levels; ideally, sound files would be publicly shared (see Jerem and Mathews 2020). As with any research field, and with anthropogenic noise studies to date, a range of data-collection approaches will continue to be beneficial because that provides a complementary understanding of different aspects of noise effects. For example, tight control of extraneous variables might only be possible in captive situations, but then there are the inherent limitations relating to restrictions on how animals can and do behave, as well as the sound field generated (especially in aquatic studies in small tanks). Field studies can provide both acoustic and ecological relevance, but are logistically often much more challenging. To isolate the importance of noise per se, experimental (or pseudo-experimental) tests are required. Ideally, there might be a combination of different approaches in the same study—the most appropriate will depend on the exact question being asked—although cost and feasibility will also play a part in the decision-making.

2.5.2 Mitigation and management

Noise pollution is unusual in the sense that it does not linger in the environment once the source is removed (compared to, for instance, chemical pollution). Also, noise levels can be potentially lessened much more quickly than other pollutants, as evidenced during lockdowns associated with the COVID-19 pandemic (Lecocq et al. 2020; March et al. 2021). Moreover, it is plausible to reduce the effects of noise pollution on a local scale (compared to, for instance, climate warming), with potentially short- and long-term benefits. Whilst there is increasing evidence that various mitigation options—both technological and behavioural—lead to less noise entering the environment, it is crucial to test explicitly whether there are improvements for wildlife and then to scale-up for longer-term and broad-scale management plans.

Technological innovations can help to lessen noise from human activities and its impact on wildlife in various ways. First, there can be reduction of noise generated by the source: for instance, engineering of quieter engines, propellors, tyres, road surfaces, and pile-driving foundations. Some experimental work has shown that this can be beneficial to animals: for example, the negative impacts of two-stroke motorboat engines on the behaviour of Ward's damselfish *Pomacentrus wardi* were lessened or eliminated by replacement with four-stroke engines, which are more fuel-efficient and quieter (McCormick et al. 2018b). Similarly, the use of electric engines rather than petrol-driven ones reduced avoidance behaviour in various terrestrial mammals (Yosef et al. 2021). A second type of technological mitigation involves the inclusion of sound barriers, such as baffles alongside roads and bubble curtains around pile-driving units, to reduce the sound energy propagated into the environment. As one example, using bubble curtains around the monopiles in a windfarm construction project resulted in an approximate halving of the displacement distance of harbour porpoises *Phocoena phocoena*, thus minimizing the temporary loss of habitat for these animals (Dähne et al. 2017). A third means by which technology can lessen anthropogenic noise impacts is through either the 'soft start' of noisy activities or the use of acoustic deterrent devices (ADDs). The purpose of both is to reduce the likelihood of animals being in the vicinity of high-intensity human-made noises, thus mitigating exposure levels without changing the sound field itself. For instance, reductions in initial piling energy and pre-exposure to ADDs reduced the likelihood of harbour porpoises being present in zones where the noise would cause injury (Thompson et al. 2020). As yet, studies directly testing such benefits are relatively rare; technological mitigation measures are also expensive and time-consuming, so alternatives need consideration.

As a potentially cheaper and faster way to reduce noise impacts, there is the option to change human behaviours such as slowing down traffic, moving sources further away from vulnerable areas, and stopping activities at crucial times (e.g. the breeding season or during migration). As with technological solutions, there is a need to test not only how such modifications decrease sound propagation but, crucially, whether there are discernible benefits to wildlife. Lockdowns used in an attempt to control the spread of COVID-19 reduced noise in urban areas considerably—for example, due to greatly reduced traffic and construction activity (Hasegawa and Lau 2022)—providing a 'natural' experiment. As a specific example of how that affected wildlife, white-crowned sparrows responded by producing higher-performance songs at lower amplitudes, effectively maximizing the salience of their vocalizations and the distance over which they communicated (Derryberry et al. 2020). Away from lockdowns, Nedelec et al. (2022) used field and laboratory experiments to test how limiting motorboat traffic, and associated noise, could potentially have a positive effect on coral reef fish reproductive success (Box 2.1a). Successfully lobbying for such changes in human behaviour is far from easy though. So, it may also be sensible to consider the potential for mitigating noise impacts on animals through the use of learning principles: using classical and operant conditioning to alter the behaviour of wildlife to minimize their risks (see Proppe et al. 2016). As one hypothetical example, which would need careful testing to determine the direct and indirect consequences, playback of a positive, biologically salient stimulus (e.g. conspecific song) could be paired with road noise to reduce fear responses induced by the latter (Proppe et al. 2016).

As we move forward with tests of mitigation measures, it must be kept in mind that the benefits might not always be immediate. Some species may have the capacity to exploit quieter soundscapes rapidly, with discernible behavioural changes. These may, in turn, translate into positive demographic effects and higher species diversity, but such consequences will take time to manifest. There may also be slower but equally important recoveries, so long-term monitoring is critical, especially when scaling-up from behavioural effects to those at the level of populations and communities (Phillips et al. 2021). Modelling of the projected impact of mitigation measures, in addition to noise propagation and its effects, is thus important too (Barber et al. 2011). Ultimately, national or international legislation limiting noise levels is required, but determining the most appropriate level is dependent on increasing our knowledge about variation in noise impacts between individuals, species, and habitats (Box 2.1b).

Box 2.1   Illustrations of how scientific studies are crucial for the design of suitable mitigation and management policies relating to noise pollution

(a) Changes in human behaviour to reduce noise pollution

Potentially the fastest way to reduce noise pollution in an area is through a change in human behaviour: a decrease in occurrence or spatial shift in the noise-producing activity. For this to be viewed as a viable option, studies are needed that test the benefits to wildlife. As a specific example, Nedelec et al. (2022) conducted a season-long field manipulation where they limited motorboat activity near some coral reefs in comparison to other reefs that received typical traffic levels. They coupled this fieldwork with a laboratory playback experiment to isolate the importance of reduced noise rather than other aspects of motorboat disturbance. Reducing noise resulted in greater reproductive success of the fish study species (spiny chromis *Acanthochromis polyacanthus*): individuals on the reduced-noise reefs were almost twice as likely as those at busier motorboat sites to have surviving offspring at the end of the breeding season, likely due to improvements in parental care. More such studies are required to build a convincing case that could then be used to persuade those engaged in a noise-generating activity (in this case, motorboat use) that a change in behaviour would be beneficial. In this particular instance, there is not necessarily the need for a human activity to cease; rather, some limiting of traffic speeds and/or proximity to vulnerable habitats would make a positive difference. Moving forward in this regard will require a delicate balancing of wildlife protection and human needs, but it will necessarily be best-informed by rigorous science.

(b) Noise levels in policy

If noise pollution is to be managed at a national or global scale, then policy-makers will need to set cumulative noise limits. One early example of this in relation to ocean anthropogenic noise was the Marine Strategy Framework Directive, which required members of the EU to attain noise levels 'that do not adversely affect the marine environment'. Of course, there is the immediate question about what those levels actually are; without defined targets, then coordinated, focused action is impossible. To manage noise pollution requires: (a) quantification of risk (e.g. production of risk maps and modelling of the population consequences arising from different noise scenarios); and (b) setting of scalable noise-budgets that are conveyed to decision-makers. Merchant et al. (2018) describe a possible framework: using noise-exposure curves to quantify the proportion of a habitat or population exposed, and thus associated exposure durations that would be deemed acceptable given current knowledge of impacts. This 'indicator' methodology can then be the basis for both location-based and ecosystem-based management measures. Merchant et al. (2018) showcase the applicability of this approach with two case studies, using data from an international assessment of cumulative impulsive noise activity in the North Sea to predict the risk to harbour porpoises *Phocoena phocoena* and to herring *Clupea harengus* spawning. One of the benefits of this risk-based approach is that it is flexible—new scientific knowledge about noise levels and their impacts on wildlife can be assimilated quickly, allowing adjustment to statutory commitments.

2.6 Conclusions

We have known for a long time that anthropogenic noise has a range of detrimental effects on humans. Research in the last two decades has demonstrated that non-human animals in all taxa also suffer from our noisy activities. These negative impacts on wildlife extend from individual and social behaviour to consequences for survival, reproductive success, population viability, community structure, and ecosystem functioning. But there is cause for cautious optimism. Animals have always lived in a world full of sound, so they are capable of adaptation when given the chance. Moreover, there are simple starting solutions that can help quickly and on a local scale. What is needed moving forward, therefore, is both an increasing understanding of the problems and increased testing and refining of mitigation and management policies. Mitigating and managing anthropogenic noise is important not only in its own right—it is a pollutant found in all ecosystems throughout the world—but because building resilience in this respect may help wildlife 'fight' against other human disturbances for which we do not necessarily have such potentially rapid solutions.

Acknowledgements

I am extremely grateful to Steve Simpson, as well as to Rick Bruintjes, Lucille Chapuis, Isla Davidson, Emma Eastcott, Harry Harding, Cathy Hobbs, Julie Kern, Amy Morris-Drake, Kieran McCloskey, Erica Morley, Sophie Nedelec, Julia Purser, Tim Lamont, Irene Voellmy, Matthew Wale, Emma Weschke, and Clemency White for invaluable insight and discussions on the topic of noise pollution across many years.

References

Barber, J.R., Burdett, C.L., Reed, S.E., et al. (2011). Anthropogenic noise exposure in protected natural areas: estimating the scale of ecological consequences. *Landscape Ecology*, 26, 1281.

Barber, J.R., Crooks, K.R., and Fristrup, K.M. (2010). The costs of chronic noise exposure for terrestrial organisms. *Trends in Ecology & Evolution*, 25, 180–189.

Bee, M.A., and Swanson, E.M. (2007). Auditory masking of anuran advertisement calls by road traffic noise. *Animal Behaviour*, 74, 1765–1776.

Bent, A.M., Ings, T.C., and Mowles, S.L. (2018). Anthropogenic noise disrupts mate searching in *Gryllus bimaculatus*. *Behavioral Ecology*, 29, 1271–1277.

Brook, B.W., Tonkyn, D.W., O'Grady, J.J., and Frankham, R. (2002). Contribution of inbreeding to extinction risk in threatened species. *Conservation Ecology*, 6, 16.

Bruintjes, R., and Radford, A.N. (2014). Context-dependent impacts of anthropogenic noise on individual and social behaviour in a cooperatively breeding fish. *Animal Behaviour*, 85, 1343–1349.

Brumm, H., and Slabbekoorn, H. (2005). Acoustic communication in noise. *Advances in the Study of Behavior*, 35, 151–209.

Bunkley, J.P., McClure, C.J.W., Kawahara, A.Y., et al. (2017). Anthropogenic noise changes arthropod abundances. *Ecology and Evolution*, 7, 2977–2985.

Buxton, R.T., McKenna, M.F., Mennitt, D., et al. (2019). Anthropogenic noise in US national parks—sources and spatial extent. *Frontiers in Ecology and the Environment*, 17, 559–564.

Candolin, U., and Wong, B.B.M. (2012). *Behavioural Responses to a Changing World: Mechanisms to Consequences*. Oxford University Press, Oxford.

Chan, A.A.Y.H., and Blumstein, D.T. (2011). Attention, noise, and implications for wildlife conservation and management. *Applied Animal Behaviour Science*, 131, 1–7.

Chan, A.A.Y.H., Giraldo-Perez, P., Smith, S., and Blumstein, D.T. (2010). Anthropogenic noise affects risk assessment and attention: the distracted prey hypothesis. *Biology Letters*, 6, 458–461.

Côté, I.M., Darling, E.S., and Brown, C.J. (2016). Interactions among ecosystem stressors and their importance in conservation. *Proceedings of the Royal Society B: Biological Sciences*, 283, 20152592.

Dähne, M., Tougaard, J., Carstensen, J., et al. (2017). Bubble curtains attenuate noise from offshore wind farm construction and reduce temporary habitat loss for harbour porpoises. *Marine Ecology Progress Series*, 580, 221–237.

Derryberry, E.P., Phillipss, J.N., Derryberry, G.E., et al. (2020). Singing in a silent spring: birds respond to a half-century soundscape reversion during the COVID-19 shutdown. *Science*, 370, 575–579.

Dominoni, D., Smit, J.A.H., Visser, M.E., and Halfwerk, W. (2020). Multisensory pollution: artificial light at night and anthropogenic noise have interactive effects on activity patterns of great tits (*Parus major*). *Environmental Pollution*, 256, 113314.

Duarte, C.M., Chapuis, L., Collin, S.P., et al. (2021). The soundscape of the Anthropocene ocean. *Science*, 371, eaba4658.

Duquette, C.A., Loss, S.R., and Hovik, T.J. (2021). A meta-analysis of the influence of anthropogenic noise on terrestrial wildlife communication strategies. *Journal of Applied Ecology*, 58, 1112–1121.

Eastcott, E., Kern, J.M., Morris-Drake, A., and Radford, A.N. (2020). Intrapopulation variation in the behavioural responses of dwarf mongooses to anthropogenic noise. *Behavioral Ecology*, 31, 680–691.

Ferrari, M.C.O., McCormick, M.I., Meekan, M.G., et al. (2018). School is out on noisy reefs: the effect of boat noise on predator learning and survival of juvenile coral reef fishes. *Proceedings of the Royal Society B: Biological Sciences*, 285, 20180033.

Fewtrell, J.L., and McCauley, R.D. (2012). Impact of air gun noise on the behaviour of marine fish and squid. *Marine Pollution Bulletin*, 64, 984–993.

Fisher, D.N., Kilgour, R.J., Siracusa, E.R., et al. (2021). Anticipated effects of abiotic environmental change on intraspecific social interactions. *Biological Reviews*, 96, 2661–2693.

Fleming, P.A., and Bateman, P.W. (2018). Novel predation opportunities in anthropogenic landscapes. *Animal Behaviour*, 138, 145–155.

Francis, C.D., and Barber, J.R. (2013). A framework for understanding noise impacts on wildlife: an urgent conservation priority. *Frontiers in Ecology and the Environment*, 11, 305–313.

Francis, C.D., Ortega, C.P., and Cruz, A. (2009). Noise pollution changes avian communities and species interactions. *Current Biology*, 19, 1415–1419.

Grace, M.K., and Noss, R.F. (2018). Evidence for selective avoidance of traffic noise by anuran amphibians. *Animal Conservation*, 21, 343–351.

Grade, A.M., and Sieving, K.E. (2016). When the birds go unheard: highway noise disrupts information transfer between bird species. *Biology Letters*, 12, 20160113.

Halfwerk, W., Holleman, L.I.M., Lessells, C., and Slabbekoorn, H. (2011). Negative impact of traffic noise on avian reproductive success. *Journal of Applied Ecology*, 48, 210–219.

Halfwerk, W., and Slabbekoorn, H. (2015). Pollution going multimodal: the complex impact of the human-altered sensory environment on animal perception and performance. *Biology Letters*, 11, 20141051.

Harding, H.R., Gordon, T.A.C., Eastcott, E., et al. (2019). Causes and consequences of intraspecific variation in animal responses to anthropogenic noise. *Behavioral Ecology*, 30, 1501–1511.

Hasegawa, Y., and Lau, S.-K. (2022). A qualitative and quantitative synthesis of the impacts of COVID-19 on soundscapes: a systematic review and meta-analysis. *Science of the Total Environment*, 844, 157223.

Herbert-Read, J.E., Kremer, L., Bruintjes, R., et al. (2017). Anthropogenic noise pollution from pile-driving disrupts the structure and dynamics of fish shoals. *Proceedings of the Royal Society B: Biological Sciences*, 284, 20170660.

Hildebrand, J.A. (2009). Anthropogenic and natural sources of ambient noise in the ocean. *Marine Ecology Progress Series*, 395, 5–20.

Holles, S., Simpson, S.D., Radford, A.N., et al. (2013). Boat noise disrupts orientation behaviour in a coral reef fish. *Marine Ecology Progress Series*, 485, 295–300.

Hubert, J., Campbell, J., van der Beek, J.G., et al. (2018). Effects of broadband sound exposure on the interaction between foraging crab and shrimp: a field study. *Environmental Pollution*, 243, 1923–1929.

Jerem, P., and Mathews, F. (2020). Trends and knowledge gaps in field research investigating effects of anthropogenic noise. *Conservation Biology*, 35, 115–129.

Kaiser, K., and Hammers, J.L. (2009). The effect of anthropogenic noise on male advertisement call rate in the neotropical treefrog, *Dendropsophus triangulum*. *Behaviour*, 146, 1053–1069.

Kavanagh, A.S., Nykänen, M., Hunt, W., et al. (2019). Seismic surveys reduces cetacean sightings across a large marine ecosystem. *Scientific Reports*, 9, 19164.

Kern, J.M., and Radford, A.N. (2016). Anthropogenic noise disrupts use of vocal information about predation risk. *Environmental Pollution*, 218, 988–995.

Kight, C.R., and Swaddle, J.P. (2011). How and why environmental noise impacts animals: an integrative, mechanistic review. *Ecology Letters*, 14, 1052–1061.

Kok, A.C.M., Berkhour, B.W., Carlson, N., et al. (2023). How chronic anthropogenic noise can affect wildlife communities. *Frontiers in Ecology and the Environment*, 11, 1130075.

Kunc, H.P., McLaughlin, K.E., and Schmidt, R. (2016). Aquatic noise pollution: implications for individuals, populations and ecosystems. *Proceedings of the Royal Society B: Biological Sciences*, 283, 20160839.

Kunc, H.P., and Schmidt, R. (2020). Species sensitivities to a global pollutant: a meta-analysis on acoustic signals in response to anthropogenic noise. *Global Change Biology*, 27, 675–688.

LaZerte, S.E., Slabbekoorn, H., and Otter, K.A. (2016). Learning to cope: vocal adjustment to urban noise is correlated with prior experience in black-capped chickadees. *Proceedings of the Royal Society B: Biological Sciences*, 283, 20161058.

Lecchini, D., Bertucci, F., Gache, C., et al. (2018). Boat noise prevents soundscape-based habitat selection by coral planulae. *Scientific Reports*, 8, 9283.

Lecocq, T., Hicks, S.P., Van Noten, K., et al. (2020). Global quieting of high-frequency seismic noise due to COVID-19 pandemic lockdown measures. *Science*, 369, 1338–1343.

Lengagne, T. (2008). Traffic noise affects communication behaviour in a breeding anuran, *Hyla arborea*. *Biological Conservation*, 141, 2023–2031.

Lucass, C., Eens, M., and Müller, W. (2016). When ambient noise impairs parent–offspring communication. *Environmental Pollution*, 212, 592–597.

Luther, D.A., and Derryberry, E.P. (2012). Birdsongs keep pace with city life: changes in song over time in an urban songbird affects communication. *Animal Behaviour*, 83, 1059–1066.

March, D., Metcalfe, K., Tintoré, J., and Godley, B.J. (2021). Tracking the global reduction of marine traffic during the COVID-19 pandemic. *Nature Communications*, 12, 2415.

Matthews, L.P., Fournet, M.E.H., Gabriele, C., et al. (2020). Acoustically advertising male harbour seals in southeast Alaska do not make biologically relevant acoustic adjustments in the presence of vessel noise. *Biology Letters*, 16, 20190795.

McClure, C.J.W., Ware, H.E., Carlisle, J.D., and Barber, J.R. (2017). Noise from a phantom road experiment alters the age structure of a community of migrating birds. *Animal Conservation*, 20, 164–172.

McCormick, M.I., Allan, B.J.M., Harding, H., and Simpson, S.D. (2018a). Boat noise impacts risk assessment in a coral reef fish but effects depend on engine type. *Scientific Reports*, 8, 3847.

McCormick, M.I., Watson, S.-A., Simpson, S.D., and Allan, B.J.M. (2018b). Effect of elevated CO_2 and small boat noise on the kinematics of predator–prey interactions. *Proceedings of the Royal Society B: Biological Sciences*, 285, 20172650.

McKenna, M.F., Shannon, G., and Fristrup, K. (2016). Characterizing anthropogenic noise to improve understanding and management of impacts to wildlife. *Endangered Species Research*, 31, 279–291.

McMahon, T.A., Rohr, J.R., and Bernal, X.E. (2017). Light and noise pollution interact to disrupt interspecific interactions. *Ecology*, 98, 1290–1299.

Merchant, N.D., Faulkner, R.C., and Martinez, R. (2018). Marine noise budgets in practice. *Conservation Letters*, 11, 1–8.

Mockford, E.J., and Marshall, R.C. (2009). Effects of urban noise on song and response behaviour in great tits. *Proceedings of the Royal Society B: Biological Sciences*, 276, 2979–2985.

Moore, B.C.J. (2012). *An Introduction to the Psychology of Hearing*. Brill, Leiden.

Morley, E.L., Jones, G., and Radford, A.N. (2014). The importance of invertebrates when considering the impacts of anthropogenic noise. *Proceedings of the Royal Society B: Biological Sciences*, 281, 20132683.

Morris-Drake, A., Bracken, A.M., Kern, J.M., and Radford, A.N. (2017). Anthropogenic noise alters dwarf mongoose responses to heterospecific alarm calls. *Environmental Pollution*, 223, 476–483.

Morris-Drake, A., Kern, J.M., and Radford, A.N. (2016). Cross-modal impacts of anthropogenic noise on information use. *Current Biology*, 26, R911–R912.

Mortensen, L.O., Chudzinska, M.E., Slabbekoorn, H., and Thomsen, F. (2021). Agent-based models to investigate sound impact on marine animals: bridging the gap between effects on individual behaviour and population level consequences. *Oikos*, 130, 1074–1086.

Mulholland, T.I., Ferraro, D.M., Boland, K.C., et al. (2018). Effects of experimental anthropogenic noise exposure on the reproductive success of secondary cavity nesting birds. *Integrative and Comparative Biology*, 58, 967–976.

Nedelec, S.L., Mills, S.C., Lecchini, D., et al. (2016). Repeated exposure to noise increases tolerance in a coral reef fish. *Environmental Pollution*, 216, 428–436.

Nedelec, S.L., Mills, S.C., Radford, A.N., et al. (2017). Motorboat noise disrupts co-operative interspecific interactions. *Scientific Reports*, 7, 6987.

Nedelec, S.L., Radford, A.N., Gatenby, P., et al. (2022). Limiting motorboat noise on coral reefs boosts fish reproductive success. *Nature Communications*, 13, 2822.

Nedelec, S.L., Radford, A.N., Pearl, L., et al. (2017). Motorboat noise impacts parental behaviour and offspring survival in a reef fish. *Proceedings of the Royal Society B: Biological Sciences*, 284, 20170143.

Nelson, D.V., Klinck, H., Carbaugh-Rutland, A., et al. (2017). Calling at the highway: the spatiotemporal constraint of road noise on Pacific chorus frog communication. *Ecology and Evolution*, 7, 429–440.

Neo, Y.Y., Hubert, J., Bolle, L.J., et al. (2018). European seabass respond more strongly to noise exposure at night and habituate over repeated trials of sound exposure. *Environmental Pollution*, 239, 367–374.

Nowacek, D.P., Thorne, L.H., Johnston, D.W., and Tyack, P.L. (2007). Responses of cetaceans to anthropogenic noise. *Mammal Review*, 37, 81–115.

Osbrink, A., Meatte, M.A., Tran, A., et al. (2021). Traffic noise inhibits cognitive performance in a songbird. *Proceedings of the Royal Society B: Biological Sciences*, 288, 20202851.

Owens, J.L., Stec, C.L., and O'Hatnick, A. (2012). The effects of extended exposure to traffic noise on parid social and risk-taking behaviour. *Behavioural Processes*, 91, 61–69.

Phillips, J.N., Termondt, S.E., and Francis, C.D. (2021). Long-term noise pollution affects seedling recruitment and community composition, with negative effects persisting after removal. *Proceedings of the Royal Society B: Biological Sciences*, 288, 20202906.

Pieniazek, R.H., Mickle, M.F., and Higgs, D.M. (2020). Comparative analysis of noise effects on wild and captive freshwater fish behaviour. *Animal Behaviour*, 168, 129–135.

Potvin, D.A., and Mulder, R.A. (2013). Immediate, independent adjustment of call pitch and amplitude in response to varying background noise by silvereyes (*Zosterops lateralis*). *Behavioral Ecology*, 24, 1363–1368.

Proppe, D.S., McMillan, N., Congdon, J.V., and Sturdy, C.B. (2016). Mitigating road impacts on animals through learning principles. *Animal Cognition*, 20, 19–31.

Purser, J., Bruintjes, R., Simpson, S.D., and Radford, A.N. (2016). Condition-dependent physiological and behavioural responses to anthropogenic noise. *Physiology & Behavior*, 155, 157–161.

Putland, R.L., Merchant, N.D., Farcas, A., and Radford, C.A. (2017). Vessel noise cuts down communication space for vocalizing fish and marine mammals. *Global Change Biology*, 24, 1708–1721.

Read, J., Jones, G., and Radford, A.N. (2014). Fitness costs as well as benefits are important when considering responses to anthropogenic noise. *Behavioral Ecology*, 25, 4–7.

Sebastianutto, L., Picciulin, M., Costantini, M., and Ferrero, E. (2011). How boat noise affects an ecologically crucial behaviour: the case of territoriality in *Gobius cruentatus* (Gobiidae). *Environmental Biology of Fishes*, 92, 207–215.

Senzaki, M., Barber, J.R., Phillips, J.N., et al. (2020a). Sensory pollutants alter bird phenology and fitness across a continent. *Nature*, 587, 605–609.

Senzaki, M., Kadoya, T., and Francis, C.D. (2020b). Direct and indirect effects of noise pollution alter biological communities in and near noise-exposed environments. *Proceedings of the Royal Society B: Biological Sciences*, 287, 20200176.

Shannon, G., Angeloni, L.M., Wittemyer, G., et al. (2014). Road traffic noise modifies behaviour of a keystone species. *Animal Behaviour*, 94, 135–141.

Shannon, G., McKenna, M.F., Angeloni, L.M., et al. (2016). A synthesis of two decades of research documenting the effects of noise on wildlife. *Biological Reviews*, 91, 982–1005.

Siemers, B.M., and Schaub, A. (2011). Hunting at the highway: traffic noise reduces foraging efficiency in acoustic predators. *Proceedings of the Royal Society B: Biological Sciences*, 278, 1646–1652.

Simpson, S.D., Radford, A.N., Nedelec, S.L., et al. (2016). Anthropogenic noise increases fish mortality by predation. *Nature Communications*, 7, 10,544.

Slabbekoorn, H., and Peet, M. (2003). Birds sing at a higher pitch in urban noise. *Nature*, 424, 267.

Templeton, C.N., Zollinger, S.A., and Brumm, H. (2016). Traffic noise drowns out great tit alarm calls. *Current Biology*, 26, R1173–R1174.

Thompson, P.M., Graham, I.M., Cheney, B., et al. (2020). Balancing risks of injury and disturbance to marine mammals when pile driving at offshore windfarms. *Ecological Solutions and Evidence*, 1, e12034.

Tuomainen, U., and Candolin, U. (2011). Behavioural responses to human-induced environmental change. *Biological Reviews*, 86, 640–657.

Voellmy, I.K., Purser, J., Flynn, D., et al. (2014). Acoustic noise reduces foraging success in two sympatric fish species via different mechanisms. *Animal Behaviour*, 89, 191–198.

Walsh, E.P., Arnott, G., and Kunc, H.P. (2017). Noise affects resource assessment in an invertebrate. *Biology Letters*, 13, 20170098.

Ware, H.E., McClure, C.J.W., Carlisle, J.D., and Barber, J.R. (2015). A phantom road experiment reveals traffic noise is an invisible source of habitat degradation. *Proceedings of the National Academy of Sciences*, 112, 12,105–12,109.

Yosef, R., Kumbhojkar, S., Sharma, S., and Morelli, F. (2021). Electric vehicles minimise disturbance to mammals. *European Journal of Wildlife Research*, 67, 74.

Zhou, Y., Radford, A.N., and Magrath, R.D. (2019). Why does noise reduce response to alarm calls? Experimental assessment of masking, distraction and greater vigilance in wild birds. *Functional Ecology*, 33, 1280–1289.

CHAPTER 3

Chemical pollution

Michael G. Bertram, Tomas Brodin, Marlene Ågerstrand, and Bob B. M. Wong

Overview

Ecosystems around the globe are increasingly being inundated with a cocktail of chemical pollutants. From antidepressant drugs found in the tissues of fish from the Niagara River to persistent organic pollutants detected in polar bears from Arctic ecosystems, pollutants are capable of affecting the development, physiology, morphology, and behaviour of wildlife. Here, we highlight the relatively young field of behavioural ecotoxicology, which has shown that exposure to even low, environmentally realistic levels of contaminants can cause a wide range of behavioural changes in animals. This is cause for major concern, given that the ability to appropriately produce and maintain behaviours is fundamental to the ecology and evolution of wild animal populations. Further, we underscore that not only is studying animal behaviour a vital component of understanding the impacts of chemical pollution, it also represents an extremely valuable—but as yet underutilized—tool for informing more effective chemicals regulation, which is urgently needed to protect wildlife living in an increasingly toxic world.

3.1 Introduction

'All things are poison, and nothing is without poison,' observed Paracelsus (1538), the founder of toxicology, half a millennium ago.

Like the world of Paracelsus, ours is a toxic one. Chemical pollution disrupts the balance of ecosystems, threatening the health of humans and wildlife everywhere (Persson et al. 2022). A key driver of this widespread contamination is the rapid proliferation of the chemicals industry, with global chemical sales reaching over US\$5.6 trillion in 2017 and projected to almost double by 2030 (UNEP 2019). Indeed, the rate of change in the production and diversification of synthetic chemicals is outstripping other well-recognized agents of global change, such as habitat destruction and rising atmospheric CO_2 concentrations (Berhardt et al. 2017). As a result, ecosystems around the world are being drenched in a cocktail of chemicals—from metals and plasticizers to pharmaceuticals and personal care products (Bertram et al. 2022a, b). Chemical pollutants enter the environment in a wide variety of ways, including the discharge of industrial and domestic wastewater, spray-drift of herbicides, pesticides, and fungicides applied to gardens and agricultural fields, migration of metals from mining operations into groundwater, and more localized sources such as chemical spills and toxic waste dumping. As a result of these highly varied sources, contaminant concentrations in the environment vary both spatially and temporally, with many anthropogenic chemicals being highly resistant to degradation and remarkably persistent over time (Cousins et al. 2019).

Once present in ecosystems, uptake of contaminants by animals can occur through direct contact with their environment (e.g. water, sediment) and/or through the food and liquid they ingest (Bertram et al. 2022a, b). Some pollutants can also be passed from mothers to offspring via placental and lactational transfer, such as polychlorinated biphenyls (PCBs) that are driving global killer whale *Orcinus orca* population declines (Desforges et al. 2018). Other contaminants are concentrated in

Michael G. Bertram, Tomas Brodin, Marlene Ågerstrand, and Bob B. M. Wong, *Chemical pollution*. In: *Behavioural Responses to a Changing World*. Edited by: Bob B. M. Wong and Ulrika Candolin, Oxford University Press. © Oxford University Press (2024). DOI: 10.1093/oso/9780192858979.003.0003

animal tissues at successively higher levels as they make their way up the food chain (i.e. biomagnification), potentially crossing ecosystem boundaries in the process. This potential to cross boundaries is seen, for example, in the case of pharmaceutical drugs that have been shown to accumulate in osprey *Pandion haliaetus* nestlings feeding on contaminated riverine fish (Bean et al. 2018). Not surprisingly given the ubiquity of anthropogenic chemicals in the environment, these contaminants are even showing up in the tissues of animals in the most remote places on Earth. This includes, for instance, the polar extremes of the Arctic (Khairy et al. 2016) and Antarctic (Garnett et al. 2022), the world's highest peak (Napper et al. 2020), and even the deepest ocean trenches (Jamieson et al. 2017).

Anthropogenic chemicals in the environment have been linked to a multitude of adverse effects in wildlife. For instance, road runoff of the ubiquitous tyre rubber-derived chemical 6PPD-quinone has been implicated in large-scale die-offs of coho salmon *Oncorhynchus kisutch* in the Pacific Northwest (Tian et al. 2021). Moreover, decades of research originating from the UK has shown that exposure to wastewater effluent containing oestrogenic pharmaceuticals found in the contraceptive pill (e.g. 17α-ethinyloestradiol [EE2]) causes feminization and widespread sexual disruption in wild fish populations (Jobling et al. 2006). Death and reproductive dysfunction aside, chemical pollution can also have profound impacts on animal behaviour. Indeed, exposure to pollutants can fundamentally alter ecologically relevant behaviours crucial for survival and reproductive success (Saaristo et al. 2018; Aulsebrook et al. 2020). In this regard, mounting evidence suggests that organismal behaviour can be especially sensitive to chemical pollution, with effects often manifesting at exposure concentrations far lower than in other endpoints that are conventionally measured in ecotoxicological tests, such as LC_{50} (i.e. the concentration at which 50% of the exposed animals are expected to die). This is significant because behaviour provides a critical link between an organism's physiological function and the environment, with consequential impacts across multiple levels of biological organization—from individuals to populations and communities, and even entire ecosystems (Figure 3.1).

In this chapter, we begin by exploring the multifarious ways that chemical pollution can affect wildlife behaviour. We firstly discuss the mechanisms through which pollutants can exert such effects, why some species may be more susceptible than others, and how long these impacts might persist. We then focus on the ecological and evolutionary implications of pollution-induced behavioural changes, before turning to key knowledge gaps and avenues for future research. We conclude by discussing the vital, yet greatly underutilized, role of behavioural knowledge in formulating regulation to limit the ongoing environmental impacts of chemical pollution.

3.2 Behavioural impacts of chemical pollution

Chemical pollution can potentially affect a wide range of ecologically important behaviours critical to organismal fitness. These behavioural changes can be induced by numerous, highly varied, direct and indirect mechanisms (reviewed in Zala and Penn 2004; Saaristo et al. 2018; Bertram et al. 2022c; Michelangeli et al. 2022). For instance, certain pollutants such as endocrine-disrupting chemicals can directly alter morphology (e.g. size, colouration, sex), with flow-on impacts on behaviour (see Gore et al. 2018; Bertram et al. 2022b). This includes the potential to alter sensory anatomy (e.g. visual, olfactory), which may reduce the ability of animals to detect conspecific and/or heterospecific cues (see Clotfelter et al. 2004). In addition to morphological changes, exposure to pollutants can cause physiological, neurological, and/or hormonal changes, which may lead to changes in behaviour. Such changes can, for example, result from exposure to chemicals that are neurotoxic, including certain insecticides (e.g. organophosphates, neonicotinoids) and metals (e.g. cadmium, lead), and those that are neuroactive, like some pharmaceuticals (e.g. anxiolytics, antidepressants) (see Kaisarevic et al. 2021). In addition, the behaviour of organisms can be disrupted indirectly by chemical pollution, for example due to pollution-induced changes to the behaviour of predator or competitor species, which

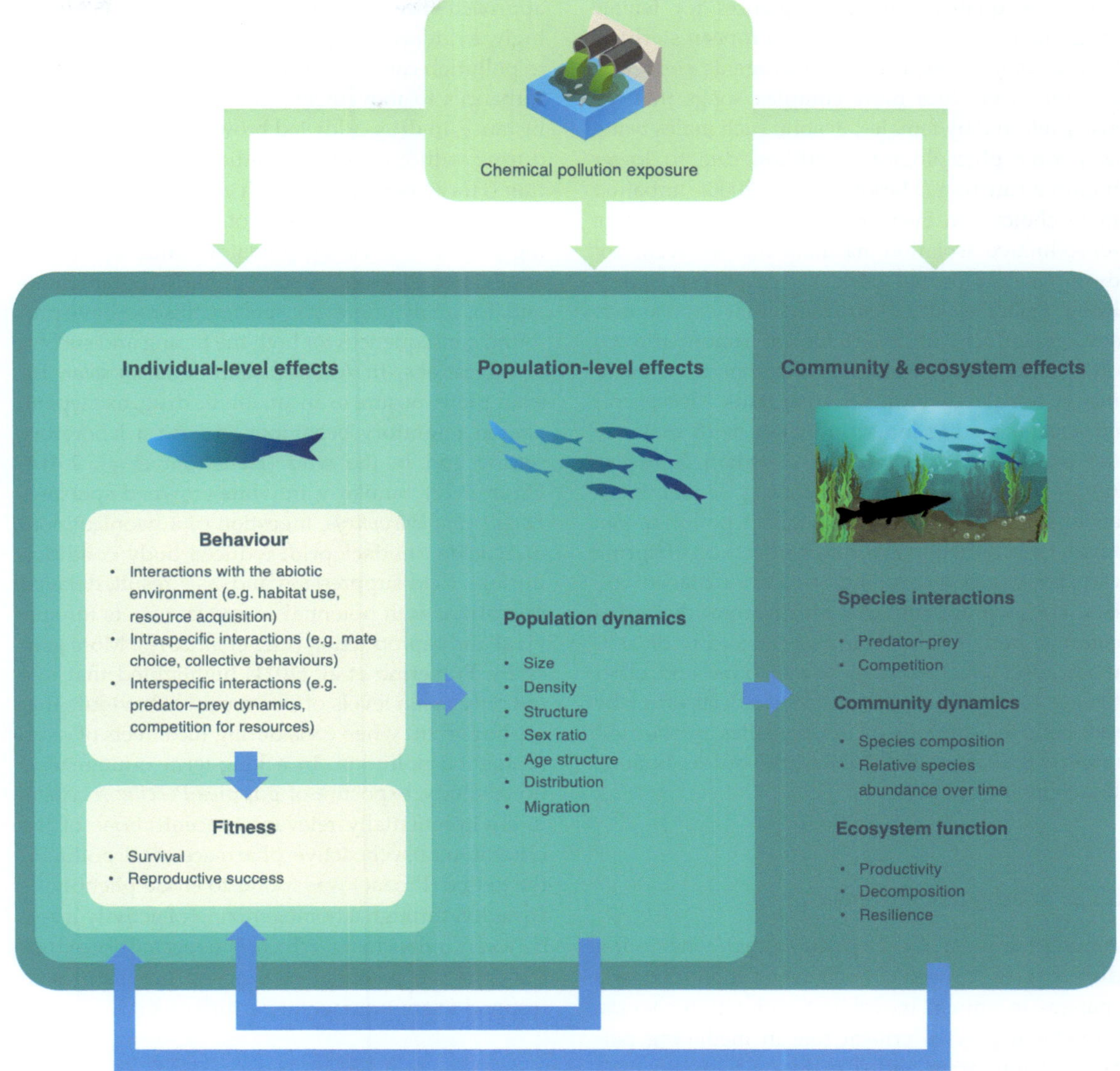

Figure 3.1 Effects of behavioural changes due to chemical pollution exposure on individuals, populations, and communities.

can have flow-on effects even for species that are not directly impacted by the pollutant (reviewed in Saaristo et al. 2018).

3.2.1 Reproductive behaviours

In a reproductive context, exposure to pollutants can affect behavioural interactions between the sexes by disrupting the production, transmission, and detection of sexual signals and displays that are vital to the location, attraction, and evaluation of potential mates (Candolin and Wong 2019). For example, exposure to PCBs impairs the development of the male larynx of African clawed frogs *Xenopus laevis*, thus preventing them from producing mate-attracting advertisement calls (Qin et al. 2007). Chemical pollution can also decouple the pivotal relationship between sexual trait expression and male quality, with consequences for female mate choice. For instance, male European starlings *Sturnus vulgaris* exposed to oestrogenic endocrine disruptors produce more complex songs that are also preferred by females, despite such males being in poorer physiological condition due to lower immune function (Markman et al. 2008). Impaired mate choice can even lead to the breakdown of reproductive isolation mechanisms, as seen, for example, in swordtail fishes *Xiphophorus* sp. Specifically, Fisher et al. (2006) found that exposure to sewage effluent and agricultural runoff affected olfactory cues important in mate recognition, resulting in female sheepshead swordtails *Xiphophorus birchmanni* mating indiscriminately with males of the closely related highland swordtail *Xiphophorus malinche*. Apart from affecting sexual interactions, another way that chemical pollution can impact reproduction is through effects on offspring care. Some urban-dwelling birds, for instance, collect and manipulate the fibres from discarded cigarette butts to line their nests as a form of 'self-medication' (Figure 3.2a). The nicotine and other toxicants, in this case, have a beneficial effect by reducing the amount of ectoparasites in the nest (Suárez-Rodríguez and Garcia 2017; but see Suárez-Rodríguez et al. 2017).

3.2.2 Activity and movement

Among the most widely documented behavioural disturbances caused by chemical pollution are changes in animal movement and activity levels. Movement plays a critical role in mediating patterns of migration and dispersal, and in shaping encounter rates with resources, such as food, shelter, and mates, as well as risks, such as competitors, predators, parasites, and disease. Movement disorders, as a result of neural dysfunction, are a common symptom of exposure to a wide variety of chemical pollutants. For example, manganese contamination from active mining operations was found to impair motor performance in an island population of an endangered marsupial, the northern quoll *Dasyurus hallucatus* (Nasir et al. 2018; Figure 3.2b). Specifically, motor tests revealed that quolls with a higher manganese burden had lower maximum approach speeds when negotiating a turn. Alarmingly, evidence suggests that even acute exposure to pollution can lead to neurologically induced disturbances to movement behaviour, as shown, for instance, in drug-addicted brown trout *Salmo trutta* during withdrawal from methamphetamine pollution (Horký et al. 2021). Such changes, in the wild, are likely to affect the success of individuals at catching prey and avoiding predators. Indeed, activity appears to be especially susceptible to contaminants and, for many migratory species, this can have profound consequences for both the timing and success of migrations. In Atlantic salmon *Salmo salar*, for example, exposure to an anxiolytic drug, oxazepam, altered migratory behaviour in both a laboratory setting and in the wild (Hellström et al. 2016a; Figure 3.2c). Similarly, in white-crowned sparrows *Zonotrichia leucophrys*, ingestion of a neonicotinoid insecticide, imidacloprid, reduced body condition through food suppression and, as a result, delayed migration, with potential carryover effects for survival and reproduction (Eng et al. 2019). More generally, Polverino et al. (2021) highlighted that it is not only mean levels of movement behaviours that are important when considering the effects of environmental pollution. In a long-term multigenerational study, exposure of guppies *Poecilia reticulata* to environmentally relevant concentrations of the ubiquitous psychoactive pharmaceutical pollutant fluoxetine (Prozac) was found to erode phenotypic variation through a homogenizing of activity levels between individuals, which is expected to reduce the adaptive potential of exposed populations to respond to environmental perturbations.

3.2.3 Anti-predator responses

Pollutants can imperil animals by impacting their anti-predator responses. This can occur, for example, by compromising the ability of prey to detect predator or conspecific alarm cues, as demonstrated, for example, in intertidal gastropods *Littorina littorea* exposed to leachates from microplastics (Seuront 2018). Chemical pollutants can also affect behaviour through impaired physiological performance. For instance, high-speed imaging has revealed that damselfly *Coenagrion hastulatum* larvae exposed to environmentally

Figure 3.2 Chemical pollution can disrupt a wide array of behaviours in wildlife belonging to a striking diversity of taxa. For instance: (a) house finches *Carpodacus mexicanus* have been found to add disassembled cigarette butts to their nests to repel ectoparasites, and are thereby exposed to nicotine and other toxicants; (b) northern quolls *Dasyurus hallucatus* contaminated with manganese from mining operations display impaired motor performance, with likely flow-on effects for prey capture and predator avoidance; (c) contamination with the widespread pharmaceutical pollutant oxazepam alters migratory behaviour in Atlantic salmon *Salmo salar*, both when tested in the laboratory and in the field; and (d) honeybees *Apis mellifera* favour food laced with the neonicotinoid pesticides imidacloprid and thiamethoxam, presenting a major hazard to foraging bees.

Photos: (a, left) iStock.com/Angel Di Bilio; (a, right) iStock.com/Olga Ihnatsyeva; (b, left) iStock.com/chameleonseye; (b, right) iStock.com/SergeyZavalnyuk; (c, left) iStock.com/Jakub Rutkiewicz; (c, right) iStock.com/Toa55; (d, left) iStock.com/bo1982, (d, right) iStock.com/alffoto

relevant concentrations of the antihistamine contaminant diphenhydramine exhibit impaired predator-escape behaviours (the C-start escape response), thereby increasing their likelihood of being captured and consumed by predators (Jonsson et al. 2019). Some pollutants, including psychoactive drugs, are deliberately designed to exert anxiolytic effects that can have unintended consequences for the anti-predator responses of non-target organisms. An example of this is seen in work carried out on the eastern mosquitofish *Gambusia holbrooki*, where fish exposed to environmentally relevant concentrations of fluoxetine displayed heightened activity levels irrespective of actual predation risk, and reduced freezing behaviour (an important anti-predator response) following a simulated predator attack (Martin et al. 2017). Intriguingly, while male mosquitofish showed a reduced freezing response at both the low and high exposure concentration, a reduction in female freezing response only occurred in the former. Such findings underscore the complexity of effects on behaviour, which can be sex-specific and also non-monotonic (i.e. effects that manifest or are more severe at lower, rather than higher, concentrations).

3.2.4 Sociality and group-level behaviours

While the vast majority of behavioural studies investigating the impacts of contaminants have tended to focus on individual-level responses, it is important to realize that such responses can also profoundly impact social interactions as well as the collective behaviour of group-living animals. Indeed, many species live in highly complex societies or gather in loose social groupings where behavioural disturbances have the potential to not only affect individual- but also group-level fitness (Michelangeli et al. 2022). Pollutants, in this regard, can interfere with social cues important in group formation by affecting either the cues themselves and/or the ability of conspecifics to be able to detect those cues, as well as the information that these cues convey. For example, exposure to 4-nonylphenol, a surfactant widely used in industrial and sewage treatment processes, was found to affect social recognition and social organization in

juvenile banded killifish *Fundulus diaphanous* (Ward et al. 2008). Evidence suggests that chemical pollutants can also affect social structure through impacts on dominance hierarchies. Such effects can arise due to asymmetries in the effects of exposure on different social phenotypes within a group. For instance, exposure to a low, environmentally relevant dose of the psychoactive pharmaceutical pollutant oxazepam resulted in subordinate juvenile brown trout *Salmo trutta* becoming more successful at securing food in a competitive setting (McCallum et al. 2021). This change was caused by a shift in behavioural dynamics across social ranks caused by differential absorption of the drug in dominant and subordinate group members (McCallum et al. 2021). Pollution-mediated shifts in the behaviour of affected individuals may even heighten the chances of exposing other members in a group or colony, with implications for collective outcomes. Foraging honeybees *Apis mellifera*, for example, lack the gustatory receptors to detect the presence of neonicotinoid pesticides in nectar and actually prefer food tainted with the toxicants due to the pharmacological effects of the compounds on nicotinic acetylcholine receptors in the brain. However, not only does exposure to these chemicals reduce foraging efficiency in the bees through impaired learning and memory, but a preference for the pollutant can result in more neonicotinoid-laced food being brought back to the hive, thus placing the survival of the entire colony at risk (Kessler et al. 2015; Figure 3.2d).

3.3 Ecological implications

The capacity for chemical pollution to affect the fitness of wild animals through changes in behaviour has important implications for population viability, species interactions, and ecosystem function. An obvious way in which pollution-induced behavioural changes can affect the dynamics of populations is through shifts in key demographic parameters, such as births, deaths, and rates of migration. For example, impaired mate choice can lead to population declines if individuals end up mating with poor-quality suitors, resulting in fewer or less-viable offspring being produced and recruited into the population (Candolin and Wong

2019). While behaviour can clearly have important population-level consequences, the reverse can also be true: changes to the density, structure, and spatial distribution of the population can influence individual behaviour. This means that behaviour may not only be sensitive to the impacts of exposure itself, but also to impacts that chemical pollution might have on population demography, such as mortality rates. It also means that changes to population dynamics as a result of altered behaviours can, through complex feedback loops, potentially reinforce or entrain further behavioural changes at the individual level. For example, reduced population size or density could make it even more difficult for individuals to encounter, and mate with, high-quality suitors (Candolin and Wong 2019).

Given the complex linkages that connect species within an ecological network, altered behaviours have the potential to influence the nature and strength of species interactions (Wong and Candolin 2015). Such a scenario can arise, for example, through so-called 'asymmetrical effects', whereby different organisms exposed to the same contaminant differ in their uptake of, and/or sensitivity to, the effects of a pollutant (Saaristo et al. 2018). For example, while exposure to oxazepam was found to make predatory European perch *Perca fluviatilis* more active, the psychoactive pollutant had no effect on the behaviour of the perch's damselfly prey *Coenagrion hastulatum* (Brodin et al. 2014). Such differences in responses across trophic levels could have important community- and ecosystem-level effects. For instance, drastic population declines in planktivorous small-bodied fishes following a whole-lake exposure to the synthetic birth control oestrogen EE2 resulted in a corresponding increase in the abundance of their zooplankton prey and a decrease in the abundance of their predator, the lake trout *Salvelinus namaycush* (Kidd et al. 2014). Similarly, population demise of *Gyps* vultures as a result of diclofenac poisoning allowed feral dog populations on the Indian subcontinent to flourish due to the absence of competition for scavengeable carcasses (Markandya et al. 2008). Evidence suggests that species interactions may even mediate the transfer of toxicants across ecosystem boundaries, as seen, for example, in riparian spiders feeding on emerging insects near wastewater treatment plants

(Richmond et al. 2018). While most studies have focused on consumer-resource interactions, such as predation, it is important to realize that such asymmetries in behavioural impacts are also expected to affect a wide range of species interactions, from competition and mutualisms, to interactions with parasites and pathogens.

3.4 Evolutionary responses

Chemical pollutants have the potential to act as a strong selective force promoting the evolution of physiological tolerance that can, in turn, affect subsequent behaviours. Alternatively, contaminants could stimulate the evolution of avoidance behaviours (e.g. altered habitat or diet choice) that reduce animal exposure to pollutants. Both of these processes—evolved tolerance and avoidance—can affect population-level success and evolutionary trajectories.

Although the evolution of physiological tolerance in wildlife to chemical pollutants has been the subject of growing research interest (see Hamilton et al. 2017), very little attention has been given to understanding how physiological tolerance might affect the behavioural responses of exposed organisms. On one hand, adaptive physiological adjustments can reduce the likelihood that downstream behaviours are maladaptive. On the other, evolved resistance can result in bioenergetic costs that can impact negatively on behavioural performance. For example, in wild Atlantic killifish *Fundulus heteroclitus*, evolved tolerance to highly toxic polycyclic aromatic hydrocarbons has been linked to altered metabolic responses, with consequences for swimming performance and habitat selection (Jayasundara et al. 2017).

Behavioural adaptations can also allow organisms to physically avoid pollutants (reviewed in Araújo et al. 2020). Examples include the avoidance of copper-contaminated water by zebrafish *Danio rerio* (Islam et al. 2019) and avoidance of soils containing polyethylene microplastics by springtails *Folsomia candida* (Ju et al. 2019). Organisms might also cope via the evolution of altered life-history traits, allowing for the avoidance of aggregated effects of pollutants in time. For instance, high adult mortality or reduced fitness later in life can result in selection acting on key life-history traits to promote faster growth and earlier reproduction (Hendry et al. 2011; Sih et al. 2011). However, it is still unclear whether pollution-induced life-history shifts might also result in corresponding behavioural changes (e.g. increased foraging to sustain faster growth, or earlier mating attempts associated with earlier maturation). More generally, evidence suggests that some species may already possess the inherent adaptive potential to respond rapidly to strong selection favouring earlier maturation and reproduction, and this, in turn, could facilitate adaptation to novel stressors (Reznick et al. 1990; but see Rolshausen et al. 2015).

Pollutant-induced changes have the potential to dramatically affect evolutionary trajectories (Bickham 2011; Saaristo et al. 2018). For example, exposure to chemical pollutants can, in some cases, result in drastic population declines (e.g. Desforges et al. 2018; Tian et al. 2021). By increasing genetic drift and inbreeding, a reduction in population size can result in the further loss of genetic diversity, creating population bottlenecks that limit the adaptive potential of populations (Blanquart et al. 2012; Falk et al. 2012), including adaptive behavioural responses. While some pollutants (i.e. genotoxicants) affect mutation rates (e.g. Rinner et al. 2011; Møller and Mousseau 2015) that, in turn, may compensate for the loss of genetic diversity during population bottlenecks (Matson et al. 2006) or otherwise enhance responses to pollutants (Whitehead et al. 2017), most of these mutations are likely to be deleterious and impose high fitness costs (Suárez-Rodríguez et al. 2017). Thus, adaptive plasticity in behaviour that shields genotypes from otherwise harsh selection imposed by pollution may allow for population persistence and the maintenance of adequate levels of standing genetic variation crucial for further adaptation (Chevin and Lande 2010). As such, behaviours involved in either the tolerance or avoidance of chemical pollutants may contribute to population persistence and subsequent evolutionary change. Such a possibility, however, warrants further investigation.

Chemical pollutants, through their effects on behaviour, can also affect the strength and targets of selection (reviewed in Gore et al. 2018). For example, in the context of sexually selected traits, exposure to the endocrine-disrupting agricultural pollutant 17β-trenbolone has been shown to alter male courtship displays and coercive behaviour in guppies (Bertram et al. 2015, 2020), and disrupt the interplay between male pre- and post-copulatory sexual traits in mosquitofish (Tan et al. 2021). While evidence suggests that a weakening of sexual selection could potentially contribute to population declines (Martínez-Ruiz and Knell 2017), it is important to point out that it may also favour the evolution of mechanisms that could allow exposed populations to better cope with pollutants. An example of this is seen in flour beetles *Tribolium castaneum*, which evolved resistance to a pyrethroid pesticide faster when sexual selection was allowed to occur than when it was experimentally precluded (Jacomb et al. 2016).

3.5 Animal behaviour and chemicals regulation

Despite the fact that animal behaviour is vital to the ecology and evolution of wildlife, and is sensitive to disruption by chemical pollution exposure, behavioural studies are almost never used to inform the regulation of chemicals. Why is this the case and what can be done to better harness behavioural information in chemical regulation?

From an applied perspective, effective management strategies are crucial in limiting or preventing the potential risks posed by chemical pollution. Examples of such measures include banning or partly restricting the use of certain chemicals, or setting limits on what concentrations in the environment are acceptable so that, when those concentrations are exceeded, they trigger risk-mitigation measures, such as improvements to wastewater technology. Management measures, in this regard, are dependent on identifying and characterizing the hazards that specific chemicals pose to human health and the environment, which is termed hazard assessment. The specific requirements of hazard assessments are set out in national or regional regulations, but the assessment processes and guidelines are similar around the world. An assessment is typically performed by governmental agencies or producers/importers of the chemical in question and, when assessing the potential hazard of a chemical, three questions are typically asked. First, what adverse effects are caused by exposure to the chemical? Second, how much of the chemical is needed to cause those adverse effects? Third, can a 'safe' dose be established; that is, a concentration at which organisms in the environment may be exposed without any expected adverse effects (European Chemicals Agency 2011)?

The process of assessing the environmental hazard of a chemical is based on ecotoxicological studies involving test organisms that are chosen to represent different trophic levels (e.g. fish, crustaceans) to reflect ecological realism (European Commission 2018). Historically, these studies have investigated effects such as mortality, reproduction, and growth, and have been performed according to accepted and validated guidelines established by, for example, the Organisation for Economic Cooperation and Development (OECD) (e.g. OECD test number 211: *Daphnia magna* Reproduction Test). Very few validated guidelines, however, investigate effects on behaviour. Thus, despite the key role that animal behaviour plays in determining how wildlife are affected by chemical pollution, current environmental assessments rarely consider behavioural impacts. Indeed, when it comes to behavioural effects, a perception among risk assessors is that behavioural studies may be less reliable or less relevant than traditional ecotoxicological endpoints used in chemical risk assessment (discussed in Ågerstrand et al. 2020; Ford et al. 2021). Reliability, in this regard, refers to the quality, robustness, and reproducibility of the study (i.e. can the results be trusted?) while relevance refers to the usefulness of the study for a particular assessment. As a striking example of

just how infrequently behavioural effects are used in chemical assessment, a recent analysis identified just six instances where behavioural studies have been used, or at least considered for use, in EU chemical regulation (Ågerstrand et al. 2020; Box 3.1).

The transfer of science to policy is not straightforward. At least two aspects will need to be carefully considered when trying to understand the reasons for the low use of behavioural studies in chemical assessments. First, the validated guidelines recommended in chemical regulations do not currently cover behavioural impacts for most test species used in chemical risk assessments (with the notable exception of bees; Ågerstrand et al. 2020). As a result, without input from scientific experts via their peer-reviewed publications, behavioural effects will continue to be ignored in environmental assessments. One argument for not including behavioural effects is the lack of clarity regarding the connection between behavioural disturbances observed in the laboratory and effects seen in the wild. However, this argument is based on outdated knowledge; as we have already explored in the chapter, a growing evidence base shows that observable behavioural effects can, and do, reflect disturbances in the wild—with potentially serious ecological and evolutionary implications. Given the wide range of behavioural disturbances caused by exposure to chemical pollutants, behaviour is a highly useful, sublethal indicator of environmental contamination, with direct fitness consequences for exposed organisms and entire populations (Ågerstrand et al. 2020; Ford et al. 2021). In this regard, a fairly straightforward way to address the low regulatory use would be to add behavioural endpoints to the

---

Box 3.1 Use of behavioural data in the regulation of chemicals

Regulatory use of studies investigating behavioural effects has been limited. Table 3.1 presents the known cases in the regulation of chemicals by the EU where behavioural studies have been considered. In only one of six cases was a study deemed to constitute key evidence in the assessment of the chemical in question (shaded green), while in two cases, the behavioural studies were considered to provide only

supportive evidence (yellow). In addition, in the remaining three cases, the studies were given low weight in the chemical assessment process (red). Those three studies were not considered key evidence because they were deemed by the assessors to be insufficiently reported and/or of insufficient reliability.

Box 3.1 *Continued*

Table 3.1 Known cases in the regulation of chemicals by the EU where behavioural studies have been considered.

Case	Behavioural endpoint and reference	Test animal	Use in EU chemical regulation
1	Avoidance behaviour (Petersen 2003)	Eel *Anguilla anguilla*	This behavioural endpoint was used as the **key study** when the solvent and fuel component methyl tertiary-butyl ether (MTBE) was assessed. Contrary to most behaviour studies, this study was performed with the specific purpose of providing evidence for the environmental risk assessment.
2	Spawning behaviour (Schoenfuss et al. 2008) Feeding, social, and aggressive behaviours (Ward et al. 2006) Reproductive behaviour (Cardinali et al. 2004)	Fathead minnow *Pimephales promelas* Rainbow trout *Oncorhynchus mykiss* Guppy *Poecilia reticulata*	These behavioural endpoints were used as **supportive evidence** when assessing 4-nonylphenol, which is used in the manufacturing of different chemical products. The assessment concluded that the substance is an endocrine disruptor, i.e., that it interferes with the normal functioning of the endocrine system, resulting in adverse effects on, for example, reproduction, the nervous system, the immune system, and metabolic function.
3	Drifting behaviour (Lauridsen and Friberg 2005) Pre-copulatory behaviour (Heckmann et al. 2005) Drifting behaviour (Nørum et al. 2010)	Mayfly *Baetis rhodani*, ant *L. cf. fusca* Amphipod *Gammarus pulex* Stonefly *Leuctra nigra*, amphipod *Gammarus pulex*, yellow may *Heptagenia sulphurea*	These behavioural endpoints were used as **supportive evidence** when the plant protection ingredient lambda-cyhalothrin was assessed. The studies were considered to have relevant results but were not assessed as key evidence for several reasons. First, the tested substance was not identical to the assessed substance. Second, the effects were observed at even the lowest tested concentration and therefore it was not established at what concentrations there would be no effects. Third, there was insufficient reporting of the research methods used. Fourth, there were concerns regarding the studies' reliability. Fifth, the studies were not performed according to a validated guideline.
4	Foraging activity (field study carried out by a chemical company)	Honeybee *Apis mellifera*	This field study reported a short-term reduction in foraging activity in bees exposed to the plant protection product alpha-cypermethrin. Because of the limited effects shown, the behavioural endpoint was evaluated to have **low weight** in the assessment. Also included in the assessment were studies showing a limited and short-term increase in mortality and no effects on the brood nest or bee brood development. The behavioural effects were not considered sufficient to alter the conclusion of these studies.
5	Predation efficiency (Woin and Larsson 1987)	Dragonfly larvae *Aeshna*	Effects on predation efficiency of dragonfly larvae were evaluated to have low reliability and were therefore assigned a **low weight** in the assessment of the plasticizer di-2-ethylhexyl phthalate (DEHP). There were two reasons for concluding that the study was of low reliability. First, only one concentration of DEHP was tested. Second, the study used high amounts of the solvent ethanol when spiking the sediment through which the larvae were exposed.
6	Avoidance behaviour (Eriksson Wiklund et al. 2006)	Amphipod *Monoporeia affinis*	The reporting of the results from this study was considered to be insufficient, as the raw data were not available, and the tested concentrations had not been analytically confirmed. Therefore, the study was assigned to have **low weight** in the assessment of the biocidal substance zinc pyrithione. Despite this, it was used to justify the selection of the assessment factor (which is used to calculate a concentration that is assumed to be safe for exposure).

validated guidelines used as the basis for chemical regulation (Ford et al. 2021).

Second, studies from the peer-reviewed literature, including those investigating behavioural effects, are often conducted and/or reported in a format that is different to the validated guidelines carried out for the purpose of chemical assessment. As a consequence, peer-review studies demonstrating behavioural effects often fall short of the regulatory testing and reporting requirements expected of studies used in such assessments, and are therefore often excluded. In addition, risk assessors may have difficulty understanding the importance of behavioural effects as they may have been investigated using experimental designs, test species, and/or traits that are not standard in regulatory toxicology and environmental risk assessment (Ågerstrand et al. 2020). To aid in the regulatory use of behavioural studies, researchers can potentially refer to existing reporting recommendations and guidelines in designing and implementing their studies, such as the Criteria for Reporting and Evaluating Ecotoxicity Data (CRED; Moermond et al. 2016), and those available via the web-tool Science in Risk Assessment and Policy (SCIRAP; www.scirap.org). These resources provide specific information about the exposure conditions, testing chambers, and feeding regimes of test subjects, as well as the connection between the investigated endpoint and effects at the population level.

For regulation to serve its purpose in protecting the environment, we must assess the potential risks posed by chemicals using studies that aim to capture the true complexity and diversity of the environment. By disregarding animal behaviour in environmental risk assessment, we exclude critical, sublethal endpoints that are highly sensitive to contaminant exposure. As such, the true environmental risks posed by chemicals may be underestimated, and the subsequent implementation of risk management measures may not be sufficient to protect wildlife.

3.6 Conclusions

Behaviour is fundamental to the ecology and evolution of wildlife populations. Therefore, research investigating whether and how chemical pollutants impact behavioural processes is vital to understanding and mitigating the effects of contaminants, which are now widespread in ecosystems globally. This issue is particularly urgent given that the threats posed by chemical pollution are being compounded by a wide range of other anthropogenic stressors (see Chapter 10 on multiple stressors). Here, sophisticated methodological approaches that capitalize on recent technological developments for studying animal behaviour represent an exciting new path for understanding contaminant-induced behavioural changes (reviewed in Bertram et al. 2022c). For instance, automated visual tracking is quickly becoming an indispensable tool for studying behaviour in the laboratory, enabling experiments that are faster and more accurate than ever before (reviewed in Panadeiro et al. 2021). At the same time, high-resolution tracking technologies such as acoustic telemetry are revolutionizing our ability to monitor the behaviour of free-roaming animals in the wild (reviewed in Hellström et al. 2022; Nathan et al. 2022). Although underutilized in behavioural ecotoxicology to date, such tracking approaches hold great promise for uncovering the real-world impacts of chemical pollution on animal behaviour (see Hellström et al. 2016b). Future research harnessing such experimental approaches is expected to provide crucial new information to both researchers seeking to understand the large-scale impacts of contaminants, as well as regulators concerned with the population-level impacts of chemical pollutants. Such information, in turn, will be crucial to developing effective solutions for better management and protection of wildlife living in an increasingly toxic world.

Acknowledgements

We acknowledge funding support from the Swedish Research Council Formas via a Mobility Grant (2020-02293 to M.G.B.) and a Research Project Grant (2018-00828 to T.B.), as well as the Australian Research Council (FT190100014, DP190100642, DP220100245 to B.B.M.W.).

References

Ågerstrand, M., Arnold, K., Balshine, S., et al. (2020). Use of behavioural endpoints in the regulation of chemicals. *Environmental Science: Processes & Impacts*, 22, 49–65.

Araújo, C.V., Laissaoui, A., Silva, D.C., et al. (2020). Not only toxic but repellent: what can organisms' responses tell us about contamination and what are the ecological consequences when they flee from an environment? *Toxics*, 8, 118.

Aulsebrook, L.C., Bertram, M.G., Martin, J.M., et al. (2020). Reproduction in a polluted world: implications for wildlife. *Reproduction*, 160, R13–R23.

Bean, T.G., Rattner, B.A., Lazarus, R.S., et al. (2018). Pharmaceuticals in water, fish and osprey nestlings in Delaware River and Bay. *Environmental Pollution*, 232, 533–545.

Bernhardt, E.S., Rosi, E.J., and Gessner, M.O. (2017). Synthetic chemicals as agents of global change. *Frontiers in Ecology and the Environment*, 15, 84–90.

Bertram, M.G., Gore, A.C., Tyler, C.R., and Brodin, T. (2022b). Endocrine-disrupting chemicals. *Current Biology*, 32, R727–R730.

Bertram, M.G., Martin, J.M., McCallum, E.S., et al. (2022c). Frontiers in quantifying wildlife behavioural responses to chemical pollution. *Biological Reviews*, 97, 1346–1364.

Bertram, M.G., Martin, J.M., Wong, B.B.M., and Brodin, T. (2022a). Micropollutants. *Current Biology*, 32, R17–R19.

Bertram, M.G., Saaristo, M., Baumgartner, J.B., et al. (2015). Sex in troubled waters: widespread agricultural contaminant disrupts reproductive behaviour in fish. *Hormones and Behavior*, 70, 85–91.

Bertram, M.G., Tomkins, P., Saaristo, M., et al. (2020). Disruption of male mating strategies in a chemically compromised environment. *Science of the Total Environment*, 703, 134991.

Bickham, J.W. (2011). The four cornerstones of evolutionary toxicology. *Ecotoxicology*, 20, 497–502.

Blanquart, F., Gandon, S., and Nuismer, S.L. (2012). The effects of migration and drift on local adaptation to a heterogeneous environment. *Journal of Evolutionary Biology*, 25, 1351–1363.

Brodin, T., Piovano, S., Fick, J., et al. (2014). Ecological effects of pharmaceuticals in aquatic systems—impacts through behavioural alterations. *Philosophical Transactions of the Royal Society B: Biological Sciences*, 369, 20130580.

Candolin, U., and Wong, B.B.M. (2019). Mate choice in a polluted world: consequences for individuals, populations and communities. *Philosophical Transactions of the Royal Society B: Biological Sciences*, 374, 20180055.

Cardinali, M., Maradonna, F., Olivotto, I., et al. (2004). Temporary impairment of reproduction in freshwater teleost exposed to nonylphenol. *Reproductive Toxicology*, 18, 597–604.

Chevin, L.-M., and Lande, R. (2010). When do adaptive plasticity and genetic evolution prevent extinction of a density-regulated population? *Evolution*, 64, 1143–1150.

Clotfelter, E.D., Bell, A.M., and Levering, K.R. (2004). The role of animal behaviour in the study of endocrine-disrupting chemicals. *Animal Behaviour*, 68, 665–676.

Cousins, I.T., Ng, C.A., Wang, Z., and Scheringer, M. (2019). Why is high persistence alone a major cause of concern? *Environmental Science: Processes & Impacts*, 21, 781–792.

Desforges, J.P., Hall, A., McConnell, B., et al. (2018). Predicting global killer whale population collapse from PCB pollution. *Science*, 361, 1373–1376.

Eng, M.L., Stutchbury, B.J., and Morrissey, C.A. (2019). A neonicotinoid insecticide reduces fueling and delays migration in songbirds. *Science*, 365, 1177–1180.

Eriksson Wiklund, A.-K., Börjesson, T., and Wiklund, S.J. (2006). Avoidance response of sediment living amphipods to zinc pyrithione as a measure of sediment toxicity. *Marine Pollution Bulletin*, 52, 96–99.

European Chemicals Agency. (2011). Guidance on Information Requirements and Chemical Safety Assessment. Part B: Hazard Assessment. https://echa.europa.eu/guidance-documents/guidance-on-information-requirements-and-chemical-safety-assessment

European Commission. (2018). Guidance No. 27—Technical Guidance for Deriving Environmental Quality Standards. https://circabc.europa.eu/ui/group/9ab5926d-bed4-4322-9aa7-9964bbe8312d/library/ba6810cd-e611-4f72-9902-f0d8867a2a6b/details

Falk, J.J., Parent, C.E., Agashe, D., and Bolnick, D.I. (2012). Drift and selection entwined: asymmetric reproductive isolation in an experimental niche shift. *Evolutionary Ecology Research*, 14, 403–423.

Fisher, H.S., Wong, B.B., and Rosenthal, G.G. (2006). Alteration of the chemical environment disrupts communication in a freshwater fish. *Proceedings of the Royal Society B: Biological Sciences*, 273, 1187–1193.

Ford, A.T., Ågerstrand, M., Brooks, B.W., et al. (2021). The role of behavioral ecotoxicology in environmental protection. *Environmental Science & Technology*, 55, 5620–5628.

Garnett, J., Halsall, C., Winton, H., et al. (2022). Increasing accumulation of perfluorocarboxylate contaminants revealed in an Antarctic firn core (1958–2017). *Environmental Science & Technology*, 56, 11246–11255.

Gore, A.C., Holley, A.M., and Crews, D. (2018). Mate choice, sexual selection, and endocrine-disrupting chemicals. *Hormones and Behavior*, 101, 3–12.

Hamilton, P.B., Rolshausen, G., Webster, T.M.U., and Tyler, C.R. (2017). Adaptive capabilities and fitness consequences associated with pollution exposure in fish. *Philosophical Transactions of the Royal Society B: Biological Sciences*, 372, 20160042.

Heckmann, L.-H., Friberg, N., and Ravn, H.W. (2005). Relationship between biochemical biomarkers and pre-copulatory behaviour and mortality in *Gammarus pulex* following pulse-exposure to lambda-cyhalothrin. *Pest Management Science*, 61, 627–635.

Hellström, G., Klaminder, J., Finn, F., et al. (2016a). GABAergic anxiolytic drug in water increases migration behaviour in salmon. *Nature Communications*, 7, 1–7.

Hellström, G., Klaminder, J., Jonsson, M., et al. (2016b). Upscaling behavioural studies to the field using acoustic telemetry. *Aquatic Toxicology*, 170, 384–389.

Hellström, G., Lennox, R.J., Bertram, M.G., and Brodin, T. (2022). Acoustic telemetry. *Current Biology*, 32, R863–R865.

Hendry, A.P., Kinnison, M.T., Heino, M., et al. (2011). Evolutionary principles and their practical application. *Evolutionary Applications*, 4, 159–183.

Horký, P., Grabic, R., Grabicová, K., et al. (2021). Methamphetamine pollution elicits addiction in wild fish. *Journal of Experimental Biology*, 224, p.jeb242145.

Islam, M.A., Blasco, J., and Araújo, C.V.M. (2019). Spatial avoidance, inhibition of recolonization and population isolation in zebrafish (*Danio rerio*) caused by copper exposure under a non-forced approach. *Science of the Total Environment*, 653, 504–511.

Jacomb, F., Marsh, J., and Holman, L. (2016). Sexual selection expedites the evolution of pesticide resistance. *Evolution*, 70, 2746–2751.

Jamieson, A.J., Malkocs, T., Piertney, S.B., et al. (2017). Bioaccumulation of persistent organic pollutants in the deepest ocean fauna. *Nature Ecology & Evolution*, 1, 1–4.

Jayasundara, N., Fernando, P.W., Osterberg, J.S., et al. (2017). Cost of tolerance: physiological consequences of evolved resistance to inhabit a polluted environment in teleost fish *Fundulus heteroclitus*. *Environmental Science & Technology*, 51, 8763–8772.

Jobling, S., Williams, R., Johnson, A., et al. (2006). Predicted exposures to steroid estrogens in UK rivers correlate with widespread sexual disruption in wild fish populations. *Environmental Health Perspectives*, 114, 32–39.

Jonsson, M., Andersson, M., Fick, J., et al. (2019). High-speed imaging reveals how antihistamine exposure affects escape behaviours in aquatic insect prey. *Science of the Total Environment*, 648, 1257–1262.

Ju, H., Zhu, D., and Qiao, M. (2019). Effects of polyethylene microplastics on the gut microbial community, reproduction and avoidance behaviors of the soil springtail, *Folsomia candida*. *Environmental Pollution*, 247, 890–897.

Kaisarevic, S., Vulin, I., Tenji, D., et al. (2021). Approaches, limitations and challenges in development of biomarker-based strategy for impact assessment of neuroactive compounds in the aquatic environment. *Environmental Sciences Europe*, 33, 1–13.

Kessler, S.C., Tiedeken, E.J., Simcock, K.L., et al. (2015). Bees prefer foods containing neonicotinoid pesticides. *Nature*, 521, 74–76.

Khairy, M.A., Luek, J.L., Dickhut, R., and Lohmann, R. (2016). Levels, sources and chemical fate of persistent organic pollutants in the atmosphere and snow along the western Antarctic Peninsula. *Environmental Pollution*, 216, 304–313.

Kidd, K.A., Paterson, M.J., Rennie, M.D., et al. (2014). Direct and indirect responses of a freshwater food web to a potent synthetic oestrogen. *Philosophical Transactions of the Royal Society B: Biological Sciences*, 369, 20130578.

Lauridsen, R.B., and Friberg, N. (2005). Stream macroinvertebrate drift response to pulsed exposure of the synthetic pyrethroid lambda-cyhalothrin. *Environmental Toxicology*, 20, 513–521.

Markandya, A., Taylor, T., Longo, A., et al. (2008). Counting the cost of vulture decline—an appraisal of the human health and other benefits of vultures in India. *Ecological Economics*, 67, 194–204.

Markman, S., Leitner, S., Catchpole, C., et al. (2008). Pollutants increase song complexity and the volume of the brain area HVC in a songbird. *PLOS ONE*, 3, e1674.

Martin, J.M., Saaristo, M., Bertram, M.G., et al. (2017). The psychoactive pollutant fluoxetine compromises antipredator behaviour in fish. *Environmental Pollution*, 222, 592–599.

Martínez-Ruiz, C., and Knell, R.J. (2017). Sexual selection can both increase and decrease extinction probability: reconciling demographic and evolutionary factors. *Journal of Animal Ecology*, 86, 117–127.

Matson, C.W., Lambert, M.M., McDonald, T.J., et al. (2006). Evolutionary toxicology: population-level effects of chronic contaminant exposure on the marsh frogs (*Rana ridibunda*) of Azerbaijan. *Environmental Health Perspectives*, 114, 547–552.

McCallum, E.S., Dey, C.J., Cerveny, D., et al. (2021). Social status modulates the behavioral and physiological consequences of a chemical pollutant in animal groups. *Ecological Applications*, 31, e02454.

Michelangeli, M., Martin, J.M., Pinter-Wollman, N., et al. (2022). Predicting the impacts of chemical pollutants on animal groups. *Trends in Ecology & Evolution*, 37, 789–802.

Moermond, C.T., Kase, R., Korkaric, M., and Ågerstrand, M. (2016). CRED: Criteria for reporting and evaluating ecotoxicity data. *Environmental Toxicology and Chemistry*, 35, 1297–1309.

Møller, A.P., and Mousseau, T.A. (2015). Strong effects of ionizing radiation from Chernobyl on mutation rates. *Scientific Reports*, 5, 8363.

Napper, I.E., Davies, B.F., Clifford, H., et al. (2020). Reaching new heights in plastic pollution—preliminary findings of microplastics on Mount Everest. *One Earth*, 3, 621–630.

Nasir, A.F.A.A., Cameron, S.F., Niehaus, A.C., et al. (2018). Manganese contamination affects the motor performance of wild northern quolls (*Dasyurus hallucatus*). *Environmental Pollution*, 241, 55–62.

Nathan, R., Monk, C., Arlinghaus, R., et al. (2022). Big-data approaches lead to an increased understanding of the ecology of animal movement. *Science*, 375, eabg1780.

Nørum, U., Friberg, N., Jensen, M.R., et al. (2010). Behavioural changes in three species of freshwater macroinvertebrates exposed to the pyrethroid lambda-cyhalothrin: laboratory and stream microcosm studies. *Aquatic Toxicology*, 98, 328–335.

Panadeiro, V., Rodriguez, A., Henry, J., et al. (2021). A review of 28 free animal-tracking software applications: current features and limitations. *Lab Animal*, 50, 246–254.

Paracelsus, A.T.H. (1538). *Septem Defensiones, Die dritte Defension*. Huser, Basel.

Persson, L., Carney Almroth, B.M., Collins, C.D., et al. (2022). Outside the safe operating space of the planetary boundary for novel entities. *Environmental Science & Technology*, 56, 1510–1521.

Petersen, G.I. (2003). Eels (*Anguilla anguilla*) avoidance test with MTBE, Report to EFOA—European Fuel Oxygenates Association. Report number DHI Project number 52294. 93.

Polverino, G., Martin, J.M., Bertram, M.G., et al. (2021). Psychoactive pollution suppresses individual differences in fish behaviour. *Proceedings of the Royal Society B: Biological Sciences*, 288, 20202294.

Qin, Z.F., Qin, X.F., Yang, L., et al. (2007). Feminizing/demasculinizing effects of polychlorinated biphenyls on the secondary sexual development of *Xenopus laevis*. *Aquatic Toxicology*, 84, 321–327.

Reznick, D.A., Bryga, H., and Endler, J.A. (1990). Experimentally induced life-history evolution in a natural population. *Nature*, 346, 357–359.

Richmond, E.K., Rosi, E.J., Walters, D.M., et al. (2018). A diverse suite of pharmaceuticals contaminates stream and riparian food webs. *Nature Communications*, 9, 1–9.

Rinner, B.P., Matson, C.W., Islamzadeh, A., et al. (2011). Evolutionary toxicology: contaminant-induced genetic mutations in mosquitofish from Sumgayit, Azerbaijan. *Ecotoxicology*, 20, 365–376.

Rolshausen, G., Phillip, D.A., Beckles, D.M., et al. (2015). Do stressful conditions make adaptation difficult? Guppies in the oil-polluted environments of southern Trinidad. *Evolutionary Applications*, 8, 854–870.

Saaristo, M., Brodin, T., Balshine, S., et al. (2018). Direct and indirect effects of chemical contaminants on the behaviour, ecology and evolution of wildlife. *Proceedings of the Royal Society B: Biological Sciences*, 285, 20181297.

Schoenfuss, H.L., Bartell, S.E., Bistodeau, T.B., et al. (2008). Impairment of the reproductive potential of male fathead minnows by environmentally relevant exposures to 4-nonylphenol. *Aquatic Toxicology*, 86, 91–98.

Seuront, L. (2018). Microplastic leachates impair behavioural vigilance and predator avoidance in a temperate intertidal gastropod. *Biology Letters*, 14, 20180453.

Sih, A., Ferrari, M.C.O., and Harris, D.J. (2011). Evolution and behavioural responses to human-induced rapid environmental change. *Evolutionary Applications*, 4, 367–387.

Suárez-Rodríguez, M., and Garcia, C.M. (2017). An experimental demonstration that house finches add cigarette butts in response to ectoparasites. *Journal of Avian Biology*, 48, 1316–1321.

Tan, H., Bertram, M.G., Martin, J.M., et al. (2021). The endocrine disruptor 17β-trenbolone alters the relationship between pre- and post-copulatory sexual traits in male mosquitofish (*Gambusia holbrooki*). *Science of the Total Environment*, 790, 148028.

Tian, Z., Zhao, H., Peter, K.T., et al. (2021). A ubiquitous tire rubber–derived chemical induces acute mortality in coho salmon. *Science*, 371, 185–189.

UNEP (United Nations Environment Programme). (2019). Global Chemicals Outlook II—From Legacies to Innovative Solutions. UNEP, Geneva, Switzerland. https:// wedocs.unep.org/bitstream/handle/20.500.11822/ 28113/GCOII.pdf?sequence=1&isAllowed=y

Ward, A.J., Duff, A.J., and Currie, S. (2006). The effects of the endocrine disrupter 4-nonylphenol on the behaviour of juvenile rainbow trout (*Oncorhynchus mykiss*). *Canadian Journal of Fisheries and Aquatic Sciences*, 63, 377–382.

Ward, A.J., Duff, A.J., Horsfall, J.S., and Currie, S. (2008). Scents and scents-ability: pollution disrupts chemical social recognition and shoaling in fish. *Proceedings of the Royal Society B: Biological Sciences*, 275, 101–105.

Whitehead, A., Clark, B.W., Reid, N.M., et al. (2017). When evolution is the solution to pollution: key principles, and lessons from rapid repeated adaptation of killifish (*Fundulus heteroclitus*) populations. *Evolutionary Applications*, 10, 762–783.

Williams, L.M., and Oleksiak, M.F. (2011). Ecologically and evolutionarily important SNPs identified in natural populations. *Molecular Biology and Evolution*, 28, 1817–1826.

Woin, P. and Larsson, P. (1987). Phthalate esters reduce predation efficiency of dragonfly larvae, *Odonata; Aeshna. Bulletin of Environmental Contamination and Toxicology*, 38, 220–225.

Wong, B.B.M., and Candolin, U. (2015). Behavioral responses to changing environments. *Behavioral Ecology*, 26, 665–673.

Zala, S.M., and Penn, D.J. (2004). Abnormal behaviours induced by chemical pollution: a review of the evidence and new challenges. *Animal Behaviour*, 68, 649–664.

CHAPTER 4

Artificial light at night

Therésa M. Jones and Kathryn B. McNamara

Overview

What is artificial light pollution, and how does it affect the behaviour of animals and, in turn, populations? In this chapter, we review key literature that has framed our understanding of the two mechanisms by which artificial light (as an individual stressor) shifts animal behaviour: altering the nocturnal photic landscape and disrupting the hormonal pathways that drive circadian behaviour. Our review highlights that behavioural research in this field has primarily focused on individual-level impacts, particularly in terrestrial systems, and that further research is required in landscape-level approaches to understand the impact of artificial light on ecosystem dynamics and health. Finally, we highlight case studies where recent research has driven effective management policy that has mitigated the behavioural impact of artificial light at night for wildlife, resulting in measurable ecological benefits.

4.1 The importance of light for life on Earth

Life on Earth has evolved under relatively reliable daily, monthly, and annual light cycles (Gaston et al. 2014). Daily cycles arise because there are periods within a 24-hour cycle when organisms receive direct intense light from the Sun (daytime) and periods when they are in the Earth's shadow and thus only illuminated by sunlight that is reflected from the Moon (nighttime). The variation in light intensity over a day can be three orders of magnitude, but nighttime light intensities have comparable levels of variation across the lunar cycle, from almost darkness (0.0001 lux for a new moon on a cloudy night) through to levels of light that are enough for the human eye to read (0.3 lux provided by a full moon on a cloudless night) (Kyba et al. 2017b). Finally, light (or more precisely photoperiod) and its spectral composition vary annually due to the tilt of the Earth's axis and its path around the Sun.

Due to their constancy, life (across all biological realms) is organized around light cycles (Gaston et al. 2017). For example, circadian or daily activity patterns of animals are defined by the intensity of natural light: diurnal species are active during daylight hours, crepuscular species during the twilight hours of dawn and dusk, and nocturnal species favour the darkness of night. However, variation in light intensity across the lunar cycle affects the likelihood of activity even within nocturnally active species: those at risk from visual predators are less likely to be active during full moon nights (Beier 2006), and those, like the willie wagtail *Rhipidura leucophrys*, whose males sing at night do so most during full moon nights, presumably to take advantage of both acoustic and visual cues in their pursuit of a suitable mate (Dickerson et al. 2022). Finally, annual shifts in light cycles associated with increasing photoperiod can drive circannual hormonal upregulation, such as the reproductive hormones in seasonally breeding birds and mammals (Dawson et al. 2001; Jimenez et al. 2015). This leads to synchronized offspring production at times when resources are plentiful. Animals may use multiple temporal cues of light, such as in the iconic mass spawning events of some coral species: spawning occurs once annually, at night, approximately three to five days after a full moon (Iluz and Dubinsky 2015).

4.2 The emergence of artificial light at night as an anthropogenic stressor

Human-led change to natural light cycles represents one of the first major anthropogenic stresses on Earth (Gaston et al. 2014; Kyba et al. 2017a). Today, the area of the Earth exposed to outdoor lighting (in multiple forms) is currently increasing by approximately 2% per year (Kyba et al. 2017a) and approximately 30% of the terrestrial environment (Gaston et al. 2014) and 22% of global coastlines (Davies et al. 2020) are affected by some form of ecological light pollution. Light pollution arises through point sources of light affecting the local environment but, due to light scattering far beyond its source, and because it also reflects upward, light from urban environments can reach tens of kilometres from its source in the form of skyglow (Kyba et al. 2012).

Artificial light changes multiple physical properties of the nocturnal environment. First, it enhances the illumination of the nocturnal environment to levels rarely, if ever, present in nature (Kyba et al. 2017a). In areas where artificial light at night is present, cloudy nights that would normally result in a darker nocturnal environment enhance night-time brightness due to scattered light reflecting back from the clouds (Kyba et al. 2017a). Second, artificial light at night typically changes the colour of the night. Recent advances in lighting technologies have seen a shift from longer-wavelength amber lighting to blue-rich white lighting. Spectral variation can have differential and species-specific effects on the physiology and behaviour of animals (van Geffen et al. 2015b; de Jong et al. 2017; Spoelstra et al. 2017b; Bennie et al. 2018); however, blue-rich nocturnal lighting is particularly problematic for aquatic environments, as it penetrates deeper into the water column (Bolton et al. 2017; Marangoni et al. 2022). Finally, artificial light masks natural light cycles, including the major daily transitions observed at sunrise and sunset, low intensity monthly variation of the lunar cycle, and annual shifts in photoperiod (Gaston et al. 2017). Given that these natural light cycles are zeitgebers of cellular processes and circadian function, it is therefore unsurprising that human-induced changes to natural light cycles result in myriad physiological and behavioural effects in both terrestrial and aquatic ecosystems (Mayer-Pinto et al. 2021; Sanders et al. 2021; Bassi et al. 2022).

4.3 Physiological changes underpinning behavioural responses

Behavioural shifts are underpinned by physiological and neurochemical changes. Artificial light at night, particularly short-wavelength blue light, interferes with the circadian system, including the synthesis of the indolamine melatonin (Tan et al. 2010; Jones et al. 2015). In animals, the circadian system is composed of a circadian clock (or clocks) and a neurochemical entrainment pathway that relays the light-induced temporal signal to the rest of the organism. Melatonin is a photosensitive chemical involved in the entrainment pathway and is believed to be ubiquitously present across much of life (Tan et al. 2010; Jones et al. 2015). While perhaps best known for its circadian role, its second function, and likely evolutionary origin, is as a powerful antioxidant (Zhao et al. 2019). As a consequence, this has implications for life-history traits, and potentially behaviour. Under natural conditions, for virtually all species in which melatonin has been studied (both nocturnal and diurnal), melatonin concentrations are typically highest at night and are reduced during the day due to light disruption of the enzyme pathways (Figure 4.1). Largely conclusive evidence now suggests that even low intensity dim artificial light at night can inhibit the synthesis of melatonin, significantly reducing circulating concentrations (Brüning et al. 2015; Grubisic et al. 2019; Figure 4.2a and b). This has inevitable direct and indirect behavioural consequences (see Sections 4.4 and 4.5).

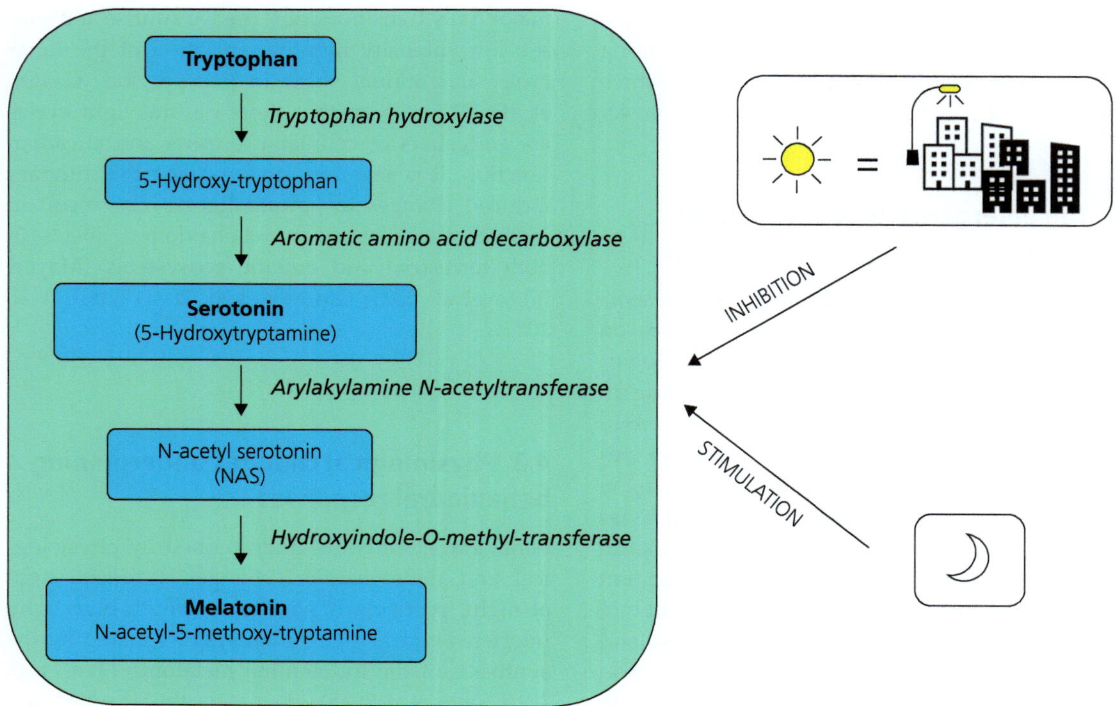

Figure 4.1 Simplified schematic of the biosynthetic pathway for melatonin synthesis in vertebrates and invertebrates (enzymes shown in italics). In vertebrates, the primary site of endogenous melatonin synthesis is the pineal gland, but it has also been found in the retina and other organs. In invertebrates, melatonin is typically synthesized in the cerebral ganglia and eyes (for a review, see Jones et al. 2015). Enzymes in the biosynthetic pathway have peak expression during darkness or night and are inhibited by either natural or anthropogenic light.

4.4 The behavioural implications of artificial light at night

Artificial light at night affects different levels of biological organization, promoting short-term or immediate behavioural responses, developmental plasticity, and longer-term adaptive evolutionary change (Swaddle et al. 2015). Biological responses are observed at multiple ecological levels and geographic scales and, even at the level of the individual, effects can be broad and cascading.

4.4.1 Attraction, avoidance, and ecological traps

Among the most obvious effects of artificial lighting is its ability to affect movement patterns of animals. References to flying animals aggregating around outdoor lighting both on land and at sea

have been documented as far back as the early 1800s (Rich and Longcore 2006). Today, there is a plethora of examples where artificial night lighting alters general movement patterns of terrestrial and aquatic animals (Rich and Longcore 2006; Marangoni et al. 2022). Recent evidence suggests its presence and spectra may have catastrophic disruptive consequences for nocturnal avian migration (Cabrera-Cruz et al. 2018), and can alter diel and other daily movement patterns of behaviour, including settlement in marine and other aquatic species (Davies et al. 2015; Bolton et al. 2017; Kuhne et al. 2021). Strictly nocturnal species, such as many species of microbats, some small mammals, or marine fishes and invertebrates, may be light-shy or completely photophobic. Such species actively avoid areas with lights or switch their activity patterns when nearing artificial lights (Rydell 2006;

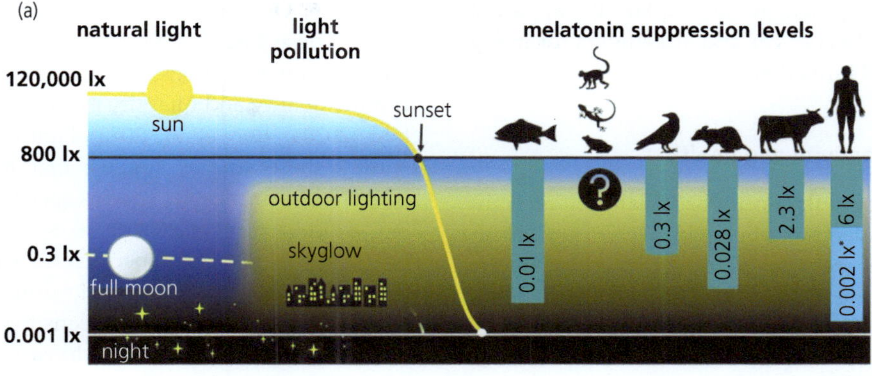

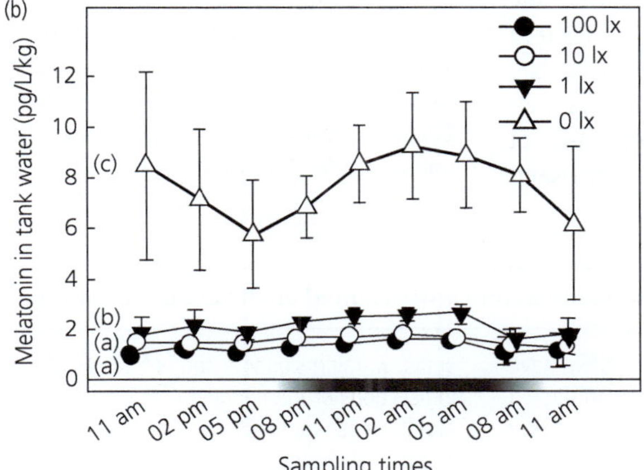

Figure 4.2 (a) The minimum reported illuminance required to suppress melatonin in various taxa (as indicated by blue bars) relative to light levels from natural and typical levels of artificial light sources; * indicates the minimum level of monochromatic light (460 nm) capable of suppressing melatonin levels in humans, under laboratory conditions. (b) Change in concentration of melatonin in tank water of European perch exposed to 0 lx, 1 lx, 10 lx, or 100 lx at night over a 24-hour period (sunset at 4 p.m. and sunrise at 7 a.m.).

(a) Image: Andreas Jechow; icons: Freepik; modified from Grubisic et al. (2019); (b) Redrawn from Brüning et al. (2015), with permission from Elsevier

Spoelstra et al. 2017a; Owens and Lewis 2018; Marangoni et al. 2022).

The precise underlying mechanism promoting an animal's immediate behavioural response to a light source is unclear. Birds and insects may mistake point sources of light as celestial cues (i.e. the Moon or stars) typically used in navigation (Rich and Longcore 2006). Scattered and polarized light from artificial lights also reflects off impervious surfaces, making them resemble water bodies, which may attract insects, such as dragonflies, mayflies, and moths, for oviposition (Frank 2006;

Horvath et al. 2014). Alternatively, animals may not be orienting to the light itself but, rather, the prey that it attracts (see Section 4.4.3). Ultimately, the presence of nocturnal lighting may result in an animal being trapped in a sub-optimal location, or an ecological trap (Hale and Swearer 2016). Once attracted, birds, bats, and insects can become temporarily blinded and/or disoriented and then fly until exhausted, until they perish, or the lights are turned off (Rodríguez et al. 2014; see Box 4.1). Animals can become functionally blind and then disorientated when exposed to even the dimmest of artificial lighting, which may explain the apparent hypnotizing effect of car and streetlights for many mammals and birds (Beier 2006).

4.4.2 Sleep activity and health

The disrupting influence of artificial light at night for sleep or sleep-like behaviour has been documented in several species of terrestrial vertebrate and invertebrate (Dominoni et al. 2016; Aulsebrook et al. 2018). However, its true impact is likely understated for most animals, other than humans, due to the current difficulty in distinguishing true sleep (Aulsebrook et al. 2016). Consequently, few field-based studies exploring the impact of light at night have measured sleep directly (but see Aulsebrook et al. 2020b). However, experimental evidence on diurnal birds indicates earlier onset of daily activity and increased nocturnal activity (Raap et al. 2016; Ouyang et al. 2017; Aulsebrook et al. 2020b). These findings have been corroborated in laboratory investigations where light at night affects both the quality and quantity of sleep (Aulsebrook et al. 2020a; Ren et al. 2022; Figure 4.4). Interestingly, these studies suggest that the effect of light at night on sleep is species-specific (Sun et al. 2017; Aulsebrook et al. 2020a), sex-specific (Ouyang et al. 2017), and spectra-specific (Aulsebrook et al. 2020a). Whether species-specific differences arise through differences in spectral sensitivities and/or perceived predation risk or due to variation in neurochemical responses to light exposure remains to be explored.

The fitness implications arising from loss of sleep are likely substantial (given comparable evidence of deleterious effects in humans; Cain et al. 2020), although we largely lack the evidence to substantiate this for most animals (Dominoni et al.

Box 4.1 Case studies of effective management and mitigation of artificial light pollution

Research into the effects of artificial light on wildlife have already led to tangible and substantive changes to human behaviour to mitigate these impacts. There are two key examples in which the implementation of research-driven management decisions has reduced the ecological impact of artificial light and improved population survival. These examples highlight both the highly collaborative nature of the management practices required, and the importance of considering the nature of light pollution (intensity, spectral distribution, and temporality) in mitigating its impacts.

The attraction of seabirds to artificial light sources, and their associated mortality, has been long known (Lewis 1927). Yet few experimental studies have quantified the impact of management mitigation actions. Phillip Island, in the state of Victoria, Australia, is a major tourism hotspot, and home to more than 1 million short-tailed shearwaters *Ardenna tenuirostris* (Figure 4.3a). Migrating shearwaters become trapped in the artificial lights, fall to the ground, and are frequently hit by cars. Fledglings are particluarly affected, with 68–99% of all birds observed (Rodríguez et al. 2014). Rodríguez et al. (2014) report a highly successful and collaborative effort between the Phillip Island National Park, the state government department with oversight of roads in Victoria, and their associated lighting contractors. Lighting intensity restriction and traffic management (speed limits, warning signage, and enforced traffic stopping) during sections of the fledging period were implemented. This multifaceted approach saw a significant decrease in the number of grounded birds, reducing the road-associated fatalities.

The importance of considering the temporal impacts of light in mitigating artificial light's effects is demonstrated in the management response to the September 11 memorial, *Tribute in Light*, in New York City (operated by the National September 11 Memorial & Museum (NSMM)). The light installation comprises 88 vertical searchlights, visible from nearly 100 km away. The *Tribute in Light* is a potent light trap to nocturnally migrating birds and bats (Figure 4.3b). Management of this memorial is part of a broader legislative push by groups, such as the advocacy group NYC Audubon (NYCA), for a reduction in all artificial nighttime lighting during peak migratory periods within the city. The NYCA and NSMM intermittently shut down the tribute for 20-minute periods when volunteer-recorded observations of bird numbers exceed 1000 individuals. Van Doren et al. (2017) demonstrated that the memorial attracted bird densities of up to 20 times baseline levels, but that this disruptive effect was immediately reversed by the lighting shutdowns, mitigating its impact.

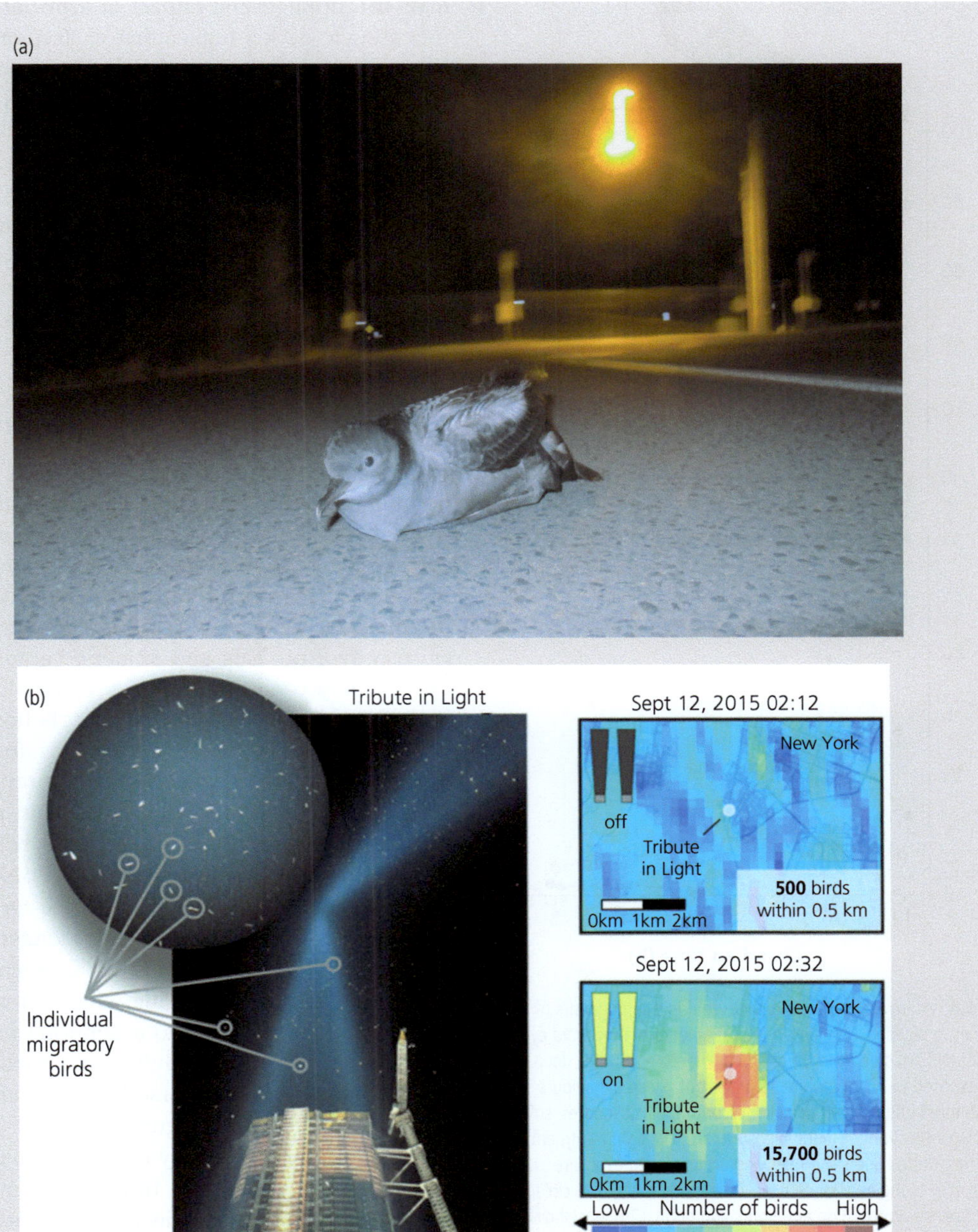

Figure 4.3 (a) A short-tailed shearwater fledgling, attracted to the park's lights and grounded on the road. (b) Migratory birds are highly attracted to the memorial, but rapidly dissipate when the lights are turned off.

(a) Photo: Airam Rodríguez; used with permission. (b) Redrawn from Van Doren et al. (2017)

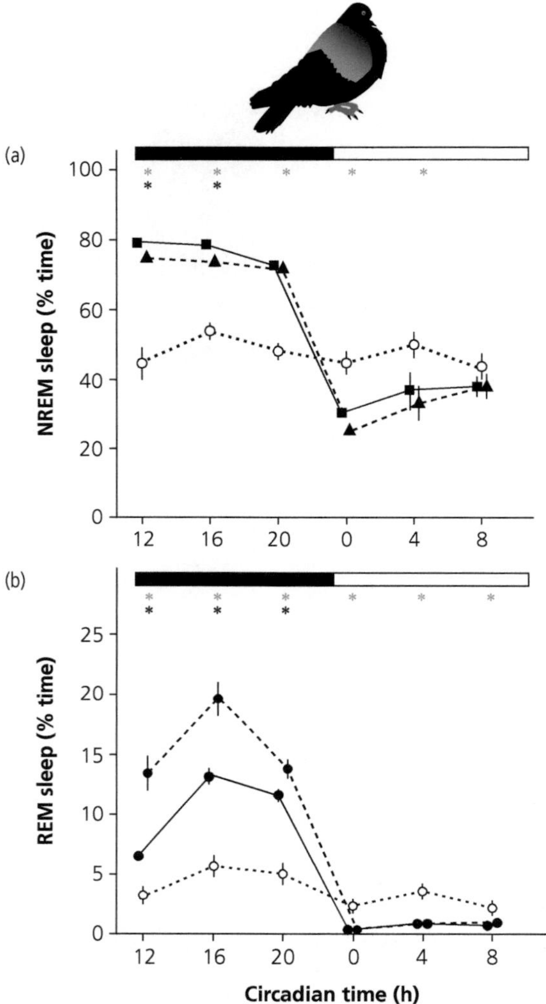

Figure 4.4 White artificial light reduces multiple components of sleep cycles in pigeons. Changes in the (a) amount of non-rapid eye movement (NREM) sleep and (b) rapid eye movement (REM) sleep in pigeons exposed to 12 h of white light at night. Sleep behaviours were monitored under no artificial light (control: black squares, solid line), during white light at night exposure (artificial light treatment: white circles, dotted line), and following artificial light exposure (recovery: grey triangles, dashed line). Time is represented as circadian time; lights were switched on at 0 hours and off at 12 hours (shown in the horizontal bar at the top of each plot). Values represent 4-hour means (± standard errors) across the 24-hour period. Significant differences between the control and white light treatment (grey asterisks) or recovery (black asterisks) are indicated.

Redrawn from Aulsebrook et al. (2020a), with permission from Elsevier. Illustration: Juliane Gaviraghi Mussoi

2016; Aulsebrook et al. 2018). It is likely that animals can mitigate the impact of artificial light by retreating to dark refugia either to sleep or to carry out nocturnal activities. Evasive behaviour such as this reduces the physiological impact of light at night and may reduce predation risk associated with visual predators (Aulsebrook et al. 2018). Alternatively, shifts in nocturnal sleeping patterns (both duration and intensity) may be mitigated through increased sleep during the day (Raap et al. 2016; Aulsebrook et al. 2020a; Aulsebrook et al. 2020b). Whether this has additional fitness implications again remains to be explored.

4.4.3 Foraging and its consequences for health

Artificial light at night can directly affect the temporality and efficacy of foraging behaviours, often through its influence on intraspecific and interspecific competitive interactions. For example, illumination of the nocturnal environment can result in enhanced foraging success for predators, such as insectivorous bats and spiders, because lights can spatially concentrate their prey (Heiling 1999; Rydell 2006; Becker et al. 2013; Bolton et al. 2017). This has obvious implications for photo-attracted prey, but even light-shy prey species may suffer reduced foraging efficacy, in part due to light-related increases in vigilance behaviours (Farnworth et al. 2018; Zhang et al. 2020). Nocturnal illumination expands the photic niche for diurnal species, allowing them to forage for longer periods (Santos et al. 2010; Russ et al. 2015). Conversely, this reduces the temporal foraging window for nocturnal species and can lead to a temporal mismatch between peak foraging activity and prey availability, as observed in the horseshoe bat *Rhinolophus pusillus* (Luo et al. 2021; Figure 4.5).

Shifts in foraging activity can have obvious and quite significant effects on fitness in animals. Elevated foraging rates, and increased caloric intake, around artificial light sources have been reported for several aquatic and terrestrial species (Rydell 2006; Minnaar et al. 2015). In the pond snail *Lymnaea stagnalis*, exposure to artificial light at night resulted in increased feeding and rates of growth. However, the number and size of eggs produced was reduced, and development and hatching of

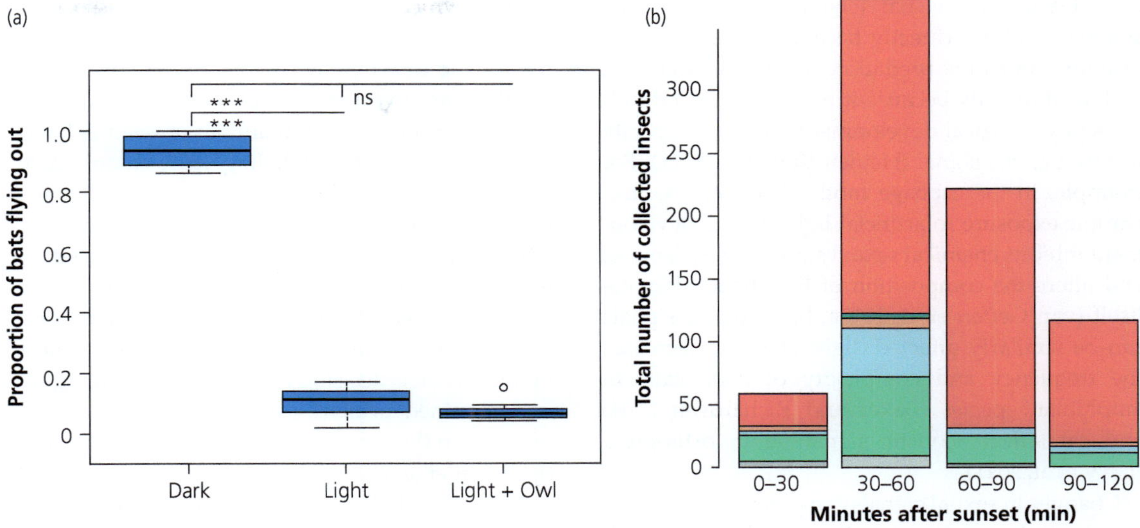

Figure 4.5 Artificial light reduces and delays the emergence of least horseshoe bats *Rhinolophus pusillus* for nocturnal foraging. (a) Most *R. pusillus* remained in their roost to avoid the LED light (Light) until the end of the experimental treatment (40–55 minutes after sunset), compared with those roosting in dark conditions (Dark). This inhibitory effect of artificial light, also present in the presence of a predator model (Light + Owl), may lead to a mismatch between foraging onset and peak food availability in this species, as (b) insect prey reached a peak of abundance 30–60 minutes after sunset (colours on bars represent different insect orders).

Redrawn from Luo et al. (2021), with permission from Elsevier

eggs delayed (Baz et al. 2022). Conversely, in the amphipod *Orchestoidea tuberculata*, the presence of artificial light reduced activity and foraging, resulting in reduced food consumption and growth rate (Luarte et al. 2016). These studies provide compelling evidence of an association between foraging behaviour and fitness. Nevertheless, such data are currently limited and frequently conflicting, suggesting this is a potentially significant gap in our understanding.

4.4.4 Mating behaviour and reproductive consequences

Due to light-related disruption to the timing of daily, lunar, and seasonal breeding circadian rhythms (Gaston et al. 2017), artificial nocturnal lighting has inevitable consequences for reproduction and mating behaviour, including the production and reception of sexual signals. Masking of daily and seasonal lighting cues can disrupt the timing of avian dawn chorusing behaviour, affect mate acquisition rates, and shift

egg-lay date (Kempenaers et al. 2010; Da Silva et al. 2016). Observed changes in behaviour can arise due to the impact of variation in light period on hormonal pathways. For example, Tammar wallabies *Macropus eugenii* exposed to streetlighting delayed and desynchronized births by up to a month (Robert et al. 2015), potentially resulting in a mismatch between offspring birth and required resources. Similarly, for the coral species *Acropora millepora* and *Acropora digitifera*, exposure to light at night resulted in delayed gametogenesis and desynchronization of gamete release (Ayalon et al. 2021). Recent evidence suggests that artificial lighting can also affect the quality of sexual signals (Baker and Richardson 2006; van Geffen et al. 2015b). The masking effect of streetlights for nocturnally produced visual signals is relatively obvious (Owens and Lewis 2018). For example, female glow-worms *Lampyris noctiluca* delay signalling under lights, but do not move away from artificial light even though it significantly reduces mate attraction, suggesting a maladaptive response (Elgert et al. 2020; Figure 4.6). However, artificial light can also disrupt non-visual (chemical and acoustic) signalling, either directly because it enhances the visibility and thus predation risk of the signaller, and/or indirectly because it influences an individual's physiological development and thus capacity to produce the signal (Heinen-Kay et al. 2021). For example, in the cabbage moth *Mamestra brassicae*, chronic exposure to artificial light during development inhibits chemical sexual signalling by females and alters the composition of the chemical signal itself (van Geffen et al. 2015a, b). Acoustic signals can be similarly affected: light at night decreases the frequency and complexity of male calls in amphibian species (Baker and Richardson 2006) and can affect acoustic signalling in orthoptera (Levy et al. 2021).

Changes in sexual signals can potentially weaken female mating preferences (Rand et al. 1997), which may have profound ecological ramifications at multiple levels. First, and most well-established, artificial light reduces individual reproductive fitness (as detailed throughout this chapter). Second, if changes in the production and reception of sexual signals leads to trait divergence, artificial light exposure may promote population divergence and subsequent speciation (Hopkins et al. 2018; Heinen-Kay et al. 2021).

4.4.5 Apparent benefits of artificial light exposure?

While there are demonstrable and quantified costs of artificial light exposure to wildlife (see examples in this chapter), there is emerging evidence that artificial light can, in some instances, benefit animals (Willmott et al. 2022). Research has focused primarily on the beneficial impacts of artificial light in predator-prey dynamics and, to a lesser extent, in mating behaviour and reproductive fitness. We highlight two recent examples below, but note that the long-term fitness implications are yet to be determined.

Perhaps the clearest beneficiaries of the presence of artificial light at night are predators whose potential prey concentrates around predictable light sources (Heiling 1999; Adams 2000; Rydell 2006; Bolton et al. 2017; Willmott et al. 2019). Empirical work suggests that such lights may act as a super-stimulus indicating areas of higher prey

density, which in turn may reduce costs associated with foraging. For example, predators such as insectivorous spiders that live in the presence of artificial lighting can invest in smaller webs and forage for shorter periods but still maintain or even enhance their foraging success (Heiling 1999; Adams 2000; Willmott et al. 2019). Whether prey availability, rather than lights per se, is the strongest determinant of foraging site preference is untested for most species. However, juvenile Australian garden orb-web spiders *Hortophora biapicata* (formerly *Eriophora*) prefer to construct their webs under artificial lights, but their decision to remain at a foraging site is determined primarily by past foraging success rather than the continued presence of artificial light (Willmott et al. 2019).

As highlighted above, the enhanced visual environment provided by artificial light at night allows normally diurnal species (particularly those that are visually reliant) to extend their periods of activity beyond natural twilight, but it also makes it easier for potential prey to detect predators, reducing the advantage of specialized nocturnal predator vision. For example, the Tammar wallaby tends to forage more and allocate less time to anti-predator vigilance when exposed to artificial light, suggesting that increased illumination was associated with increased perception of safety (Biebouw and Blumstein 2003). However, the ostensible benefits associated with enhanced foraging success or increased visual ability may mask underlying physiological impacts of nocturnal lights. While Australian garden orb-web spiders accrue foraging benefits if they build webs under lights (Willmott et al. 2019), when resources were held constant, light-exposed spiders developed more rapidly, were smaller as adults, and had a 70% reduction in female fecundity (Willmott et al. 2018). This has significant implications for individuals that live in ephemeral environments, where resources may be less reliable, and may be of particular concern for insectivorous species in the face of global declines in invertebrate biodiversity (Knop et al. 2017; Owens and Lewis 2018).

4.5 Light-induced shifts to community structure and function

As highlighted above, behavioural shifts associated with the presence of artificial light at night can

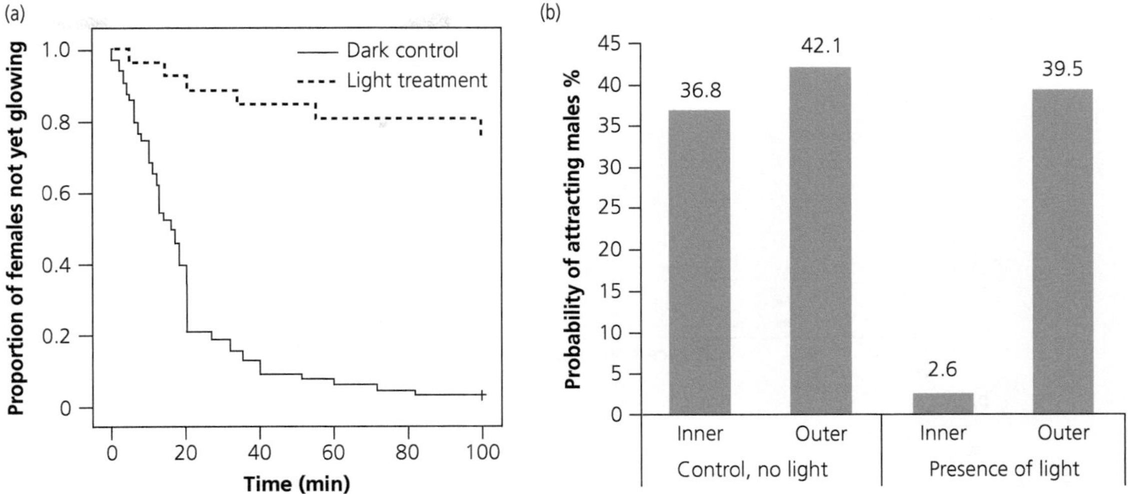

Figure 4.6 (a) The likelihood of commencing sexual signalling in the female common glow-worm *Lampyris noctiluca* is reduced in the presence of artificial light (black dashed line), compared with dark-at-night conditions (black solid line). Females do not avoid artificial light, which is maladaptive as (b) the probability of a female successfully attracting a male increased dramatically when there was no artificial light (left two bars) or when the female was further from the source of artificial light (Outer) compared with females located under the source of artificial light (Inner). Redrawn from Elgert et al. (2020)

have cascading cross-species effects, and accumulating evidence suggests this can lead to substantial shifts in community structure and ecosystem function, both within (Rydell 2006; Davies et al. 2012; Lockett et al. 2021; Marangoni et al. 2022) and across realms (Manfrin et al. 2017; Mayer-Pinto et al. 2021). Observed structural changes within communities can be influenced by top-down effects, such as predatory species exploiting behavioural shifts in potential prey (Davies et al. 2012; Manfrin et al. 2017), or through bottom-up trophic effects resulting in resource limitation (Bennie et al. 2018). However, community shifts can also arise through species-specific differences in attraction or repulsion to the presence of light if this drives differential immigration or emigration (Davies et al. 2015; Lockett et al. 2021). For example, following experimental deployment of LED lighting (three treatments: natural darkness, 19 lux, or 30 lux) within a marine environment, Davies et al. (2015) report species-specific variation in the recruitment of a range of sessile and mobile marine invertebrates, and thus significant shifts in the epifaunal marine invertebrate community. Of concern is that community-level impacts of streetlighting may have spillover effects into both neighbouring unlit environments and the

subsequent diurnal period (Davies et al. 2012; Knop et al. 2017). Moreover, these community impacts may affect spatial niches (i.e. terrestrial versus aerial communities) differently (Lockett et al. 2021).

4.6 Evolutionary consequences of artificial light

Although the phenotypic effects of artificial light at night are increasingly well established, there is little empirical evidence for the capacity for artificial light exposure to promote evolutionary change (Hopkins et al. 2018; Desouhant et al. 2019; see also Chapter 14), although evidence from the field of urban ecology would suggest it is highly likely (Johnson and Munshi-South 2017). In principle, artificial light has a significant capacity to alter evolution through its effects on direct selection (through differential survival or reproductive output of light-affected individuals), reproductive isolation (artificial light can alter sexual signalling, potentially creating distinct sub-populations, which can be maintained through assortative mating), and gene flow and genetic drift (artificial light fragments the landscape, altering the distribution and movements of organisms) (Hopkins et al. 2018; Figure 4.7). Phenotypic plasticity

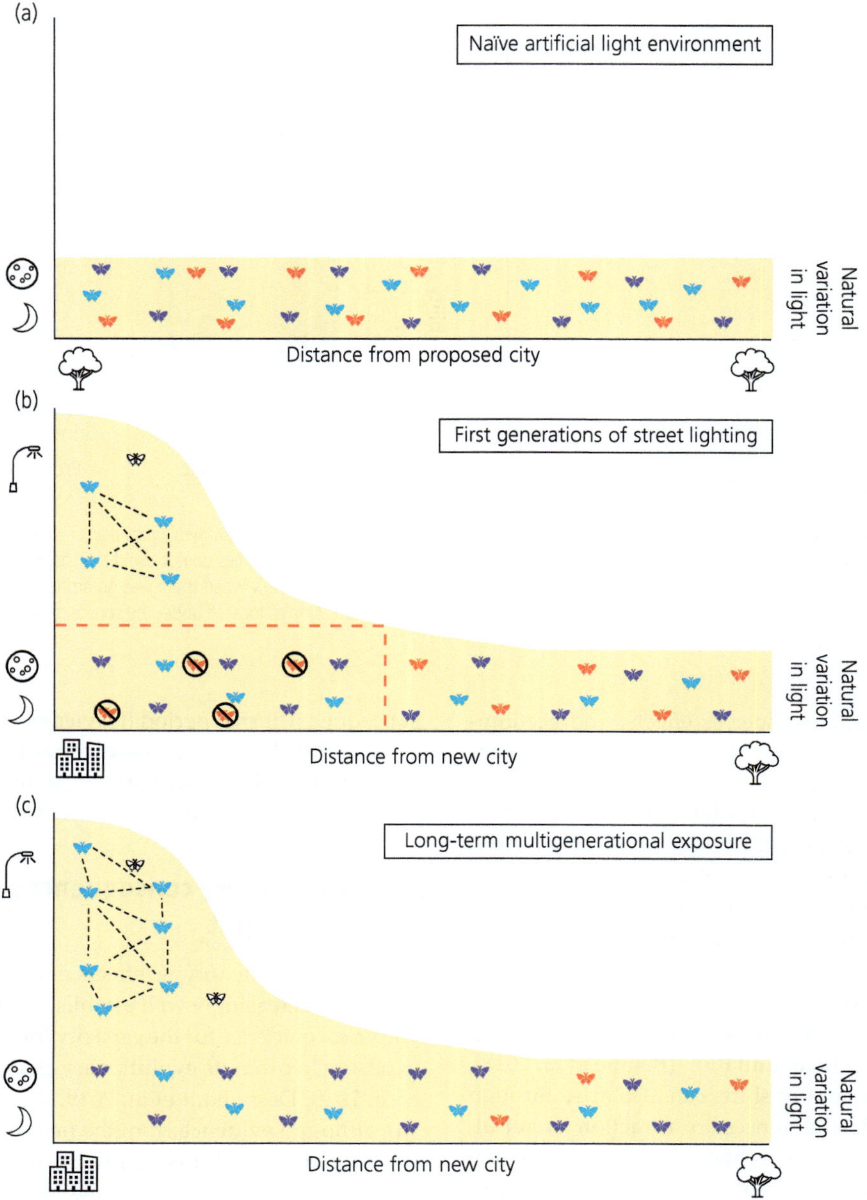

Figure 4.7 Conceptual illustration of artificial light's role in driving evolutionary change in an urban population. (a) This panel represents the distribution and abundance of allele types (red, blue, and purple) within the naturally occurring photic environment of the lunar cycle, prior to the introduction of artificial light (here, city lights). (b) The appearance of city lights creates a strong selection pressure on red alleles, which can only tolerate near-natural night lighting, limiting their distribution (thresholds demarcated by red dashed lines); this reduces their abundance and overall reduces gene flow across the gradient. Blue alleles are attracted to artificial light and undergo a distributional shift towards the city to fill this niche. Assortative mating (black dashed lines) occurs within this sub-population, creating the potential for reproductive isolation. Purple alleles are unaffected by artificial light and retain their initial abundance and distribution. Finally, artificial light may have mutagenic effects (depicted by novel white alleles), introducing new variation into the population. (c) The resulting population after generational exposure to artificial light. The red alleles' abundance is further diminished by stochastic process (genetic drift) in favour of purple alleles and they are restricted to non-lit environments. The blue alleles thrive under artificial light, increasing their presence in highly urbanized areas. The changes in seasonal reproductive behaviour of this urbanized population further enhances reproductive isolation of this sub-population. White mutant alleles may continue to increase in abundance. The allele frequencies under artificial light are altered, demonstrating evolution.

in response to artificial light may become fixed through genetic accommodation and assimilation, potentially leading to heritable change (Levis and Pfennig 2016). However, current studies rarely distinguish between plastic or heritable genetic responses in light-affected populations.

Direct evidence of evolutionary changes in behavioural responses to artificial light are scarce (Swaddle et al. 2015; Hopkins et al. 2018). Indirect evidence of local adaptation in response to artificial light is found in the heritable preference of spiders to build webs in artificially lit areas, irrespective of their developmental exposure (Heiling 1999) and the reduced flight to light behaviour of ermine moths *Yponomeuta cagnagella* collected as larvae from light-affected populations (Altermatt and Ebert 2016). However, neither study was able to demonstrate conclusively evolved differences specifically in relation to light at night. Moreover, responses to artificial light in the ermine moth were sex- and behaviour-specific, as there was no clear evidence of comparable evolution in feeding behaviours or female sexual signalling (Cieraad et al. 2022).

4.7 The future

Our chapter reviews a snapshot of the accumulating evidence for the impacts of artificial light on individuals through to populations. However, there are several key gaps in our knowledge. First, our understanding of the impacts of light pollution on communities and ecosystems is hampered by a lack of fundamental research. Similarly, research to date has primarily focused on single-realm terrestrial systems. This is likely problematic given we know that there is significant interspecific variation in melatonin suppression under light exposure, and biological realms may be differentially affected (Mayer-Pinto et al. 2021). A wider approach is required to elucidate the impacts on ecosystems and population health and to quantify the broad-scale impacts of artificial light pollution. Second, artificial light is inextricably linked with other anthropogenic stressors (chemical, noise, and heat). Indeed, the degree to which artificial light is solely responsible for the observed declines in biodiversity remains to be properly investigated.

Studies investigating the effects of artificial light in a combined stressor context are rare, but have critically demonstrated non-additive effects with heat (Miller et al. 2017) and noise (Dominoni et al. 2020; see also Chapter 10). Interpretation of artificial light's effects in isolation may therefore lead to an underestimation of its 'real-world' impacts, limiting our capacity for mitigation. Third, we lack insight into the long-term consequences of exposure to artificial light at night, which compromises our understanding of its potential evolutionary impact. Targeted longitudinal field studies or large-scale experimental evolution trials would provide much needed data. Such studies would allow us to assess the relative importance of plastic and evolutionary responses to light at night. Finally, our understanding of the behavioural impacts of artificial light has largely been empirically tested using a single constant daytime light intensity, at least one intensity (or spectrum) of dim light at night as the light treatment, and an absence of light (i.e. darkness) as the experimental control. This is highly problematic, as the difference between day and night lighting rarely matches that in nature; spectral sensitivities of animals are highly species-specific and the impact of different spectra appears to be varied (van Geffen et al. 2015a; Spoelstra et al. 2017a; Aulsebrook et al. 2020a; Aulsebrook et al. 2020b; Ren et al. 2022); and, perhaps most importantly, no light at night may, in itself, be a biological stressor (Aulsebrook et al. 2022). Ultimately, while lab-based and semi-natural field-based experiments often yield some of the cleanest data, and both authors are strong advocates for such approaches, experimental lighting specifications should be carefully considered to effectively translate the biological relevance of our findings.

References

Adams, M. (2000). Choosing hunting sites: web site preferences of the orb weaver spider, *Neoscona crucifera*, relative to light cues. *Journal of Insect Behavior*, 13, 299–305.

Altermatt, F., and Ebert, D. (2016). Reduced flight-to-light behaviour of moth populations exposed to long-term urban light pollution. *Biology Letters*, 12, 20160111.

Aulsebrook, A.E., Connelly, F., Johnsson, R.D., et al. (2020a). White and amber light at night disrupt sleep physiology in birds. *Current Biology*, 30, 0220035.

Aulsebrook, A.E., Jechow, A., Krop-Benesch, A., et al. (2022). Nocturnal lighting in animal research should be replicable and reflect relevant ecological conditions. *Biology Letters*, 18, 20220035.

Aulsebrook, A.E., Jones, T.M., Mulder, R.A., and Lesku, J.A. (2018). Impacts of artificial light at night on sleep: a review and prospectus. *Journal of Experimental Zoology Part A—Ecological and Integrative Physiology*, 329, 409–418.

Aulsebrook, A.E., Jones, T.M., Rattenborg, N.C., et al. (2016). Sleep ecophysiology: integrating neuroscience and ecology. *Trends in Ecology & Evolution*, 31, 590–599.

Aulsebrook, A.E., Lesku, J.A., Mulder, R.A., et al. (2020b). Streetlights disrupt night-time sleep in urban black swans. *Frontiers in Ecology and Evolution*, 8, 13.

Ayalon, I., Rosenberg, Y., Benichou, J.I.C., et al. (2021). Coral gametogenesis collapse under artificial light pollution. *Current Biology*, 31, 413–419.

Baker, B.J., and Richardson, J.M.L. (2006). The effect of artificial light on male breeding-season behaviour in green frogs, *Rana clamitans melanota*. *Canadian Journal of Zoology*, 84, 1528–1532.

Bassi, A., Love, O.P., Cooke, S.J., et al. (2022). Effects of artificial light at night on fishes: a synthesis with future research priorities. *Fish and Fisheries*, 23, 631–647.

Baz, E., Hussein, A.A.A., Vreeker, E.M.T., Soliman, M.F.M., Tadros, M.M., and El-Shenawy, N.S., Koene, J.M. (2022). Consequences of artificial light at night on behavior, reproduction, and development of *Lymnaea stagnalis*. *Environmental Pollution*, 307.

Becker, A., Whitfield, A.K., Cowley, P.D., et al. (2013). Potential effects of artificial light associated with anthropogenic infrastructure on the abundance and foraging behaviour of estuary-associated fishes. *Journal of Applied Ecology*, 50, 43–50.

Beier, P. (2006). Effects of artificial night lighting on terrestrial mammals. In: C. Rich and T. Longcore (eds), *Ecological Consequences of Artificial Night Lighting*, pp. 19–42. Island Press, Washington, DC.

Bennie, J., Davies, T.W., Cruse, D., et al. (2018). Artificial light at night causes top-down and bottom-up trophic effects on invertebrate populations. *Journal of Applied Ecology*, 55, 2698–2706.

Biebouw, K., and Blumstein, D.T. (2003). Tammar wallabies (*Macropus eugenii*) associate safety with higher levels of nocturnal illumination. *Ethology Ecology and Evolution*, 15, 159–172.

Bolton, D., Mayer-Pinto, M., Clark, G.F., et al. (2017). Coastal urban lighting has ecological consequences for multiple trophic levels under the sea. *Science of the Total Environment*, 576, 1–9.

Brüning, A., Hölker, F., Franke, S., et al. (2015). Spotlight on fish: light pollution affects circadian rhythms of European perch but does not cause stress. *Science of the Total Environment*, 511, 516–522.

Cabrera-Cruz, S.A., Smolinsky, J.A., and Buler, J.J. (2018). Light pollution is greatest within migration passage areas for nocturnally-migrating birds around the world. *Scientific Reports*, 8, 8.

Cain, S.W., McGlashan, E.M., Vidafar, P., et al. (2020). Evening home lighting adversely impacts the circadian system and sleep. *Scientific Reports*, 10, 19,110.

Cieraad, E., van Grunsven, R.H.A., van der Sman, F., et al. (2022). Lack of local adaptation of feeding and calling behaviours by *Yponomeuta cagnagellus* moths in response to artificial light at night. *Insect Conservation and Diversity*, 15, 445–452.

Da Silva, A., Valcu, M., and Kempenaers, B. (2016). Behavioural plasticity in the onset of dawn song under intermittent experimental night lighting. *Animal Behaviour*, 117, 155–165.

Davies, T.W., Bennie, J., and Gaston, K.J. (2012). Street lighting changes the composition of invertebrate communities. *Biology Letters*, 8, 764–767.

Davies, T.W., Coleman, M., Griffith, K.M., and Jenkins, S.R. (2015). Night-time lighting alters the composition of marine epifaunal communities. *Biology Letters*, 11, 20150080.

Davies, T.W., McKee, D., Fishwick, J., et al. (2020). Biologically important artificial light at night on the seafloor. *Scientific Reports*, 10, 10.

Dawson, A., King, V.M., Bentley, G.E., and Ball, G.F. (2001). Photoperiodic control of seasonality in birds. *Journal of Biological Rhythms*, 16, 365–380.

de Jong, M., Caro, S.P., Gienapp, P., et al. (2017). Early birds by light at night: effects of light color and intensity on daily activity patterns in blue tits. *Journal of Biological Rhythms*, 32, 323–333.

Desouhant, E., Gomes, E., Mondy, N., and Amat, I. (2019). Mechanistic, ecological, and evolutionary consequences of artificial light at night for insects: review and prospective. *Entomologia Experimentalis et Applicata*, 167, 37–58.

Dickerson, A.L., Hall, M.L., and Jones, T.M. (2022). The effect of natural and artificial light at night on nocturnal song in the diurnal willie wagtail. *Science of the Total Environment*, 808, 151986.

Dominoni, D.M., Borniger, J.C., and Nelson, R.J. (2016). Light at night, clocks and health: from humans to wild organisms. *Biology Letters*, 12, 4.

Dominoni, D., Smit, J.A.H., Visser, M.E., and Halfwerk, W. (2020). Multisensory pollution: artificial light at night and anthropogenic noise have interactive effects on activity patterns of great tits (*Parus major*). *Environmental Pollution*, 256, 113314.

Elgert, C., Hopkins, J., Kaitala, A., and Candolin, U. (2020). Reproduction under light pollution: maladaptive response to spatial variation in artificial light in a glow-worm. *Proceedings of the Royal Society B: Biological Sciences*, 287, 20200806.

El-Sayed, B., Ahmed, A.A., Hussein, E.M.T., et al. (2022). Consequences of artificial light at night on behavior, reproduction, and development of *Lymnaea stagnalis*. *Environmental Pollution*, 307, 119507.

Farnworth, B., Innes, J., Kelly, C., et al. (2018). Photons and foraging: artificial light at night generates avoidance behaviour in male, but not female, New Zealand weta. *Environmental Pollution*, 236, 82–90.

Frank, K. (2006). Effect of artificial night lighting on moths. In: C. Rich and T. Longcore (eds), *Ecological Consequences of Artificial Light Pollution*, pp. 305–344. Island Press, Washington, DC.

Gaston, K.J., Davies, T.W., Nedelec, S.L., and Holt, L.A. (2017). Impacts of artificial light at night on biological timings. In: D.J. Futuyma (ed.), *Annual Review of Ecology, Evolution, and Systematics, Vol 48*, pp. 49–68. Annual Reviews, Palo Alto, CA.

Gaston, K.J., Duffy, J.P., Gaston, S., et al. (2014). Human alteration of natural light cycles: causes and ecological consequences. *Oecologia*, 176, 917–931.

Grubisic, M., Haim, A., Bhusal, P., et al. (2019). Light pollution, circadian photoreception, and melatonin in vertebrates. *Sustainability*, 11, 136–141.

Hale, R., and Swearer, S.E. (2016). Ecological traps: current evidence and future directions. *Proceedings of the Royal Society B: Biological Sciences*, 283, 8.

Heiling, A.M. (1999). Why do nocturnal orb-web spiders (Araneidae) search for light? *Behavioral Ecology and Sociobiology*, 46, 43–49.

Heinen-Kay, J.L., Kay, A.D., and Zuk, M. (2021). How urbanization affects sexual communication. *Ecology and Evolution*, 11, 17,625–17,650.

Hopkins, G.R., Gaston, K.J., Visser, M.E., et al. (2018). Artificial light at night as a driver of evolution across urban–rural landscapes. *Frontiers in Ecology and the Environment*, 16, 472–479.

Horvath, G., Kriska, G., and Robertson, B. (2014). *Anthropogenic Polarization and Polarized Light Pollution Inducing Polarized Ecological Traps*. Springer-Verlag Berlin, Berlin.

Iluz, D., and Dubinsky, Z. (2015). Coral photobiology: new light on old views. *Zoology*, 118, 71–78.

Jimenez, R., Burgos, M., and Barrionuevo, F.J. (2015). Circannual testis changes in seasonally breeding mammals. *Sexual Development*, 9, 205–215.

Johnson, M.T.J., and Munshi-South, J. (2017). Evolution of life in urban environments. *Science*, 358, 6363.

Jones, T.M., Durrant, J., Michaelides, E.B., and Green, M.P. (2015). Melatonin: a possible link between the presence of artificial light at night and reductions in biological fitness. *Philosophical Transactions of the Royal Society B: Biological Sciences*, 370, 10.

Kempenaers, B., Borgström, P., Loës, P., et al. (2010). Artificial night lighting affects dawn song, extra-pair siring success, and lay date in songbirds. *Current Biology*, 20, 1735–1739.

Knop, E., Zoller, L., Ryser, R., et al. (2017). Artificial light at night as a new threat to pollination. *Nature*, 548, 206–209.

Kuhne, J.L., van Grunsven, R.H.A., Jechow, A., and Holker, F. (2021). Impact of different wavelengths of artificial light at night on phototaxis in aquatic insects. *Integrative and Comparative Biology*, 61, 1182–1190.

Kyba, C.C.M., Kuester, T., de Miguel, A.S., et al. (2017a). Artificially lit surface of Earth at night increasing in radiance and extent. *Science Advances*, 3, 8.

Kyba, C.C.M., Mohar, A., and Posch, T. (2017b). How bright is moonlight? *Astronomy and Geophysics*, 58, 1.31–31.32.

Kyba, C.C.M., Ruhtz, T., Fischer, J., and Holker, F. (2012). Red is the new black: how the colour of urban skyglow varies with cloud cover. *Monthly Notices of the Royal Astronomical Society*, 425, 701–708.

Levis, N.A., and Pfennig, D.W. (2016). Evaluating 'plasticity-first' evolution in nature: key criteria and empirical approaches. *Trends in Ecology & Evolution*, 31, 563–574.

Levy, K., Wegrzyn, Y., Efronny, R., et al. (2021). Lifelong exposure to artificial light at night impacts stridulation and locomotion activity patterns in the cricket *Gryllus bimaculatus*. *Proceedings of the Royal Society B: Biological Sciences*, 288, 20211626.

Lewis, H.F. (1927). Destruction of birds by lighthouses in the provinces of Ontario and Quebec. *Canadian Field Naturalist*, 41, 55–58.

Lockett, M.T., Jones, T.M., Elgar, M.A., et al. (2021). Urban street lighting differentially affects community attributes of airborne and ground-dwelling invertebrate assemblages. *Journal of Applied Ecology*, 58, 2329–2339.

Luarte, T., Bonta, C.C., Silva-Rodriguez, E.A., et al. (2016). Light pollution reduces activity, food consumption and growth rates in a sandy beach invertebrate. *Environmental Pollution*, 218, 1147–1153.

Luo, B., Xu, R., Li, Y., et al. (2021). Artificial light reduces foraging opportunities in wild least horseshoe bats. *Environmental Pollution*, 288, 117765.

Manfrin, A., Singer, G., Larsen, S., et al. (2017). Artificial light at night affects organism flux across ecosystem boundaries and drives community structure in the recipient ecosystem. *Frontiers in Environmental Science*, 5, 14.

Marangoni, L.F.B., Davies, T., Smyth, T., et al. (2022). Impacts of artificial light at night in marine ecosystems—a review. *Global Change Biology*, 28, 5346–5367.

Mayer-Pinto, M., Jones, T.M., Swearer, S.E., et al. (2021). Light pollution: a landscape-scale issue requiring cross-realm consideration. *UCL Open: Environment*, 29, 4:e036.

Miller, C.R., Barton, B.T., Zhu, L., et al. (2017). Combined effects of night warming and light pollution on predator-prey interactions. *Proceedings of the Royal Society B: Biological Sciences*, 284, 20171195.

Minnaar, C., Boyles, J.G., Minnaar, I.A., et al. (2015). Stacking the odds: light pollution may shift the balance in an ancient predator–prey arms race. *Journal of Applied Ecology*, 52, 522–531.

Ouyang, J.Q., de Jong, M., van Grunsven, R.H.A., et al. (2017). Restless roosts: light pollution affects behavior, sleep, and physiology in a free-living songbird. *Global Change Biology*, 23, 4987–4994.

Owens, A.C.S., and Lewis, S.M. (2018). The impact of artificial light at night on nocturnal insects: a review and synthesis. *Ecology and Evolution*, 8, 11,337–11,358.

Raap, T., Pinxten, R., and Eens, M. (2016). Artificial light at night disrupts sleep in female great tits (*Parus major*) during the nestling period, and is followed by a sleep rebound. *Environmental Pollution*, 215, 125–134.

Rand, A.S., Bridarolli, M.E., Dries, L., and Ryan, M.J. (1997). Light levels influence female choice in Túngara frogs: predation risk assessment? *Copeia*, 1997, 447–450.

Ren, Z.F., Chen, Y.Q., Liu, F.B., et al. (2022). Effects of artificial light with different wavelengths and irradiances on the sleep behaviors of chestnut buntings (*Emberiza rutila*). *Biological Rhythm Research*, 53, 1454–1473.

Rich, C., and Longcore, T. (2006). *Ecological Consequences of Artificial Night Lighting*. Island Press, Washington, DC.

Robert, K.A., Lesku, J.A., Partecke, J., and Chambers, B. (2015). Artificial light at night desynchronizes strictly seasonal reproduction in a wild mammal. *Proceedings of the Royal Society B: Biological Sciences*, 282, 20151745.

Rodríguez, A., Burgan, G., Dann, P., et al. (2014). Fatal attraction of short-tailed shearwaters to artificial lights. *PLOS ONE*, 9, e110114.

Russ, A., Rüger, A., and Klenke, R. (2015). Seize the night: European blackbirds (*Turdus merula*) extend their foraging activity under artificial illumination. *Journal of Ornithology*, 156, 123–131.

Rydell, J. (2006). Bats and their insect prey at streetlights. In: C. Rich and T. Longcore (eds), *Ecological Consequences of Artificial Night Lighting*, pp. 43–61. Island Press, Washington, DC.

Sanders, D., Frago, E., Kehoe, R., et al. (2021). A meta-analysis of biological impacts of artificial light at night. *Nature Ecology & Evolution*, 5, 10.

Santos, C.D., Miranda, A.C., Granadeiro, J.P., et al. (2010). Effects of artificial illumination on the nocturnal foraging of waders. *Acta Oecologica*, 36, 166–172.

Spoelstra, K., van Grunsven, R.H.A., Ramakers, J.J.C., et al. (2017a). Response of bats to light with different spectra: light-shy and agile bat presence is affected by white and green, but not red light. *Proceedings of the Royal Society B: Biological Sciences*, 284, 8.

Spoelstra, K., van Grunsven, R.H.A., Ramakers, J.J.C., et al. (2017b). Response of bats to light with different spectra: light-shy and agile bat presence is affected by white and green, but not red light. *Proceedings of the Royal Society B: Biological Sciences*, 284, 20170075.

Sun, J.C., Raap, T., Pinxten, R., and Eens, M. (2017). Artificial light at night affects sleep behaviour differently in two closely related songbird species. *Environmental Pollution*, 231, 882–889.

Swaddle, J.P., Francis, C.D., Barber, J.R., et al. (2015). A framework to assess evolutionary responses to anthropogenic light and sound. *Trends in Ecology & Evolution*, 30, 550–560.

Tan, D.X., Hardeland, R., Manchester, L.C., et al. (2010). The changing biological roles of melatonin during evolution: from an antioxidant to signals of darkness, sexual selection and fitness. *Biological Reviews of the Cambridge Philosophical Society*, 85, 7–623.

Van Doren, B.M., Horton, K.G., Dokter, A.M., et al. (2017). High-intensity urban light installation dramatically alters nocturnal bird migration. *Proceedings of the National Academy of Sciences*, 114, 11,175–11,180.

van Geffen, K.G., Groot, A.T., Van Grunsven, R.H.A., et al. (2015a). Artificial night lighting disrupts sex pheromone in a noctuid moth. *Ecological Entomology*, 40, 401–408.

van Geffen, K.G., van Eck, E., de Boer, R.A., et al. (2015b). Artificial light at night inhibits mating in a Geometrid moth. *Insect Conservation and Diversity*, 8, 282–287.

Willmott, N.J., Henneken, J., Elgar, M.A., and Jones, T.M. (2019). Guiding lights: foraging responses of juvenile nocturnal orb-web spiders to the presence of artificial light at night. *Ethology*, 125, 289–297.

Willmott, N.J., Henneken, J., Selleck, C.J., and Jones, T.M. (2018). Artificial light at night alters life history in a nocturnal orb-web spider. *PeerJ*, 6, 17.

Willmott, N., Wong, B., Lowe, E., et al. (2022). Wildlife exploitation of anthropogenic change: interactions and consequences. *The Quarterly Review of Biology*, 97, 15–35.

Zhang, F.S., Wang, Y., Wu, K., et al. (2020). Effects of artificial light at night on foraging behavior and vigilance in a nocturnal rodent. *Science of the Total Environment*, 724, 138271.

Zhao, D., Yu, Y., Shen, Y., et al. (2019). Melatonin synthesis and function: evolutionary history in animals and plants. *Frontiers in Endocrinology*, 10, 16.

CHAPTER 5

Ocean acidification

Timothy D. Clark, Jeff C. Clements, Fredrik Jutfelt, and Josefin Sundin

Overview

Ocean acidification describes the decline in pH of marine environments as they continue to absorb anthropogenically derived carbon dioxide (CO_2). Research over the past ~15 years has reported that levels of ocean acidification forecasted for the end of the century (CO_2 ~800–1000 µatm; pH ~7.6–7.7) can severely impair behaviours of marine animals, including fishes and invertebrates. Impaired behaviours of most concern are those linked with neural and sensory systems, such as the capacity to respond appropriately to predators. However, after an initial proliferation of studies reporting dire behavioural disturbances, studies finding negligible effects of end-of-century ocean acidification began to accumulate. We have now reached a point where there is little consensus on whether, and how much, ocean acidification will impact animal behaviour. Here, we outline existing knowledge regarding the effects of ocean acidification on animal behaviour, discuss the chronology of discoveries and controversies in the field, and provide guidance for improving rigour and transparency in behavioural ecology more broadly.

5.1 Introduction

Climate change is impacting—and will increasingly impact—animals at multiple levels of biological organization. While much of the research on the behavioural responses of wildlife to climate change has focused on the effects of warming (see Chapter 1), there has also been considerable interest in other behavioural impacts, including those associated with ocean acidification.

'Ocean acidification' is a term used to describe the decline in oceanic pH resulting from the uptake of anthropogenically derived carbon dioxide (CO_2) from the atmosphere. From average pre-industrial levels of ~280 µatm through to current levels of ~420 µatm, the partial pressure of CO_2 (pCO_2) of the open ocean is expected to reach 800–1000 µatm by the end of the century if emissions remain unabated, resulting in a pH decline from current levels of ~8.0 to ~7.6–7.7 (IPCC 2022). Such pCO_2 levels are thought to be higher than oceans have experienced in the past 30 million years (Luthi et al. 2008; Hönisch et al. 2012). When CO_2 is absorbed by seawater, it reacts with water molecules (H_2O) and forms carbonic acid (H_2CO_3), which then breaks down into bicarbonate (HCO_3^-) and a hydrogen ion (H^+), the latter causing a decline in pH. Existing carbonate ions (CO_3^{2-}) in seawater buffer the pH by combining with the extra hydrogen ions to form bicarbonate, which consequently lowers the concentration of carbonate ions and the saturation state of aragonite (a soluble form of calcium carbonate).

While there can be dramatic daily and seasonal fluctuations in pCO_2 in some ecosystems (e.g. coral reefs, intertidal zones), the continuing increase in pCO_2, and decrease in aragonite saturation state, in all marine environments may be cause for concern. For example, animals that build calcified body structures like shells could be negatively affected, as the deposition of calcium carbonate is argued to become less efficient under ocean acidification conditions (Orr et al. 2005; but see Leung et al.

Timothy D. Clark, Jeff C. Clements, Fredrik Jutfelt, and Josefin Sundin, *Ocean acidification*. In: *Behavioural Responses to a Changing World*. Edited by: Bob B. M. Wong and Ulrika Candolin, Oxford University Press. © Oxford University Press (2024). DOI: 10.1093/oso/9780192858979.003.0005

2022). Moreover, species that have poor physiological regulation of their internal pH (i.e. poor acid-base regulation) may suffer from ocean acidification because cellular processes can be impaired when pH deviates from optimal levels (Melzner et al. 2009). Around 2007, reports began to emerge about marine invertebrate behaviours being affected by elevated pCO_2, including predator avoidance, settlement, burrowing, and activity (reviewed in Clements and Hunt 2015). Compared with invertebrates, fishes have relatively robust acid-base regulation and there is a rich history of studies that have documented physiological resilience to pCO_2 conditions many times greater than expected by the end of the century (e.g. ~15,000 µatm; Toews et al. 1983; Pörtner et al. 2004; Nelson 2015). Thus, it was

surprising when a series of studies in 2009–2010 reported serious behavioural disturbances in coral reef fishes after a few days of exposure to end-of-century ocean acidification conditions of ~1000 µatm pCO_2 (Munday et al. 2009; Dixson et al. 2010; Munday et al. 2010), despite the fact that coral reefs can already experience daily/seasonal pCO_2 fluctuations that can exceed 1000 µatm (Shaw et al. 2012). The studies on coral reef fishes are particularly important to the topic of ocean acidification and animal behaviour, as they are widely considered to have kick-started the field, resulting in a dramatic rise in related studies over the past ~15 years (Figure 5.1).

Our understanding of ocean acidification and animal behaviour almost exclusively relies on

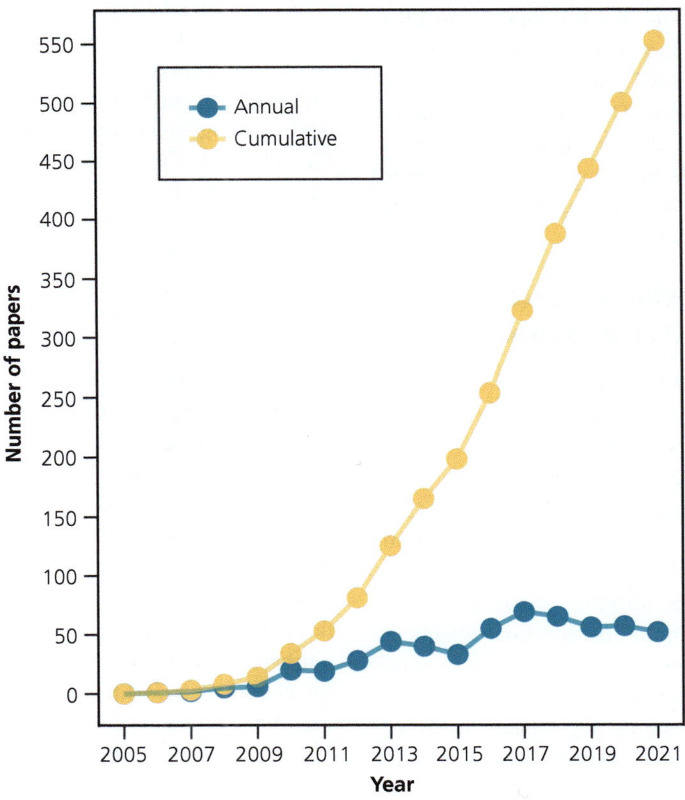

Figure 5.1 Estimated temporal trends in the number of papers related to ocean acidification and behaviour from 2005 to 2021. Data were generated by conducting a Scopus search for the keyword string 'ocean acidification behaviour' using the software Publish or Perish (Harzing 2016). Searches were conducted individually for each year.

laboratory-based manipulations of pCO_2 (or pH) to expose ectothermic animals to forecasted ocean acidification conditions. Thus, this chapter necessarily focuses on studies that have acclimated ectothermic animals to simulated ocean acidification conditions for relatively short periods of time relative to the rate of change in average pCO_2 levels in the oceans. Since chronic acclimation over time (years) is more relevant to progressive acidification in the world's oceans, as it can illuminate the capacity of species to acclimate or adapt, we have noted where studies have conducted multigenerational experiments. For simplicity, we primarily use the term 'CO_2' rather than 'pCO_2' herein, and we use the term 'ocean acidification' bearing in mind that almost all studies have been conducted by *simulating* ocean acidification conditions in aquaria. First, this chapter reviews the range of behaviours reported to be impacted by ocean acidification, explores the potential for species- and life stage-specific effects, and discusses the proposed mechanisms underlying acidification-induced behavioural disturbances. The chapter then highlights controversies in this field, and suggests a range of improvements in behavioural research to enhance transparency and maximize repeatability across independent laboratories.

5.2 Reported behavioural/sensory consequences of ocean acidification

Some of the most prominent reported effects of ocean acidification are those on behaviours related to the use of sensory systems (Ashur et al. 2017; Tigert and Porteus 2023). Given that the sensory systems of animals play an essential role in lifetime survival and fitness, effects on sensory function have the potential to influence predator-prey interactions, social dynamics, foraging, dispersal, activity levels, mating, and more. Behavioural effects of CO_2 have been most extensively studied in relation to the chemosensory system, but effects relating to other sensory systems (e.g. vision, hearing) and neural processes have also been investigated to some extent. Below, we outline the state of knowledge for some of the most extensively measured behaviours (see Table 5.1 for an overview).

5.2.1 Chemosensory behaviours

Of the myriad behavioural impairments reported in aquatic ectotherms when faced with end-of-century levels of CO_2, those associated with the chemosensory system (olfaction/gustation) have been the most profound. Most of these studies tested animals in choice flumes (Figure 5.2; see Jutfelt et al. 2017) or injected chemical cues into aquaria containing the focal animals (Lönnstedt et al. 2013). Acclimation times were generally short (e.g. 4 days) and experimental protocols were rapid (e.g. < 10 minutes). The first study to report extreme acidification-induced chemosensory impairments in fish was published in 2009, where elevated CO_2 was said to impair olfactory discrimination and homing ability in settlement-stage larvae of the coral reef clownfish *Amphiprion percula* (Munday et al. 2009). This paper spawned a great number of studies (Figure 5.1) and continues to be one of the most cited papers in the field (Clements et al. 2022). Many of the follow-up papers reported effects of elevated CO_2 on olfactory-mediated predator-prey interactions and homing behaviours (reviewed in: Heuer and Grosell 2014; Clements and Hunt 2015). Coral reef fishes emerged as the most sensitive to ocean acidification, as the sensory and behavioural impairments reported for those species were often orders of magnitude more severe than reported for temperate species (Porteus et al. 2018). For example, larval coral reef clownfish *A. percula* and settlement-stage damselfish *Pomacentrus wardi* were reported to exhibit extreme attraction to predator chemical cues when held for > 2 days in forecasted CO_2 levels of ~850 µatm (Munday et al. 2010). In an even more extreme example, it was reported that settlement-stage clownfish *A. percula* under current-day CO_2 levels completely avoided the chemical cues of predators (with no inter-individual variation), whereas conspecifics exposed to CO_2 of ~1000 µatm showed complete preference for the predator chemical cues (again with no inter-individual variation) (Dixson et al. 2010).

The authors of the 2009–2010 studies published many subsequent papers reporting similar impairments, highlighting that the extreme impacts of end-of-century ocean acidification were ubiquitous

Table 5.1 Summary table of behaviours studied in the context of ocean acidification that have been highlighted in the text of this chapter, including the type of behaviour assessed, examples of endpoints measured in individual studies, examples of invertebrate and vertebrate taxa studied, the documented outcomes reported (positive, +; negative, −; null, ≠), and the general conclusions that can be drawn for each type of behaviour based on current literature. Note that results across studies are mixed and inconsistent for all behaviour types and broad conclusions are not possible for the majority.

Behaviour type	Example endpoints measured	Example taxa studied		Documented outcomes	General conclusions
		Invertebrates	Vertebrates		
Chemosensory	Predator detection and avoidance, homing ability, chemical/odour detection	Crustaceans (crabs/lobsters), gastropods, bivalves	Teleosts (coral reef + temperate), sharks	+ − ≠	Varied and unpredictable for invertebrates; contemporary studies suggest negligible impacts for fishes
Vision	Response to visual threat/cue, phototaxis, retinal flick frequency, eye morphology, visual navigation	Crustaceans (crabs), cephalopods	Teleosts (coral reef + temperate)	+ − ≠	Insufficient data preclude broad conclusions
Hearing	Otolith morphology, vestibulo-ocular reflex	−	Teleosts (coral reef + temperate)	+ − ≠	Insufficient data preclude broad conclusions
Learning	Naïve vs. learned behavioural responses	−	Teleosts (coral reef + temperate)	+ − ≠	Insufficient data and inconsistent findings preclude broad conclusions; contradictory results between studies relating to basic biological principles of learning drive large uncertainty in any reported effects
Activity and feeding	Activity levels, anxiety, swimming/movement patterns, food strikes/bites, feeding rates	Cephalopods, gastropods, bivalves, echinoderms, crustaceans, polychaetes	Teleosts (coral reef + temperate), sharks	+ − ≠	Insufficient data and inconsistent findings preclude broad conclusions; swimming activity appears generally robust to ocean acidification; feeding in invertebrates may be affected, but dependent on ontogeny and phylogeny
Behavioural lateralization	Right-left turning preference	Gastropods	Teleosts (coral reef + temperate), sharks	+ − ≠	Inconsistent findings preclude broad conclusions; non-repeatable behaviour with varied functional ramifications, and highly sensitive to experimental set-up, thus potentially poor behavioural indicator
Mating and reproduction	Courtship behaviour, ornamentation, egg fanning, nest-building, parental care, mate guarding, egg shedding	Amphipods, copepods	Teleosts (coral reef + temperate)	+ − ≠	Insufficient data preclude broad conclusions; parental exposure does not improve effects on subsequent generations for invertebrates

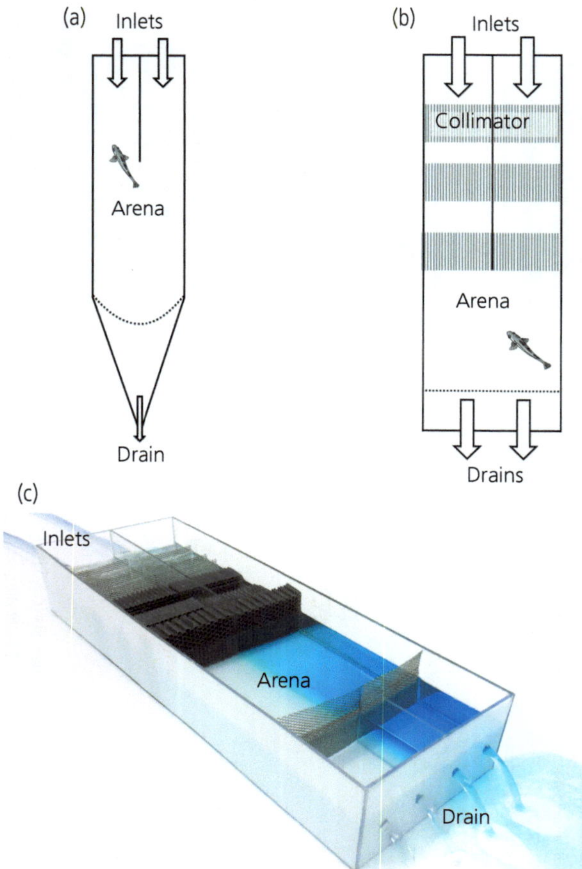

Figure 5.2 Examples of two-current choice flumes used in ocean acidification research, adapted from Jutfelt et al. (2017). Water with different properties typically flows from two header tanks (e.g. one may contain the chemical cue of interest; header tanks not shown in figure) and enters the flume through the inlets. Many studies have used flumes with no collimators and a tapered drain end, with a physical barrier dividing parts of the choice arena (a), while some have used collimators to remove turbulence and straighten the flow, with a non-tapered drain end, and no physical barrier dividing any of the choice arena (b, c). To ensure that laminar flow has been achieved, dye can be used, as in (c), where blue dye was added to one of the header tanks to show how the water sources remain separated as they flow through the arena despite the absence of a physical barrier. The time spent on either side of the arena is quantified and used to determine preference/avoidance of the chemical cue of interest.

across species of coral reef fishes. It also became clear that the sensory behaviour of later life stages was just as vulnerable to elevated CO_2 as larvae and early-stage juveniles (Figure 5.3; Table 5.2). The olfactory discrimination of fish living on natural CO_2 seeps in Papua New Guinea was reported to be equally impaired (Munday et al. 2014), and even transgenerational acclimation to end-of-century ocean acidification was reportedly unable to attenuate the negative impacts (Welch et al. 2014). The prospect of prey fishes being strongly attracted to the chemical cues of their predators later this century appeared catastrophic for coral reef ecosystems, but concerns were tempered by the emergence of studies from a broader range of research groups showing smaller/negligible behavioural impairments (discussed in Section 5.4).

While not as extensively studied as marine fishes, chemosensory abilities in marine invertebrates can potentially be sensitive to elevated CO_2. For example, the gastropod *Littorina littorea* exhibited the opposite response to that reported for coral

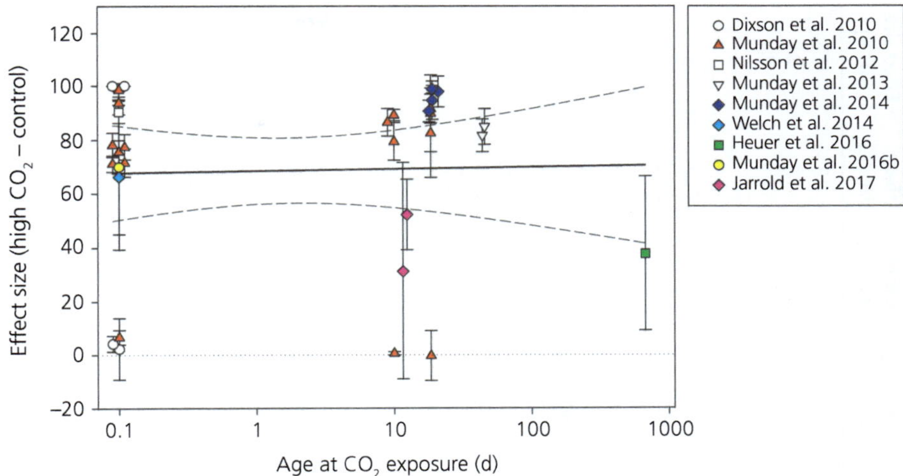

Figure 5.3 Literature-derived effect sizes (mean of high CO_2 group minus mean of control group, determined from two-current choice flumes; Jutfelt et al. 2017) for the percentage of time that coral reef fishes spent in water containing predator chemical cues or conspecific chemical alarm cues, as a function of age at exposure to high CO_2 (references given in legend). Error bars are SD, calculated as SD = $\sqrt{(SD_{control\ group}^2 + SD_{CO2\ group}^2)}$. An effect size of 100% indicates that fish in the high CO_2 (simulated ocean acidification) treatment spent 100% of their time on the side of the choice flume containing the predator/alarm cue, and fish in the control treatment spent 0% of their time on that side. An exposure age of 0.1 days was given to fish that were exposed to high CO_2 immediately after hatching. Note that the age at exposure does not always approximate the age of the fish at behavioural testing (see Table 5.2). For example, if a fish was placed into high CO_2 at one day of age and the behavioural test was performed two weeks later, the age at exposure is one day. Linear regression (solid line) and 95% confidence intervals (dashed lines) have been fitted to all data in the figure. Minor horizontal jitter was applied to some data points to minimize overlap of symbols and error bars. All points around an effect size of 0% were from treatment groups that had spent less than one day post-hatch at high CO_2 prior to the choice flume test. Age of fish in Heuer et al. (2016) was estimated from fish mass by applying a polynomial regression to age vs mass data in Munday et al. (2008). See Table 5.2 for further details.

reef fishes—an increased predator cue avoidance behaviour under extreme ocean acidification conditions (pH 6.6) (Bibby et al. 2007). In similarly extreme ocean acidification conditions (pH < 7), hermit crabs were reported to exhibit a reduced ability to track odours (de la Haye et al. 2012). Altered predator avoidance behaviour under ocean acidification is also reported to be a consequence of impaired chemosensory ability in marine snails (Manríquez et al. 2014; Watson et al. 2014). In contrast, Clements et al. (2020) reported that ocean acidification does not appear to influence the chemosensory ability of epibenthic marine mussels to detect and respond to chemical alarm cues (dead conspecifics) via shell closures. Overall, the effects of ocean acidification on invertebrate chemosensory abilities appear varied and unpredictable, which may be more likely due to differences in methodology rather than species-specific biological phenomena.

5.2.2 Vision

The effects of elevated CO_2 on vision have mainly been investigated in the context of feeding, survival, and predator-prey interactions. Some studies have reported that elevated CO_2 alters visual risk assessment, causing fishes and cephalopods to respond inappropriately to an apparent threat (Ferrari et al. 2012b; Spady et al. 2018), while other studies have reported no such effects (Lönnstedt et al. 2013; Moura et al. 2019).

The preference of larval fishes to swim toward or away from a light source (positive or negative phototaxis, respectively) is essential for feeding and survival. Again, some studies have reported a CO_2-induced increase in positive phototaxis (Forsgren et al. 2013), while others report no effects of CO_2 exposure on phototaxis or startle response (Munday et al. 2016a). Likewise, while studies on invertebrates are limited, elevated CO_2 (1600 and 3200

Table 5.2 Data used in Figure 5.3 concerning the preference/avoidance of coral reef fishes to predation-related chemical cues in ocean acidification experiments using two-current choice flumes. Columns include: associated references, relevant figures within the reference, species tested, chemical cue type, post-hatch age at first exposure to high CO_2, duration in high CO_2 before being tested, age of the fish at testing, percentage time spent in the cue in each of the high CO_2 and control treatment groups, effect size (high CO_2 minus control), and the standard deviation associated with the effect size (calculated as $SD = \sqrt{(SD_{control\ group}^2 + SD_{CO2\ group}^2)}$). In studies with multiple CO_2 treatment groups, the group closest to $CO_2 = 1000$ μatm was selected.

Reference	Figures in ref.	Species	Cue type	Age at CO_2 exposure (d)	Duration in CO_2 (d)	Age at test (d)	High CO_2 mean (%)	Control mean (%)	Effect size (%)	Effect size SD (%)
Dixson et al. (2010)	Fig. 1a & 2a	*Amphiprion percula*	Predator sp. 1	0*	1*	1	11.1	8.8	2.3	11.5
Dixson et al. (2010)	Fig. 1a & 2a	*Amphiprion percula*	Predator sp. 2	0*	1*	1	13.4	9.2	4.2	2.9
Dixson et al. (2010)	Fig. 1b & 2b	*Amphiprion percula*	Predator sp. 1	0*	11*	11[Ψ]	100	0	100	0.0
Dixson et al. (2010)	Fig. 1b & 2b	*Amphiprion percula*	Predator sp. 2	0*	11*	11[Ψ]	100	0	100	0.0
Munday et al. (2010)	Fig. 1	*Amphiprion percula*	Predator	0	1	1	18.5	11.9	6.6	2.7
Munday et al. (2010)	Fig. 1	*Amphiprion percula*	Predator	0	2	2	82.8	11.6	71.2	3.1
Munday et al. (2010)	Fig. 1	*Amphiprion percula*	Predator	0	3	3	83	11.4	71.6	5.3
Munday et al. (2010)	Fig. 1	*Amphiprion percula*	Predator	0	4	4	87.8	9.6	78.2	4.5
Munday et al. (2010)	Fig. 1	*Amphiprion percula*	Predator	0	5	5	87.6	11.9	75.7	4.2
Munday et al. (2010)	Fig. 1	*Amphiprion percula*	Predator	0	6	6	87.1	9.6	77.5	4.7
Munday et al. (2010)	Fig. 1	*Amphiprion percula*	Predator	0	8	8	98.5	0	98.5	2.5
Munday et al. (2010)	Fig. 1	*Amphiprion percula*	Predator	0	10	10[Ψ]	93.7	0	93.7	4.3
Munday et al. (2010)	Fig. 2a	*Amphiprion percula*	Predator	10[Ψ]	1	11	0.6	0	0.6	0.6
Munday et al. (2010)	Fig. 2a	*Amphiprion percula*	Predator	10[Ψ]	2	12	79.4	0	79.4	7.0
Munday et al. (2010)	Fig. 2a	*Amphiprion percula*	Predator	10[Ψ]	3	13	86.5	0	86.5	5.1
Munday et al. (2010)	Fig. 2a	*Amphiprion percula*	Predator	10[Ψ]	4	14	89.1	0	89.1	2.1
Munday et al. (2010)	Fig. 2b	*Pomacentrus wardi*	Predator	18.5[Ψ]	1	19.5	1.7	2	−0.3	9.3
Munday et al. (2010)	Fig. 2b	*Pomacentrus wardi*	Predator	18.5[Ψ]	2	20.5	85.6	3.1	82.5	16.6
Munday et al. (2010)	Fig. 2b	*Pomacentrus wardi*	Predator	18.5[Ψ]	3	21.5	91.8	2	89.8	14.2
Munday et al. (2010)	Fig. 2b	*Pomacentrus wardi*	Predator	18.5[Ψ]	4	22.5	93.2	1.7	91.5	5.3
Nilsson et al. (2012)	Fig. 2a	*Amphiprion percula*	Predator	0	11	11[Ψ]	91.5	1.1	90.4	4.1
Munday et al. (2013)	Fig. 1	*Plectropomus leopardus*	Predator	45	4	49	91.1	6.2	84.9	6.6
Munday et al. (2013)	Fig. 1	*Plectropomus leopardus*	Predator	45	28	73	90.8	9.2	81.6	6.0
Munday et al. (2014)	Fig. 1a	*Dascyllus aruanus*	Predator	21[Ψ†]	25	46	97.8	0	97.8	5.7
Munday et al. (2014)	Fig. 1a	*Pomacentrus moluccensis*	Predator	18[Ψ†]	38	56	90.6	0	90.6	4.0
Munday et al. (2014)	Fig. 1a	*Apogon cyanosoma*	Predator	19[Ψ†]	22	41	94.6	0	94.6	7.1
Munday et al. (2014)	Fig. 1a	*Cheilodipterus quinquelineatus*	Predator	19[Ψ†]	20	39	98.7	0	98.7	1.4
Welch et al. (2014)	Fig. 1	*Acanthochromis polyacanthus*	Alarm	0	42.5	42.5	77	10.8	66.2	26.8
Heuer et al. (2016)	Fig. 2	*Acanthochromis polyacanthus*	Alarm	673[#]	4	677	53.1	15.3	37.8	28.6
Munday et al. (2016b)	Fig. 1	*Amphiprion percula*	Predator	0	13.5[Ψ]	13.5	79	9	70	25.0
Jarrold et al. (2017)	Fig. 1c	*Amphiprion percula*	Predator	12[Ψ]	7	19	57.2	4.9	52.3	13.0
Jarrold et al. (2017)	Fig. 2c	*Amphiprion percula*	Predator	12[Ψ]	7	19	45.5	14.2	31.3	40.3

* Fish were exposed to high CO_2 throughout their ~8-day incubation prior to hatch; [#] estimated from fish mass by applying a polynomial regression to age vs mass data in Munday et al. (2008); [Ψ] age at settlement (approx.); [†] fish were naturally exposed to high CO_2 when they settled on reefs at CO_2 seeps.

μatm) had no significant effect on phototaxis in zoea larvae of Dungeness crabs *Metacarcinus magister* (Roberts 2013).

In search of a mechanism to explain the acidification-induced visual impairments reported in some studies of fishes, Chung et al. (2014) measured the maximal flicker frequency of the retina and reported that it was reduced by continuous exposure to elevated CO_2, potentially impairing the capacity of fish to react to fast events. In addition, some studies documented developmental damage to the eyes of fish that could potentially impair vision (Frommel et al. 2016). Despite these reports, Clark et al. (2020a) found that coral reef fish acclimated to elevated CO_2 were equally as proficient as control fish at visually determining the shortest way around an obstacle when coaxed along a narrow channel. Clearly, insufficient data exist to form a consensus on the effects of ocean acidification on animal vision, but it is clear that our knowledge of this topic is compromised by a lack of standardization and consistency across studies and species.

5.2.3 Hearing

In aquatic environments, the ability to respond to auditory cues is evidently present in mammals, but it has also been observed in fish (Popper and Fay 1973), arthropods (Lovell et al. 2005), cephalopods (Kaifu et al. 2008), and cnidaria (Vermeij et al. 2010), where hearing is defined as the ability to detect non-contact vibrational stimuli. While we are not aware of any studies examining the effects of ocean acidification on 'hearing' in marine invertebrates, ocean acidification has been reported to decrease hearing sensitivity in fish (Radford et al. 2021) and to alter the ability of fish larvae to navigate using acoustic cues, affecting both the timing of settlement and the attraction to habitat cues (Simpson et al. 2011; Rossi et al. 2018).

Auditory capabilities of fishes are dependent on internal structures, such as the inner ear and otoliths. Elevated CO_2 has been suggested to impact otolith structure because aragonite—of which otoliths are composed—is predicted to become less abundant as CO_2 levels in the oceans increase (Orr et al. 2005). However, it has more

recently been argued that otolith growth may actually be enhanced by the mechanisms used to compensate extracellular pH (Kwan and Tresguerres 2022). Several studies have investigated whether CO_2 alters the shape and/or size of otoliths, but the results are not congruent. It is tempting to assume that differences between studies could be ascribed to different CO_2 exposure durations, but the likelihood of finding an effect is not clearly related to exposure durations ranging from a few days through to transgenerational exposure (Franke and Clemmesen 2011; Hurst et al. 2012; Schade et al. 2014; Réveillac et al. 2015). While the weight of evidence suggests that otolith size increases at CO_2 levels exceeding end-of-century forecasts, it has been suggested that the level of CO_2 exposure needed to increase otolith size is higher than levels needed to induce altered auditory perception, such that behavioural effects are expected to occur before otolith enlargement (Ashur et al. 2017).

Mathematical modelling suggests that changes in ocean pH are unlikely to affect sound velocity (Sehgal et al. 2010), but absorption-based attenuation of acoustic signals can decrease with decreasing pH due to pH-dependent chemical relaxations in the $B(OH)_3/B(OH)_4^-$ and HCO_3^-/CO_3^{2-} systems (Hester et al. 2008). Such reduced attenuation, especially at frequencies below ~10 kHz, could make oceans noisier places if acidification continues (Hester et al. 2008; Sehgal et al. 2010). It has been suggested that a change in sound absorption might be one of the reasons for a reported change in the frequency of blue whale *Balaenoptera musculus* song (Sehgal et al. 2010; noise pollution is discussed in Chapter 2). However, out of several hypotheses examined (changes in sexual selection, increasing ocean noise, increasing whale body size post-whaling, global warming, interference from other animal sounds, and post-whaling increases in abundance), none were found to provide a satisfactory explanation for the observed change (McDonald et al. 2009).

5.2.4 Learning

It has been argued, again for coral reef fishes, that ocean acidification impacts the capacity to learn from chemical and visual information. Ferrari et al. (2012a) and Chivers et al. (2014a) reported that pre-settlement juveniles of the tropical fish *Pomacentrus amboinensis* failed to learn to respond appropriately to common predators when held under ocean acidification conditions. The effects of ocean acidification on learning in marine invertebrates remains unstudied. One study that may have relevance to learning reported that seawater pH did not affect short-term habituation in valve closure responses to tactile predator cues in marine mussels *Mytilus* sp. (Clements et al. 2021).

Whilst learning capacity in the context of ocean acidification has received scant research attention, the Ferrari et al. (2012a) and Chivers et al. (2014a) studies mentioned above raise a larger issue that has resulted in significant confusion in the literature: do larval/pre-settlement fishes need to learn how to respond to the chemical and visual cues of predators/conspecifics, or is the response innate? Ferrari et al. (2012a) and Chivers et al. (2014a) state that coral reef fishes must learn how to respond to chemical and visual stimuli (i.e. it is not an innate response), otherwise they lack appropriate responses to predators. In other studies with overlapping authors, however, predator-naïve fish reportedly avoided predator cues without any prior learning (e.g. Dixson et al. 2010; Munday et al. 2010; Vail and McCormick 2011; Munday et al. 2012). Adding to the complexity, it has been reported that learning can take place at the embryo stage prior to hatching (Atherton and McCormick 2015). With little cross-citation between papers to explain this contradictory narrative between innate versus learned avoidance of predators, there is an opportunity for future work by independent lab groups to address the topic using standardized and transparent techniques.

5.2.5 Activity and feeding

A range of studies have examined various activities and feeding behaviours of marine animals in the context of ocean acidification. Marked hyperactivity (up to a 9000% increase) has been reported for coral reef fishes in response to elevated CO_2 (Munday et al. 2010; Munday et al. 2013; Munday et al. 2014), yet other studies on temperate species were not able to corroborate these findings (Maneja et al. 2015; Sundin and Jutfelt 2016). This again sparked concern that coral reef species were most sensitive to ocean acidification and that they would face a substantial activity-induced burden on their metabolism and energy requirements (Munday et al. 2013). It was proposed that changes in activity patterns were the result of CO_2-induced effects on the central nervous system, leading to 'anxiety-like' behaviour (Hamilton et al. 2014). Nevertheless, as more scientists searched for ocean acidification effects in coral reef fishes, it became clear that the hyperactivity reported in some papers was not common (Bignami et al. 2014; Sundin et al. 2017a; Raby et al. 2018), and a review by Clements and Hunt (2015) concluded that swimming behaviour appeared relatively unaffected by ocean acidification.

Effects of ocean acidification on the routine activities of marine invertebrates are well documented for various taxa (Clements and Hunt 2015). For example, ocean acidification has been reported to increase activity in multiple species of squid (Spady et al. 2018). In gastropods, ocean acidification effects on movement speed are highly variable, with studies reporting increased (Fields 2013; Watson et al. 2017), decreased (Queirós et al. 2015), and unaltered (Schram et al. 2014) speeds under ocean acidification conditions. Taxonomic differences appear to explain some of the effect size variability in ocean acidification studies on shellfish prey defence behaviours, with bivalves exhibiting responses to ocean acidification perceived to be detrimental, cephalopods exhibiting responses perceived to be beneficial, and gastropods, echinoderms, and malacostracan crustaceans appearing unaffected (Clements and Comeau 2019). Equally, a meta-analysis by Clements and Darrow (2018) reported that effect sizes pertaining to the feeding behaviour of calcifying marine invertebrates were predominantly moderated by phylum, life stage, and CO_2 exposure time. Therein, the feeding behaviour of suspension feeding molluscs and grazing echinoderms appeared to be affected by ocean acidification while arthropods were generally unaffected, and larvae and juveniles were more susceptible to ocean acidification effects on feeding than adults; however, only CO_2 concentrations > 1000 µatm yielded a significant mean effect size. Overall, the effects of ocean acidification on activity, movement, and suspension feeding in marine invertebrates appears varied and mechanisms are not well understood.

5.2.6 Behavioural lateralization

The term 'behavioural lateralization' refers to asymmetric behaviour and relies on the assumption that behavioural left or right biases reflect underlying asymmetries in the functioning of the nervous system (Vallortigara and Rogers 2005). Adaptive values of behavioural lateralization are often explained in terms of possible advantages of asymmetric cognitive control, such as benefits associated with enabling multiple stimuli to be processed simultaneously (Vallortigara and Rogers 2005). Disadvantages of lateralization have also been reported, such as interference with exploratory behaviour in fish (Dadda et al. 2009). Given the potential for both positive and negative effects relating to individual fitness, it is perhaps not surprising that great variation in the degree of lateralization has been reported both between and within species, with many species not being detectably lateralized at all (Vallortigara and Rogers 2005; Clark et al. 2020a; Roche et al. 2020).

Like the behaviours already discussed, the literature regarding the effects of CO_2 on behavioural lateralization is mixed. Several studies of fishes have reported effects, where the general trend is for weaker behavioural lateralization under elevated CO_2 (Domenici et al. 2012; Nilsson et al. 2012; Green and Jutfelt 2014; Welch et al. 2014; Jarrold et al. 2017). Other studies have found small or mixed effects (Sundin and Jutfelt 2016; Schmidt et al. 2017), or no effects at all (Clark et al. 2020a; Jarvis et al. 2022). Explanations for these discrepancies undoubtedly include insufficient sample sizes (e.g. < 30 per treatment group; Clements et al. 2022), and also the fact that measurements of behavioural lateralization can be particularly sensitive to the experimental set-up (Sundin and Jutfelt 2016; Clark et al. 2020a; Penry-Williams et al. 2022). Moreover, behavioural lateralization was shown to have low repeatability in a comprehensive study of several temperate and tropical fish species, raising questions about its usefulness as a behavioural performance metric (Roche et al. 2020).

5.2.7 Mating and reproduction

Some of the least studied behaviours in the context of elevated CO_2 are those related to mating and reproduction. A study of three-spined stickleback *Gasterosteus aculeatus* reported that males under both control and elevated CO_2 treatments developed normal sexual ornaments, pursued normal nest-building activities, exhibited similar levels of courtship behaviours and displacement fanning, and had the same mating probability and reproductive output (Sundin et al. 2017b). Other studies have reported increased reproductive investment in fish near CO_2 vents, relative to control CO_2 levels, mediated by increased energy intake via intensified foraging on more abundant prey (Nagelkerken et al. 2021). A reduction in time spent on parental care behaviour at volcanic CO_2 seeps compared with control CO_2 sites has also been reported (Spatafora et al. 2021). While CO_2 seeps can act as a window to the future in some respects, there are likely to be many other factors that vary around CO_2 vents that could influence ecology and complicate direct assessments of CO_2 effects.

The mixed findings for reproductive behaviours are reflected in mixed findings for reproductive output, with studies reporting stimulated reproductive output at elevated CO_2 (Miller et al. 2013; Schade et al. 2014; Welch and Munday 2016), no effects (Forsgren et al. 2013; Sundin et al. 2013; Miller et al. 2015; Milazzo et al. 2016), or a decrease in reproduction (Welch and Munday 2016). Thus, yet again, there is no consensus when summarizing existing knowledge. Papers examining transgenerational effects of CO_2 on tropical fishes have shown that successful mating and viable offspring are possible at end-of-century levels of CO_2 (Welch et al. 2014; Schunter et al. 2016; Schunter et al. 2018), showing that current-day fishes can have the inherent capacity to maintain at least some level of lifetime fitness when faced with end-of-century ocean acidification.

Very few studies have assessed the effects of ocean acidification on mating and reproductive behaviours in marine invertebrates. Borges et al. (2018) reported that ocean acidification exposure reduced the time spent engaging in mate guarding behaviour in amphipods. In the copepod *Calanipeda aquaedulcis*, gravid females (i.e. females carrying eggs) exhibited increased shedding behaviour of unhatched eggs under ocean acidification conditions; increased shedding would have negative functional ramifications, as unhatched eggs would not develop into offspring (Bhuiyan et al. 2022). In both aforementioned papers, parental exposure to ocean acidification conditions did not improve mating behaviour responses in subsequent generations.

5.3 Mechanisms of behavioural/sensory impairments

With the field of ocean acidification and animal behaviour growing at a rapid rate (Figure 5.1), interest arose in the underlying physiological mechanisms and the reasons why coral reef species were most sensitive. The reported effects of ocean acidification on the behaviour of aquatic ectotherms gave some clues as to the potential mechanisms. The switch from avoidance to attraction to predator and chemical alarm cues reported in coral reef fishes suggested that it could not be a simple reduction or malfunction of chemical sensory cells, as originally hypothesized by fish sensory physiologist Kjell Døving and colleagues (Munday et al. 2009). Instead, the maintained high precision of the behavioural response to chemical cues suggested a very specific, directed mechanism. The most common mechanistic explanations are discussed below.

5.3.1 The GABA hypothesis

The gamma-aminobutyric acid (GABA) system is the main inhibitory system of the brains of vertebrates (Figure 5.4), and in the peripheral and central nervous systems of invertebrates (Lummis 1990; Lunt 1991). The GABA neurotransmitter hyperpolarizes the inside of the post-synaptic cell membrane, meaning that the connected cell is electrically further away from a future action potential (i.e. it has reduced chance for activity). Higher activity of the GABA system can therefore reduce the activity of the brain, and conversely, a reduced function in the GABA system can lead to hyperactivity of the brain and subsequently of the animal's behaviour (Nelson et al. 2002). The $GABA_A$ receptor is a chloride (Cl^-) channel and when it opens, Cl^- enters the cell; by bringing in the negative Cl^- ions, the $GABA_A$ receptor causes the hyperpolarization.

Using coral reef fish, Nilsson et al. (2012) hypothesized a potential connection between the GABA system and the tissue pH regulation that occurs when animals encounter acidified water. Fish in water with reduced pH often buffer their blood and interstitial fluid pH by accumulating bicarbonate (HCO_3^-). This leads to a reduced concentration of Cl^- and an increased concentration of HCO_3^- in the blood and interstitial fluids. That can mean, according to Nilsson et al. (2012), that when a $GABA_A$ channel opens, there is less of a chemical gradient for Cl^- to enter the neuron, leading to a reduction in Cl^- influx and consequently reduced inhibitory function of the GABA system. It is also possible that more HCO_3^- can pass through the channel and interfere with its function, possibly in the direction out from the neuron. If the concentrations of Cl^- and HCO_3^- are substantially altered by the pH regulation seen in elevated water CO_2 conditions, there is even the potential for reversal of $GABA_A$ channel function, meaning the $GABA_A$ system may change from inhibitory to excitatory. There is some evidence for the European sea bass *Dicentrarchus labrax* that gene regulation in the olfactory rosette (the peripheral olfactory organ) is sensitive to ocean acidification, including genes involved in ion transport and GABAergic signalling (Cohen-Rengifo et al. 2022). Likewise, ocean acidification was reported to induce substantial up-regulation of many genes associated with the nervous system, including $GABA_A$, in pteropods *Heliconoides inflatus* (Moya et al. 2016) and swimming crabs *Portunus trituberculatus* (Ren et al. 2018). It is prudent to note that mRNA levels do not necessarily reflect protein levels, and mRNA levels can be highly heterogeneous in the various cell types that form complex organs such as the brain and the olfactory bulb. Thus, caution should be applied when forming conclusions from transcriptomic analyses.

The involvement of the $GABA_A$ receptor was first tested in coral reef clownfish *A. percula* and damselfish *Neopomacentrus azysron* exposed to ~900 µatm CO_2 by bathing them in 4 mg/L of the $GABA_A$ receptor antagonist drug gabazine prior to behavioural testing (Nilsson et al. 2012). As predicted by the authors, this treatment 'cured' the fish of their acidification-induced behavioural disturbances (Nilsson et al. 2012). Since then, several studies—most by the same research group—have similarly reported that gabazine can remove ocean

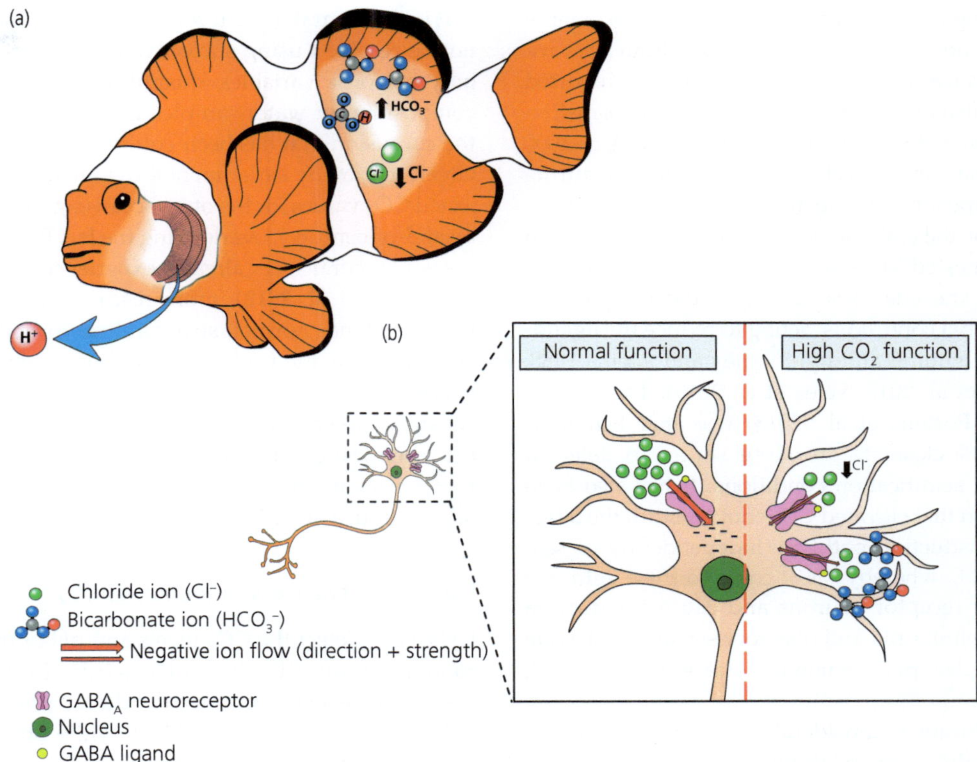

Figure 5.4 The proposed GABA hypothesis for behavioural disturbance in aquatic animals exposed to elevated CO_2 levels. In (a), a marine teleost fish in elevated water CO_2 levels maintains its internal pH by accumulating HCO_3^-. That results in elevated levels of HCO_3^- and reduced levels of Cl^- in the blood and the interstitial fluids surrounding the cells. This may affect neurons such as the GABAergic neuron in (b). During normal function in current-day CO_2 levels, the GABA neurotransmitter binds to a $GABA_A$ receptor, which opens a Cl^- channel whereby Cl^- enters the cell and hyperpolarizes it. In high CO_2 conditions, the reduced concentration of Cl^- may reduce the inflow of Cl^- into the cell during $GABA_A$ opening. Additionally, elevated levels of HCO_3^- in the interstitial fluid may also pass out through the $GABA_A$ ion channel and depolarize the cell. These changes to the $GABA_A$ function have been proposed to cause behavioural disturbances in animals exposed to future ocean acidification conditions.

acidification-induced impairments and restore normal behaviour in fishes and invertebrates (e.g. Watson et al. 2014; Thomas et al. 2020; Thomas et al. 2021).

Curiously, many studies also report that the control animals were unaffected by the gabazine drug (reviewed in Tresguerres and Hamilton 2017), which is unexpected given that reduced $GABA_A$ functioning in gabazine-treated animals is predicted to result in hyperactivity and even seizures (Tresguerres and Hamilton 2017). Additionally, it has been estimated that reversal of the $GABA_A$ receptor function may require plasma HCO_3^- concentrations of around 15 mM (Esbaugh 2018), while ocean acidification conditions are not thought to elevate plasma HCO_3^- concentrations to more than ~7 mM (Tresguerres and Hamilton 2017).

Some independent attempts to replicate the benefits of gabazine on fishes and invertebrates have failed (Charpentier and Cohen 2016; Abboud et al. 2019; Sundin et al. 2019), adding further questions about the veracity of previous reports and the role of $GABA_A$ receptors in the reported behavioural effects linked with ocean acidification.

5.3.2 Direct chemical effects of low pH and/or elevated CO_2

Separately from the GABA system, ocean acidification conditions have been suggested to alter behavioural responses to chemical cues through direct effects of pH and/or CO_2 on chemosensation (Roggatz et al. 2016; Porteus et al. 2018; Velez et al. 2019; Velez et al. 2021). For example, by measuring electrical activity of the olfactory nerve while exposing the olfactory epithelium to various chemical cues, Porteus et al. (2018) reported that the fish *Dicentrarchus labrax* had lower sensitivity to some chemical cues in elevated CO_2 (~1000 µatm) than in control water (~400 µatm) during acute exposure. As postulated by Roggatz et al. (2016) for the crab *Carcinus maenas*, the mechanism was suggested to be direct, through chemical alteration of the cue molecules and the receptors in reduced pH conditions, acting to reduce the ligand-receptor binding affinity (Tierney and Atema 1988; Porteus et al. 2018; Velez et al. 2019). The conclusion by Porteus et al. (2018) was that fish need to be 42% closer to an odour source to detect it in ocean acidification conditions. It has also been suggested that elevated CO_2 could disturb the olfactory transduction pathway independently of seawater pH, whereby CO_2 could diffuse into the olfactory receptor neurons and cause a reduction of intracellular pH and olfactory sensitivity despite extracellular pH remaining stable (Velez et al. 2021).

An additional consideration is the potential for ocean acidification to influence the degradation rate of chemical cues. Chivers et al. (2013) reported that chemical alarm cues degrade rapidly, with cues aged for 30 minutes no longer evoking anti-predator responses. The research group further reported that exposing the cues to CO_2 causes degradation within 15 minutes (Chivers et al. 2014b). This would mean that the lack of response to chemical cues reported for animals under ocean acidification conditions could be a consequence of the chemical cues degrading too quickly to elicit a response. Despite this possibility, studies by the same authors have used batches of alarm cue in ocean acidification studies where the experimental duration far exceeded the apparent degradation time, with no reported changes in animal responses to the chemical cue through time (e.g. Welch et al. 2014). Moreover, a hypothesized degradation of chemical cues under elevated CO_2 could not explain the complete reversal of preference/avoidance behaviours reported in several studies, as cue degradation would simply result in impartiality rather than extreme attraction/avoidance (e.g. Dixson et al. 2010; Munday et al. 2014).

It is apparent that such contradictory reports cannot be resolved using subjective behaviours as the only response variables; quantifying the chemical composition of water samples is required to solve the confusion that has been cultivated in this field. Given that many experimental treatments use projected CO_2 values for the open ocean (e.g. 1000 µatm by end-of-century), overlooking the fact that coastal areas like coral reefs already experience such levels (Shaw et al. 2012), more research is required to understand 'normal' structural conformation of chemical cue molecules under current-day CO_2 variability. In fact, there is a need for behavioural ecologists to collaborate with chemists to identify the 'chemical cue molecules' that are said to elicit preference/avoidance responses in the first place (e.g. Mathuru et al. 2012; Velez et al. 2019).

5.3.3 Direct sensory detection of CO_2

Fishes can detect the CO_2 levels and pH of the surrounding water (Jutfelt and Hedgärde 2013). Some species have been reported to detect prey using pH sensing (Caprio et al. 2014), while others can detect and avoid areas with elevated CO_2 levels (Jutfelt and Hedgärde 2013; Suski 2020; Bzonek and Mandrak 2022). Such direct sensing of local pH and CO_2 levels can therefore cause behavioural responses in fishes. Thus, behavioural responses to direct CO_2 detection may be confounded with chronic behavioural impacts, as the two can be difficult to disentangle in experimental set-ups. Locally reduced pH and elevated CO_2 can be used as proxies for high prey biomass and act as an attractant to predators (Caprio et al. 2014), but may also imply a risk of predators, or be used as a proxy for poor water quality (e.g. hypoxic waters can have high pCO_2; Pörtner et al. 2005). In the small-spotted catshark *Scyliorhinus canicula*, direct detection of CO_2 was suggested to cause increased swimming activity at night in elevated CO_2 levels compared with control conditions (Green and Jutfelt 2014). While fish reared under, and acclimated to, elevated CO_2 levels may be expected to habituate to the conditions, six weeks of exposure to elevated CO_2 did not reduce the strong CO_2 avoidance behaviour of juvenile Atlantic cod *Gadus morhua* (Jutfelt and Hedgärde 2013). It also appears that invertebrates can detect localized pH and CO_2 conditions as well—at least for conditions within sediments—and use this information to avoid burrowing into low pH sediments, which may result in shell dissolution during early life stages (Clements and Hunt 2017). This suggests that this potential mechanism of behavioural changes in fishes and invertebrates may persist and be the cause of some of the reported effects of elevated CO_2 on fish behaviour (e.g. hyperactivity as a form of avoidance behaviour), and can be adaptively beneficial. Nonetheless, this mechanism of direct CO_2 detection cannot be responsible for the more profound behavioural disturbances reported, such as attraction of fishes to predator chemical cues.

5.4 Anomalies and controversies

Following the early proliferation of studies reporting severe behavioural disturbances in Great Barrier Reef fishes exposed to elevated CO_2 (Munday et al. 2009; Dixson et al. 2010; Munday et al. 2010), a range of studies of coral reef fishes began to emerge that reported negligible effects, including for some of the same species that had been reported to suffer negative impacts (Sundin et al. 2017a; Raby et al. 2018; Sundin et al. 2019). The leading mechanism to explain behavioural/sensory impairments (a malfunction of the $GABA_A$ receptors in the brain) also came into question (Charpentier and Cohen 2016; Abboud et al. 2019; Sundin et al. 2019).

Approximately a decade after the field of acidification-induced behavioural disturbances was initiated, a large multi-year effort attempted to replicate some of the clearest impairments reported for coral reef fishes. Despite using overlapping species, life stages, and locations, Clark et al. (2020a) found no evidence for the previous claims, in that ocean acidification did not cause coral reef fishes to be (1) attracted to predator chemical cues, (2) consistently more active, (3) less behaviourally lateralized, or (4) visually impaired. Moreover, a data bootstrapping approach in Clark et al. (2020a) demonstrated that the within-group variance was implausibly small in most of the previous studies reporting strong ocean acidification effects in coral reef fishes. These findings were met with confusion and controversy throughout the field, causing

division within the scientific community (Clark et al. 2020b). The controversy highlighted the lack of objective evidence in the field of acidification-induced behavioural impacts; indeed, more than a decade of behavioural observations had been published without any accompanying video evidence to support the remarkable claims (see Clark 2017).

A subsequent meta-analysis of acidification-induced behavioural effects on fishes revealed that the extreme effect sizes in the initial studies (2009–2010) declined to near-zero by 2015 and remained around zero for the subsequent four years included in the meta-analysis (Clements et al. 2022). This suggests that the effects of ocean acidification were initially highly exaggerated, and the field did not uncover the true effect sizes until after many years of study and much investment of resources.

While the potential for a 'decline effect' has yet to be examined in invertebrates, studies suggest that ocean acidification effects on invertebrate behaviour are not robust. For example, while earlier studies reported that ocean acidification affects predator escape responses in some molluscs (Manríquez et al. 2014; Watson et al. 2014), recent studies of other molluscs have found that such behaviours are resilient to ocean acidification (Clements et al. 2020; Clements et al. 2021). Likewise, while early studies reported effects on self-righting in various species spanning multiple taxonomic groups (Manríquez et al. 2013; Zittier et al. 2013; Appelhans et al. 2014), more recent studies fail to document effects of ocean acidification (as a single stressor) on self-righting (Zlatkin and Heuer 2019; Duarte et al. 2022). Such declining effects through time can have several causes, which are explored in the next section.

5.5 Mechanisms of effect size inflation and the 'decline effect'

The decline effect phenomenon has been shown in several scientific fields in the past, but perhaps not to the same extent as shown by Clements et al. (2022) in the field of acidification-induced behavioural disturbances in fishes (Figure 5.5). There can be general causes underlying decline effects, but also specific causes linked to particular fields (Box 5.1).

Box 5.1 The plight of behavioural ecology and the need for enhanced transparency

Many people are deeply captivated when observing animal behaviour. This can be through direct observation of common behaviours of wild animals or pets, or the smorgasbord of extreme animal behaviours observed through our screens, from TikTok clips to high production value documentaries. Many documentaries include scenes with the behavioural ecologists who research animal behaviours, and the NOVA documentary *Lethal Seas* featured researchers measuring ocean acidification-induced alterations in fish behaviour (YouTube 2017). Animal behaviour research helps the public relate to the natural world and has led to many successful conservation campaigns (Berger-Tal et al. 2016; Greggor et al. 2019). Likewise, the positive framing of biologists in the public eye has likely benefitted research efforts on animal behaviour. Therefore, how the public views research in behavioural ecology is of great importance to scientific and conservation efforts. This is particularly important in light of global anthropogenic change and associated impacts on the environment.

Unfortunately, the lustre of behavioural ecology has been tarnished by several scientific misconduct cases, some of which are still unravelling. For example, the field of spider sociality was rocked by revelations in 2020 of data fabrication in many papers; at the time of writing this chapter, 26 papers had been retracted or given Expressions of Concern (Wikipedia 2023). In another recent and ongoing case, data duplications were found in several papers investigating animal behaviour and ecology, leading to at least 11 retractions, Expressions of Concern, or corrections (PubPeer 2023). Furthermore, two early-career behavioural ecologists separately published papers in *Science* that were later retracted due to data fabrication (Berg 2017; Enserink 2017; Enserink 2022; Thorp 2022), and their papers on ocean acidification and fish behaviour are amongst those that remain under investigation (Enserink 2021).

These and other cases have caused many to ponder if behavioural ecology is more susceptible to unreliable data than are other fields of biology. In some sense, behavioural ecology can be, especially if not done right. The type of experiments that often underlie animal behaviour findings are prone to several biases, such as observer bias (Tuyttens et al. 2014; Jutfelt et al. 2017; Clements et al. 2022). Many experiments also use small-scale and custom-built equipment, and are performed in remote and unique field settings, making replications challenging and increasing the difficulty of detecting scientific misconduct (Clark et al. 2016; Clark et al. 2020a). These issues and pitfalls have likely contributed to the recent cases of scientific misconduct that threaten the reputation of behavioural ecology. Furthermore, these issues have led to a poor understanding of acidification-induced effects on animal behaviour and a severe lack of predictive capacity, despite nearly two decades of research on the topic.

Fortunately, there are well-described paths to remedy these issues. Some of the solutions to minimize bias and increase replicability have been known for a long time. For example, the benefits of blinding of treatments to reduce observer biases have been known for decades (Beatty 1972), and automated quantification of animal behaviour from video recordings has been in use since the 1980s (Randelov et al. 1986). More recently, the 'Open Science' movement (e.g. Center for Open Science: COS 2023) is rapidly gaining traction, continuously improving transparency in science and providing protocols, tools, and platforms for sharing of data, scripts, videos, and other research outputs (e.g. Roche et al. 2022; SORTEE 2023). Journals, editors, and reviewers have an obligation to ensure results based on unreliable and potentially biased methods are detected before publication. Together, these factors give us the tools to ensure that behavioural ecology retains a solid reputation within academia, as well as providing a captivating glimpse of biological science for the public. This will keep people excited about animal behaviour and conservation, now and into the future.

The general causes of the early inflation of effect sizes include a range of biases such as publication bias (e.g. publication of the most dramatic findings, while leaving less dramatic findings unpublished) and confirmation bias (e.g. publishing results that support previous reports, to capitalize on the receptivity of journals to new 'phenomena'). These factors undoubtedly contributed to the decline effect in acidification-behaviour research, but there are specific factors identified by Clements et al. (2022) that appear to be larger drivers. Below, we outline the chronology and mechanisms of the decline effect in acidification-fish behaviour research throughout the past ~15 years.

Initial studies in the field (Munday et al. 2009; Dixson et al. 2010; Munday et al. 2010) reported effect sizes that were consistently greater than observed in nearly any other study in ecology and evolution (Jennions and Møller 2002). These studies were published in high-profile journals, which generated much attention in the media and throughout the scientific community, and even led to a presentation at the White House (Roberts 2015). As expected, other scientists became interested in this phenomenon and commenced similar experiments in their own study systems, which generally were not in coral reef environments. Given that studies finding an ocean acidification effect were more likely to be published in higher-profile journals than studies finding no effect (Clements et al. 2022), a strong incentive existed to report an effect of acidification. Selective reporting undoubtedly played a big role, whereby studies measured a swathe of behaviours but only reported the ones that showed a 'significant' treatment effect, without appropriately adjusting statistical thresholds (Rosenthal 1979; Simonsohn et al. 2014). Indeed, the circumstances were ripe for various analytical and statistical biases to come into play, including 'P-hacking' and 'HARKing' (Parker et al. 2016). Subsequent years saw a flurry of publications reporting ocean acidification effects on the behaviour of various aquatic animals (Figure 5.1), noting that sample sizes were often low and effect sizes were never of the same magnitude as those reported by the authors of the initial studies (Figure 5.5; Clements et al. 2022).

As the novelty of the ocean acidification-behaviour phenomenon wore off and high-profile journals became less receptive, an increasing number of studies reporting no effects were published in lower-ranking journals. Indeed, there was even a Special Issue of *ICES Journal of Marine Science* targeted at publishing those 'filed away' studies that found little effect of ocean acidification (Browman 2016). The progressive publication of these studies contributed to the decline effect in the field, yet other factors were clearly at play. How is it possible that very large effect sizes with small or non-existent among-individual variability in initial studies could decline to near-zero over time? In their meta-analysis, Clements et al. (2022) confirmed that the decline effect was not a result of biological factors that may have resulted from

an increasing proportion of studies examining (1) coldwater species, (2) nonolfactory-associated behaviours, or (3) nonlarval life stages. Intriguingly, when papers by the authors of the initial studies were removed from the meta-analysis dataset, there was essentially no behavioural effect of ocean acidification on fishes throughout the decade (Figure 5.5; Clements et al. 2022). This role of 'authorship' represents an underappreciated bias in science (Moulin and Amaral 2020), but one that must be addressed in parallel with more recognized biases.

Alarmingly, a misconduct investigation in 2022 concluded widespread data fabrication and falsification in published papers by one of the scientists who had co-founded the field of acidification-induced behavioural impairments in coral reef fishes (Enserink 2022). So far, two papers have been retracted as a result of the investigation outcomes (Thorp 2022; Barrett 2023), and associated investigations are ongoing. The misconduct revelation came as little surprise to scientists familiar with the behaviour of aquatic ectotherms, as the large effect sizes and small or non-existent variances reported in many of the studies by the co-founding scientists were implausible from biological and statistical standpoints (see Baird et al. 2014; Clark et al. 2020a). How did the field come to this point, and what can we do to minimize the potential for it to happen again?

5.6 Minimizing bias and improving transparency in behavioural ecology

Transparency and reproducibility are cornerstones of science, yet they are given scant attention. Recent cases have highlighted that disciplines relying heavily on subjective observations—like behavioural ecology—can be particularly vulnerable to scientific misconduct. Indeed, serious issues such as data fabrication (Box 5.1), and misleading claims of conducting experiments 'blind' (Berg 2017), have emphasized the need for an overhaul of standard practices in behavioural research.

Several approaches can be used to improve transparency and reproducibility. First, behavioural experiments should be filmed and made available

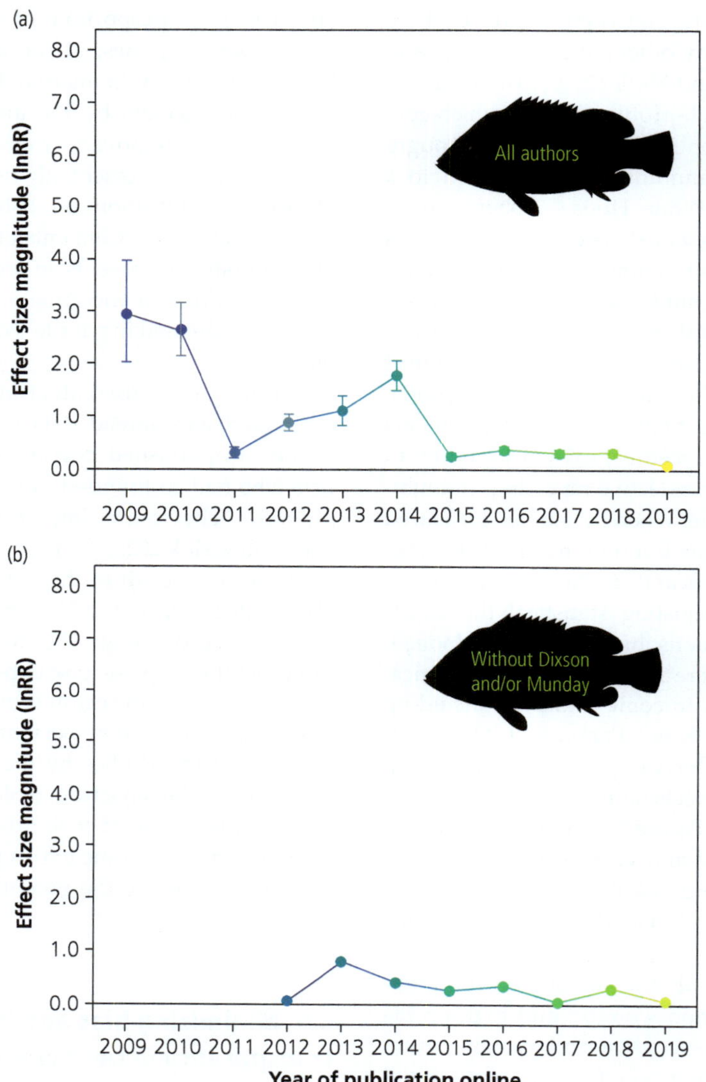

Figure 5.5 The decline effect in ocean acidification studies on fish behaviour. Data show mean effect sizes and their 95% confidence bounds for each year from 2009 to 2019 for (a) all studies and (b) for only those studies in which Dixson and/or Munday were not authors. Note that Dixson and Munday were the main authors of the initial studies that largely spawned the field, and all studies before 2012 were authored by Dixson and/or Munday. The decline effect in this field is one of the most striking examples of this phenomenon in all of ecology and evolution.

This figure is reprinted from Clements et al. (2023)

(Clark 2017). This means filming *all* behavioural trials and making them available, not just a 'representative' subset. For full transparency, videos should preferably start by showing a note in display of the camera at the commencement of each trial, where the note indicates the trial number, animal number, treatment history, and other important details of the animal being placed in the experimental apparatus (Clark et al. 2020a). In experiments using a choice flume, or any apparatus using chemical cues, a signalling stick can be used within frame of the camera to highlight which side of the apparatus a particular chemical stimulus is on, and/or the time point when the stimulus is added (Clark et al. 2020a). Such techniques could be refined further and be applied as mandatory standards across behavioural sciences. Interestingly, 20 online videos were highlighted by the founders of the field of ocean acidification and fish behaviour, which claim to provide representative examples of coral reef fish responses to ocean acidification and chemical alarm cues in choice flume experiments (Welch and Munday 2013). A recent analysis of the behavioural variance in those videos highlights that the expected variance (based on visual evidence) is much higher than reported in most published papers by the same research group (where no visual evidence exists) (Clements et al. 2023), highlighting how visual evidence can help to uncover anomalous data. The relatively small amount of time spent refining the set-up at the commencement of experiments is a worthwhile investment to improve data reliability and provide visual transparency.

Second, sample sizes should be increased dramatically as a general rule, given the particularly high inter-individual variability that is typical of animal behaviour (Jennions and Møller 2003; Yang et al. 2022). Clements et al. (2022) showed that most studies of ocean acidification and fish behaviour reporting large effect sizes used sample sizes below 30 animals per treatment group, identifying this value as a potential threshold to aim for. Sample sizes below this threshold can give rise to statistical artefacts due to low power and increase the risk that the resulting dataset is manipulated using questionable research practices (Parker et al. 2016). Moreover, most behavioural studies on the topic of ocean acidification have used very short measurement durations (e.g. < 10 minutes), providing little opportunity for animals to recover from handling stress. Given the evidence of non-biological data in this field of study (Clements et al. 2023), statistical power based on meta-analyses may not be accurate and, thus, there is a need for targeted, transparent experiments to provide guidance on robust sample sizes and measurement durations.

Third, it is critical for the sake of transparency that authors provide raw data and analysis code (Roche et al. 2022). These should be accessible to reviewers at the time of first submission (e.g. by sharing a link to a data repository), and then made publicly available once the paper is published. While it is encouraging to see that the proportion of papers providing 'open data' is increasing through time (Roche et al. 2022), it has been estimated that the majority of available datasets are incomplete and/or of a form that does not enable their reuse (Roche et al. 2015). Raw data availability assists with evaluating the quality of a study, and has been valuable in many cases to detect misconduct (Box 5.1).

Last, but certainly not least, scientists should collaborate on projects as much as possible with an aim to encompass diverse opinions and interpretations. Collaboration across disciplines, in particular, can foster a healthy level of scepticism, avoid 'groupthink', and dramatically improve the robustness of experimental approaches to maximize objectivity in the measurements.

5.7 Conclusions and future directions

Given the contradictory results across studies, the finding of an extreme decline effect in this field over the past > 10 years, and the revelation of misconduct (data fabrication) by one of the authors who founded the field of ocean acidification and coral reef fish behaviour, there are several challenging questions that must be addressed. What do we know about ocean acidification and animal behaviour? How much of the literature is independently replicable? Can the publications in this field be used to forecast any potential impacts of ocean acidification on marine animals? What research needs to be conducted to improve the current state of knowledge, and how do we standardize experimental approaches to ensure robust comparisons across studies and species?

Robust and transparent experiments and syntheses suggest that ocean acidification will have little direct adverse effect on the behaviour of fishes. Data and syntheses for marine invertebrates are lagging, but are needed to better understand acidification effects on these animals. In all future efforts to reach a true understanding of any potential impacts of ocean acidification on animal behaviour, experiments should be transparent, systematically documented (including with video footage), robust, and statistically powerful (Roche et al. 2015; Parker et al. 2016; Clark 2017; Roche et al. 2022). Previously claimed 'phenomena' that prove to be irreproducible should be retracted from the literature to prevent any further investment of resources into flawed narratives. It is imperative that additional information gets us closer to the truth, rather than further clouding the field.

It is important to note that the unfortunate occurrences in the field of ocean acidification and animal behaviour should not tarnish the field of climate change biology as a whole. Negative effects of climate warming, for example, have been shown consistently by many independent research groups across many biological systems (Pecl et al. 2017). Climate change continues to be one of the most pervasive threats facing animals worldwide, so research into the consequences must be paralleled by mitigation efforts and global action on cutting greenhouse gas emissions (see also Chapter 1). Research on animal behaviour plays an important role in relaying the wonders and challenges of the natural world to the general public (Box 5.1), and we look forward to a continuation of this association as the scientific techniques in behavioural ecology evolve and strengthen.

References

Abboud, J.-C., Bartolome, E.A., Blanco, M., et al. (2019). Carbon dioxide enrichment alters predator avoidance and sex determination but only sex is mediated by $GABA_A$ receptors. *Hydrobiologia*, 829, 307–322.

Appelhans, Y.S., Thomsen, J., Opitz, S., et al. (2014). Juvenile sea stars exposed to acidification decrease feeding and growth with no acclimation potential. *Marine Ecology Progress Series*, 509, 227–239.

Ashur, M.M., Johnston, N.K., and Dixson, D.L. (2017). Impacts of ocean acidification on sensory function in marine organisms. *Integrative and Comparative Biology*, 57, 63–80.

Atherton, J.A., and McCormick, M.I. (2015). Active in the sac: damselfish embryos use innate recognition of odours to learn predation risk before hatching. *Animal Behaviour*, 103, 1–6.

Baird, A.H., Cumbo, V.R., Figueiredo, J., et al. (2014). Comment on 'Chemically mediated behavior of recruiting corals and fishes: a tipping point that may limit reef recovery'. *PeerJ*, 2, e628v1.

Barrett, S.C.H. (2023). Retraction of: 'Reef fishes can recognize bleached habitat during settlement: sea anemone bleaching alters anemonefish host selection' (2016) by Scott and Dixson. *Proceedings of the Royal Society B: Biological Sciences*, 290, 20152694.

Beatty, W.W. (1972). How blind is blind? A simple procedure for estimating observer naiveté. *Psychological Bulletin*, 78, 70–71.

Berg, J. (2017). Addendum to 'Editorial retraction of the report "Environmentally relevant concentrations of microplastic particles influence larval fish ecology", by O. M. Lönnstedt and P. Eklöv'. *Science*, 358. https://www.science.org/doi/abs/10.1126/science.aar7766.

Berger-Tal, O., Blumstein, D.T., Carroll, S., et al. (2016). A systematic survey of the integration of animal behavior into conservation. *Conservation Biology*, 30, 744–753.

Bhuiyan, M.K.A., Billah, M.M., DelValls, T.Á., and Conradi, M. (2022). Intergenerational effects of ocean acidification on reproductive traits of an estuarine copepod. *Journal of Experimental Marine Biology and Ecology*, 557, 151799.

Bibby, R., Cleall-Harding, P., Rundle, S., et al. (2007). Ocean acidification disrupts induced defences in the intertidal gastropod *Littorina littorea*. *Biology Letters*, 3, 699–701.

Bignami, S., Sponaugle, S., and Cowen, R.K. (2014). Effects of ocean acidification on the larvae of a high-value pelagic fisheries species, mahi-mahi *Coryphaena hippurus*. *Aquatic Biology*, 21, 249–260.

Borges, F.O., Figueiredo, C., Sampaio, E., et al. (2018). Transgenerational deleterious effects of ocean acidification on the reproductive success of a keystone crustacean (*Gammarus locusta*). *Marine Environmental Research*, 138, 55–64.

Browman, H.I. (2016). Applying organized scepticism to ocean acidification research. *ICES Journal of Marine Science*, 73, 529–536.

Bzonek, P.A., and Mandrak, N.E. (2022). Wetland fishes avoid a carbon dioxide deterrent deployed in the field. *Conservation Physiology*, 10, coac021.

Caprio, J., Shimohara, M., Marui, T., et al. (2014). Marine teleost locates live prey through pH sensing. *Science*, 344, 1154–1156.

Charpentier, C.L., and Cohen, J.H. (2016). Acidification and γ-aminobutyric acid independently alter kairomone-induced behaviour. *Royal Society Open Science*, 3, 160311.

Chivers, D.P., Dixson, D.L., White, J.R., et al. (2013). Degradation of chemical alarm cues and assessment of risk throughout the day. *Ecology and Evolution*, 3, 3925–3934.

Chivers, D.P., McCormick, M.I., Nilsson, G.E., et al. (2014a). Impaired learning of predators and lower prey survival under elevated CO_2: a consequence of neurotransmitter interference. *Global Change Biology*, 20, 515–522.

Chivers, D.P., Ramasamy, R.A., McCormick, M.I., et al. (2014b). Temporal constraints on predation risk assessment in a changing world. *Science of the Total Environment*, 500–501, 332–338.

Chung, W.-S., Marshall, N.J., Watson, S.-A., et al. (2014). Ocean acidification slows retinal function in a damselfish through interference with $GABA_A$ receptors. *Journal of Experimental Biology*, 217, 323–326.

Clark, T.D. (2017). Science, lies and video-taped experiments. *Nature: World View*, 542, 139.

Clark, T.D., Binning, S.A., Raby, G.D., et al. (2016). Scientific misconduct: the elephant in the lab. *Trends in Ecology & Evolution*, 31, 899–900.

Clark, T.D., Raby, G.D., Roche, D.G., et al. (2020a). Ocean acidification does not impair the behaviour of coral reef fishes. *Nature*, 577, 370–375.

Clark, T.D., Raby, G.D., Roche, D.G., et al. (2020b). Reply to: 'Methods matter in repeating ocean acidification studies'. *Nature*, 586, E25–E27.

Clements, J.C., and Comeau, L.A. (2019). Behavioral defenses of shellfish prey under ocean acidification. *Journal of Shellfish Research*, 38, 725–742.

Clements, J.C., and Darrow, E.S. (2018). Eating in an acidifying ocean: a quantitative review of elevated CO_2 effects on the feeding rates of calcifying marine invertebrates. *Hydrobiologia*, 820, 1–21.

Clements, J.C., and Hunt, H.L. (2015). Marine animal behaviour in a high CO_2 ocean. *Marine Ecology Progress Series*, 536, 259–279.

Clements, J.C., and Hunt, H.L. (2017). Effects of CO_2-driven sediment acidification on infaunal marine bivalves: a synthesis. *Marine Pollution Bulletin*, 117, 6–16.

Clements, J.C., Poirier, L.A., Pérez, F.F., et al. (2020). Behavioural responses to predators in Mediterranean mussels (*Mytilus galloprovincialis*) are unaffected by elevated pCO_2. *Marine Environmental Research*, 161, 105148.

Clements, J.C., Ramesh, K., Nysveen, J., et al. (2021). Animal size and sea water temperature, but not pH, influence a repeatable startle response behaviour in a wide-ranging marine mollusc. *Animal Behaviour*, 173, 191–205.

Clements, J.C., Sundin, J., Clark, T.D., and Jutfelt, F. (2022). Meta-analysis reveals an extreme 'decline effect' in the impacts of ocean acidification on fish behavior. *PLOS Biology*, 20, e3001511.

Clements, J.C., Sundin, J., Clark, T.D., and Jutfelt, F. (2023). Extreme original data yield extreme decline effects. *PLOS Biology*, 21, e3001996.

Cohen-Rengifo, M., Danion, M., Gonzalez, A.A., et al. (2022). The extensive transgenerational transcriptomic effects of ocean acidification on the olfactory epithelium of a marine fish are associated with a better viral resistance. *BMC Genomics*, 23, 448.

COS. (2023). Center for Open Science. [online]. Available from: https://www.cos.io [accessed 1 September 2023].

Dadda, M., Zandonà, E., Agrillo, C., and Bisazza, A. (2009). The costs of hemispheric specialization in a fish. *Proceedings of the Royal Society B: Biological Sciences*, 276, 4399–4407.

de la Haye, K.L., Spicer, J.I., Widdicombe, S., and Briffa, M. (2012). Reduced pH sea water disrupts chemoresponsive behaviour in an intertidal crustacean. *Journal of Experimental Marine Biology and Ecology*, 412, 134–140.

Dixson, D.L., Munday, P.L., and Jones, G.P. (2010). Ocean acidification disrupts the innate ability of fish to detect predator olfactory cues. *Ecology Letters*, 13, 68–75.

Domenici, P., Allan, B., McCormick, M.I., and Munday, P.L. (2012). Elevated carbon dioxide affects behavioural lateralization in a coral reef fish. *Biology Letters*, 8, 78–81.

Duarte, C., Jahnsen-Guzmán, N., Quijón, P.A., et al. (2022). Morphological, physiological and behavioral responses of an intertidal snail, *Acanthina monodon* (Pallas), to projected ocean acidification and cooling water conditions in upwelling ecosystems. *Environmental Pollution*, 293, 118481.

Enserink, M. (2017). Researcher in Swedish fraud case speaks out: 'I'm very disappointed by my colleague'. *Science*. https://www.science.org/content/article/researcher-swedish-fraud-case-speaks-out-i-m-very-disappointed-my-colleague.

Enserink, M. (2021). Sea of doubts. *Science*, 372, 560–565. https://www.science.org/doi/full/10.1126/science.372.6542.560.

Enserink, M. (2022). Star marine ecologist committed misconduct, university says. *Science*, 377.

Esbaugh, A.J. (2018). Physiological implications of ocean acidification for marine fish: emerging patterns and new insights. *Journal of Comparative Physiology B*, 188, 1–13.

Ferrari, M.C.O., Manassa, R.P., Dixson, D.L., et al. (2012a). Effects of ocean acidification on learning in coral reef fishes. *PLOS ONE*, 7, e31478.

Ferrari, M.C.O., McCormick, M.I., Munday, P.L., et al. (2012b). Effects of ocean acidification on visual risk assessment in coral reef fishes. *Functional Ecology*, 26, 553–558.

Fields, J. (2013). Effects of ocean acidification on the behavior of two marine invertebrates: a study of predator-prey responses of the molluscs *Conus marmoreus* and *Strombus luhuanus* at elevated-CO_2 conditions. *Independent Study Project (ISP) Collection*, 1750, https://digitalcollections.sit.edu/isp_collection/1750.

Forsgren, E., Dupont, S., Jutfelt, F., and Amundsen, T. (2013). Elevated CO_2 affects embryonic development and larval phototaxis in a temperate marine fish. *Ecology and Evolution*, 3, 3637–3646.

Franke, A., and Clemmesen, C. (2011). Effect of ocean acidification on early life stages of Atlantic herring (*Clupea harengus* L.). *Biogeosciences*, 8, 3697–3707.

Frommel, A.Y., Margulies, D., Wexler, J.B., et al. (2016). Ocean acidification has lethal and sub-lethal effects on larval development of yellowfin tuna, *Thunnus albacares*. *Journal of Experimental Marine Biology and Ecology*, 482, 18–24.

Green, L., and Jutfelt, F. (2014). Elevated carbon dioxide alters the plasma composition and behaviour of a shark. *Biology Letters*, 10, 20140538.

Greggor, A.L., Blumstein, D.T., Wong, B.B.M., and Berger-Tal, O. (2019). Using animal behavior in conservation management: a series of systematic reviews and maps. *Environmental Evidence*, 8, 23.

Hamilton, T.J., Holcombe, A., and Tresguerres, M. (2014). CO_2-induced ocean acidification increases anxiety in Rockfish via alteration of $GABA_A$ receptor functioning. *Proceedings of the Royal Society B: Biological Sciences*, 281, 20132509.

Harzing, A.W. (2016). *The Publish or Perish tutorial: 80 easy tips to get the best out of the Publish or Perish software.* Tarma Software Research, London.

Hester, K.C., Peltzer, E.T., Kirkwood, W.J., and Brewer, P.G. (2008). Unanticipated consequences of ocean acidification: a noisier ocean at lower pH. *Geophysical Research Letters*, 35, L19601.

Heuer, R.M., and Grosell, M. (2014). Physiological impacts of elevated carbon dioxide and ocean acidification on fish. *American Journal of Physiology: Regulatory, Integrative and Comparative Physiology*, 307, R1061–R1084.

Heuer, R.M., Welch, M.J., Rummer, J.L., et al. (2016). Altered brain ion gradients following compensation for elevated CO_2 are linked to behavioural alterations in a coral reef fish. *Scientific Reports*, 6, 33216.

Hönisch, B., Ridgwell, A., Schmidt, D.N., et al. (2012). The geological record of ocean acidification. *Science*, 335, 1058–1063.

Hurst, T.P., Fernandez, E.R., Mathis, J.T., et al. (2012). Resiliency of juvenile walleye pollock to projected levels of ocean acidification. *Aquatic Biology*, 17, 247–259.

IPCC. (2022). Climate change 2022: impacts, adaptation and vulnerability. In: H.-O. Pörtner, D.C. Roberts, M. Tignor, et al. (eds), *Contribution of Working Group II to the Sixth Assessment Report of the Intergovernmental Panel on Climate Change*, pp. 3056. Cambridge University Press, Cambridge.

Jarrold, M.D., Humphrey, C., McCormick, M.I., and Munday, P.L. (2017). Diel CO_2 cycles reduce severity of behavioural abnormalities in coral reef fish under ocean acidification. *Scientific Reports*, 7, 10153.

Jarvis, D.M., Pope, E.C., Duteil, M., et al. (2022). Elevated CO_2 does not alter behavioural lateralization in free-swimming juvenile European sea bass (Dicentrarchus labrax) tested in groups. *Journal of Fish Biology*, 101, 1361–1365.

Jennions, M.D., and Møller, A.P. (2002). Relationships fade with time: a meta-analysis of temporal trends in publication in ecology and evolution. *Proceedings of the Royal Society B: Biological Sciences*, 269, 43–48.

Jennions, M.D., and Møller, A.P. (2003). A survey of the statistical power of research in behavioral ecology and animal behavior. *Behavioral Ecology*, 14, 438–445.

Jutfelt, F., and Hedgärde, M. (2013). Atlantic cod actively avoid CO_2 and predator odour, even after long-term CO_2 exposure. *Frontiers in Zoology*, 10, 81.

Jutfelt, F., Sundin, J., Raby, G.D., et al. (2017). Two-current choice flumes for testing avoidance and preference in aquatic animals. *Methods in Ecology and Evolution*, 8, 379–390.

Kaifu, K., Akamatsu, T., and Segawa, S. (2008). Underwater sound detection by cephalopod statocyst. *Fisheries Science*, 74, 781–786.

Kwan, G.T., and Tresguerres, M. (2022). Elucidating the acid-base mechanisms underlying otolith overgrowth in fish exposed to ocean acidification. *Science of the Total Environment*, 823, 153690.

Leung, J.Y.S., Zhang, S., and Connell, S.D. (2022). Is ocean acidification really a threat to marine calcifiers? A systematic review and meta-analysis of 980+ studies spanning two decades. *Small*, 18, 2107407.

Lönnstedt, O.M., Munday, P.L., McCormick, M.I., et al. (2013). Ocean acidification and responses to predators: can sensory redundancy reduce the apparent impacts of elevated CO_2 on fish? *Ecology and Evolution*, 3, 3565–3575.

Lovell, J.M., Findlay, M.M., Moate, R.M., and Yan, H.Y. (2005). The hearing abilities of the prawn *Palaemon serratus*. *Comparative Biochemistry and Physiology Part A: Molecular & Integrative Physiology*, 140, 89–100.

Lummis, S.C.R. (1990). GABA receptors in insects. *Comparative Biochemistry and Physiology Part C: Comparative Pharmacology*, 95, 1–8.

Lunt, G.G. (1991). GABA and GABA receptors in invertebrates. *Seminars in Neuroscience*, 3, 251–258.

Luthi, D., Le Floch, M., Bereiter, B., et al. (2008). High-resolution carbon dioxide concentration record 650,000–800,000 years before present. *Nature*, 453, 379–382.

Maneja, R.H., Frommel, A.Y., Browman, H.I., et al. (2015). The swimming kinematics and foraging behavior of larval Atlantic herring (*Clupea harengus* L.) are unaffected by elevated pCO_2. *Journal of Experimental Marine Biology and Ecology*, 466, 42–48.

Manríquez, P.H., Jara, M.E., Mardones, M.L., et al. (2013). Ocean acidification disrupts prey responses to predator cues but not net prey shell growth in *Concholepas concholepas* (loco). *PLOS ONE*, 8, e68643.

Manríquez, P.H., Jara, M.E., Mardones, M.L., et al. (2014). Ocean acidification affects predator avoidance behaviour but not prey detection in the early ontogeny of a keystone species. *Marine Ecology Progress Series*, 502, 157–167.

Mathuru, A.S., Kibat, C., Cheong, W.F., et al. (2012). Chondroitin fragments are odorants that trigger fear behavior in fish. *Current Biology*, 22, 538–544.

McDonald, M.A., Hildebrand, J.A., and Mesnick, S. (2009). Worldwide decline in tonal frequencies of blue whale songs. *Endangered Species Research*, 9, 13–21.

Melzner, F., Gutowska, M.A., Langenbuch, M., et al. (2009). Physiological basis for high CO_2 tolerance in marine ectothermic animals: pre-adaptation through lifestyle and ontogeny? *Biogeosciences*, 6, 2313–2331.

Milazzo, M., Cattano, C., Alonzo, S.H., et al. (2016). Ocean acidification affects fish spawning but not paternity at CO_2 seeps. *Proceedings of the Royal Society B: Biological Sciences*, 283, 20161021.

Miller, G.M., Kroon, F.J., Metcalfe, S., and Munday, P.L. (2015). Temperature is the evil twin: effects of increased temperature and ocean acidification on reproduction in a reef fish. *Ecological Applications*, 25, 603–620.

Miller, G.M., Watson, S.-A., McCormick, M.I., and Munday, P.L. (2013). Increased CO_2 stimulates reproduction in a coral reef fish. *Global Change Biology*, 19, 3037–3045.

Moulin, T.C., and Amaral, O.B. (2020). Using collaboration networks to identify authorship dependence in meta-analysis results. *Research Synthesis Methods*, 11, 655–668.

Moura, É., Pimentel, M., Santos, C.P., et al. (2019). Cuttlefish early development and behavior under future high CO_2 conditions. *Frontiers in Physiology*, 10, 975.

Moya, A., Howes, E.L., Lacoue-Labarthe, T., et al. (2016). Near-future pH conditions severely impact calcification, metabolism and the nervous system in the pteropod *Heliconoides inflatus*. *Global Change Biology*, 22, 3888–3900.

Munday, P.L., Cheal, A.J., Dixson, D.L., et al. (2014). Behavioural impairment in reef fishes caused by ocean acidification at CO_2 seeps. *Nature Climate Change*, 4, 487–492.

Munday, P.L., Dixson, D.L., Donelson, J.M., et al. (2009). Ocean acidification impairs olfactory discrimination and homing ability of a marine fish. *Proceedings of the National Academy of Sciences*, 106, 1848–1852.

Munday, P.L., Dixson, D.L., McCormick, M.I., et al. (2010). Replenishment of fish populations is threatened by ocean acidification. *Proceedings of the National Academy of Sciences*, 107, 12,930–12,934.

Munday, P., Kingsford, M., O'Callaghan, M., and Donelson, J. (2008). Elevated temperature restricts growth potential of the coral reef fish *Acanthochromis polyacanthus*. *Coral Reefs*, 27, 927–931.

Munday, P.L., McCormick, M.I., Meekan, M., et al. (2012). Selective mortality associated with variation in CO_2 tolerance in a marine fish. *Ocean Acidification*, 1, 1–5.

Munday, P.L., Pratchett, M.S., Dixson, D.L., et al. (2013). Elevated CO_2 affects the behavior of an ecologically and economically important coral reef fish. *Marine Biology*, 160, 2137–2144.

Munday, P.L., Watson, S.-A., Parsons, D.M., et al. (2016a). Effects of elevated CO_2 on early life history development of the yellowtail kingfish, *Seriola lalandi*, a large pelagic fish. *ICES Journal of Marine Science*, 73, 641–649.

Munday, P.L., Welch, M.J., Allan, B.J.M., et al. (2016b). Effects of elevated CO_2 on predator avoidance behaviour by reef fishes is not altered by experimental test water. *PeerJ*, 4, e2501.

Nagelkerken, I., Alemany, T., Anquetin, J.M., et al. (2021). Ocean acidification boosts reproduction in fish via indirect effects. *PLOS Biology*, 19, e3001033.

Nelson, J.A. (2015). Pickled fish anyone? The physiological ecology of fish from naturally acidic waters. In: R. Riesch, M. Tobler, and M. Plath (eds), *Extremophile Fishes: Ecology, Evolution, and Physiology of Teleosts in Extreme Environments*, pp. 193–215. Springer, Heidelberg.

Nelson, L.E., Guo, T.Z., Lu, J., et al. (2002). The sedative component of anesthesia is mediated by $GABA_A$ receptors in an endogenous sleep pathway. *Nature Neuroscience*, 5, 979–984.

Nilsson, G.E., Dixson, D.L., Domenici, P., et al. (2012). Near-future carbon dioxide levels alter fish behaviour by interfering with neurotransmitter function. *Nature Climate Change*, 2, 201–204.

Orr, J.C., Fabry, V.J., Aumont, O., et al. (2005). Anthropogenic ocean acidification over the twenty-first century and its impact on calcifying organisms. *Nature*, 437, 681–686.

Parker, T.H., Forstmeier, W., Koricheva, J., et al. (2016). Transparency in ecology and evolution: real problems, real solutions. *Trends in Ecology & Evolution*, 31, 711–719.

Pecl, G.T., Araújo, M.B., Bell, J.D., et al. (2017). Biodiversity redistribution under climate change: impacts on ecosystems and human well-being. *Science*, 355, eaai9214.

Penry-Williams, I.L., Brown, C., and Ioannou, C.C. (2022). Detecting behavioural lateralisation in *Poecilia reticulata* is strongly dependent on experimental design. *Behavioral Ecology and Sociobiology*, 76, 25.

Popper, A.N., and Fay, R.R. (1973). Sound detection and processing by teleost fishes: a critical review. *The Journal of the Acoustical Society of America*, 53, 1515–1529.

Porteus, C.S., Hubbard, P.C., Uren Webster, T.M., et al. (2018). Near-future CO_2 levels impair the olfactory system of a marine fish. *Nature Climate Change*, 8, 737–743.

Pörtner, H.O., Langenbuch, M., and Michaelidis, B. (2005). Synergistic effects of temperature extremes, hypoxia, and increases in CO_2 on marine animals: from Earth history to global change. *Journal of Geophysical Research: Oceans*, 110, C09S10.

Pörtner, H., Langenbuch, M., and Reipschläger, A. (2004). Biological impact of elevated ocean CO_2 concentrations: lessons from animal physiology and earth history. *Journal of Oceanography*, 60, 705–718.

PubPeer. (2023). Denon Start. [online]. Available from: https://pubpeer.com/search?q=Denon+start [accessed 15 September 2023].

Queirós, A.M., Fernandes, J.A., Faulwetter, S., et al. (2015). Scaling up experimental ocean acidification and warming research: from individuals to the ecosystem. *Global Change Biology*, 21, 130–143.

Raby, G.D., Sundin, J., Jutfelt, F., et al. (2018). Exposure to elevated carbon dioxide does not impair short-term swimming behaviour or shelter-seeking in a predatory coral-reef fish. *Journal of Fish Biology*, 93, 138–142.

Radford, C.A., Collins, S.P., Munday, P.L., and Parsons, D. (2021). Ocean acidification effects on fish hearing. *Proceedings of the Royal Society B: Biological Sciences*, 288, 20202754.

Randelov, A., Poulsen, E., and Pedersen, B. (1986). *Flugtadfaerd hos ål, skrubber og hesterejer*. Miljostyrelsenand COWIconsult, Ringkobing, Denmark.

Ren, Z., Mu, C., Li, R., et al. (2018). Characterization of a γ-aminobutyrate type A receptor-associated protein gene, which is involved in the response of *Portunus trituberculatus* to CO_2-induced ocean acidification. *Aquaculture Research*, 49, 2393–2403.

Réveillac, E., Lacoue-Labarthe, T., Oberhänsli, F., et al. (2015). Ocean acidification reshapes the otolith-body allometry of growth in juvenile sea bream. *Journal of Experimental Marine Biology and Ecology*, 463, 87–94.

Roberts, C.P. (2013). *Phototaxis of Dungeness Crab Zoeae in High-CO_2 Seawater: Implications for Coastal Ecosystems in an Acidified Ocean, vol. Master of Environmental Studies (MES) thesis*, p. 64, The Evergreen State College, Olympia, Washington.

Roberts, K.B. (2015). Sea change: UD's Dixson discusses ocean acidification at White House briefing. *UDaily*, http://www1.udel.edu/udaily/2016/dec/ocean-acidification-120415.html [December 2016 issue; accessed 1 February 2023].

Roche, D.G., Amcoff, M., Morgan, R., et al. (2020). Behavioural lateralization in a detour test is not repeatable in fishes. *Animal Behaviour*, 167, 55–64.

Roche, D.G., Kruuk, L.E.B., Lanfear, R., and Binning, S.A. (2015). Public data archiving in ecology and evolution: how well are we doing? *PLOS Biology*, 13, e1002295.

Roche, D.G., Raby, G.D., Norin, T., et al. (2022). Paths towards greater consensus building in experimental biology. *Journal of Experimental Biology*, 225, jeb243559.

Roggatz, C.C., Lorch, M., Hardege, J.D., and Benoit, D.M. (2016). Ocean acidification affects marine chemical communication by changing structure and function of peptide signalling molecules. *Global Change Biology*, 22, 3914–3926.

Rosenthal, R. (1979). The file drawer problem and tolerance for null results. *Psychological Bulletin*, 86, 638–641.

Rossi, T., Pistevos, J.C.A., Connell, S.D., and Nagelkerken, I. (2018). On the wrong track: ocean acidification attracts larval fish to irrelevant environmental cues. *Scientific Reports*, 8, 5840.

Schade, F.M., Clemmesen, C., and Mathias Wegner, K. (2014). Within- and transgenerational effects of ocean acidification on life history of marine three-spined stickleback (*Gasterosteus aculeatus*). *Marine Biology*, 161, 1667–1676.

Schmidt, M., Gerlach, G., Leo, E., et al. (2017). Impact of ocean warming and acidification on the behaviour of two co-occurring gadid species, *Boreogadus saida* and *Gadus morhua*, from Svalbard. *Marine Ecology Progress Series*, 571, 183–191.

Schram, J.B., Schoenrock, K.M., McClintock, J.B., et al. (2014). Multiple stressor effects of near-future elevated seawater temperature and decreased pH on righting and escape behaviors of two common Antarctic gastropods. *Journal of Experimental Marine Biology and Ecology*, 457, 90–96.

Schunter, C., Welch, M.J., Nilsson, G.E., et al. (2018). An interplay between plasticity and parental phenotype determines impacts of ocean acidification on a reef fish. *Nature Ecology & Evolution*, 2, 334–342.

Schunter, C., Welch, M.J., Ryu, T., et al. (2016). Molecular signatures of transgenerational response to ocean acidification in a species of reef fish. *Nature Climate Change*, 6, 1014–1018.

Sehgal, A., Tumar, I., and Schönwälder, J. (2010). Effects of climate change and anthropogenic ocean acidification on underwater acoustic communications. In: OCEANS'10, IEEE, Sydney, pp. 1–6.

Shaw, E.C., McNeil, B.I., and Tilbrook, B. (2012). Impacts of ocean acidification in naturally variable coral reef flat ecosystems. *Journal of Geophysical Research: Oceans*, 117, C03038.

Simonsohn, U., Nelson, L.D., and Simmons, J.P. (2014). P-curve and effect size: correcting for publication bias using only significant results. *Perspectives on Psychological Science*, 9, 666–681.

Simpson, S.D., Munday, P.L., Wittenrich, M.L., et al. (2011). Ocean acidification erodes crucial auditory behaviour in a marine fish. *Biology Letters*, 7, 917–920.

SORTEE. (2023). Society for Open, Reliable, and Transparent Ecology and Evolutionary biology. [online]. Available from: https://www.sortee.org [accessed 23 November 2023].

Spady, B.L., Munday, P.L., and Watson, S.-A. (2018). Predatory strategies and behaviours in cephalopods are altered by elevated CO_2. *Global Change Biology*, 24, 2585–2596.

Spatafora, D., Quattrocchi, F., Cattano, C., et al. (2021). Nest guarding behaviour of a temperate wrasse differs between sites off Mediterranean CO_2 seeps. *Science of the Total Environment*, 799, 149376.

Sundin, J., Amcoff, M., Mateos-González, F., et al. (2017a). Long-term exposure to elevated carbon dioxide does not alter activity levels of a coral reef fish in response to predator chemical cues. *Behavioral Ecology and Sociobiology*, 71, 108.

Sundin, J., Amcoff, M., Mateos-González, F., et al. (2019). Long-term acclimation to near-future ocean acidification has negligible effects on energetic attributes in a juvenile coral reef fish. *Oecologia*, 190, 689–702.

Sundin, J., and Jutfelt, F. (2016). 9–28 d of exposure to elevated pCO_2 reduces avoidance of predator odour but had no effect on behavioural lateralization or swimming activity in a temperate wrasse (*Ctenolabrus rupestris*). *ICES Journal of Marine Science*, 73, 620–632.

Sundin, J., Rosenqvist, G., and Berglund, A. (2013). Altered oceanic pH impairs mating propensity in a pipefish. *Ethology*, 119, 86–93.

Sundin, J., Vossen, L.E., Nilsson-Sköld, H., and Jutfelt, F. (2017b). No effect of elevated carbon dioxide on reproductive behaviors in the three-spined stickleback. *Behavioral Ecology*, 28, 1482–1491.

Suski, C.D. (2020). Development of carbon dioxide barriers to deter invasive fishes: insights and lessons learned from bigheaded carp. *Fishes*, 5, 25.

Thomas, J.T., Munday, P.L., and Watson, S.-A. (2020). Toward a mechanistic understanding of marine invertebrate behavior at elevated CO_2. *Frontiers in Marine Science*, 7:345.

Thomas, J.T., Spady, B.L., Munday, P.L., and Watson, S.-A. (2021). The role of ligand-gated chloride channels in behavioural alterations at elevated CO_2 in a cephalopod. *Journal of Experimental Biology*, 224, jeb242335.

Thorp, H.H. (2022). Editorial retraction. *Science*, 377, 826–826.

Tierney, A.J., and Atema, T. (1988). Amino acid chemoreception: effects of pH on receptors and stimuli. *Journal of Chemical Ecology*, 14, 135–141.

Tigert, L.R., and Porteus, C.S. (2023). Invited review—the effects of anthropogenic abiotic stressors on the sensory systems of fishes. *Comparative Biochemistry and Physiology Part A: Molecular & Integrative Physiology*, 277, 111366.

Toews, D.P., Holeton, G.F., and Heisler, N. (1983). Regulation of the acid-base status during environmental hypercapnia in the marine teleost fish *Conger conger*. *Journal of Experimental Biology*, 107, 9–20.

Tresguerres, M., and Hamilton, T.J. (2017). Acid–base physiology, neurobiology and behaviour in relation to CO_2-induced ocean acidification. *Journal of Experimental Biology*, 220, 2136–2148.

Tuyttens, F.A.M., de Graaf, S., Heerkens, J.L.T., et al. (2014). Observer bias in animal behaviour research: can we believe what we score, if we score what we believe? *Animal Behaviour*, 90, 273–280.

Vail, A.L., and McCormick, M.I. (2011). Metamorphosing reef fishes avoid predator scent when choosing a home. *Biology Letters*, 7, 921–924.

Vallortigara, G., and Rogers, L.J. (2005). Survival with an asymmetrical brain: advantages and disadvantages of cerebral lateralization. *Behavioral and Brain Sciences*, 28, 575–633.

Velez, Z., Costa, R.A., Wang, W., and Hubbard, P.C. (2021). Independent effects of seawater pH and high PCO_2 on olfactory sensitivity in fish: possible role of carbonic anhydrase. *Journal of Experimental Biology*, 224, jeb238485.

Velez, Z., Roggatz, C.C., Benoit, D.M., et al. (2019). Short- and medium-term exposure to ocean acidification reduces olfactory sensitivity in gilthead seabream. *Frontiers in Physiology*, 10, 731.

Vermeij, M.J.A., Marhaver, K.L., Huijbers, C.M., et al. (2010). Coral larvae move toward reef sounds. *PLOS ONE*, 5, e10660.

Watson, S.-A., Fields, J.B., and Munday, P.L. (2017). Ocean acidification alters predator behaviour and reduces predation rate. *Biology Letters*, 13, 20160797.

Watson, S.-A., Lefevre, S., McCormick, M.I., et al. (2014). Marine mollusc predator-escape behaviour altered by near-future carbon dioxide levels. *Proceedings of the Royal Society B: Biological Sciences*, 281, 20132377.

Welch, M.J., and Munday, P.L. (2013). Raw data for olfactory response of *Acanthochromis polyacanthus* in a Y-maze flume (dataset). Database: Research Data JCU.

[online]. Available from: https://doi.org/10.4225/28/5add60af3a267 [accessed 26 June 2020].

Welch, M.J., and Munday, P.L. (2016). Contrasting effects of ocean acidification on reproduction in reef fishes. *Coral Reefs*, 35, 485–493.

Welch, M.J., Watson, S.-A., Welsh, J.Q., et al. (2014). Effects of elevated CO_2 on fish behaviour undiminished by transgenerational acclimation. *Nature Climate Change*, 4, 1086–1089.

Wikipedia. (2023). Jonathan Pruitt. [online]. Available from: https://en.wikipedia.org/wiki/Jonathan_Pruitt [accessed 23 November 2023].

Yang, Y., Hillebrand, H., Lagisz, M., et al. (2022). Low statistical power and overestimated anthropogenic impacts, exacerbated by publication bias, dominate field studies in global change biology. *Global Change Biology*, 28, 969–989.

YouTube. (2017). Lethal Seas. Public Broadcasting Service NOVA Documentary, https://youtu.be/LQY-uHazpLQ.

Zittier, Z.M.C., Hirse, T., and Pörtner, H.-O. (2013). The synergistic effects of increasing temperature and CO_2 levels on activity capacity and acid–base balance in the spider crab, *Hyas araneus*. *Marine Biology*, 160, 2049–2062.

Zlatkin, R.L., and Heuer, R.M. (2019). Ocean acidification affects acid-base physiology and behaviour in a model invertebrate, the California sea hare (*Aplysia californica*). *Royal Society Open Science*, 6, 191041.

CHAPTER 6

Biological invasion

Lauren M. Pintor and Lindsey S. Reisinger

Overview

Biological invasions can result in pronounced changes in animal behaviour, both for invading species and for organisms in the recipient community. Here, we describe behavioural responses to invasion of both native and non-native species and how these responses depend on ecological interactions. Specifically, changes in competition and predation typically result in different behavioural responses by non-native and native species. Changes in competition often result in changes in habitat use, resource use, or aggression, whereas changes in predation often result in changes in movement or activity for prey species and changes in foraging for predators. We then focus on the mechanisms underlying these behavioural changes, including learning and evolution. Finally, we describe how the management of invasive species has impacted animal behaviour. Lethal control, for example, may select for shy or bold phenotypes depending on the capture method. We conclude that ecological context is key for understanding how invasions impact behaviour and behavioural changes can alter the effectiveness of invasive species control.

6.1 Introduction

Biological invasion remains a strong driver of environmental change by altering species composition, community dynamics, and ecosystem processes (Simberloff et al. 2013). Behavioural variation plays a key role for both non-native and native species throughout the invasion process.

Although only the non-native species experiences all stages of the invasion process (i.e. transport, arrival/introduction, establishment, spread), breaking down the process into these stages is a powerful framework to identify the selective pressures facing both non-native and native species in recipient communities. Furthermore, evidence suggests that different behaviours facilitate success at each stage (Chapple et al. 2012; Phillips and Suarez 2012; Chapple and Wong 2016; Figure 6.1). Boldness and exploration have been associated with the successful transport of species outside of the native range, but aggressiveness has been shown to be more important for processes associated with establishment. For example, bolder, more exploratory species often live in closer proximity to humans and may be more likely to get entrenched into an unintentional transport vector, such as cargo or freight (Chapple et al. 2012). However, following introduction and during establishment, higher levels of interspecific aggressiveness have explained the success of non-native species in outcompeting native species for shared resources (Hudina et al. 2014). Additionally, at the establishment stage, the pressures associated with invasion (e.g. increased competition, predation) begin to apply to native species in the recipient community. Finally, behaviours such as dispersal tendency have been shown to facilitate the spread stage of the invasion process for non-native species. Although several of these behaviours (boldness, exploration, aggressiveness, dispersal) are often correlated with each other, these behaviours may vary in relative importance during each invasion stage.

Lauren M. Pintor and Lindsey S. Reisinger, *Biological invasion*. In: *Behavioural Responses to a Changing World*. Edited by: Bob B. M. Wong and Ulrika Candolin, Oxford University Press. © Oxford University Press (2024). DOI: 10.1093/oso/9780192858979.003.0006

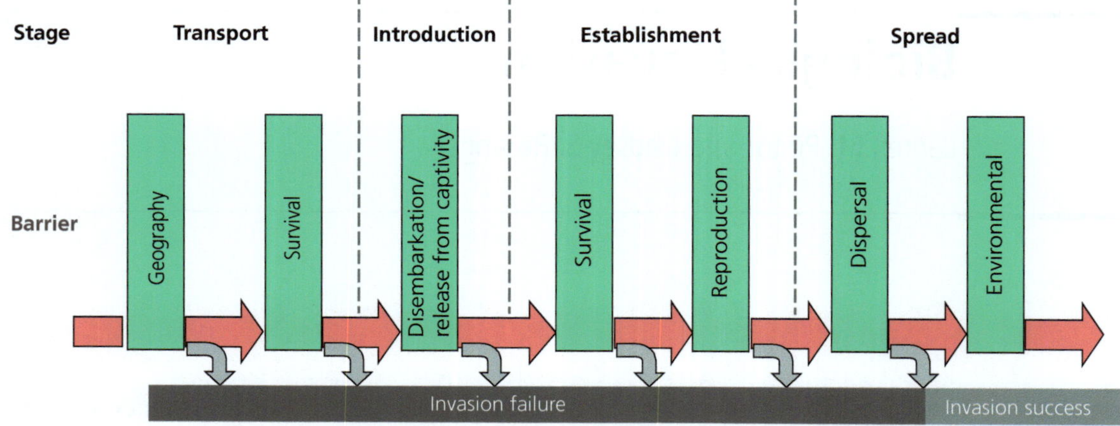

Figure 6.1 Outline of the invasion process and the four stages of invasion a species must pass to successfully invade. Each stage has different barriers that a non-native species must navigate in order to become invasive. Native species in the recipient community only experience the pressures that may drive behavioural change during the establishment and spread phases.

Image adapted from Chapple and Wong (2016)

Behavioural responses of non-native and native species to invasion are also important to understand and integrate into the management of invasive species. Invasive species management often involves the removal of non-native species and can have similar effects on behaviour as the harvest of managed game species (Diaz Pauli et al. 2015; Diaz Pauli and Sih 2017; see also Chapter 7). Specifically, harvest-induced selection of managed native species has resulted in life-history, behavioural, and morphological trait changes within a population. Behavioural change in response to invasive species management will also likely have important impacts on species interactions and can have a cascading impact on the recipient food web and ecosystem functions associated with it.

Finally, the behavioural response of a non-native species does not occur in isolation from native species within the recipient community. There is increasing evidence that the outcome of an interaction is the result of a 'back-and-forth' response (i.e. reciprocal behavioural feedback; Figure 6.2) where the behaviour of an individual of species A changes the behaviour of an individual of species B, and vice versa (McGhee et al. 2013; Belgrad and Griffen 2016). As such, if we are to understand the behavioural impacts of invasive species in recipient communities, it is

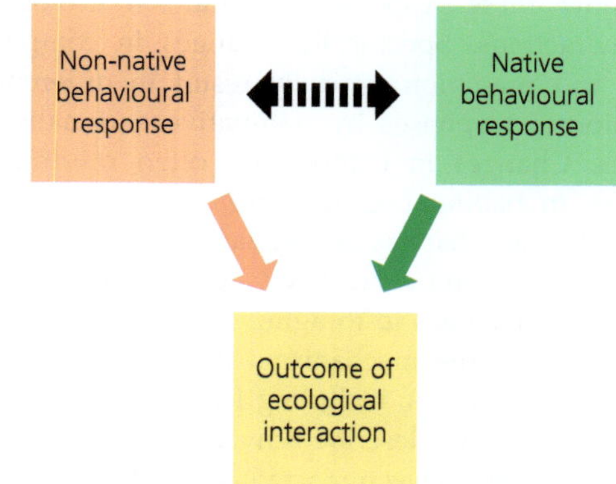

Figure 6.2 Solid arrows indicate a direct effect of the native or non-native species behaviour on the outcome of an interaction. Dashed, double-headed arrow indicates where reciprocal behavioural plasticity might occur.

important to examine the response of both native and non-native species to the biological invasion. Often behavioural responses by non-native and native species will depend on ecological interactions, such as competition and predation. As such, studying the behavioural responses in the context of trophic interactions can provide insights into how behaviour affects population and community-level consequences of invasion.

In this chapter, we discuss the current patterns of how native species and non-native species respond to biological invasion. We focus on behavioural change both in response to within trophic level (e.g. response to a new competitor) and across trophic level (e.g. new prey, new predator) species interactions. Next, we discuss the mechanisms that drive the behavioural response to invasion; learned versus evolutionary responses of species to invasion. Finally, we discuss the behavioural changes observed in response to invasive species management programmes and future research opportunities to test predictions about how the behavioural response to eradication may alter management success and food web interactions.

6.2 Patterns of behavioural change

6.2.1 Behavioural changes in response to competition

When the competitor community's composition changes for either a native or an invasive species, behaviours associated with resource requirements may also need to change for the species to co-occur in the short or long term. For example, behavioural shifts in habitat use have been frequently documented in native species. Many shift their habitat use to a different, less favourable microhabitat (e.g. physical or thermal) to coexist with an invasive competitor. For example, native virile crayfish *Faxonius virilis* shift their habitat use in lakes from locations with rocky substrate to locations with dense vegetation and organic matter in the presence of invasive rusty crayfish *Faxonius rusticus*, even though the vegetated locations often have low dissolved oxygen (Peters and Lodge 2013). Similarly, native Atlantic salmon *Salmo salar* and brook trout *Salvelinus fontinalis* in rivers shift their habitat use from habitats with cover (e.g. riparian vegetation) to open habitats in the presence of invasive rainbow trout *Oncorhynchus mykiss* (Thibault and Dodson 2013). Further, native black garden ants *Lasius niger* were able to persist at high densities in the presence of invasive ants *Tapinoma magnum* by shifting their foraging to a habitat with a different microclimate (Gippet et al. 2022). Occupying less favourable habitat may, to some extent, be detrimental to the native

species, but in many cases, these habitat shifts are thought to be key for the persistence of the native species in invaded areas (Peters and Lodge 2013).

Some native species exhibit changes in aggression in the face of an invasive competitor. For example, native topminnows *Fundulus notatus* exhibit high rates of agonistic behaviours (e.g. chases and nips) towards conspecifics, but only when in the presence of invasive western mosquitofish *Gambusia affinis* (Sutton et al. 2012). Likewise, native anoles *Anolis oculatus* exhibit more aggressive displays towards both conspecifics and heterospecifics in the presence of invasive anoles *Anolis cristatellus* (Dufour et al. 2020). As another example, native virile crayfish that co-occur with invasive rusty crayfish are more aggressive towards the invader than virile crayfish from naïve populations (Hayes et al. 2009). In some cases, changes in aggression in response to the invasion of a competitor may be beneficial, allowing a native species to better compete with an invader (Hayes et al. 2009).

Fewer studies have shown shifts in food/diet in comparison to behavioural changes in aggression or habitat use. However, the native day gecko *Phelsuma ornata* exhibits higher rates of cannibalism to compensate for reduced food availability caused by the invasion of the house gecko *Hemidactylus frenatus*, a superior competitor for shared prey (Cole and Harris 2011). In addition, brown anoles *Anolis sagrei* change their diet to include less arboreal prey in response to the colonization of islands by an arboreal competitor, the green anole *Anolis smaragdinus* (Pringle et al. 2019). Although these examples show native species changing their behaviour in response to a competitor's invasion, it is important to also consider potential changes in the invasive species' behaviour.

Native competitors in the invaded range may alter the behaviour of invasive species, particularly in cases where the invader is an inferior competitor compared to the native species. For example, the invasive gecko *Hemidactylus frenatus* is smaller and competitively subordinate to the larger native gecko *Gehyra australis*. Where the two taxa co-occur, the invasive gecko shifts its habitat use to a microhabitat not used by the native species, that is, a vacant niche (Yang et al. 2012). The vacant niche was an early leading hypothesis of invasion success, whereby an invader was more successful in establishing an invaded site if there was an 'empty' or 'vacant' niche not occupied by any of the native species in that environment (Herbold and Moyle 1986). Therefore, invasive species that can shift their behaviour to occupy a vacant niche may be more successful than those that compete for resources with native species. Research demonstrating different resource usage by invasive and native species is often not sufficient, of itself, to determine whether this is the result of a behavioural change. Indeed, determining whether behaviour has changed following invasion can be challenging, since the behaviour of the individuals that were initially introduced is often unknown. However, some research suggests that changes in behaviour are important for the success of invasive species. For example, bird species with greater behavioural flexibility (species with relatively large brains or a high frequency of foraging innovations) are more likely to be successful invaders than species without these traits (Sol et al. 2002). Behavioural flexibility may be advantageous for non-native species for several reasons, including allowing non-native species to use novel, unexploited resources and to avoid strong competition from native species.

6.2.2 Behavioural changes in response to predation

Many of the most dramatic examples of impacts of invasive species have come from studies illustrating the naiveté of prey to invasions by non-native predators (i.e. cases in which prey do not change their behaviour to respond appropriately to novel predators). For example, the invasion of the brown tree snake *Boiga irregularis* in Guam has caused either a dramatic reduction in the abundance or, in some cases, the extirpation of many of the island's bird species (Wiles et al. 2003). Similarly, several studies have found that Atlantic prey fishes often do not recognize invasive lionfish *Pterois* sp. as a predator, which is likely to enhance its success as a predator (Cote and Smith 2018). Although these examples highlight the potential for invasive predators to have devastating impacts when native prey do not change their behaviour, many prey species do show a behavioural response to invasion, so that,

over time, changes in behaviour may allow native prey species to persist.

Invasion may drive behavioural change for prey exposed to new predators in a recipient community/food web, regardless of the prey's status as native or non-native. Changes in movement or activity patterns of both native and non-native prey species have been frequently observed in response to invasion. This behavioural response is typically presumed to help a prey species to either avoid detection by a predator or to escape a predator upon detection. For example, several different species of native amphibian tadpoles readily reduce their movement or activity in response to non-native predators, such as crayfish (Nunes et al. 2014; Polo-Cavia and Gomez-Mestre 2014). Likewise, non-native prey often exhibits the same response to new (native) predators in the recipient community. For example, invasive tadpoles of the Mediterranean painted frog *Discoglossus pictus* reduce their activity 35–40% in the presence of native predators (Pujol-Buxo et al. 2013). Yet, prey do not always reduce activity in response to a new predator. Some native and non-native prey respond to a new predator by fleeing (i.e. increased activity). For example, in Australia, native common ringtail possums *Pseudochirus peregrinus* are more alert and flee in response to predator odours from the invasive European red fox *Vulpes vulpes* (Anson and Dickman 2013). Interestingly, this is the same response common ringtail possums exhibit to their native predator, the lace monitor *Varanus varius* (Anson and Dickman 2013). These examples illustrate the potential for prey to respond appropriately to novel predators. In fact, in several studies reviewed by Ruland and Jeschke (2020), the behavioural response of a prey species (native or non-native) towards a predator is often the same, regardless of whether the predator and prey share a long evolutionary history. This suggests that prey are using and responding to a general cue (e.g. scent of a dead conspecific, faecal odour of a predator) that alerts them to a looming predation risk, or that prey that exhibit behavioural changes in activity or movement in response to invasion are those that use a general predator cue. Introduced prey that fail to respond to novel predators may be less likely to become established or move further in the invasion process.

New predator-prey interactions that arise due to invasion can also drive prey to shift to new microhabitats. These shifts in microhabitat can occur in space and/or time, and sometimes cause prey to be displaced into more marginal habitat to avoid the new predator. Typically, these shifts have been observed in native species responding to an invasive predator, probably because native prey behaviour can be compared between invaded and uninvaded sites, or before and after the invasion, while invasions do not lend themselves as readily to testing shifts in invasive prey behaviour. Native brown anoles, for instance, change their habitat use to become more arboreal in response to island colonization by ground-dwelling curly-tailed lizards *Leiocephalus carinatus* (Pringle et al. 2019). Diets of brown anoles also shifted to include more arboreal prey in response to colonization by curly-tailed lizards (Pringle et al. 2019). As another example, native coots *Fulica atra* and great crested grebes *Podiceps cristatus* change their nesting locations and behaviour in response to the invasion of American mink *Neogale vison*. Specifically, to avoid mink, coots and grebes change their nesting locations to areas near human settlements (Brezinski et al. 2012). Furthermore, following invasion of the mink, a higher proportion of both coots and grebes exhibit colonial breeding to further reduce predation risk by mink. Similarly, in lake ecosystems, several native copepod species respond to an invasive cladoceran predator *Bythothrephes longimanus* by shifting their vertical distribution in the water column (Bourdeau et al. 2011). Typically, during the day, most copepod species inhabit the upper strata of Lake Michigan where there is typically a higher abundance of phytoplankton (i.e. food resources) than deeper in the water column. However, in the presence of high abundances of *Bythothrephes*, many copepod species inhabit a lower/deeper position in the water column to reduce spatial overlap with the invader. Interestingly, not all native copepods respond by shifting vertical distribution, and evidence suggests that the species that do not shift their distribution do not respond to water-borne cues from *Bythothrephes*. Thus, the type of cue used by prey to detect predation risk might be particularly important for predicting whether native prey respond to invasion by a predator (Sih et al. 2010).

Although the examples described here are drawn from cases of native prey responding to a non-native predator, similar behavioural shifts are also displayed by non-native prey responding to new predators in the recipient community. However, the examples are too few to suggest a general pattern.

Predators also often respond behaviourally to invasion. There is ample evidence that many native and non-native predator species change their diet and foraging behaviour following a successful invasion. In Ruland and Jeschke's (2020) literature review, approximately 85% of the studies they reviewed on a native predator's response to a non-native prey showed predators changing their diet. For example, the diet of several freshwater fish species in the Great Lakes ecosystem now includes several non-native prey species. Invasive round goby *Neogobius melanostomus* now comprises close to 90% of the diet of predatory smallmouth bass *Micropterus dolomieu*, and round gobies are also readily consumed by native water snakes and other piscivorous native fish (Steinhart et al. 2004; Johnson et al. 2005). Similarly, following the invasion of the red swamp crayfish *Procambarus clarkii* into the Guadalquivir marshes in southwestern Spain, over 60% of the 41 predators in the marsh had incorporated red swamp crayfish into their diet (Tablado et al. 2010).

Invasive species must use novel food resources in the invaded range to survive (unless species with which they share evolutionary history are also present), but the behavioural mechanisms that might generate a change in diet in a native predator are less clear. Foraging theory provides several hypotheses that might explain the behavioural mechanisms behind a native predator consuming a new, non-native prey. If predators forage optimally, then they should maximize the energy gained per unit handling time (E/h) needed to consume a prey. Thus, a non-native prey may have greater energy or lower handling time relative to native prey and drive native predators to change their diet. For example, native ant lions *Myrmeleon* spp. grow more quickly and weigh more on a diet of invasive Argentine ants *Linepithema humile* in comparison to native ant prey (Glenn and Holway 2008). Argentine ants are more susceptible to capture by ant lions in comparison to native ant prey, which suggests a lower handling time and likely a greater E/h for the invasive prey. Yet, equally important is the relative abundance of alternative prey. Although initially rare, non-native prey can often become extremely abundant. Even if their relative E/h is lower than that of a native prey, a native predator might change its diet and consume non-native prey if native prey becomes rare. For example, in nearshore oyster reef habitats, the predatory common mud crab *Panopeus herbsti* exhibits a persistent preference for native prey, and only increases its consumption of the non-native green porcelain crab *Petrolisthes armatus* when the latter's abundance, relative to native prey, increases in the environment (Kinney et al. 2023).

Native predators also sometimes respond to non-native prey by reducing their feeding intensity or attack rate of the prey, but this almost always appears to be because the prey was toxic. The most well-known and well-studied examples of this are the responses of native predators to the invasion of the toxic cane toad *Rhinella marina* in Australia. Toxins in the cane toad are different to those found in native frogs preyed upon by native vertebrate predators. As a result, many native predators die after attacking or consuming the cane toad (Webb et al. 2005; Smith and Phillips 2006). However, several studies have demonstrated that native predators can learn to avoid consuming cane toads (Box 6.1).

Box 6.1 The spread and management of cane toads in northern Australia

The cane toad *Rhinella marina* (Figure 6.3a) invasion in Australia provides an illustrative example of the behavioural responses of animals to invasion and how these behavioural responses can inform management aimed at mitigating invasion impacts. Cane toads were introduced to the northeast coast of Australia in 1935 as a control for insect pests, and they have since spread throughout a substantial portion of tropical and subtropical Australia (Urban et al. 2008). Cane toads are toxic and have had devastating impacts on native species, particularly large predators that eat anurans, such as the northern quoll *Dasyurus hallucatus*, freshwater crocodiles *Crocodylus johnstoni*, and large monitor lizards (Figure 6.3b) in the family Varanidae (Pettit et al. 2021). Many aspects of this invasion have been intensively studied, including the traits of cane toads and native species in recipient communities and techniques to mitigate cane toad impacts.

Cane toad traits, including behavioural traits, have changed as a result of the invasion process. Cane toads from invasion-front populations are more exploratory and have a higher propensity to take risks than conspecifics from long-established populations (Gruber et al. 2017). These behavioural differences were maintained in toads reared in a common environment, suggesting that evolutionary processes may underlie these behavioural changes (Gruber et al. 2017). In addition, cane toads from invasion-front populations had greater dispersal rates (travelled further, were more highly directional, and re-used refugia less frequently) than conspecifics from long-established populations (Pizzatto et al. 2017).

These changes in behaviour have implications for management of cane toads. For example, population control through manual collection or trapping may be more successful in regions where cane toads have been present for a long period of time, since areas near the invasion front will be recolonized more quickly by individuals with high dispersal rates (Phillips 2016). Phillips et al. (2016) also used a simulation model to investigate whether a 'genetic backburn' could be used as a strategy to prevent individuals with high-dispersal phenotypes from overcoming a barrier to dispersal. This strategy involves translocating invasion-core individuals with low-dispersal phenotypes to the nearside of a barrier ahead of the arrival of the invasion front. Their model indicated that this approach could enhance the effectiveness of a barrier by preventing individuals with high-dispersal phenotypes from reaching the barrier. Other mitigation strategies that take advantage of the behaviour of the invader have also been proposed. For example, alarm cues from injured conspecifics elicit a behavioural response in cane toad tadpoles, and these tadpoles have reduced rates of survival and growth; thus, applying the chemicals involved in alarm cues to waterbodies could potentially be used to reduce cane toad recruitment (Crossland et al. 2019).

The cane toad invasion has also altered the behaviour of native predators. Native predator populations exposed to cane toads have developed learned behavioural responses to the toads or genetically based changes in feeding behaviour. For example, lace monitors *Varanus varius* and Australian marbled frogs *Limnodynastes convexiusculus* that consume a toad that has enough toxin to elicit illness—but not

Box 6.1 *Continued*

enough to kill the predator—have learned not to consume toads in future encounters (Greenlees et al. 2010; Jolly et al. 2016). Further, predators collected from populations that persist in cane toad invaded areas are less likely to consume toads than those collected from naïve populations (Kelly and Phillips 2017). Scientists have used information about the learned behavioural responses of predators to develop protocols that could enhance predator survival after cane toads arrive. Specifically, studies have tested the efficacy of conditioned taste aversion training to teach naïve predators to

avoid cane toads (Tingley et al. 2017). This training involves exposing naïve predators to small toads that have lower levels of toxin or 'toad aversion baits' (i.e. cane toad tissue containing a nausea-inducing chemical) ahead of the arrival of the cane toad invasion to reduce predator mortality (Tingley et al. 2017). Overall, the invasion of cane toads in Australia exemplifies, not only the role of invasions in changing animal behaviour, but also how knowledge of those behavioural changes can potentially be harnessed to address invasive species impacts.

Figure 6.3 (a) Cane toad *Rhinella marina* (b) Australian monitor lizard *Varanus varius*.

Photos: iStock

6.2.3 Behavioural changes in response to facilitation

In addition to altering competition, prey availability, and predation, invasions are also likely to result in new positive heterospecific interactions, and these interactions may also cause changes in behaviour (Camacho-Cervantes et al. 2023). For instance, native Mexican topminnows and invasive Trinidadian guppies *Poecilia reticulata* associate with one another, which is likely to provide benefits for each species, such as reduced predation risk (Camacho-Cervantes et al. 2018). As another example, invasive Italian wall lizards *Podarcis sicula* can use social information from novel heterospecifics to increase their foraging success (Damas Moriera et al. 2018). While these positive interactions are less studied than competition and predation, they may also be important for invasion success and behavioural changes following invasion.

6.3 Mechanisms underlying behavioural changes

6.3.1 Role of learning and evolutionary mechanisms

Both native and non-native species can change their behaviour in response to invasion, and this can occur via several different evolutionary and cognitive mechanisms (Ruland and Jeschke 2020; Stewart et al. 2021). Epigenetics, maternal effects, and hormonal mechanisms can also drive behavioural change (Ledón-Rettig et al. 2013; Stewart et al. 2021). However, most published studies, to date, commonly report rapid genetic change or cognitive mechanisms, such as learning, as driving the response to invasion (reviewed in Ruland and Jeschke 2020). But learning and evolutionary change are not mutually exclusive mechanisms. Individual learning and plasticity can often work to modify existing behaviours or create new ones, allowing evolutionary mechanisms (e.g. selection) to then drive changes in behavioural traits/diversity at the population level. For example, invasion can drive behavioural change in either a non-native or native species if (1) a non-native individual modifies or creates a behaviour to survive through the invasion process and/or a native individual modifies or creates a behaviour to persist after a successful invasion, (2) the behaviours of the individuals (either native and/or non-native) are heritable, and (3) the behaviours are favoured by natural selection and, as a result, are able to spread in the population (i.e. Simpson-Baldwin effect: Baldwin 1896, 1902; Simpson 1953).

Although both learning and evolutionary mechanisms drive behavioural change in response to invasion, empirical evidence suggests that learning allows for faster behavioural adjustments and often precedes trait change via evolutionary mechanisms (Baldwin 1896, 1902; Simpson 1953; Zuk et al. 2014; Ruland and Jeschke 2020). For example, shifts in anti-predator behaviour of individual native fence lizards have increased survival from attacks by invasive predatory fire ants (Langkilde 2009; Langkilde and Freidenfelds 2010; Freidenfelds et al. 2012; Graham et al. 2012). Following this behavioural change, fence lizards have evolved longer hindlimbs, which enables them to reach and remove more predatory fire ants from their body (Langkilde 2009). In general, learning has been suggested to explain more invasion-driven behavioural changes in native species, like the fence lizard, in comparison to changes in invasive species (Ruland and Jeschke 2020). Ruland and Jeschke's (2020) review of the literature indicated that learning accounts for 75% (89 of 118 records) of behavioural changes in native species, and only 46% (30 of 65 records) in invasive species reported. In contrast, rapid genetic adaptation explained more of the behavioural change in invasive species (42%, 27 of 65 records) as opposed to native species (24%, 28 of 118 records). However, the data available for non-native species are dominated by three taxonomic groups (i.e. insects, molluscs, and amphibians) and should be interpreted with caution (Ruland and Jeschke 2020).

6.3.2 Stage of invasion and mechanisms of behavioural change

The mechanism of behavioural change for non-native and native species may also be related to differences in the stressors experienced during each stage of the invasion process. For example, human-mediated transport of a non-native species has often been viewed as a selective filter, whereby the few individuals that survive often harsh physical and abiotic conditions are introduced into a new environment, driving rapid evolutionary change in the non-native population (e.g. evolutionary bottleneck: Baños-Villalba et al. 2020; Chappell et al. 2022). Data on non-native species during the transport stages of invasion are lacking (Briski et al. 2018). Future work might wish to explore the relative importance of learning during the transport and arrival stages of the invasion and test whether behaviour mediates adaptation (West-Eberhard 2003; Duckworth 2008), or if in-situ selection during transport is non-random (Briski et al. 2018). This is critical because behavioural changes observed following introduction (i.e. during establishment) might be associated with physiological or morphological traits previously selected upon while entrained in the transport pathway. For example, non-native plankton species transported in the ballast water of ships experience high, but not total, mortality during entrainment. Those that survive would have been exposed to extreme changes in temperature, salinity, and dissolved oxygen levels. These factors, among others (e.g. exposure to toxic chemicals leached from metal and paint), are suggested to select for individuals tolerant of harsh conditions. Behaviour, physiology, and life-history traits are often highly correlated (Biro and Stamps 2008; Reale et al. 2010). Therefore, non-native species may have different behaviour upon arrival in the invaded range as a result of non-random selection of a suite of correlated traits, and this trait selection during the transport process may explain why evolutionary mechanisms (as opposed to learning) are responsible for most behavioural changes in invasive species.

Following introduction and during establishment, learning is thought to be the dominant mechanism driving change in non-native and native species (i.e. the adaptive flexibility hypothesis; Wright et al. 2010). Wright et al. (2010) suggest that innovation will be favoured as non-native individuals navigate and establish a population in a new environment, and native species respond to a change in community composition. Indeed, several studies reviewed by Ruland and Jeschke (2020) indicated that learning facilitates avoidance behaviour in both native and non-native species in response to a new competitor. For example, American mink *Neovison vison* invaded the UK during a time when two native mustelid competitors were largely absent (Harrington et al. 2009). As the two native mustelids began to recover in the areas now occupied by the non-native mink, the latter shifted from being primarily nocturnal to being primarily diurnal following the re-establishment of native competitors. The rapid expansion of sea urchins *Centrostephanus rodgersii* on the east coast of Tasmania, Australia, has similarly increased competition with the black-lipped abalone *Haliotis rubra* (Strain et al. 2013). In-situ experimental manipulations of urchins with abalone indicated that abalone increased activity and shifted microhabitats to avoid urchins and reduce competition for resources via short-term learning.

Learning plays a larger role than rapid genetic change for observed changes in behaviour of both native prey (native prey: 73% or 40 of 55 studies) and non-native prey (non-native prey: 73% or 11 of 15 studies) in response to a new predator (Ruland and Jeschke 2020). However, in response to a new prey, native predators and non-native predators respond differently. Specifically, changes in the feeding behaviour of native predators following invasion of a non-native prey are driven more by learning (84% or 32 of 38 studies) rather than genetic change (16% or six of 38 studies). But changes in the feeding behaviour of non-native predators following invasion are driven more by rapid genetic change (56% or five of nine studies) than by learning (33% or three of nine studies). Although behavioural plasticity (e.g. learning) is clearly an

important mechanism driving behavioural change, behaviour is increasingly being recognized as playing a key role in evolutionary processes as well. For behavioural changes that take place in response to invasion, it will be key that researchers consider the stage of the invasion process in which the changes have occurred to understand the ecological context and pressures being experienced by individuals as the invasion unfolds.

6.4 Behavioural response to invasive species management programmes

In the same way that the invasion process can act as a selective filter on behavioural traits, eradication of invasive species from an invaded ecosystem can equally select either for or against certain behaviours and, in so doing, drive phenotypic change. Lethal control of invasive species may operate in the same way as harvest-induced selection on behavioural, morphological, and life-history traits, which has been documented for several exploited species (Heino et al. 2015; Kuparinen and Festa-Bianchet 2017; Chapter 7). Many examples come from the fish literature. Here, not only can harvesting, in general, induce selection, but different harvesting methods can select for different behavioural traits in the species being exploited (i.e. through removal of either bold or shy individuals; reviewed in Diaz Pauli and Sih 2017; Chapter 7). In a similar vein, when applied to the control of invasive species, employing management methods that remove bolder individuals with higher reproductive effort can help to slow or curb an invasion, whereas methods that remove shy individuals with low reproductive effort might not impact the population growth of non-native individuals that remain in the invaded habitat.

However, management strategies used to cull or remove non-native species can also drive behavioural responses that, over time, may decrease the effectiveness of a particular approach or gear. Spearfishing has been one of the primary strategies used to cull invasive lionfish in the Bahamas. Over time, however, lionfish on culled reefs have become less active and hide deeper within the reef during

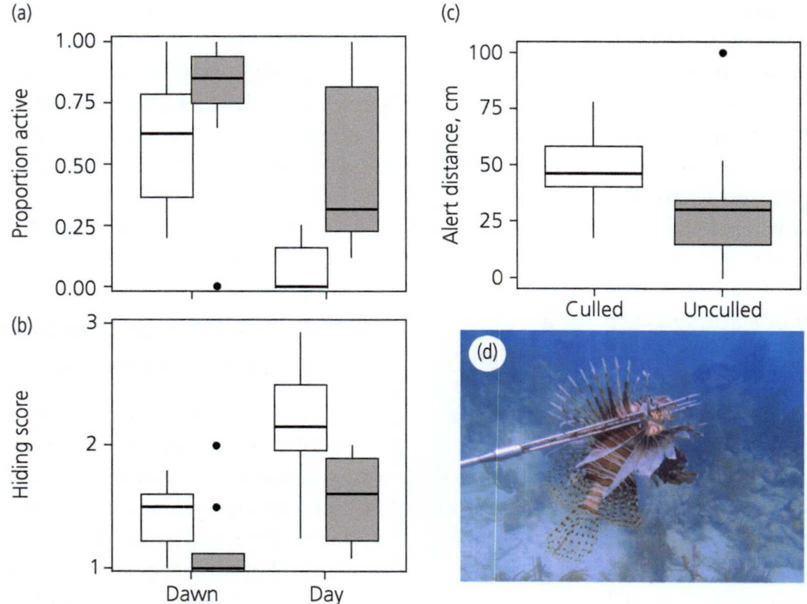

Figure 6.4 (a) Proportion of lionfish that were active, and (b) extent of concealment, as measured on a three-point scale, with three being the most hidden on the culled (open bars) and unculled (shaded bars) coral reef patches. (c) Alert distance of lionfish to an approaching diver on culled and unculled coral reef patches. Larger distances denote increased wariness. Boxplots show medians, first and third quartiles, and 95% confidence intervals, along with outliers. (d) Image of lionfish culling.

Images (a) to (c) adapted from Cote et al. (2014); Photo (d): iStock

the day (Figure 6.4), thus making them more difficult to spear during culling attempts (Cote et al. 2014). Thus, any management strategy that targets individuals with behaviours that make them more vulnerable to capture can also potentially drive behavioural change in the invasive species. Additionally, in the case of non-native lionfish, those inhabiting fished areas have a higher flight initiation distance (i.e. the distance between the diver and the location of the fish when it flees) when compared to unfished areas (Figure 6.4; Januchowski-Hartley et al. 2012; Cote et al. 2014). It is not known if this is a learned response or one driven by evolutionary change, but it is important to consider how the different mechanisms might influence management. For example, if a change in reaction distance in a culled reef is a learned behaviour that spreads quickly through a population through social learning, a management strategy centred on culling might become less effective more quickly than if the behavioural

change is driven by an evolutionary response. However, learned responses can be forgotten if a management technique is not continually applied to the population. Together these points highlight that behavioural responses and the mechanisms driving them can influence the effectiveness of management strategies over time.

Finally, there is growing evidence that encounters with eradication attempts may cause animals to experience high levels of stress and drive long-lasting behavioural changes in individuals that successfully escape capture. For example, in many animal communities, humans have now taken the role of 'super predator', either through direct killing or by eliciting fear responses in animals, such as large carnivores (Darimont et al. 2015; see also Chapter 13 on managing human-wildlife conflicts). Pumas, for instance, exhibit much higher flee responses and lower foraging rates when exposed to human predator playbacks as opposed to non-predator (frog) sounds at puma

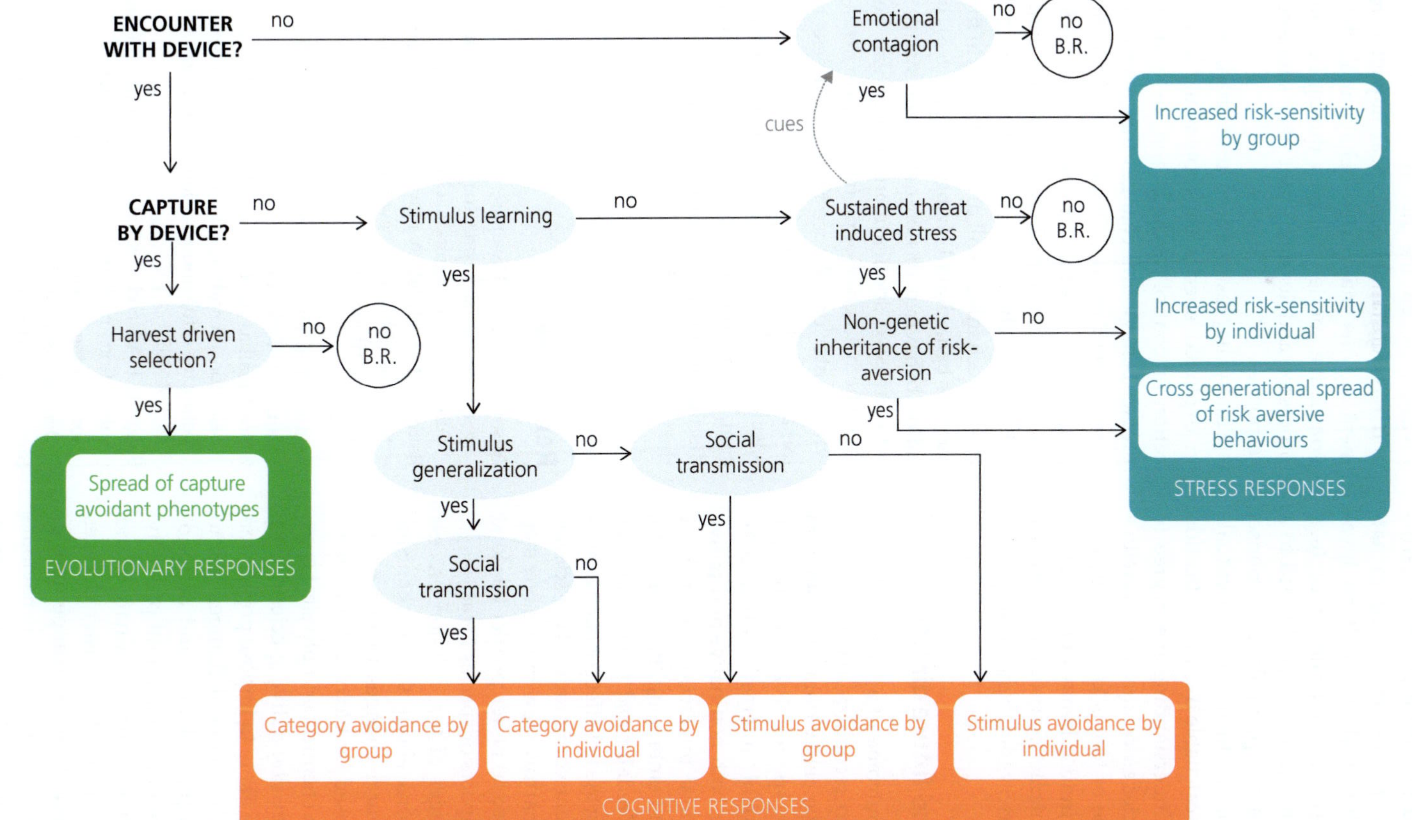

Figure 6.5 Framework linking interactions of individuals with management devices (in bold black capitals), to mechanisms (in grey circles), to the different types of behavioural responses possible displayed by individuals and/or populations (in coloured boxes). Behavioural responses are grouped into three categories (evolutionary, cognitive, and stress responses); in some cases, combinations of events and mechanisms do not elicit a behavioural response (white circles with 'no B.R.' indicate no behavioural response).

Image adapted from Diquelou and Griffin (2020)

feeding sites (Smith et al. 2017). Similarly, nuisance badgers exposed to control efforts on farms exhibit greater levels of stress compared to natural predators (Clinchy et al. 2016). If an invasive species associates an eradication tool (e.g. traps or poison) with humans, this could lead to a generalized aversion or a hypervigilance response. The ensuing association can then cause individuals to be less detectable or catchable during eradication or control attempts. Understanding the degree to which an invasive species associates humans with a management technique or tool is key to predicting how often stress can induce behavioural changes and, in so doing, counteract control measures (see also Chapter 11 on conservation behaviour).

6.5 Conclusions and future directions

To better predict the outcome of a biological invasion, it is important to recognize that invasion is a stressor that can drive behavioural shifts in both non-native and native species. Post-introduction, invasions will alter trophic interactions (e.g. competition, predation) for both non-native and native species. Evaluating those interactions as a 'back and forth' response and embracing the behaviours of both players (e.g. predator *and* prey) brings us closer to understanding the net effect of behavioural changes on the recipient community. Two-species interactions are still a narrow view but are one step closer to what is needed to better predict the food web and ecosystem impacts of species invasions.

Additionally, the stages of the invasion process can improve predictions about the mechanisms that underlie behavioural changes in non-native and native species caused by biological invasions. For example, the harsh chemical and physical environment experienced by non-native species during transport may be more likely to impose population bottlenecks and drive behavioural changes due to selection on correlated physiological and morphological traits, rather than on behavioural plasticity. However, during establishment, non-native and native species may need to exhibit faster changes in behaviour in response to novel competitors as well as novel predators or prey. Thus, behaviours associated with species interactions between native and non-native species may be more likely to

change via learning. During the later stages of an invasion (when the non-native species spreads and has substantial ecological impacts), management actions may drive behavioural changes in non-native species, not only through evolutionary processes and learning, but also stress-hormonal responses.

Finally, harvest-induced trait changes observed in exploited species can potentially guide eradication or control methods and, in so doing, help predict behavioural changes that may accompany management efforts. Diquelou and Griffin (2020) proposed a useful framework for understanding the relationships between the mechanism and behavioural change, but also for predicting how different mechanisms will drive variation in the prevalence of behavioural changes (Figure 6.5). Incorporating behaviour of non-native and native species into invasive species management can help predict whether behavioural responses to management will counter or enhance the effectiveness and sustainability of control efforts. Yet, given that native species are also clearly changing in response to invasion, equal consideration needs to be given to the disruption that eradication and control can have on the rest of the food web.

References

Anson, J.R., and Dickman, C.R. (2013). Behavioural responses of native prey to disparate predators: naiveté and predator recognition. *Oecologia*, 171, 367–377.

Baldwin, J.M. (1896). A new factor in evolution. Adaptive individuals in evolving populations: models and algorithms. *The American Naturalist*, 30, 441–451, 536–553.

Baldwin, J.M. (1902). *Development and Evolution*. Macmillan, New York, NY.

Baños-Villalba, A., Carrete, M., Tella, J.L., et al. (2020). Selection on individuals of introduced species starts before the actual introduction. *Evolutionary Applications*, 14, 781–793.

Belgrad, B.A., and Griffen, B.D. (2016). Predator–prey interactions mediated by prey personality and predator hunting mode. *Proceedings of the Royal Society B: Biological Sciences*, 283, 20160408.

Biro, P.A., and Stamps, J.A. (2008). Are animal personality traits linked to life-history productivity? *Trends in Ecology & Evolution*, 23, 361–368.

Bourdeau, P.E., Pangle, K.L., and Peacor, S.D. (2011). The invasive predator Bythotrephes induces changes in the vertical distribution of native copepods in Lake Michigan. *Biological Invasions*, 13, 2533–2545.

Briski, E., Chan, F.T., Darling, J.A., et al. (2018). Beyond propagule pressure: importance of selection during the transport stage of biological invasions. *Frontiers in Ecology and the Environment*, 16, 345–353.

Brzeziński, M., Natorff, M., Zalewski, A., and Żmihorski, M. (2012). Numerical and behavioural responses of waterfowl to the invasive American mink: a conservation paradox. *Biological Conservation*, 147, 68–78.

Camacho-Cervantes, M., Keller, R.P., and Vila, M. (2023). Could non-native species boost their chances of invasion success by socializing with natives? *Philosophical Transactions of the Royal Society B: Biological Sciences*, 378, 20220106.

Camacho-Cervantes, M., Ojanguren, A.F., Dominguez-Dominguez, O., and Magurran, A.E. (2018). Sociability between invasive guppies and native topminnows. *PLOS ONE*, 13, e0192539.

Chappell, C.R., Dhami, M.K., Bitter, M.C., et al. (2022). Wide-ranging consequences of priority effects governed by an overarching factor. *eLife*, 11, e79647.

Chapple, D.G., Simmonds, S.M., and Wong, B.B.M. (2012). Can behavioural and personality traits influence the success of unintentional species introductions? *Trends in Ecology & Evolution*, 27, 57–64.

Chapple, D.G., and Wong, B.B.M. (2016). The role of behavioural variation across different stages of the introduction process. In: D. Sol and J.S. Weis (eds), *Biological Invasions and Animal Behaviour*, pp. 7–25. Cambridge University Press, Cambridge.

Clinchy, M., Zanette, L.Y., Roberts, D., et al. (2016). Fear of the human 'super predator' far exceeds the fear of large carnivores in a model mesocarnivore. *Behavioral Ecology*, 27, 1826–1832.

Cole, N.C., and Harris, S. (2011). Environmentally-induced shifts in behaviour intensify indirect competition by an invasive gecko in Mauritius. *Biological Invasions*, 13, 2063–2075.

Cote, I.M., Darling, E.S., Malpica-Cruz, L., et al. (2014). What doesn't kill you makes you wary? Effect of repeated culling on the behaviour of an invasive predator. *PLOS ONE*, 9, e94248.

Cote, I.M., and Smith, N.S. (2018). The lionfish *Pterois* sp. invasion: has the worst-case scenario come to pass? *Journal of Fish Biology*, 92, 660–689.

Crossland, M.R., Salim, A.A., Capon, R.J., and Shine, R. (2019). The effects of conspecific alarm cues on larval cane toads (*Rhinella marina*). *Journal of Chemical Ecology*, 45, 838–848.

Damas-Moreira, I., Oliveria, D., Santos, J.L., et al. (2018). Learning from others: an invasive lizard uses social information from both conspecifics and heterospecifics. *Biology Letters*, 14, 20180532.

Darimont, C.T., Fox, C.H., Bryan, H.M., and Reimchen, T.E. (2015). The unique ecology of human predators. *Science*, 349, 858–860.

Diaz Pauli, B., and Sih, A. (2017). Behavioural responses to human-induced change: why fishing should not be ignored. *Evolutionary Applications*, 10, 231–240.

Diaz Pauli, B., Wiech, M., Heino, M., and Utne-Palm, A.C. (2015). Opposite selection on behavioural types by active and passive fishing gears in a simulated guppy *Poecilia reticulata* fishery. *Journal of Fish Biology*, 86, 1030–1045.

Diquelou, M.C., and Griffin, A.S. (2020). Behavioural responses of invasive and nuisance vertebrates to harvesting: a mechanistic framework. *Frontiers in Ecology and Evolution*, 8, 177.

Duckworth, R.A. (2008). Adaptive dispersal strategies and the dynamics of a range expansion. *The American Naturalist*, 172, S4–S17.

Dufour, C.M., Clark, D.L., Herrel, A., and Losos, J.B. (2020). Recent biological invasion shapes species recognition and aggressive behaviour in a native species: a behavioural experiment using robots in the field. *Journal of Animal Ecology*, 89, 1604–1614.

Freidenfelds, N.A., Robbins, T.R., and Langkilde, T. (2012). Evading invaders: the effectiveness of a behavioural response acquired through lifetime exposure. *Behavioral Ecology*, 23, 659–664.

Gippet, J.M., George, L., and Bertelsmeier, C. (2022). Local coexistence of native and invasive ant species is associated with micro-spatial shifts in foraging activity. *Biological Invasions*, 24, 761–773.

Glenn, S., and Holway, D. (2008). Consumption of introduced prey by native predators: Argentine ants and pit-building ant lions. *Biological Invasions*, 10, 273–280.

Graham, S.P., Freidenfelds, N.A., McCormick, G.L., and Langkilde, T. (2012). The impacts of invaders: basal and acute stress glucocorticoid profiles and immune function in native lizards threatened by invasive ants. *General and Comparative Endocrinology*, 176, 400–408.

Greenlees, M.J., Phillips, B.L., and Shine, R. (2010). Adjusting to a toxic invader: native Australian frogs learn not to prey on cane toads. *Behavioral Ecology*, 21, 966–971.

Gruber, J., Brown, G., Whiting, M.J., and Shine, R. (2017). Is the behavioural divergence between range-core and range-edge populations of cane toads (*Rhinella marina*) due to evolutionary change or developmental plasticity? *Royal Society Open Science*, 4, 170789.

Harrington, L.A., Harrington, A.L., Yamaguchi, N., et al. (2009). The impact of native competitors on an alien invasive: temporal niche shifts to avoid interspecific aggression? *Ecology*, 90, 1207–1216.

Hayes, N.M., Butkas, K.J., Olden, J.D., and Vander Zanden, J.M. (2009). Behavioural and growth differences between experienced and naïve populations of a native crayfish in the presence of invasive rusty crayfish. *Freshwater Biology*, 54, 1876–1887.

Heino, M., Diaz Pauli, B., and Dieckmann, U. (2015). Fisheries-induced evolution. *Annual Review of Ecology and Systematics*, 46, 461–480.

Herbold, B., and Moyle, P.B. (1986). Introduced species and vacant niches. *The American Naturalist*, 128, 751–760.

Hudina, S., Hock, K., and Žganec, K. (2014). The role of aggression in range expansion and biological invasions. *Current Zoology*, 60, 401–409.

Januchowski-Hartley, F.A., Nash, K.L., and Lawton, R.J. (2012). Influence of spear guns, dive gear and observers on estimating fish flight initiation distance on coral reefs. *Marine Ecology Progress Series*, 469, 113–119.

Johnson, T.B., Bunnell, D.B., and Knight, C.T. (2005). A potential new energy pathway in central Lake Erie: the round goby connection. *Journal of Great Lakes Research*, 31, 238–251.

Jolly, C.J., Shine, R., and Greenlees, M.J. (2016). The impacts of a toxic invasive prey species (the cane toad, *Rhinella marina*) on a vulnerable predator (the lace monitor, *Varanus varius*). *Biological Invasions*, 18, 1499–1509.

Kelly, E., and Phillips, B.L. (2017). Get smart: native mammal develops toad-smart behavior in response to a toxic invader. *Behavioral Ecology*, 28, 854–858.

Kinney, K., Pintor, L.M., Mell, A., and Byers, J.E. (2023). Density-dependent predation and predator preference for native prey may facilitate an invasive crab's escape from natural enemies. *Biological Invasions*, 25, 1–10.

Kuparinen, A., and Festa-Bianchet, M. (2017). Harvest-induced evolution: insights from aquatic and terrestrial systems. *Philosophical Transactions of the Royal Society B: Biological Sciences*, 372, 20160036.

Langkilde, T. (2009). Invasive fire ants alter behaviour and morphology of native lizards. *Ecology*, 90, 208–217.

Langkilde, T., and Freidenfelds, N. (2010). Consequences of envenomation: red imported fire ants have delayed effects on survival but not growth of native fence lizards. *Wildlife Research*, 37, 566.

Ledón-Rettig, C.C., Richards, C.L., and Martin, L.B. (2013). Epigenetics for behavioural ecologists. *Behavioral Ecology*, 24, 311–324.

McGhee, K.E., Pintor, L.M., and Bell, A.M. (2013). Reciprocal behavioral plasticity and behavioral types during predator-prey interactions. *The American Naturalist*, 182, 704–717.

Nunes, A.L., Orizaola, G., Laurila, A., and Rebelo, R. (2014). Rapid evolution of constitutive and inducible defenses against an invasive predator. *Ecology*, 95, 1520–1530.

Peters, J.A., and Lodge, D.M. (2013). Habitat, predation, and coexistence between invasive and native crayfishes: prioritizing lakes for invasion prevention. *Biological Invasions*, 15, 2489–2502.

Pettit, L., Somaweera, R., Kaiser, S., et al. (2021). The impact of invasive toads (Bufonidae) on monitor lizards (Varanidae): an overview and prospectus. *Quarterly Review of Biology*, 96, 105–125.

Phillips, B.L. (2016). Behaviour on invasion fronts, and the behaviour of invasion fronts. In: J.S. Weis and D. Sol (eds), *Biological Invasions and Animal Behaviour*, pp. 82–95. Cambridge University Press, Cambridge.

Phillips, B.L., Shine, R., and Tingley, R. (2016). The genetic backburn: using rapid evolution to halt invasions. *Proceedings of the Royal Society B: Biological Sciences*, 283, 1–9.

Phillips, B.L., and Suarez, A.V. (2012). The role of behavioural variation in the invasion of new areas. In: U. Candolin and B.B.M. Wong (eds), *Behavioural Responses to a Changing World: Mechanisms and Consequences*, pp. 190–196. Oxford University Press, Oxford.

Pizzatto, L., Both, C., Brown, G., and Shine, R. (2017). The accelerating invasion: dispersal rates of cane toads at an invasion front compared to an already-colonized location. *Evolutionary Ecology*, 31, 533–545.

Polo-Cavia, N., and Gomez-Mestre, I. (2014). Learned recognition of introduced predators determines survival of tadpole prey. *Functional Ecology*, 28, 432–439.

Pringle, R.M., Kartzinel, T.R., Palmer, T.M., et al. (2019). Predator-induced collapse of niche structure and species coexistence. *Nature*, 570, 58–64.

Pujol-Buxó, E., San Sebastián, O., Garriga, N., and Llorente, G.A. (2013). How does the invasive/native nature of species influence tadpoles' plastic responses to predators? *Oikos*, 122, 19–29.

Reale, D., Garant, D., Humphries, M.M., et al. (2010). Personality and the emergence of the pace-of-life syndrome concept at the population level. *Philosophical Transactions of the Royal Society B: Biological Sciences*, 365, 4051–4063.

Ruland, F., and Jeschke, J.M. (2020). How biological invasions affect animal behaviour: a global, cross-taxonomic analysis. *Journal of Animal Ecology*, 89, 2531–2541.

Sih, A., Bolnick, D.I., Luttbeg, B., et al. (2010). Predator-prey naïveté, antipredator behaviour, and the ecology of predator invasions. *Oikos*, 119, 610–621.

Simberloff, D., Martin, J.L., Genovesi, P., et al. (2013). Impacts of biological invasions: what's what and the way forward. *Trends in Ecology & Evolution*, 28, 58–66.

Simpson, G.G. (1953). The Baldwin effect. *Evolution*, 7, 110–117.

Smith, J., and Phillips, B. (2006). Toxic tucker: the potential impact of cane toads on Australian reptiles. *Pacific Conservation Biology*, 12, 40–49.

Smith, J.A., Suraci, J.P., Clinchy, M., et al. (2017). Fear of the human 'super predator' reduces feeding time in large carnivores. *Proceedings of the Royal Society B: Biological Sciences*, 284, 20170433.

Sol, D., Timmermans, S., and Lefebvre, L. (2002). Behavioural flexibility and invasion success in birds. *Animal Behaviour*, 63, 495–502.

Steinhart, G.B., Stein, R.A., and Marschall, E.A. (2004). High growth rate of young-of-the-year smallmouth bass in Lake Erie: a result of the round goby invasion? *Journal of Great Lakes Research*, 30, 381–389.

Stewart, P.S., Hill, R.A., Stephens, P.A., et al. (2021). Impacts of invasive plants on animal behaviour. *Ecology Letters*, 24, 891–907.

Strain, E.M.A., Johnson, C.R., and Thomson, R.J. (2013). Effects of a range-expanding sea urchin on behaviour of commercially fished abalone. *PLOS ONE*, 8, e73477.

Sutton, T.M., Zeiber, R.A., and Fisher, B.E. (2012). Behavioural interactions between blackstripe topminnow and other native Indiana topminnows. *Proceedings of the Indiana Academy of Science*, 121, 62–70.

Tablado, Z., Tella, J.L., Sanchez-Zapata, J.A., and Hiraldo, F. (2010). The paradox of the long-term positive effects of a North American crayfish on a European community of predators. *Conservation Biology*, 24, 1230–1238.

Thibault, I., and Dodson, J. (2013). Impacts of exotic rainbow trout on habitat use by native juvenile salmonid species at an early invasive stage. *Transactions of the American Fisheries Society*, 142, 1141–1150.

Tingley, R., Ward-Fear, G., Schwarzkopf, L., et al. (2017). New weapons in the toad toolkit: a review of methods to control and mitigate the biodiversity impacts of invasive cane toads (*Rhinella marina*). *Quarterly Review of Biology*, 92, 123–149.

Urban, M.C., Phillips, B.L., Skelly, D.K., and Shine, R. (2008). A toad more travelled: the heterogeneous invasion dynamics of cane toads in Australia. *The American Naturalist*, 171, E134–E148.

Webb, J.K., Shine, R., and Christian, K.A. (2005). Does intraspecific niche partitioning in a native predator influence its response to an invasion by a toxic prey species? *Austral Ecology*, 30, 201–209.

West-Eberhard, M.J. (2003). *Developmental Plasticity and Evolution*. Oxford University Press, Oxford.

Wiles, G.J., Bart, J., Beck, R.E., and Aguon, C.F. (2003). Impacts of the brown tree snake: patterns of decline and species persistence in Guam's avifauna. *Conservation Biology*, 17, 1350–1360.

Wright, T.F., Eberhard, J.R., Hobson, E.A., et al. (2010). Behavioural flexibility and species invasions: the adaptive flexibility hypothesis. *Ethology, Ecology and Evolution*, 22, 393–404.

Yang, D., Gonzalez-Bernal, E., Greenlees, M., and Shine, R. (2012). Interactions between native and invasive gecko lizards in tropical Australia. *Austral Ecology*, 37, 592–599.

Zuk, M., Bastiaans, E.J., Langkilde, T., and Swanger, E. (2014). The role of behaviour in the establishment of novel traits. *Animal Behaviour*, 92, 333–344.

CHAPTER 7

Wildlife harvesting

Beatriz Díaz Pauli and Marco Festa-Bianchet

Overview

Harvesting of wildlife usually decreases the abundance of populations and can be selective for certain morphological traits, such as body or weapon size, sex, or age class. Harvesting also affects behaviour, either because some behavioural traits make an individual more vulnerable to harvesting or because harvesting favours behaviours that increase the chance of avoiding capture. We review studies from the wild showing that behavioural traits related to spatial and temporal movement patterns, foraging, sexual selection, parental care, and personality make some individuals more prone to being harvested than others. Most harvest-induced behavioural changes appear to involve a plastic response, although few studies have addressed a possible genetic component. A few studies show that harvest-induced behavioural changes can involve microevolution, and some identified the genes involved. Independently of the mechanism underlying the behavioural change, there is evidence that such change has implications at the population and ecosystem levels and, hence, should be considered in management plans.

7.1 Introduction

Wildlife harvesting refers here to any kind of exploitation of wildlife through hunting and fishing. Harvesting takes many forms and has a range of motivations, from providing food or materials to maintaining cultural traditions, recreation, and pest control. Harvesting is ubiquitous in marine, freshwater, and terrestrial ecosystems. Wildlife and fisheries managers have long been concerned with the ecological sustainability of harvest: how many can we take and be able to continue exploiting the population? More recently, some scientists, managers, and harvesters have become concerned about possible effects on the behaviour of harvested species (Box 7.1). This chapter will examine how animal behaviour can be modified by harvest, what the adaptive and maladaptive consequences of those behavioural changes are, whether changes are plastic or partly driven by evolutionary changes, and what the consequences of behavioural changes are for populations, communities, and ecosystems. Harvesting affects a wide range of taxa, including all classes of vertebrates, as well as invertebrates, such as crustaceans, molluscs, and insects. We will concentrate on fish and terrestrial mammals, because those two groups have been the most studied (Figure 7.1). We will mention examples from other taxa when relevant information is available. In addition, we will mainly consider studies that have been carried out in nature.

The potential effects of harvesting on animal behaviour are varied. Behaviour can change because individuals adopt strategies that reduce their chance of being killed, because harvests change the age or size structure, sex ratio, or density of a population, or because behaviours that reduce the probability of harvest are correlated with changes in habitat use, circadian rhythms, appetite, foraging behaviour, predator avoidance, and other traits. Those changes can have cascading effects on community structure. Some changes may be maladaptive in the face of natural selective pressures and will be maintained only if avoiding

Beatriz Díaz Pauli and Marco Festa-Bianchet, *Wildlife harvesting*. In: *Behavioural Responses to a Changing World*. Edited by: Bob B. M. Wong and Ulrika Candolin, Oxford University Press. © Oxford University Press (2024). DOI: 10.1093/oso/9780192858979.003.0007

Box 7.1 Animal behaviour and wildlife harvesting: management implications

Currently, most harvest regulations and practices do not consider their potential impacts on animal behaviour, but many select for behavioural traits, or for morphological traits that can be linked to behaviour. Management plans typically seek to limit the demographic impact of harvests, often by protecting animals at specific phases of their life history, such as reproduction and juvenile phases, which are in turn linked to behaviours including migration, mating tactics, and parental care. When management plans do consider behaviour, they commonly seek to avoid bycatch or lower harvest efficiency, rather than reduce selectivity on behaviour. When behavioural consequences of harvest are undesirable, management plans should consider different aspects of behaviour, including space use and habitat selection, personality, and the landscape of fear, as in conservation (Chapter 11) and plans to reduce human-wildlife conflicts (Chapter 13).

Protected areas have been a mainstay of conservation for decades. Increasingly, management plans include the establishment of refuge areas to mitigate the demographic effects of harvesting. Although protected areas were not originally designed to protect certain behavioural phenotypes, they have allowed behavioural ecologists to study the behavioural consequences of harvesting. Some of these refuges provide effective protection, including the maintenance of certain behavioural traits that mitigate harvest-induced selection. Some protected areas, however, may impose selective pressures themselves, for instance favouring reduced movements, home ranges, or dispersal. Habituation to humans in these areas can lead to increased conflict. Therefore, the effectiveness of protected areas in mitigating harvest-induced changes and conserving ecosystems is highly context-specific, not only depending on the taxa and population, but also on the size, design, and enforcement of the protected area.

Inclusion of behavioural considerations in harvest management strategies should expand beyond considering single populations affected by a single stressor and examine how the rapidly changing nature of the Anthropocene, with multiple human-produced stressors, affects behaviour in multiple interacting species.

---

harvest was the most important determinant of fitness. Generally, the greater the harvest rate, the greater the potential that maladaptive behaviours could be maintained. That is an intuitive observation, but its quantification can help avoid unwanted behavioural consequences of harvest. We conclude the chapter by considering what management strategies could minimize behavioural changes in harvested populations, especially when behavioural changes are undesirable, from ecological, social, or economic viewpoints.

7.2 The selective nature of harvesting

Harvests can be selective because of regulations, cultural preferences, and because of the gear used. Some regulations specify the species, size, sex, age, and other characteristics of individuals that can and cannot be harvested. Size regulations can stipulate, for example, the minimum number of tines in deer antlers, size of bovid horns, and/or set minimum size or slot sizes of fish that can be killed. Other regulations can limit the harvest to mature males (e.g. crustaceans and some ungulates and game birds managed to limit demographic impacts on populations), or can direct much of the harvest to young of the year, as increasingly adopted for cervids in parts of Europe (Milner et al. 2011). In addition, regulations commonly limit the number or total weight of harvested individuals, and effort (number of hunters, fishers, or boats, and type and number of fishing gears or hunting weapons). Finally, harvesting regulations may control the time and areas where harvesting is allowed, which can further direct the harvest to, or away from, individuals with specific traits, such as, for example, prohibiting harvest during breeding seasons or in spawning areas. Individual preferences may affect harvest selectivity: some harvesters seek males with large horns, antlers, or tusks, or they may prefer larger individuals, avoid taking females followed by young of the year, or may release captured animals, particularly fish, based on size (Table 7.1).

Harvesting regulations and preferences do not directly target behaviour, but behaviour may be the key determinant of vulnerability to harvest. The timing of harvest can affect migratory species

Table 7.1 Examples of how regulations, harvest gear, and harvester preferences may potentially affect animal behaviour.

Harvest variable	Potential behavioural consequences
Timing of harvest season	Timing of migration drives harvest risk
Bait, lures	Hunger or prey preference
Type of weapon/gear	Escape distance, habitat selection, movement patterns
Decoys	Gregariousness, attraction to conspecifics
Traps	Exploratory behaviour
Daylight harvest only	Nocturnal behaviour
Use of dogs	Anti-predator response
Trophy hunting	Mating tactics of younger males
Long-range rifles	Group size if larger groups are more easily seen

by changing the risk of harvest for early or late migrants. Harvest can be restricted to daylight hours, be limited to certain areas, or rely on baits, lures, imitations of calls, decoys, or scents. Moreover, in the case of fish, behaviour can drive the design of ever more effective and selective gears, as well as regulations on the depth at which fishing gear can be deployed to avoid undesirable bycatch. Finally, harvest can be habitat-specific if it depends on visibility, access, or ability to deploy and recover certain types of gear. All these characteristics of harvest can put individual animals at either greater or lower risk of being killed depending on their behaviour (Table 7.1).

7.3 How behaviour affects the risk of harvest

In this section, we consider situations where harvesting is intense, so that avoiding getting shot, trapped, hooked, or netted overrides many natural selective pressures. In other cases, harvest may be light overall, with only a very small proportion of the population being targeted, but can be extremely intense for certain kinds of individuals. For example, trophy-hunted ungulates may experience overall harvest rates of less than 2%, but, for trophy-sized males in those same population, the annual risk of harvest can be well over 50%.

For most species, harvest can favour behaviours that decrease the probability of being detected and caught. Several behavioural tactics can lower the risk of detection. Animals may change movement patterns to reduce the risk of encountering harvesters or could use areas where harvester access is limited/prohibited and where visibility is impaired (Table 7.2). Animals could also avoid areas with greater risk of harvesting—such as roads, trails, or other areas where hunters are more likely to be encountered—or they could become more nocturnal (Table 7.3). All these behavioural changes may lower the risk of harvest but can also have consequences for individual fitness. They can also lead to broader ecological consequences by changing risk of predation, foraging behaviour and diet selection, and pressure on local resources. Detection risk varies depending on the taxon and the harvesting method. Hunters of mammals and birds often detect animals through sight or sound. The use of hunting dogs, which can detect prey species by scent, changes what behaviours affect the risk of hunting mortality. For fish, avoiding detection is only an option in shallow, clear water, since sonars and echosounders can easily detect fish. Avoidance of nets, bait, and other gear is then key to reducing the chances of being caught. Currently, knowledge is limited on how harvesting affects these behaviours, whether any changes in such behaviours are plastic and/or genetic, or whether they are driven by other mechanisms (see Section 7.6). Plastic behavioural changes likely occur faster and are easier to reverse, while evolutionary changes would occur more slowly and may be more difficult to reverse when harvesting halts. Next, we present evidence of how harvesting can affect a range of ecologically important behaviours and attempt to specify—in cases where the information is available—the mechanisms underpinning these behavioural changes (Figure 7.1).

7.3.1 Movement patterns

Animals can modify their movement patterns to reduce their vulnerability to capture in three key ways: (1) as an immediate response to hunters or to fishing gear, (2) via changes in daily movements and home ranges, and (3) through temporal and spatial changes in migration patterns.

Intuitively, high activity seems linked to higher encounter rate with harvesters, but this relationship is actually much more complex. Among terrestrial mammals, for example, evidence suggests that higher movement rate can either increase or

Figure 7.1 Case studies on the behavioural consequences of harvesting. (a) Two blacktail deer *Odocoileus hemionus* in an urban environment. In areas where hunting is not allowed, deer can become habituated to humans, sometimes leading to conflict. Escape distance from humans, habitat use, and movement rates can all be affected by hunting activities. (b) Atlantic cod *Gadus morhua*. Individuals in shallow water are more likely to be caught in various fishing gears. (c) *Labrus bergylta*. Active individuals are more vulnerable to fishing. (d) Wild boar *Sus scrofa*. Individuals increase or decrease movement depending on whether dogs are used for hunting or not. (e) Three adult bighorn sheep *Ovis canadensis* rams. Based on hunting regulations in Alberta, Canada, only the ram on the left could be legally harvested. Removal of most large-horned rams through trophy hunting may affect the behaviour of surviving rams, and the migration of adult rams from and to protected areas during the breeding season. (f) European lobster *Homarus gammarus*. Female mate preference is relaxed in fished populations. (g) Brown bear *Ursus arctos*. Active individuals are better at avoiding hunters and their dogs. (h) Largemouth bass *Micropterus salmoides*. Nesting males are vulnerable to angling.

Photos: (a, e) Marco Festa-Bianchet; (b) Lise D. Sivle; (c) David Villegas Ríos; (d) Håkan Lindgren; (f) Tonje K. Sørdalen; (g) Charles J. Sharp, CC BY-SA 4.0; (h) Cory Suski

decrease vulnerability to hunting. For instance, in brown bears *Ursus arctos*, Leclerc et al. (2019) found that individuals that had low movement rates during hunting hours were more likely to be shot. That result could be explained by the ability of more active bears to avoid hunters and especially hunting dogs. In contrast, Ciuti et al. (2012) found that higher movement rates by elk *Cervus canadensis* of both sexes actually increased the risk of harvest, likely by increasing the probability of encountering hunters. Although greater movement rates may increase hunting risk, in open habitats, ungulates may have no choice but to move away from hunters, even if by doing so they may encounter other hunters (Proffitt et al. 2009). Moreover, movement rate is increased or decreased in wild boar *Sus scrofa* whether the hunting dogs are used or not (Thurfjell et al. 2013). Therefore, habitat characteristics, hunter density, and hunting method affect behaviour of hunted mammals. Finally, disturbance by hunters can itself affect movement rates and patterns—and not only in mammals. For example, northern bobwhite quail *Colinus virginianus* reduced movement by 38% in the presence of rabbit hunters, and their tracks were straighter relative to when hunters were not present (Mohlman et al. 2019).

In fish, simply encountering fishing gears does not necessarily lead to being harvested. In the case of both passive (angling and traps) and active gears, an 'active choice' of avoiding the gear can reduce vulnerability. This active choice may depend on an individual's personality, willingness to forage, or could be learned from experience with gear (Diaz Pauli and Sih 2017, and references therein). Even for gillnets, where vulnerability does increase with activity, individual differences in activity may be driven by other behaviours, such as foraging and mating. For instance, in Ballan wrasse *Labrus bergylta*, foraging increases activity, while mating reduces it; thus vulnerability to gillnets peaks in spring-summer when foraging rate is highest (Villegas-Ríos et al. 2014). In several fish species, individuals that presented higher variation in speed and erratic movements were less vulnerable to the trawl (Kim and Wardle 2003). Larger variation in speed may increase trawl avoidance also in mesopelagic fishes (Underwood et al.

2020). Most evidence of active gear avoidance comes from studies testing the efficiency of the trawl with behavioural observations of fish already in the trawl. Thus, we know very little about behaviour earlier in the capture process.

Harvesting also affects daily movements. For instance, both passive and active fishing gears selectively captured cod *Gadus morhua* that occupied shallow waters and displayed extensive diel vertical migration, increasing the abundance of deepwater individuals (Jakobsdóttir et al. 2011; Olsen et al. 2012). Home range size appears to influence the risk of harvesting differently according to species and to presence of protected areas. Villegas-Ríos et al. (2021) studied several fish species along the Norwegian coast and concluded that the selectivity of fishing depended on whether the home range was inside or outside a marine protected area. Individuals with home range centroids inside the protected area experienced higher vulnerability to fishing with increasing home range size. The opposite occurred to individuals outside the protected area. However, for European lobster *Homarus gammarus*, individuals with smaller home ranges were more vulnerable to traps (Moland et al. 2019). The home range sizes of individual lobsters were highly consistent over time, suggesting a possible genetic component. Thus, home range size may evolve in response to fishing selection (Moland et al. 2019).

There is evidence that harvesting affects seasonal migrations. For example, bighorn sheep *Ovis canadensis* rams that undertake breeding migrations in late October looking for mating opportunities become vulnerable to trophy hunters, especially if they leave protected areas (Poisson et al. 2020). Red deer *Cervus elaphus* in Norway were more likely to migrate in autumn immediately after the start of the hunting season (Rivrud et al. 2016). Late migrant sockeye salmon *O. nerka* experienced higher vulnerability to fishing, and selection, in turn, led to the migration occurring earlier (Quinn et al. 2007). The opposite pattern was shown for Atlantic salmon *Salmo salar* in Ireland (Quinn et al. 2006). The Northeast Arctic stock of Atlantic cod currently has shorter migration routes and is no longer reaching its historical southernmost spawning areas along the Norwegian coast (Opdal and Jørgensen 2015).

These shorter migrations have been linked to fishing selection on size, as small individuals are not able to migrate long distances and large fish have been removed by fishing (Opdal and Jørgensen 2015). In certain species, timing of migration might be very plastic, depending on environmental changes, such as temperature. For some salmonids, however, a strong genetic component underpins the timing of migration for reproduction—thus raising the possibility for evolutionary changes to accrue over time (Tillotson and Quinn 2018).

7.3.2 Foraging behaviour

Foraging behaviour may be the main driver of the probability of being harvested, as the trade-off between acquiring food and predator/harvesting avoidance is the key determinant of space use. In mammals and birds, evidence that harvesting reduces foraging behaviour comes from studies that considered non-consumptive effects of harvesting. For instance, fear of hunters (i.e. perceived hunting risk) reduced foraging in brown bears and northern bobwhite (Hertel et al. 2016; McGrath et al. 2018). However, no study of terrestrial vertebrates has directly tested whether hunters selectively harvest individuals with increased foraging behaviour after accounting for movement rates. In fish, on the other hand, vulnerability to fishing changes with seasonal differences in foraging activity and feeding motivation (Fernö et al. 1986; Villegas-Ríos et al. 2014). Studies considering baited gears have concluded that vulnerable individuals tend to be in poor body condition (Heermann et al. 2013) and starving (Fernö et al. 1986; Keiling et al. 2020). Hessenauer et al. (2015) revealed that populations of largemouth bass *Micropterus salmoides* exposed to angling had lower resting metabolic rates compared to unfished populations, suggesting that angling selectively removed individuals with higher metabolic rates, which also would have a higher feeding rate. Nash et al. (2013) studied fishing vulnerability indirectly (using an index based on life history and ecological characteristics) and linked it to foraging movements in several herbivorous reef fishes with different functional roles (browsers, farmers, grazer/detritivores, and scraper/excavators). Their index suggested that foraging distance was correlated to vulnerability to fishing and thus farming species that moved short foraging distances were less vulnerable than browsers that moved large foraging distances.

The greater evidence that foraging behaviour affects vulnerability to harvest in fish—compared to mammals or birds—may simply be due to greater research effort in this area for fish. It may also, however, be related to the likely greater metabolic flexibility of fish compared to birds and mammals; for example, based largely on food availability, the mass of landlocked Atlantic salmon can vary 400-fold depending on prey availability (Hutchings et al. 2019). It therefore appears that, while fish can modify foraging behaviour to reduce the risk of harvesting, mammals and birds may be much more constrained by the risk of starvation.

7.3.3 Sexual selection, parental care, sex allocation

Size-selective harvests can affect reproductive success, mating behaviour, and the strength of sexual selection by inducing changes in size or age structure and sex ratio. Size- and/or behaviour-selective harvesting may also be sex-biased—but which sex is favoured is likely to be highly species- and context-specific. For instance, harvesting may selectively remove more males in hunted mammals that have horns, antlers, or tusks, in quails, some grouse, and fiddler crabs, as these are preferred by harvesters. Regulations result in more male than female European lobsters being harvested due to the ban on fishing egg-bearing females or minimum size limits. These minimum size limits also result in male black sea bream *Spondyliosoma cantharus* being more vulnerable to angling (Pinder et al. 2017) but in perch *Perca fluviatilis*, females may be selectively removed as their higher energy demands lead to higher foraging activities and, hence, higher motivation to strike baits compared to males (Heermann et al. 2013). Moreover, sex-biased fishing could particularly affect sex-changing fishes, where the 'later' sex is more vulnerable, due to larger size. Thus, sex-selective harvesting may result in skewed sex ratios. Even non-lethal sex-selective harvest can affect operational sex ratio. In fiddler crabs *Uca tangeri*, for example, the major (enlarged) claw of the males is removed before releasing the animal back into the wild to regenerate their claws. Evidence suggests, however, that other fiddler crabs behave towards these clawless males as if they were females, with implications for male-male competition and female mate choice (Oliveira et al. 2000).

Harvest-induced changes in sex ratio may reduce the skew in male reproductive success. For example, in bighorn sheep and fallow deer *Dama dama*, a greater proportion of males, including younger males, may have access to mating when older males are removed (McElligott and Hayden 2000; Martin et al. 2016). Mating behaviour and sexual selection assessed for European lobsters, both inside and outside a marine protected area, showed that females preferred males that were larger than themselves in both areas, but selection differentials in male size and claw size were only significant in the protected area (Sørdalen et al. 2018). Multiple mating by males was more common in the protected area, suggesting a higher skew in male reproductive success. Male size also differed between fished and non-fished lakes in bluegill *Lepomis macrochirus*, a fish with two reproductive strategies: parental males that look after the offspring and cuckolders (Drake et al. 1997). Differences in size in parental males were due to size-selectivity of the fishery, which favoured males that matured early at small body size. It would be interesting to test whether the relaxed female preference towards larger males reported for lobsters also occurred for bluegills. Fishing size-selectivity may be counteracted by female preference for larger males. However, if sexual selection is relaxed, fishing-induced evolution on size could accelerate (Sørdalen et al. 2018). The fished populations of bluegill had a higher proportion of cuckolders, showing that fishing not only affects sexual selection, but also the relative frequency of the two reproductive strategies (Drake et al. 1997). In sex-changing fishes, evidence shows that sex change occurs at smaller size or younger ages when harvesting is sex-biased (Provost 2013). The consequences of altered skew in reproductive success or timing of sex change for the population and the ecosystem would depend on the mating system, social interactions, and sex-dependent movement patterns and, hence, will be taxon- and case-specific (see Section 7.7).

Parental care and maternal allocation can be affected by selective harvests in different taxa. In several ungulates, regulations or cultural preferences protect lactating females. If hunting mortality is high, harvest selection may favour greater maternal allocation than natural selection, because females that do not reproduce are more likely to be shot (Nilsen and Solberg 2006; Rughetti and Festa-Bianchet 2014). In brown bears, regulations protecting family groups, combined with high harvest mortality, may favour females that nurse their young for two years, while a strategy of nursing the young for a single year appears to be favoured by natural selection (Van de Walle et al. 2018). Finally, largemouth bass differed in their parental behaviours depending on whether or not they were exposed to fishing. Specifically, nesting males not exposed to fishing performed better parental care, were more aggressive towards nest predators, and patrolled larger areas, compared to nesting males that were exposed to fishing. Nesting males inside the protected area, however, were more likely to strike at lures and be caught (Philipp et al. 2015; Twardek et al. 2017). Results from experimental ponds confirmed that fishing selectively targets males that are preferred by females, better at parental care, and have higher reproductive success, and that the behavioural shift might be evolutionary (Philipp et al. 2015), thus suggesting a behavioural trade-off between paternal care and the risk of being fished.

7.3.4 Personality

Much recent research on behaviours that affect vulnerability to fishing has focused on personality traits, such as boldness, exploration, activity, aggression, and sociability (Arlinghaus et al. 2016b; Diaz Pauli and Sih 2017; Gunn et al. 2022). Boldness and exploration are the behaviours most studied, while sociability only recently received some attention from evolutionary ecologists. In the context of harvesting, research interest on animal personality is justified by two important considerations. First, it seems reasonable to expect that 'bolder' individuals—which are more willing to take risks, explore larger areas, and are more active—may be at greater risk of being caught. Second, some 'personality' traits have a partly genetic basis.

Consequently, if fishing selectively removes individuals of a certain personality, there could be an evolutionary change in fished populations.

A recent meta-analysis by Gunn et al. (2022) suggests that fish and birds tend to increase in boldness and exploration under harvesting, urbanization, and noise. As pointed out by Arlinghaus et al. (2016a), consumptive and non-consumptive human impacts may result in different behavioural responses. Even only considering harvesting, however, which personality traits may be favoured is likely to be context-dependent, differing with species, past evolutionary history, and harvesting techniques or gear (Diaz Pauli and Sih 2017; Gunn et al. 2022).

The most studied harvesting method linked with personality traits is angling, which captures bold individuals more frequently, so that heavily fished populations become shyer. Angling also captures more aggressive individuals, particularly in species with paternal care (see examples in Arlinghaus et al. 2016b; Diaz Pauli and Sih 2017). Less is known about how personality may affect the risk of being caught in the context of active fishing gears, like trawls and seines. In contrast to angling, active fishing gears might favour bold and exploratory individuals that are more likely to escape them (Diaz Pauli and Sih 2017). Field studies suggest that more active individuals, or those with more variable movements, are also more likely to escape the trawl (Kim and Wardle 2003; Underwood et al. 2015; Underwood et al. 2020). Interestingly, the removal of a predator species by angling can increase boldness in prey species, through reduction in predation risk (Madin et al. 2010). The mere release from predation risk may therefore explain why direct human impact (sensu Gunn et al. 2022) resulted in higher boldness. In contrast to fish, however, less attention has been paid to the role of personality in affecting the risk of harvest in terrestrial mammals. Among the notable exceptions, Ciuti et al. (2012) found that individual differences in behaviour that affected the risk of harvest in elk were already present before the onset of the hunting season and attributed these differences to individual personality.

Several studies suggest that harvesting should affect animal sociability and their tendency to form groups. Larger groups are commonly targeted by harvesters, as they are easier to find. This is particularly true for fishing, as schooling fishes are disproportionately targeted by fisheries relative to non-aggregating species, because the former are easier to catch in large numbers and hence more profitable (Parrish 1999). Schooling behaviour is not

Table 7.2 Behaviours that affect the risk of harvest.

Species	Behaviour	Reference
Ballan wrasse	Activity level increases vulnerability to gillnets	Villegas-Ríos et al. 2014
Yellowtail flounders	Swimming direction: greater chance of being caught in trawls if swimming downwards	Underwood et al. 2015
Haddock, saithe, mackerel, cod, flatfish	Higher variation in speed and erratic movements reduce vulnerability to trawl	Kim and Wardle 2003
Cod	Individuals in shallow water more likely to be caught in various fishing gears	Jakobsdóttir et al. 2011; Olsen et al. 2012
Sockeye salmon	Late migration increases risk of harvest	Quinn et al. 2007
Atlantic salmon	Late migration decreases risk of harvest	Quinn et al. 2006
Multiple species	Hunger and poor body condition increase risk of capture	Fernö et al. 1986; Villegas-Ríos et al. 2014
European lobster	Home range size negatively correlated with trappability	Moland et al. 2019
Brown bear	Risk of being shot decreases with activity and movement rates	Leclerc et al. 2019
Elk	Risk of being shot increases with movement rates, use of open habitats	Ciuti et al. 2012
Red deer	Use of forest cover decreases risk of harvest	Lone et al. 2015
Bighorn sheep	Migrating individuals have higher risk of harvest	Poisson et al. 2020
Northern bobwhite	Reduced movement in the presence of hunters	Mohlman et al. 2019

only used to locate the fish, but schooling tendencies are exploited to create more effective fishing gears (Parrish 1999). In addition, angling selectively captures more social individuals in bluegill (Louison et al. 2019) and other fishes. Sociability, however, may also help some species avoid being caught. Rosen et al. (2012) showed that larger shoals of cod had faster diving rates and when in deep water they were less vulnerable to pelagic trawling. Moreover, saithe *Pollachius virens* seemed to have longer residence times in front of the trawl when forming groups, and this, in turn, might increase escape ability (Underwood et al. 2018). In minnow *Phoxinus phoxinus*, anaerobic capability and strong swimming performance determined vulnerability to trawling in laboratory conditions, but different social contexts (group coordination and cohesion) could modulate such links (Hollins et al. 2019). Therefore, the effect of harvesting on sociability appears highly context-dependent and sociability could be an extrinsic factor modulating vulnerability. In addition, groups are often heterogeneous, with bold individuals leading and shy ones following. Thus, disrupting such variability by selectively removing one behavioural type may, in turn, disrupt group cohesion and coordination (Palkovacs et al. 2018). A laboratory study with zebrafish *Danio rerio*, for example, showed that trawls selectively captured more social individuals and concluded that once an individual enters a trap or a trawl, others may follow (Crespel et al. 2021). Thus, it might be the behaviour of leaders that determines whether or not a shoal is captured (Crespel et al. 2021) or escapes (Rosen et al. 2012; Underwood et al. 2018).

7.4 Behavioural changes induced by harvesting

7.4.1 Avoiding risky places

One obvious way to reduce risk of being harvested is to avoid habitats with high visibility. Several studies of mammals hunted with rifles report greater use of forest in hunted populations, sometimes coinciding with the onset of the hunting season. For example, in Norway, red deer males that survived the hunting season increased their use of forest cover during the hunting season, while males

that did not shift their habitat—and were thus using open areas with equal frequency before and during the hunting season—were shot and killed (Lone et al. 2015). Once the hunting season starts, hunted ungulates may move into areas where hunters are not allowed, as reported for wild boars (Tolon et al. 2009) and female elk (Proffitt et al. 2013). Avoidance of roads is frequently reported during the hunting season, as hunters are often near vehicles (Proffitt et al. 2009). Harvesting could lead animals to move into protected areas. Proffitt et al. (2013) found that female elk made greater use of areas where hunting was not allowed during the hunting season, and elk selected non-hunted areas more than secure areas. Similarly, Tolon et al. (2009) found that wild boars, whose home ranges overlapped a small non-hunting reserve, increased their use of the reserve during the hunting season. This effect, however, was limited to boars with home ranges less than 2.1 km from the reserve boundary.

7.4.2 Avoiding risky times

A meta-analysis by Gaynor et al. (2018) found that human activity led to an increase in nocturnality for many mammals. Regulated sport hunting is usually only allowed during the day, so a decrease in diurnal activity, compensated for by an increase in nocturnality, should lower the risk of harvest. Greater nocturnal use of waterholes has been reported for African ungulates in hunted compared to non-hunted areas, despite the greater predation risk near waterholes at night (Crosmary et al. 2012). Wolves *Canis lupus* subject to aerial shooting shifted activity from day to night hours, which in turn changed their probability of encountering different prey species (Frey et al. 2022).

Fish may also modify their diel activities due to harvesting; however, whether they become more nocturnal or more diurnal depends on the species and the type of harvest. For instance, in the coral reefs of Line Island in the Pacific Ocean, areas exposed to fishing pressure had fewer large diurnal predators targeted by fishing, relative to unfished reefs (McCauley et al. 2012). Absence of large predators, in turn, increased the diurnal activity of fishes at lower trophic levels, particularly in species that are normally nocturnal (McCauley et al. 2012).

Table 7.3 Behaviours induced by harvesting.

Species	Behaviour	Reference
Elk	Movement rate increase, group size decrease	Proffitt et al. 2009
Wild boar	Movement rate increase when hunters use dogs, decrease when no dogs involved	Thurfjell et al. 2013
Wild boar	Move to protected area	Tolon et al. 2009
Elk	Avoid roads, move to non-hunted areas	Proffitt et al. 2013
African ungulates	Greater nighttime use of waterholes	Crosmary et al. 2012
Wolf	Greater nocturnal activity	Frey et al. 2022
Roe deer	Decrease use of open habitat	Gehr et al. 2017
Red deer; Atlantic cod	Migrate when hunting starts Shorter migration	Rivrud et al. 2016 Opdal and Jørgensen 2015

Most fishes are active at night and perform vertical and horizontal migrations to feed and reproduce. Recreational and commercial fishers exploit these behaviours, although most fishing effort occurs during the day. Whether or not nocturnal fishing alters the diel activity of fishes is largely unknown, but the modified movement patterns of Atlantic cod—with fewer diel vertical migrations and greater use of deep waters (Jakobsdóttir et al. 2011; Olsen et al. 2012)—might indicate that they maintain their diurnal, but reduce their nocturnal, activities.

7.5 Are behavioural changes maladaptive?

Harvest-induced changes in behaviour are expected to increase survival in harvested populations, but some selective pressures imposed by harvesting may be opposite to those favoured by natural and sexual selection. For instance, social behaviour may be selected differently by natural and harvest selection, as living in groups commonly reduces predation risk but groups may be more vulnerable to harvesting. In the hermaphroditic Hokkai shrimp *Pandalus latirostris*, the adaptive sex ratio is slightly male-biased, as is typical in protandrous hermaphrodites. Size- and female-selective fishing, however, skews it further towards males. The Hokkai shrimp can change sex adaptively based on population sex ratio, but this plastic response becomes maladaptive when females are selectively fished, because the sex change is determined by the sex ratio before the fishing season (Chiba et al.

2013). Therefore, harvest-induced behaviours may be maladaptive when harvests cease, either temporally because harvests are limited to certain seasons, or spatially in a mosaic of harvested and non-harvested areas.

In addition, behaviours favoured by harvesting may be disadvantageous when facing natural selective pressures or may be correlated to other behaviours that are disadvantageous. In these cases, the interplay between increased fitness due to harvest avoidance and decreased fitness through the correlated behaviours will likely determine whether the multi-trait response is maladaptive or not. Avoidance of hunters may decrease the ability to escape natural predators if hunting modes are different. For example, roe deer decrease their use of open habitat during the hunting season, facing a greater risk of predation by Eurasian lynx *Lynx lynx* in closed habitats (Gehr et al. 2017). There is little evidence of a trade-off between harvesting and predation risk for wild fish, but Olsen et al. (2012) found that natural selection on cod behaviour was relatively weaker than harvesting selection. Hence, the harvest-induced change in depth use that we discussed in the context of daily movements (Section 7.3.1) might be adaptive. Laboratory experiments concluded that size-selective fishing entailed reduced parental care or lower willingness to forage (see Diaz Pauli and Sih 2017 for examples). It is not clear, however, how these results may translate to wild environments. More research should therefore focus on multi-trait responses to harvesting and compare those responses to natural selective pressures.

Finally, habituation can be adaptive when it allows exploitation of habitats despite human presence, but maladaptive when it decreases escape responses from harvesting. 'Adaptive' habituation is particularly challenging when stimuli are inconsistent, such as in patchworks of protected and harvested areas, when harvesting activities are seasonal, or when harvesters and other human users of an area overlap. For example, Courbin et al. (2022) found that chamois *Rupicapra rupicapra* that habituated to people in areas of high hiker use appeared less effective at avoiding hunters. Similarly, dolphins and primates habituated to tourism are more vulnerable to fishing and poaching (Ménard et al. 2014). A literature review by Stankowich (2008) confirmed that ungulate flight distances are greater in hunted than in non-hunted populations, but the effect size was small (r = 0.17) and responses varied substantially among populations, possibly because of the effects of habitat, species, harvesting technique, and relative sizes of protected and harvested areas.

7.6 Are harvest-induced changes in behaviour plastic or genetic?

Behaviour often has a genetic component but estimates of heritability of behaviour in the wild are extremely rare (see also Chapter 14 on plasticity and adaptation). For instance, behaviours targeted by harvesting, such as timing of migration and personality, correlate with molecular markers in fish and mammals (Bubac et al. 2020). Brown bears in heavily hunted populations may have evolved to coexist with humans, partly by becoming more nocturnal and less aggressive, and by having larger litter sizes after accounting for maternal mass (Zedrosser et al. 2011). However, the relative contributions of plasticity and evolutionary change to these apparent adaptations are unknown.

Most harvest-induced changes appear to be plastic and likely involve learning. Vulnerability to fishing is initially reduced through learned gear avoidance. For mammals, there is some evidence of selection on behaviour. For example, deer that reduce movement rates during the hunting season are less likely to be shot (Ciuti et al. 2012; Lone et al. 2015). Age-specific behaviour of female elk suggests those that survived many hunting seasons moved less

than those that were subsequently shot, with the former also making more use of rugged terrain (Thurfjell et al. 2017); as a result, females aged nine years and older were almost never shot. A study of roe deer also suggested that older deer increased their use of safer habitat to avoid hunters (Padié et al. 2015). For both roe deer and elk, however, the relative contribution of learning and selection remain unclear: older animals may have learned to better avoid hunters, or animals that were better at avoiding hunters may have been more likely to survive to older ages. Similarly in fish, some studies show a genetic basis for vulnerability to fishing in semi-natural pond experiments (Philipp et al. 2015).

The only molecular evidence for genetic changes in behaviour due to harvesting is that of Icelandic cod, where fishing selectively removed individuals of the shallow-water and coastal ecotypes, leading to lower frequency of AA alleles of the pantophysin gene *Pan I* (Jakobsdóttir et al. 2011). A few genetic studies of wild populations focused on size-selectivity rather than behaviour and reached different conclusions. For instance, a whole-genome study of two other Atlantic cod populations showed no evidence of selection attributable to contemporary harvesting (Pinsky et al. 2021). Genetic changes might become evident if studies focused on extensive local sampling harvesting (Pinsky et al. 2021), as observed in Icelandic cod and Atlantic salmon. Changes in allele frequencies of the *vgll3* locus, explaining 40% of variation in age at maturity in Atlantic salmon, have been linked to fishing pressure in the Teno river in northern Europe, where different passive fishing gears had opposite effects on genotypes (Czorlich et al. 2022).

It is important to realize that genetic changes due to harvest selectivity in the wild and in laboratory conditions might not be comparable. The laboratory study by Crespel et al. (2021) suggested that trawling selectively targeted zebrafish with slow growth, low aerobic metabolic scope, high aggression, and high sociability. Trawling also showed selectivity at the genomic level, mainly on brain function and neurogenesis. The genes selected, however, differed between density conditions. Therefore, an interaction between harvest-induced phenotype selection and reduced density may lead to the observed phenotypic and genetic changes (Crespel et al. 2021). Pinsky et al. (2021) argued that the lack of selection in Atlantic cod suggested by whole-genome analyses was due to density-dependent plasticity and/or polygenic selection resulting in subtle allele changes. Both processes also played an important role in zebrafish laboratory trawling (Crespel et al. 2021).

7.7 Do behavioural responses to harvest affect population dynamics, community structure, and ecosystem function?

It was originally thought that any ecological or ecosystem consequences of harvesting would be due to the removal of large numbers of individuals. However, a meta-analysis showed that changes in trait variation within a population might cause equivalent, or even larger, ecological effects than the mere removal of individuals (Des Roches et al. 2018). Another meta-analysis revealed that body size, mouth gape size, and diet are key traits linking fish responses to fishing and their effect on ecosystem processes (Hadj-Hammou et al. 2021).

At the population level, harvest selection may lead to a substantial redistribution of male mating success in species where siring success is largely driven by the same phenotypic traits that make large males desirable as trophies, or more vulnerable to fishing, such as aggressive behaviour or size. For example, in bighorn sheep or fallow deer, a few males, often with large horns or antlers, have very high reproductive success (McElligott and Hayden 2000; Martin et al. 2016). If one of those males is shot, up to 20–25% of paternities in that population would be redistributed among other males in the following rut. Similar results are expected for largemouth bass—where fewer than 25% of males build nests (Reed 2022) and these males are preferentially removed by angling (Philipp et al. 2015)—and probably also for European lobster. In other species with a lower skew in male reproductive success, such as eastern grey kangaroos *Macropus giganteus* (Montana et al. 2022) or white-tailed deer *Odocoileus virginianus* (Newbolt et al. 2016), selective removal of a large, dominant male may only lead to a redistribution of 5–10% of paternities.

Therefore, mating system and male reproductive skew in unexploited populations are important considerations when estimating the possible genetic consequences of selective hunting. If harvest had strong effects on male age structure and on the variance in competitive ability—for example, if most mature males are harvested—the mating system could also be affected. Sexual selection could also be weakened if the skew in male reproductive success decreased following age truncation, but we know little about this possibility because individual siring success has been measured in very few wild species, and mostly in unexploited populations. In brown bears, high hunting intensity did appear to reduce the skew in male siring success (Frank et al. 2020). In sex-changing fishes, dominant individuals in leks and harems are more aggressive and may be more vulnerable than individuals in pairs, while individuals that aggregate are expected to be more vulnerable than those that do not (Provost 2013).

At the ecosystem level, harvesting may affect top-down and bottom-up regulations. There are several instances where harvest reduced the numbers of top predators (e.g. fish in Indo-Pacific reefs), leading to increases in prey (starfish), which, in turn, decreased the primary producer trophic level (coral reefs; Kroon et al. 2021). However, top-down regulation may be altered, not only by reducing the number of predators, but also by disrupting trait variation within higher trophic levels. For example, Shackell et al. (2010) showed that fishing of predatory groundfish in western Scotian Shelf doubled their prey densities, which, in turn, led to lower herbivorous zooplankton and more abundant phytoplankton. These trophic cascades were due to size-selective fishing: the smaller predators probably had impeded burst swimming speed, which reduced their predatory efficiency. In mammals, behavioural changes induced by hunting could lower grazing pressure in habitats where risk of harvest is higher and increase pressure in safer habitats. Similarly, earlier timing of migration as a response to the opening of the hunting season could reduce foraging pressure in some seasonal habitats. Neither of these possibilities, however, appears to have been investigated specifically in the context of behavioural reactions to harvest. Finally, harvesting can affect bottom-up regulation too. For instance, in a mangrove estuary, energetic models estimated that behaviour-selective harvest would reduce the supply of nutrients to the ecosystem to a much larger degree than size-selective or non-selective harvest (Allgeier et al. 2020). Large herbivores and birds are also key players in nutrient recycling and supply, as well as seed dispersion, in their ecosystems (Ripple et al. 2015). The interplay between individual size, diet, and activity is an important determinant of nutrient supply. For instance, small individuals excrete nitrogen faster due to higher metabolism, but excrete less phosphorous due to a low phosphorous diet, relative to big individuals. More active individuals would excrete higher nitrogen, not only affecting the supply locally, but also the translocation of nutrients (Allgeier et al. 2020).

To our knowledge, no study has looked directly at how harvesting-induced behavioural changes affect ecosystems. However, all behaviours mentioned above that are affected by harvesting, movement patterns, foraging, and personality, are known to affect prey consumption rate, nutrient storage, and nutrient excretion (Palkovacs et al. 2018). Therefore, harvesting selection on behaviour could affect ecosystems—both through top-down and bottom-up effects. Selective harvest of individuals

that are more willingly foragers or more active may remove individuals that impose a greater top-down and bottom-up control on the ecosystems. Moreover, size-selective mortality, together with reductions in number of individuals, may reduce foraging abilities and diet breadth, as observed in mesocosm experiments (Evangelista et al. 2021) and in the wild (see examples in Palkovacs et al. 2018). These ecosystem changes can, in turn, modify the selective pressures on harvested species, closing an eco-evolutionary feedback loop (Figure 7.2). Ecosystem effects due to changes in behaviour are likely to be much more important for fish than for terrestrial homeotherms. Fish (and possibly reptiles) can survive very large differences in feeding rate and simply change growth or reproductive rates, while mammals and birds will starve and die if they reduce feeding to the same extent. Moreover, in mammals, the hunting season may coincide with important phenological periods—such as the hyperphagia period in bears, rutting season in ungulates, and migratory season in birds—when reducing feeding rate might be particularly detrimental (Ordiz et al. 2012).

Some behavioural changes imposed by harvesting could include indirect effects, such as on fear of predators. By reducing predator size or numbers,

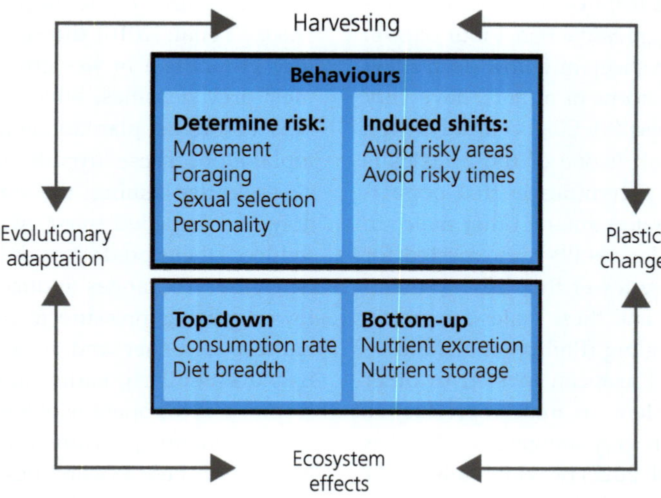

Figure 7.2 How behaviour-selective harvesting can affect populations and ecosystems.

fishing changed anti-predator behaviour in non-target species, which then performed longer foraging trips (Madin et al. 2010). Cromsigt et al. (2013) suggested that hunting could be a management tool, creating a landscape of fear where animals may reduce their use of areas or resources that generate conflicts with people. For example, the 'conservation' spring hunt of snow geese *Anser caerulescens* in Canada and the USA seeks to reduce habitat degradation in the Arctic nesting grounds. Hunters in feeding areas kill a few staging geese but disturb thousands. The spring hunt reduces breeding success because of poorer body conditions caused by disturbance from hunters and lower feeding efficiency of geese (Juillet et al. 2012). In summary, even though harvest-induced changes might be adaptive and result in increased survival population, such changes may entail larger-scale consequences at the ecosystem level and be linked to loss of ecosystem resilience towards global warming and other anthropogenic stressors (Chapters 1–10).

7.8 Conclusions and future directions

The impacts of harvest on populations and ecosystems increase with its intensity and selectivity. Those impacts can thus be avoided by reducing harvest rate and, in some cases, selectivity. Appropriate management, however, will require more information about which behavioural traits are under selection. So far, selectivity on behavioural traits has received little attention. For example, gillnets and pots are thought to be less size-selective than trawls or other fishing gear as they target medium-size fish. However, as we have discussed earlier, they may also select for certain types of behaviour. One strategy to mitigate fishing selectivity would be to employ a combination of fishing gears, for instance passive and active gears, that might target opposite behavioural traits and, hence, reduce selectivity. Reduced selectivity, however, may conflict with management to reduce bycatch. Better knowledge on gear avoidance behaviour among species would improve the development of more selective gears to avoid bycatch.

In addition to reducing harvest intensity and selectivity, in some cases, a regime of protected areas, the location of which can be shifted or rotated over time, could mitigate some of the effects of selectivity outside protected areas (Twardek et al. 2017; Palkovacs et al. 2018). However, small protected areas could create selectivity for certain behaviours, for instance individuals with large home ranges would be more vulnerable to fishing (Villegas-Ríos et al. 2021). Thus, whether protected areas are permanent or temporally variable will depend on management objectives. Harvesting season should be regulated taking into account possible behavioural impacts and should avoid phenologically important periods for the target species (Ordiz et al. 2012).

There are increasing calls for management plans to consider the conservation of individual trait variation as a major component of diversity (Arlinghaus et al. 2016b; Diaz Pauli and Sih 2017; Des Roches et al. 2018; Palkovacs et al. 2018). One way to accomplish this goal may be to monitor variance in certain traits that are targets of harvest as early warning signals of changes that may have important population or ecosystem consequences (Clements and Ozgul 2016). For terrestrial animals, one important goal is to establish which hunting-induced behavioural changes may benefit conservation and which are detrimental. For example, greater fear of humans may reduce human-wildlife conflicts in some situations, but avoidance of certain habitats or areas may reduce opportunities for population growth (see also Chapter 13). However, to fully understand the consequences of harvesting, research should not only focus on the targeted trait. Moving forward, we argue that it is vital to assess multi-trait responses to harvesting on the target species—these responses could reveal unexpected selective pressures—followed by determining how such responses might propagate up and down the food web and, thus, the entire ecosystem.

Acknowledgements

M.F.B. acknowledges the long-term support of the Natural Sciences and Engineering Research Council of Canada for his research programme and B.D.P. the Research Council of Norway (project number 275125). We thank an anonymous reviewer, Shaun Killen, Bob Wong, and Ulrika Candolin for helpful and constructive comments on earlier drafts of this chapter. We also thank Lise D. Sivle, Tonje

K. Sørdalen, David Villegas Ríos, Håkan Lindgren, Henrik Thurfjell, and Cory Suski for contributing photos.

References

Allgeier, J.E., Cline, T.J., Walsworth, T.E., et al. (2020). Individual behavior drives ecosystem function and the impacts of harvest. *Science Advances*, 6, eaax8329.

Arlinghaus, R., Alós, J., Klefoth, T., et al. (2016a). Consumptive tourism causes timidity, rather than boldness, syndromes: a response to Geffroy et al. *Trends in Ecology & Evolution*, 31, 92–94.

Arlinghaus, R., Laskowski, K.L., Alós, J., et al. (2016b). Passive gear-induced timidity syndrome in wild fish populations and its potential ecological and managerial implications. *Fish and Fisheries*, 18, 360–373.

Bubac, C.M., Miller, J.M., and Coltman, D.W. (2020). The genetic basis of animal behavioural diversity in natural populations. *Molecular Ecology*, 29, 1957–1971.

Chiba, S., Yoshino, K., Kanaiwa, M., et al. (2013). Maladaptive sex ratio adjustment by a sex-changing shrimp in selective-fishing environments. *Journal of Animal Ecology*, 82, 632–641.

Ciuti, S., Muhly, T.B., Paton, D.G., et al. (2012). Human selection of elk behavioural traits in a landscape of fear. *Proceedings of the Royal Society B: Biological Sciences*, 279, 4407–4416.

Clements, C.F., and Ozgul, A. (2016). Including trait-based early warning signals helps predict population collapse. *Nature Communications*, 7, 10984.

Courbin, N., Garel, M., Marchand, P., et al. (2022). Interacting lethal and nonlethal human activities shape complex risk tolerance behaviors in a mountain herbivore. *Ecological Applications*, 32, e2640.

Crespel, A., Schneider, K., Miller, T., et al. (2021). Genomic basis of fishing-associated selection varies with population density. *Proceedings of the National Academy of Sciences*, 118, e2020833118.

Cromsigt, J.P.G.M., Kuijper, D.P.J., Adam, M., et al. (2013). Hunting for fear: innovating management of human-wildlife conflicts. *Journal of Applied Ecology*, 50, 544–549.

Crosmary, W.-G., Valeix, M., Fritz, H., et al. (2012). African ungulates and their drinking problems: hunting and predation risks constrain access to water. *Animal Behaviour*, 83, 145–153.

Czorlich, Y., Aykanat, T., Erkinaro, J., et al. (2022). Rapid evolution in salmon life history induced by direct and indirect effects of fishing. *Science*, 376, 420–423.

Des Roches, S., Post, D.M., Turley, N.E., et al. (2018). The ecological importance of intraspecific variation. *Nature Ecology & Evolution*, 2, 57–64.

Diaz Pauli, B., and Sih, A. (2017). Behavioural responses to human-induced change: why fishing should not be ignored. *Evolutionary Applications*, 10, 231–240.

Drake, M.T., Claussen, J.E., Philipp, D.P., and Pereira, D.L. (1997). A comparison of bluegill reproductive strategies and growth among lakes with different fishing intensities. *North American Journal of Fisheries Management*, 17, 496–507.

Evangelista, C., Dupeu, J., Sandkjenn, J., et al. (2021). Ecological ramifications of adaptation to size-selective mortality. *Royal Society Open Science*, 8, 210842.

Fernö, A., Solemdal, P., and Tilseth, S. (1986). Field studies on the behaviour of whiting (*Gadus imerlangus* L.) towards baited hooks. *Fiskeridirektoratets Skrifter Serie Havundersøkelser*, 18, 83–95.

Frank, S.C., Pelletier, F., Kopatz, A., et al. (2020). Harvest is associated with the disruption of social and fine-scale genetic structure among matrilines of a solitary large carnivore. *Evolutionary Applications*, 14, 1023–1035.

Frey, S., Tejero, D., Baillie-David, K., et al. (2022). Predator control alters wolf interactions with prey and competitor species over the diel cycle. *Oikos*, 2022, e08821.

Gaynor, K.M., Hojnowski, C.E., Carter, N.H., and Brashares, J.S. (2018). The influence of human disturbance on wildlife nocturnality. *Science*, 360, 1232–1235.

Gehr, B., Hofer, E.J., Pewsner, M., et al. (2017). Hunting-mediated predator facilitation and superadditive mortality in a European ungulate. *Ecology and Evolution*, 8, 109–119.

Gunn, R.L., Hartley, I.R., Algar, A.C., et al. (2022). Understanding behavioural responses to human-induced rapid environmental change: a meta-analysis. *Oikos*, 2022, e08366.

Hadj-Hammou, J., Mouillot, D., and Graham, N.A.J. (2021). Response and effect traits of coral reef fish. *Frontiers in Marine Science*, 8, 249.

Heermann, L., Emmrich, M., Heynen, M., et al. (2013). Explaining recreational angling catch rates of Eurasian perch, *Perca fluviatilis*: the role of natural and fishing-related environmental factors. *Fisheries Management and Ecology*, 20, 187–200.

Hertel, A.G., Zedrosser, A., Mysterud, A., et al. (2016). Temporal effects of hunting on foraging behavior of an apex predator: do bears forego foraging when risk is high? *Oecologia*, 182, 1019–1029.

Hessenauer, J.-M., Vokoun, J.C., Suski, C.D., et al. (2015). Differences in the metabolic rates of exploited and unexploited fish populations: a signature of recreational fisheries induced evolution? *PLOS ONE*, 10, e0128336.

Hollins, J.P.W., Thambithurai, D., Van Leeuwen, T.E., et al. (2019). Shoal familiarity modulates effects of individual metabolism on vulnerability to capture by trawling. *Conservation Physiology*, 7.

Hutchings, J.A., Ardren, W.R., Barlaup, B.T., et al. (2019). Life-history variability and conservation status of landlocked Atlantic salmon: an overview. *Canadian Journal of Fisheries and Aquatic Sciences*, 76, 1697–1708.

Jakobsdóttir, K.B., Pardoe, H., Magnusson, Á., et al. (2011). Historical changes in genotypic frequencies at the pantophysin locus in Atlantic cod (*Gadus morhua*) in Icelandic waters: evidence of fisheries-induced selection? *Evolutionary Applications*, 4, 562–573.

Juillet, C., Choquet, R., Gauthier, G., et al. (2012). Carry-over effects of spring hunt and climate on recruitment to the natal colony in a migratory species. *Journal of Applied Ecology*, 49, 1237–1246.

Keiling, T.D., Louison, M.J., and Suski, C.D. (2020). Big, hungry fish get the lure: size and food availability determine capture over boldness and exploratory behaviors. *Fisheries Research*, 227, 105554.

Kim, Y.-H., and Wardle, C.S. (2003). Optomotor response and erratic response: quantitative analysis of fish reaction to towed fishing gears. *Fisheries Research*, 60, 455–470.

Kroon, F.J., Barneche, D.R., and Emslie, M.J. (2021). Fish predators control outbreaks of crown-of-thorns starfish. *Nature Communications*, 12, 6986.

Leclerc, M., Zedrosser, A., Swenson, J.E., and Pelletier, F. (2019). Hunters select for behavioral traits in a large carnivore. *Scientific Reports*, 9, 12,371.

Lone, K., Loe, L.E., Meisingset, E.L., et al. (2015). An adaptive behavioural response to hunting: surviving male red deer shift habitat at the onset of the hunting season. *Animal Behaviour*, 102, 127–138.

Louison, M.J., Stein, J.A., and Suski, C.D. (2019). The role of social network behavior, swimming performance, and fish size in the determination of angling vulnerability in bluegill. *Behavioral Ecology and Sociobiology*, 73, 139.

Madin, E.M.P., Gaines, S.D., and Warner, R.R. (2010). Field evidence for pervasive indirect effects of fishing on prey foraging behavior. *Ecology*, 91, 3563–3571.

Martin, A.M., Festa-Bianchet, M., Coltman, D.W., and Pelletier, F. (2016). Demographic drivers of age-dependent sexual selection. *Journal of Evolutionary Biology*, 29, 1437–1446.

McCauley, D.J., Hoffmann, E., Young, H.S., and Micheli, F. (2012). Night shift: expansion of temporal niche use following reductions in predator density. *PLOS ONE*, 7, e38871.

McElligott, A.G., and Hayden, T.J. (2000). Lifetime mating success, sexual selection and life history of fallow bucks (*Dama dama*). *Behavioral Ecology and Sociobiology*, 48, 203–210.

McGrath, D.J., Terhune II, T.M., and Martin, J.A. (2018). Northern bobwhite foraging response to hunting. *The Journal of Wildlife Management*, 82, 966–976.

Ménard, N., Foulquier, A., Vallet, D., et al. (2014). How tourism and pastoralism influence population demographic changes in a threatened large mammal species. *Animal Conservation*, 17, 115–124.

Milner, J.M., Bonenfant, C., and Mysterud, A. (2011). Hunting Bambi—evaluating the basis for selective harvesting of juveniles. *European Journal of Wildlife Research*, 57, 565–574.

Mohlman, J.L., Gardner, R.R., Parnell, I.B., et al. (2019). Nonconsumptive effects of hunting on a nontarget game bird. *Ecology and Evolution*, 9, 9324–9333.

Moland, E., Carlson, S.M., Villegas-Ríos, D., et al. (2019). Harvest selection on multiple traits in the wild revealed by aquatic animal telemetry. *Ecology and Evolution*, 9, 6480–6491.

Montana, L., King, W.J., Coulson, G., et al. (2022). Large eastern grey kangaroo males are dominant but do not monopolize matings. *Behavioral Ecology and Sociobiology*, 76.

Nash, K.L., Graham, N.A.J., and Bellwood, D.R. (2013). Fish foraging patterns, vulnerability to fishing, and implications for the management of ecosystem function across scales. *Ecological Applications*, 23, 1632–1644.

Newbolt, C.H., Acker, P.K., Neuman, T.J., et al. (2016). Factors influencing reproductive success in male white-tailed deer. *The Journal of Wildlife Management*, 81, 206–217.

Nilsen, E.B., and Solberg, E.J. (2006). Patterns of hunting mortality in Norwegian moose (*Alces alces*) populations. *European Journal of Wildlife Research*, 52, 153–163.

Oliveira, R.F., Machado, J.L., Jordão, J.M., et al. (2000). Human exploitation of male fiddler crab claws: behavioural consequences and implications for conservation. *Animal Conservation*, 3, 1–5.

Olsen, E.M., Heupel, M.R., Simpfendorfer, C.A., and Moland, E. (2012). Harvest selection on Atlantic cod behavioral traits: implications for spatial management. *Ecology and Evolution*, 2, 1549–1562.

Opdal, A.F., and Jørgensen, C. (2015). Long-term change in a behavioural trait: truncated spawning distribution and demography in Northeast Arctic cod. *Global Change Biology*, 21, 1521–1530.

Ordiz, A., Støen, O.-G., Sæbø, S., et al. (2012). Do bears know they are being hunted? *Biological Conservation*, 152, 21–28.

Padié, S., Morellet, N., Hewison, A.J.M., et al. (2015). Roe deer at risk: teasing apart habitat selection and landscape constraints in risk exposure at multiple scales. *Oikos*, 124, 1536–1546.

Palkovacs, E.P., Moritsch, M.M., Contolini, G.M., and Pelletier, F. (2018). Ecology of harvest-driven trait changes and implications for ecosystem management. *Frontiers in Ecology and the Environment*, 16, 20–28.

Parrish, J.K. (1999). Using behavior and ecology to exploit schooling fishes. *Environmental Biology of Fishes*, 55, 157–181.

Philipp, D.P., Claussen, J.E., Koppelman, J.B., et al. (2015). Fisheries-induced evolution in largemouth bass: linking vulnerability to angling, parental care, and fitness. *American Fisheries Society Symposium*, 82, 223–234.

Pinder, A.C., Velterop, R., Cooke, S.J., and Britton, J.R. (2017). Consequences of catch-and-release angling for black bream *Spondyliosoma cantharus*, during the parental care period: implications for management. *ICES Journal of Marine Science*, 74, 254–262.

Pinsky, M.L., Eikeset, A.M., Helmerson, C., et al. (2021). Genomic stability through time despite decades of exploitation in cod on both sides of the Atlantic. *Proceedings of the National Academy of Sciences*, 118, e2025453118.

Poisson, Y., Festa-Bianchet, M., and Pelletier, F. (2020). Testing the importance of harvest refuges for phenotypic rescue of trophy-hunted populations. *Journal of Applied Ecology*, 57, 526–535.

Proffitt, K.M., Grigg, J.L., Hamlin, K.L., and Garrott, R.A. (2009). Contrasting effects of wolves and human hunters on elk behavioral responses to predation risk. *Journal of Wildlife Management*, 73, 345–356.

Proffitt, K.M., Gude, J.A., Hamlin, K.L., and Messer, M.A. (2013). Effects of hunter access and habitat security on elk habitat selection in landscapes with a public and private land matrix. *The Journal of Wildlife Management*, 77, 514–524.

Provost, M.M. (2013). *Understanding sex change in exploited fish populations: a review of east coast fish stocks and assessment of selectivity and sex change in black sea bass (Centropristis striata) in New Jersey*. Master thesis, New Brunswick Rutgers, The State University of New Jersey.

Quinn, T.P., Hodgson, S., Flynn, L., et al. (2007). Directional selection by fisheries and the timing of sockeye salmon (*Oncorhynchus nerka*) migrations. *Ecological Applications*, 17, 731–739.

Quinn, T.P., McGinnity, P., and Cross, T.F. (2006). Long-term declines in body size and shifts in run timing of Atlantic salmon in Ireland. *Journal of Fish Biology*, 68, 1713–1730.

Reed, J. (2022). Potential influence of male nesting levels on largemouth bass *Micropterus salmoides* populations. *Minnesota Department of Natural Resources Investigational Report*, 577, 10.

Ripple, W.J., Newsome, T.M., Wolf, C., et al. (2015). Collapse of the world's largest herbivores. *Science Advances*, 1, e1400103.

Rivrud, I.M., Bischof, R., Meisingset, E., et al. (2016). Leave before it's too late: anthropogenic and environmental triggers of autumn migration in a hunted ungulate population. *Ecology*, 97, 1058–1068.

Rosen, S., Engås, A., Fernö, A., and Jørgensen, T. (2012). The reactions of shoaling adult cod to a pelagic trawl: implications for commercial trawling. *ICES Journal of Marine Science*, 69, 303–312.

Rughetti, M., and Festa-Bianchet, M. (2014). Effects of selective harvest of non-lactating females on chamois population dynamics. *Journal of Applied Ecology*, 51, 1075–1084.

Shackell, N.L., Frank, K.T., Fisher, J.A.D., et al. (2010). Decline in top predator body size and changing climate alter trophic structure in an oceanic ecosystem. *Proceedings of the Royal Society B: Biological Sciences*, 277, 1353–1360.

Sørdalen, T.K., Halvorsen, K.T., Harrison, H.B., et al. (2018). Harvesting changes mating behaviour in European lobster. *Evolutionary Applications*, 11, 963–977.

Stankowich, T. (2008). Ungulate flight responses to human disturbance: a review and meta-analysis. *Biological Conservation*, 141, 2159–2173.

Thurfjell, H., Ciuti, S., and Boyce, M.S. (2017). Learning from the mistakes of others: how female elk (*Cervus elaphus*) adjust behaviour with age to avoid hunters. *PLOS ONE*, 12, e0178082–e0178082.

Thurfjell, H., Spong, G., and Ericsson, G. (2013). Effects of hunting on wild boar *Sus scrofa* behaviour. *Wildlife Biology*, 19, 87–93.

Tillotson, M.D., and Quinn, T.P. (2018). Selection on the timing of migration and breeding: a neglected aspect of fishing-induced evolution and trait change. *Fish and Fisheries*, 19, 170–181.

Tolon, V., Dray, S., Loison, A., et al. (2009). Responding to spatial and temporal variations in predation risk: space use of a game species in a changing landscape of fear. *Canadian Journal of Zoology*, 87, 1129–1137.

Twardek, W.M., Elvidge, C.K., Wilson, A.D.M., et al. (2017). Do protected areas mitigate the effects of fisheries-induced evolution on parental care behaviour of a teleost fish? *Aquatic Conservation: Marine and Freshwater Ecosystems*, 27, 789–796.

Underwood, M.J., García-Seoane, E., Klevjer, T.A., et al. (2020). An acoustic method to observe the distribution and behaviour of mesopelagic organisms in front of a trawl. *Deep Sea Research Part II: Topical Studies in Oceanography*, 180, 104873. Special issue: Structure and functioning of the Norwegian, Iceland, Irminger and Labrador Seas ecosystems: a comparative study.

Underwood, M.J., Rosen, S., Engås, A., et al. (2018). Species-specific residence times in the aft part of a pelagic survey trawl: implications for inference of pre-capture spatial distribution using the Deep Vision system. *ICES Journal of Marine Science*, 75, 1393–1404.

Underwood, M., Winger, P.D., Fernö, A., and Engås, A. (2015). Behavior-dependent selectivity of yellowtail flounder (*Limanda ferruginea*) in the mouth of a commercial bottom trawl. *Fishery Bulletin*, 113, 430–441.

Van de Walle, J., Pigeon, G., Zedrosser, A., et al. (2018). Hunting regulation favors slow life histories in a large carnivore. *Nature Communications*, 9, 1100–1100.

Villegas-Ríos, D., Alós, J., Palmer, M., et al. (2014). Life-history and activity shape catchability in a sedentary fish. *Marine Ecology Progress Series*, 515, 239–250.

Villegas-Ríos, D., Claudet, J., Freitas, C., et al. (2021). Time at risk: individual spatial behaviour drives effectiveness of marine protected areas and fitness. *Biological Conservation*, 263, 109333.

Zedrosser, A., Steyaert, S.M.J.G., Gossow, H., and Swenson, J.E. (2011). Brown bear conservation and the ghost of persecution past. *Biological Conservation*, 144, 2163–2170.

CHAPTER 8

Habitat loss and fragmentation

Michael Sievers, Stephen E. Swearer, and Robin Hale

Overview

Habitat loss and fragmentation represent major alterations to natural environments and are key contributors to the global biodiversity crisis. The first response of many animals to altered environmental conditions is via behavioural adjustments. This chapter illustrates how animal behaviour can be modified in response to habitat loss and fragmentation, the implications of these changes for individual fitness, and how behavioural responses can vary both within and among species. It then describes the consequences of changes in behaviour for populations, communities, and ecosystems. Lastly, it focuses on outlining how the impacts of habitat loss and fragmentation on animals can be mitigated in the context of animal behaviour, and proposes knowledge gaps and research priorities for behavioural ecologists and environmental managers.

8.1 Introduction

The loss and fragmentation of wildlife habitat is globally widespread and affects a diverse range of ecosystems, including terrestrial forests (Keenan et al. 2015; Taubert et al. 2018), mangroves (Hamilton and Casey 2016; Bryan-Brown et al. 2020), rivers (Grill et al. 2015), seagrass meadows (Dunic et al. 2021), shellfish reefs (Gillies et al. 2018), and freshwater wetlands (Davidson 2014). The drivers of habitat loss and fragmentation are diverse and often ecosystem specific. For instance, large-scale farming, logging, and mining have cleared and fragmented forests (Laurance 2014), aquaculture and

the production of palm oil and rice continue to drive the clearing of mangroves (Friess et al. 2019), and dams disrupt longitudinal connectivity within streams (Barbarossa et al. 2020).

While habitat loss is often associated with habitat fragmentation (the division of a habitat into multiple areas, embedded within a matrix of 'non-habitat'), the latter can occur with or without a loss in absolute habitat area (Figure 8.1). The spatial configuration of habitat, for instance, can change independently of the amount of habitat available, in a process known as 'habitat fragmentation per se' (Fahrig 2003). The net effect of habitat fragmentation on individual species and biodiversity, and how to test for such effects, has been debated over the last several decades (e.g. Fahrig 2003; Ewers and Didham 2006; Haddad et al. 2015; Fletcher Jr et al. 2018; Fahrig et al. 2019). Despite this contention, and several empirical and modelling studies that highlight positive and neutral effects of habitat fragmentation (Ewers and Didham 2006; Fahrig 2017; Fahrig et al. 2019; Rybicki et al. 2020), there is strong consensus that habitat loss, and in most situations habitat fragmentation, have negative ecological effects. Supporting this, habitat loss and fragmentation have been shown to be powerful driving forces of species extinction (Püttker et al. 2020), and Haddad et al. (2015) found consistently negative effects of habitat fragmentation on biodiversity and ecosystem functions, including an ecosystem function debt (i.e. delayed changes in ecosystem function due to reduced fragment size or increased isolation).

How habitat loss and fragmentation affect animals is dependent on a suite of abiotic and biotic

Michael Sievers, Stephen E. Swearer, and Robin Hale, *Habitat loss and fragmentation*. In: *Behavioural Responses to a Changing World*. Edited by: Bob B. M. Wong and Ulrika Candolin, Oxford University Press. © Oxford University Press (2024). DOI: 10.1093/oso/9780192858979.003.0008

Figure 8.1 Basic conceptualization of habitat loss and fragmentation.

Modified from Bryan-Brown et al. (2020)

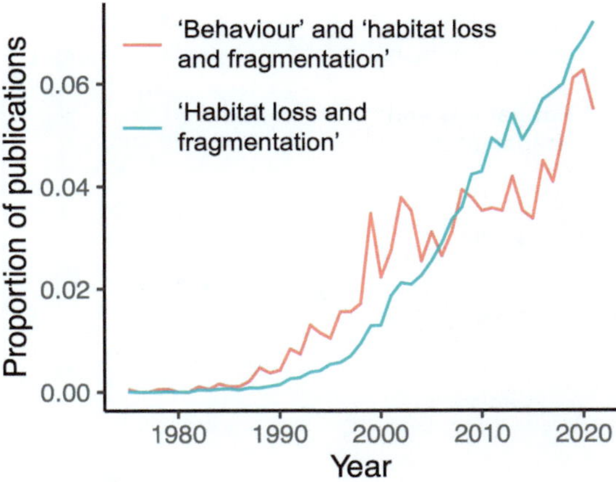

Figure 8.2 The proportion of published papers on animal behaviour and habitat loss and fragmentation (red), and on habitat loss and fragmentation more broadly (blue; ISI Web of Science search 1900–2021, inclusive). Contrasting these two trendlines provides an indication of the relative research interest in animal behaviour in the study of habitat loss and fragmentation.

factors (which can interact), and will partly be driven by how they respond behaviourally (Tuomainen and Candolin 2011; Fisher et al. 2021; Blumstein et al. 2022). Following a surge in behavioural research with respect to habitat loss and fragmentation in the 1990s, the subsequent two decades saw behavioural studies arguably underrepresented, followed by a potential resurgence in the last five years (Figure 8.2). This renewed interest may stem from acknowledgement that a better understanding of behavioural responses to habitat loss and fragmentation can provide important early warning signs of impact. For example, the ecotoxicological literature has repeatedly shown that animal behaviours are rapidly altered by human-induced rapid environmental change (see Chapter 3), and manipulating animal behaviour may be a useful management strategy for mitigating the effects of habitat loss and degradation (Sih 2013; Hale et al. 2020). This chapter considers how animal behaviour is influenced by habitat loss and fragmentation and how these behavioural changes affect individuals. We explain why individuals and species differ in their sensitivities to habitat loss and fragmentation, and describe the consequences of these behavioural responses for populations, communities, and ecosystems. We then discuss the important role of behavioural knowledge in developing effective management strategies to reduce the negative impacts of habitat loss and fragmentation on animals.

8.2 How does habitat loss and fragmentation influence animal behaviour and affect fitness?

Habitat loss and fragmentation influence animal behaviour via a complex culmination of interrelated cause and effect pathways, all of which have the potential to influence individual fitness (Figure 8.3). Landscapes that have less habitat and more fragmented habitats likely have a reduced amount and diversity of resources (e.g. nesting sites, mates, predators, food, refuges; Kupfer et al. 2006). Reduced resources can lead to altered movement patterns, higher levels of starvation or predation, altered social interactions, and reduced survival and reproductive success (Tuomainen and Candolin 2011; Valiente-Banuet et al. 2015; Laurent et al. 2020; Wilson et al. 2020). There are thus indirect effects of habitat loss and fragmentation based on how animals behaviourally respond, such as species changing foraging behaviours in ways that increase predation risk. We discuss these below, and while we delineate behavioural categories into separate subsections for clarity, these are inherently linked and rarely occur in isolation (e.g. altered sociality can affect movement and reproductive behaviours; Cousseau et al. 2020).

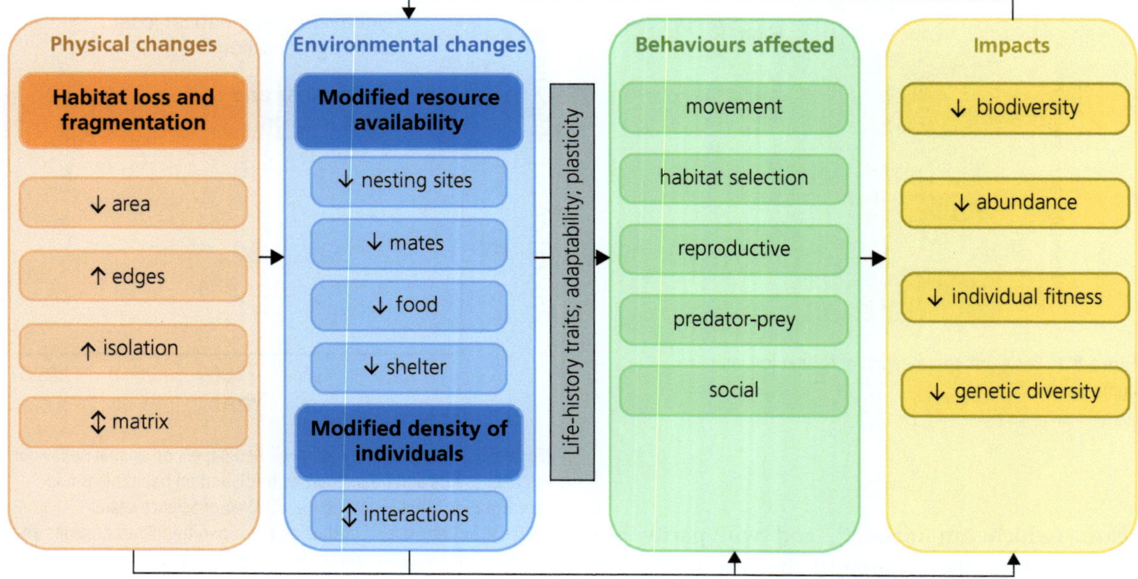

Figure 8.3 Simplified conceptual model of the key pathways through which habitat loss and fragmentation affect the environment and animal behaviour, and subsequently the impacts of these on animals. Up (increase) and down (decrease) arrows are indicative of generalized responses and do not reflect all scenarios.

8.2.1 Movement and habitat selection

Habitat loss and fragmentation can strongly influence movement and habitat selection, which are inherently related and form the basis of species' space-use patterns (Cattarino et al. 2016; Van Moorter et al. 2016; Spiegel et al. 2017; Tucker et al. 2018). Habitat loss and fragmentation, and their drivers, create significant physical barriers and inhospitable matrices that limit movement. Greater prairie chickens *Tympanuchus cupido*, for instance, generally avoid, and alter their movement rates when near, anthropogenic structures such as roads (Londe et al. 2022). In rivers, culverts can cause fragmentation, impeding the movement of fish by acting as structural and hydraulic barriers, but also by altering the light environment causing fish avoidance behaviours (Figure 8.4a; Jones and Hale 2020). Wastewater outfalls can create chemical barriers in rivers, whereby animals actively avoid polluted waters (Fuller et al. 2015). These examples show that habitats can become fragmented not only by structural changes, but also by other changes, such as noise (Chapter 2), chemical (Chapter 3), and light

(Chapter 4) pollution, with a suite of effects on animal behaviour.

Habitat loss and fragmentation also drives changes in animal movement modalities. Bale monkeys *Chlorocebus djamdjamensis* in fragmentated forests, for example, spend more time walking and galloping on the ground and in the understorey, compared to monkeys in continuous forests, which display more natural climbing behaviour (Figure 8.4a; Mekonnen et al. 2018). These changes in behaviour are a direct result of habitat fragmentation, and put bale monkeys at a greater risk of predation and conflict with humans (Mekonnen et al. 2018; see also Chapter 13). Modelling has proven useful in determining how movement modalities and habitat selection can modulate the impacts of habitat loss and fragmentation. For instance, Cattarino et al. (2016) developed a demographic model for individuals that adopt different modes related to movement, and applied the model within a spatially explicit simulation framework that had varying levels of habitat fragmentation (amongst other varied parameters). Behaviours, such as whether individuals adopt short and tortuous movements or long and straight movements, strongly influenced how fragmentation (namely the spatial scale of fragmentation; within or between foraging areas) impacted individual fitness.

Habitat selection can be influenced by habitat loss and fragmentation via changes to the matrix and the size and quality of remaining patches (e.g. for birds in the lark family Alaudidae; Morgado et al. 2010; Bosco et al. 2021), and also when lost habitats are replaced by ones created by humans that then attract animals. For example, stormwater wetlands have been created in cities around the world—primarily to capture and treat stormwater, but also to replace natural wetlands and support wildlife—and these artificial wetlands are becoming the dominant aquatic habitat in urban areas (Kentula et al. 2004; Parris 2016). Some animals prefer these artificial habitats, and there is evidence they can be ecological traps by attracting animals that colonize and subsequently suffer fitness costs due to high pollution loads (Figure 8.4b; Hale et al. 2018; Sievers et al. 2018). Research has also shown that marshland restoration can cause ecological traps by attracting butterflies into habitats that subsequently flood, reducing reproductive success (Severns 2011). Therefore, habitat loss can indirectly affect habitat selection and individual fitness via the creation or restoration of habitats that are unsuitable but contain the positive cues animals use when selecting habitats (Robertson and Hutto 2006; Hale and Swearer 2016).

8.2.2 Reproductive behaviour

Reproductive behaviours can be altered by habitat loss and fragmentation, due in part to reductions in resource availability. The forest butterfly *Salamis parhassus*, for instance, alters its mate location strategy from perching to patrolling in forests affected by logging and deforestation, in an attempt to more quickly occupy increasingly rare and valuable light gaps (Figure 8.4c; Bonte and Van Dyck 2009). Males of a cooperatively breeding bird, the placid greenbul *Phyllastrephus placidus*, from fragmented Afrotropical cloud forests reproduce earlier (due to dispersing earlier) and mostly settle within their natal patch, relative to those from continuous

forests (Cousseau et al. 2020). This outcome is predicted to be driven by both reduced natal habitat quality and enhanced reproductive opportunities in fragmented forests (Cousseau et al. 2020). Mating behaviour of water striders *Aquarius remigis* differs between large and small pools (a proxy for habitat loss) whereby striders exhibit scramble promiscuity with intense sexual conflict in large pools, and harem polygyny in smaller pools, with less frequent and shorter matings (Sih et al. 2017). Habitat loss and fragmentation, and its impact on nest availability, can also force animals to change their reproductive behaviours and accept novel nesting resources, such as artificial nesting boxes (Carstens et al. 2019) or urban buildings (Charter et al. 2007; see also Chapter 9).

8.2.3 Predator-prey interactions and foraging behaviour

Habitat loss and fragmentation can strongly influence predator-prey interactions and foraging behaviours. A review on the effect of habitat fragmentation on plant-animal interactions, for example, found that the most consistent response across species was a reduction in herbivory (Brudvig et al. 2015). Predators, on the other hand, can become more efficient in the short term in highly fragmented habitats because higher densities of prey in the remaining patches can increase encounter rates (McWilliams et al. 2019). Increasing encounter rates between predators and prey can also elicit stronger predator-avoidance behaviours in prey and, as a result, behaviours related to perceived predation risk and 'fear' are often modified by habitat loss and fragmentation. For instance, Atlantic cod *Gadus morhua* spent less time foraging within more fragmented seagrass patches; however, total prey consumption was unaffected, indicating that predatory behaviours were more efficient and/or anti-predator behaviours were less efficient (Figure 8.4d; Laurent et al. 2020).

Pollinator foraging behaviours are also susceptible to habitat loss and fragmentation, with subsequent implications for pollination success and plant reproductive output (Hadley and Betts 2012; Brudvig et al. 2015; Harrison and Winfree 2015; Hermansen et al. 2017; see also Chapter 16). For

(a) Dispersal and movement

Culverts fragment rivers and act as a barrier to movement, both from a physical point of view and due to altering the light environment[1]

Forest fragmentation alters movement modalities: bale monkeys in fragments are more likely to spend time walking and galloping on the ground and in the understorey[2]

(b) Habitat selection

Habitat loss forcing animals to prefer novel habitats, such as stormwater wetlands. Alterations to habitat selection can have deleterious effects if the novel habitats function as ecological traps[3]

(c) Reproductive behaviour

The forest butterfly alters its mate location strategy from perching to patrolling in forests affected by logging and deforestation to more quickly occupy increasingly rare and valuable light gaps[4]

(d) Predator-prey interactions

Enhanced predator efficiency: Atlantic cod in fragmented seagrass spent less time hunting but consumed the same amount of prey[5]

(e) Social behaviour and Communication

Habitat loss and fragmentation can increase aggression due to increasing the density of individuals and decreasing the density of resources[6]

Figure 8.4 Case studies of how habitat loss and fragmentation can affect animal behaviour, including impacts on (a) dispersal and movement, (b) habitat selection, (c) reproductive behaviour, (d) predator-prey interactions, and (e) social behaviour and communication.

[1] Jones and Hale (2020); [2] Mekonnen et al. (2018); [3] Sievers et al. (2018); [4] Bonte and Van Dyck (2009); [5] Laurent et al. (2020), [6] Macdonald et al. (2004). Symbols courtesy of the Integration and Application Network (ian.umces.edu/media-library), licensed under Attribution-ShareAlike 4.0 International (CC BY-SA 4.0)

instance, fragmentation but not the amount of habitat significantly reduced bumblebee *Bombus t. terrestris* foraging activity (Maurer et al. 2020), whilst flowers are often visited less frequently by pollinators and have higher selfing rates in small, fragmented urban populations (Cheptou and Avendaño 2006). However, pollinators in these fragmented habitats may be more fully exploiting resources from each visit, as the time spent at each flower can be longer, suggesting animals may be able to modify feeding behaviour to recoup energy spent flying farther (Andrieu et al. 2009; Harrison and Winfree 2015).

8.2.4 Social behaviour

Habitat loss and fragmentation can influence social interactions, structure, and hierarchies (Wang and Schreiber 2001; Banks et al. 2007; Fisher et al. 2021). For instance, increases in the density of Eurasian badgers *Meles meles* and their resources within fragmented habitats strengthens competition and causes high rates of intraspecific aggression (Figure 8.4e; Macdonald et al. 2004), male tree lizards *Urosaurus ornatus* display more aggressive interactions in more patchy habitats (Lattanzio and Miles 2014), and mosquitofish *Gambusia holbrooki* are more aggressive when water levels decline (i.e. a loss of habitat), leading to more intraspecific conflict (Flood and Wong 2017). While experimentally manipulated habitat loss altered brown trout *Salmo trutta* dominance hierarchies (Sloman et al. 2001), habitat loss and fragmentation can actually lead to more structured and stable hierarchies; for example, when conflicts induce high costs or when the level of group relatedness increases (Mathot and Giraldeau 2010; Fisher et al. 2021).

Positive adjustments to social behaviours, such as changes to patterns of den-sharing and cooperative resource use, can help mediate the impacts of resource availability on individual fitness in the wake of habitat loss and fragmentation. For example, den-sharing among hollow-dependent Australian mountain brushtail possums *Trichosurus cunninghami* was reduced when there were fewer dens available, suggesting that, under resource competition, territoriality increases (Banks et al. 2011). However, as dens became less available,

possums switched from avoiding kin to increasing den-sharing among siblings; a behavioural adjustment that can have profound impacts on individual fitness via subsequent changes in resource acquisition, inbreeding, and pathogen transmission (Banks et al. 2011).

8.3 Why do animals respond as they do, and why do species differ in their responses?

The initial response of animals to habitat loss and fragmentation is often behavioural, and those individuals and species that can adaptively modify their behaviour, for instance by shifting to utilize beneficial artificial resources (e.g. nest boxes) or by sharing resources (e.g. kin sharing of hollows), or those that have inherent behavioural traits that lead to higher fitness in disturbed habitats (e.g. urban adaptors; Chapter 9), are likely to cope better with the loss and fragmentation of habitats. Understanding which functional traits (including behavioural traits) are associated with 'winners' versus 'losers' in modified landscapes is essential for effective conservation planning (Farneda et al. 2015), especially given that specialist species and those of conservation concern are most likely to be negatively affected by habitat loss and fragmentation (Pfeifer et al. 2017). If there are differences in behavioural responses among species and/or individuals, such differences have the potential to mediate the consequences of habitat loss and fragmentation for affected animals.

One of the most striking ways in which individuals, both within and among species, can vary in their responses to habitat loss and fragmentation is the propensity to disperse among remnant habitat patches (Desrochers and Hannon 1997; Bélisle and Desrochers 2002). For example, individuals may differ in personality-dependent dispersal, which, in the context of habitat fragmentation, could mean that isolated patches are colonized by individuals with behavioural syndromes characterized by being more bold, exploratory, aggressive, and asocial (Cote et al. 2010). Consistent with this hypothesis, Cornelius et al. (2017) found that individual white-shouldered fire-eyes *Pyriglena leucoptera* from fragmented Neotropical rainforests, while more resistant to crossing boundaries, were more successful at crossing the matrix than individuals from continuous forest, possibly because they were more thorough in assessing risk. Dispersal propensity can also vary among species. This is often evidenced in comparisons between habitat generalists and specialists, with the latter typically exhibiting more limited dispersal when the costs of dispersing among sparsely distributed resource patches are high (Stevens et al. 2014). In the case of European land snails (superfamily Helicoidea), this difference is, in part, behaviourally linked, as specialists were less likely to cross boundaries between familiar and unfamiliar habitats (Dahirel et al. 2015).

Although species that are habitat or resource specialists are typically the 'losers' when habitats are lost or fragmented (Filgueiras et al. 2021), those that exhibit some degree of behavioural plasticity, and can change their resource dependencies, are likely to have greater capacity to adjust to habitat change (Chapter 14). Such adaptive behavioural plasticity can be seen, for example, in the earlier dispersal by male placid greenbuls *Phyllastrephus placidus* born in fragmented forests, compared to males born in continuous forests, as a strategy to optimize reproduction-survival trade-offs (Cousseau et al. 2020), and in the changes in foraging mode by the habitat-specialist peacock grouper *Cephalopholis argus* in response to coral reef degradation (Karkarey et al. 2017). While behavioural plasticity alone may be insufficient to cope with the magnitude of human perturbation (van Baaren and Candolin 2018), improvements to survival could be sufficient to facilitate evolutionary adaptation by providing more time for genetic changes to accrue (Tuomainen and Candolin 2011).

Assessing whether flexibility in behavioural responses is adaptive—and how responses are shaped by the interactive effects of genetics and past environmental conditions—can help illuminate the eco-evolutionary dynamics that can play out under habitat loss and fragmentation and, in so doing, allow us to predict the persistence of impacted populations (Mazza et al. 2020; Szulkin et al. 2020). Further insights can also be gained through a greater appreciation of how cognition and learning might affect the ability

of animals to respond and adapt to habitat loss and fragmentation (Shettleworth 2009). Sol et al. (2020) outline two ways in which cognition has the potential to impact adaptive evolution in an urban context that strongly relates to habitat loss and fragmentation (see also Chapter 9); first, by allowing individuals to select habitats and resources that align with their specific phenotypes, and second, by helping animals to develop learned responses to novel or unfamiliar challenges. Where species and individuals utilize cognition and learning, these processes can enhance adaptive evolution by minimizing the risk of population extinction and help ease individuals into the new environmental conditions that they are confronting (Sol et al. 2020).

8.4 What are the broader consequences of behavioural responses to habitat loss and fragmentation?

In this section, we focus on how the individual-level impacts discussed earlier can scale up to the population or community level. We previously outlined the potential negative outcomes of habitat loss and fragmentation for individual behaviour and fitness. While behavioural studies allude to broader implications at the level of populations, communities, and ecosystems, there is limited empirical investigation, which is hindering management efforts (Wilson et al. 2020). Despite this limitation, the available direct evidence shows that behavioural adjustments can manifest into impacts at higher levels of biological organization.

8.4.1 Movement and habitat selection

Many studies have inferred that changes in movement and habitat selection resulting from habitat loss and fragmentation can affect individual fitness, with flow-on consequences for populations and communities. For instance, modification of movement modalities in bale monkeys in forest fragments was speculated to increase conflicts with humans, and the risk of parasitic infection and predation, with subsequent implications for long-term population persistence (Mekonnen et al. 2018). However, positive outcomes from behavioural plasticity, such as enhanced foraging on nutrient-dense food resources and increased gene flow, highlight that empirical studies are required to fully understand the population-level implications of behavioural adjustments. As an example, Brooker and Brooker (2002) studied how fragmentation affects reproduction, survival, dispersal, and recruitment in blue-breasted fairy-wrens *Malurus pulcherrimus*, and found that the degree of habitat connectivity affected individual movement, the establishment of breeding territories, and post-dispersal survivorship. This resulted in a much higher proportion of individuals dying during dispersal in fragmented areas, which, in turn, led to declines in population size. Further, modelling of how the movement of individuals in response to habitat edges affects population persistence and spatial spread showed that including edge behaviour is needed to correctly understand population-level patterns (Maciel and Lutscher 2013).

Habitat fragmentation can also influence survival through changes in habitat selection decisions. For instance, pumas *Puma concolor* in the fragmented Santa Cruz Mountains in California preferentially hunt in human-dominated areas, where they suffer greater retaliatory killings by people following livestock loss despite little evidence that livestock are actually preyed on by the pumas (Nisi et al. 2022). While such potential ecological traps are defined by their impacts to individual fitness, theoretical work has shown that traps can scale up to compromise metapopulation persistence at regional scales (Hale et al. 2015b) and increase extinction risk (Battin 2004).

Adjustments to movement and habitat selection in response to habitat loss and fragmentation can also have longer-term evolutionary consequences. For example, habitat fragmentation has led to reductions in allelic diversity and heterozygosity in terrestrial mammals, with stronger effects in species whose movement is most susceptible to the effects of fragmentation (Lino et al. 2019). Such effects can have population-level consequences. Méndez et al. (2011) showed that reduced movement, coupled with resource reductions (e.g. mates), can restrict gene flow and genetic drift, with

subsequent reductions in individual fitness and population viability. In such cases, selection likely favours populations that consist of individuals with behavioural responses that limit inbreeding (Tuomainen and Candolin 2011; but see Szulkin et al. 2013) and, indeed, many species have evolved various behavioural patterns that avoid or reduce this risk (Szulkin et al. 2013).

8.4.2 Reproductive behaviour

There is considerable potential for habitat loss and fragmentation to result in changes to reproductive behaviour, which can have important population-level impacts. For example, fragmentation often leads to spatially clumped resources and, as a consequence, spatially clumped females, which increases the likelihood for mating systems to transition towards defence polygyny, although this will depend on female home range size relative to the size of remnant patches (Lane et al. 2011). Furthermore, the costs of habitat fragmentation for movement can differ between males and females, leading to sex-biased dispersal with flow-on effects for operational sex ratios (Banks et al. 2005). Changes in population density and connectivity can also influence the likelihood of finding or having to compete for mates, or having offspring killed by incoming males (see Banks et al. 2007; Giuntini and Pedruzzi 2022). Ultimately, changes in reproductive behaviour and the associated costs of inbreeding, reduced pairing success, and restricted gene flow are likely to reduce population viability, although recent reviews argue that more evidence is still needed (Banks et al. 2007; Giuntini and Pedruzzi 2022).

8.4.3 Predator-prey interactions and foraging behaviour

Changes in resource distribution can also impact on predator-prey interactions and community structure due to changes in how prey perceive predation risk. For example, the recovery of wolf *Canis lupus* populations in Wisconsin, USA resulted in a behavioural shift in deer *Odocoileus virginianus*, which reduced their foraging along roadsides because of fear of predation, resulting in a 24% reduction in deer-vehicle collisions (Raynor et al. 2021). By contrast, recent spatially explicit modelling has shown that such fear can actually intensify reductions in prey diversity as a result of habitat loss and fragmentation influencing species interactions, as well as shifting prey community composition from large to small animals (Teckentrup et al. 2019).

Habitat loss and fragmentation can also influence the movement decisions of animals foraging on flowering plants, with important implications for plant-pollinator interactions and plant mating systems and communities (see also Chapter 16). Insect pollinators, in comparison to vertebrate pollinators such as birds and bats, which have greater flying potential, are particularly vulnerable to habitat fragmentation (Aguilar et al. 2019) and can have lower diversity in areas with reduced/fragmented habitat (see Ewers and Didham 2006 and references therein). This can lead to decreased pollination service and resulting declines in plant fecundity, progeny quality, and long-term population viability (Aguilar et al. 2006; Aguilar et al. 2019).

8.4.4 Social behaviour

Habitat fragmentation, as well as other structural changes to the habitat and the surrounding matrix, can influence social interactions by altering signal transmission, limiting movement, and increasing competition for resources (Fisher et al. 2021; Blumstein et al. 2022). Such effects can scale up to have important impacts on population viability, such as reduced sexual signalling in wolf spiders *Hygrolycosa rubrofasciata* (Ahtiainen et al. 2004) and reduced individual song diversity in Dupont's lark *Chersophilus duponti* (Laiolo et al. 2008) being associated with lower viability of small populations. While these examples provide evidence for how behavioural adjustments can lead to broader ecological consequences, ultimately, more evidence is needed for how behavioural adjustments link to population demography and genetics, if we are to gain a better understanding of when and where mitigation of habitat loss and fragmentation should be prioritized.

8.5 How can the effects of habitat loss and fragmentation be mitigated and exploited?

8.5.1 Harnessing behaviour to manage habitat loss and fragmentation

The remainder of our chapter focuses on outlining how the impacts of habitat loss and fragmentation to wildlife can be mitigated, and how knowledge of animal behaviour can be applied to guide these efforts. We conclude by highlighting knowledge gaps and research priorities for behavioural ecologists and environmental managers.

There are different options available to mitigate the effects of habitat loss, and which one is most appropriate depends on what the objective is and for which species, and the scale at which impacts need to be ameliorated. For example, landscape design, aimed at increasing functional connectivity at large spatial extents, can involve revegetation (Thomson et al. 2009), establishing habitat corridors (Banks-Leite et al. 2020), building new or modifying existing structures to enable movement (e.g. wildlife crossings for roads; Lister et al. 2015; fishways for streams; Silva et al. 2018), or removing infrastructure entirely (e.g. barrier removal in streams; Sun et al. 2022). By contrast, in-situ habitat restoration is often undertaken at smaller spatial scales and targeted at specific locations, such as through establishing breeding populations of particular target species in terrestrial systems (Selwood et al. 2009), revegetating at specific locations to reconnect habitats in marine systems, and reinstating channel and bank habitats in freshwater systems (Hale et al. 2019). As both the actions and the scales at which they are implemented can vary between landscape-scale design and in-situ restoration, we might expect animals to respond in different ways via different types of behaviour.

Knowledge of animal behaviour can help improve efforts to mitigate habitat loss and fragmentation in several ways, including guiding actions to reconnect habitats across the landscape and restoring habitats for local scale benefits for animals. An understanding of animal behaviour can also be used as informative indicators of responses by individual animals, and to assess the wider consequences of these responses for populations, communities, and ultimately ecosystems. Such knowledge can help identify when and why unintended responses to management actions occur, and how we can minimize any undesired impacts.

8.5.2 Behavioural knowledge to reconnect fragmented habitats

Many animals move to access better resources or mates, or avoid predators, competitors, or unfavourable environmental conditions, ultimately increasing their fitness (Bowler and Benton 2005). Movement therefore has a major influence on the structure and dynamics of populations, communities, and ecosystems (Nathan et al. 2008). Animals also often move between, and use, different habitats throughout their life cycle, and these habitats will vary in terms of their ecological value; for instance, in terms of food resources, the degree to which they can provide refuge from unfavourable environmental conditions or biotic interactions, and their importance as part of dispersal pathways (e.g. stopover sites) or within metapopulations (e.g. sources; Gilby et al. 2018).

Knowledge of movement is an important component of conservation in highly fragmented landscapes (Doherty and Driscoll 2018; Jones et al. 2021). There are four mechanistic components to movement: internal state ('why move'), motion ('how to move'), navigation ('when and where to move'), and external factors affecting the first three (Nathan et al. 2008). This 'movement ecology' paradigm can be applied to understand and manage the impacts of altered movement in highly disturbed and fragmented landscapes (Doherty and Driscoll 2018). While management has often been focused on improving connectivity by providing habitat corridors or stepping-stones, more detailed information about animal behaviour and movement can help guide more targeted interventions (Box 8.1). This could include, for example, novel management of the matrix (e.g. orientating plantation rows to direct dispersing animals towards suitable habitat) based on an understanding of an organism's perceptual range and what factors drive their movements (Doherty and Driscoll 2018). Behavioural

knowledge can be equally important in guiding efforts to enhance connectivity in aquatic systems. For instance, fishways are structures placed on or around constructed barriers, such as dams, to allow fish to move through. Whether such fishways are effective in meeting ecological or conservation goals, however, will depend on whether fish are actually attracted to them (Cooke and Hinch 2013), whether they move through them (Goerig and Castro-Santos 2017), and how they respond behaviourally to altered environmental cues, such as light (Jones and Hale 2020).

Knowledge of movement ecology is important for managing the effects of habitat modification, but so too is a greater mechanistic understanding of how animals perceive and experience their habitats (Van Dyck 2012; Jones et al. 2021). For instance, landscape ecology has traditionally considered resource configurations (i.e. how habitat quality might vary) as static, but such an approach may be too simplistic, given that some animals track phenological variation in resources across space (Abrahms et al. 2021). In this regard, a better understanding from the animal's perspective of which habitats across the landscape are important—and why—might be more informative in helping us to prioritize where management actions should be undertaken to best improve connectivity.

8.5.3 Behavioural knowledge to guide in-situ habitat restoration

Behavioural knowledge is an important component of in-situ habitat restoration. Best practice restoration projects should set goals and implement restoration actions that produce desired changes to habitat structure, and result in the successful colonization and occupation of restored sites by target species that are able to maintain self-sustaining populations through successful reproduction (Hale et al. 2020). Behaviour is important in terms of how animals move around the landscape and also whether animals can feasibly encounter restored sites (Tonkin et al. 2014). Once animals encounter restored sites, they need to colonize and occupy them, and those sites need to provide the resources that animals require (Hale and Swearer 2017). Consequently, information about habitat selection and fitness is vital for assessing restoration success (Hale et al. 2019). This is best demonstrated by two contrasting situations where animals exhibit unexpected responses to restoration, driven by the decoupling of habitat preferences and habitat quality. The first, and perhaps most dramatic, situation is when restoration creates ecological traps (Robertson and Hutto 2006), whereby animals are attracted to restored sites but suffer compromised fitness once they are there (Hale et al. 2015a; Hale and Swearer 2017). Planting vegetation, such as during afforestation, for example, could attract predators that compromise populations of target species (Hawlena et al. 2010; Belder et al. 2018). The second situation is when animals fail to recolonize restored sites, regardless of habitat suitability. For instance, many birds select breeding sites on the basis of cues from calling conspecifics (Valente et al. 2021), and will not colonize restored locations when these cues are absent (Schofield et al. 2018). Interestingly, species may respond differently to the same management action so that the outcomes above, along with adaptive habitat selection, can be observed simultaneously (Komyakova et al. 2021).

The examples above illustrate the need to understand habitat selection and its consequences in the context of mitigating unexpected responses to restoration following habitat loss and fragmentation. Habitat selection is a complex process, and some animals may respond to a range of different cues, potentially with multiple sensory modalities (e.g. Huijbers et al. 2012). When cues are missing and animals avoid restored sites, it may be possible to simply seed sites with missing cues, such as using acoustic playbacks (e.g. birds: Schofield et al. 2018; coral reef fish: Gordon et al. 2019), ultimately exploiting behavioural preferences to improve outcomes. Alternatively, if cues are misleading, it may be possible to remove them to prevent animals settling into unfavourable locations (Greggor et al. 2020). However, a range of complex factors can influence habitat selection, such as the propensity for an animal to prefer habitats similar to their natal location (Davis and Stamps 2004), or when cues from other locations in the landscape affect how animals assess habitat quality (Resetarits and Binckley 2009; Resetarits and Silberbush 2016). We need

to firstly understand when and why these factors are occurring, and then identify what management approaches might be needed to produce effective management and conservation outcomes. For example, it may be possible to reduce the likelihood of newly introduced animals dispersing away from a restored site by exposing them to cues from that restored site when they are still in captivity prior to their release into the wild, or by providing suitable and sufficient resources within the restored site that encourage individuals to stay (Berger-Tal et al. 2020; see also Chapter 12).

Restoration of lost or fragmented habitats will fail if animals colonize sites that do not provide the resources they require. Often it is presumed that 'resources' here simply reflects things like food and shelter, but a broader view is required, guided by how animals view and respond to these habitats (Van Dyck 2012). Behavioural knowledge can help us to make more informative comparisons of habitat quality based on behavioural variables that are linked to fitness outcomes, and to identify resources that either make habitats suitable or unsuitable for target species (Lindell 2008). For example, prey species may benefit from predator-alarm signals produced by other species, and if these other species are not present, a habitat may be less suitable (Gil et al. 2016). Some changes in behaviour (e.g. increases in feeding rate, courtship behaviour) can also be indicators that fitness is likely to increase in restored habitats (Lindell 2008; Berger-Tal et al. 2011). However, despite the importance of behavioural information in restoration projects being highlighted more than a decade ago (Lindell 2008), responses of animals to restoration actions are most commonly measured using population- and community-level metrics, such as abundance (Hale et al. 2019), which, in isolation, can be misleading indicators of habitat quality (Hale and Swearer 2017). To that end, a more animal-centric view of restoration is needed (Hale et al. 2020; Sievers et al. 2022) so that we can better plan and evaluate projects using behavioural knowledge. For instance, supplementary feeding is often used during reintroductions without a plan for how supplementary feeding will be eventually stopped; by understanding dietary preferences, it may be possible to restore habitats (e.g. by planting species known to provide food) to ultimately generate natural food resources, and reduce the reliance on supplementary feeding for those species (Maggs et al. 2019; see also Chapter 12).

8.6 Managing and mitigating the effects of habitat loss and fragmentation

To encourage greater use of behavioural knowledge, we highlight avenues for future research as well

---

Box 8.1 Using knowledge of animal behaviour in the design and implementation of wildlife crossings and rope bridges

Here we briefly describe a case study where behaviour has been incorporated in a conservation/management setting. Understanding movement behaviour has benefitted the design and implementation of structures, such as wildlife crossings and rope bridges, that assist animal movement in fragmentated landscapes (Clevenger and Huijser 2011; Forman 2012; Smith et al. 2015). These structures can provide safe passage for animals, allowing them to cross highways or other barriers without risking injury or death (Figure 8.5; Soanes et al. 2018). By taking into account these behavioural preferences, designers can create crossings and bridges that are more likely to be used by the target species, thereby increasing their effectiveness in promoting connectivity in otherwise 'fragmented' landscapes

(Kintsch et al. 2015). For instance, species or individuals that live in structurally complex environments, such as dense forests, are likely to prefer or only use crossings that provide sufficient cover, whereas those species that inhabit more open environments may prefer to use larger and more open structures for passage. Kintsch et al. (2015) provide a framework (the 'Wildlife Crossing Guilds') to classify wildlife into classes defined by physiological and behavioural needs that underlie a species' willingness to use wildlife crossing structures. These needs are related to movement capacity and modalities, the size of the animal, their desire for cover when moving, specialized habitat needs, and factors related to anti-risk behaviour (Kintsch et al. 2015).

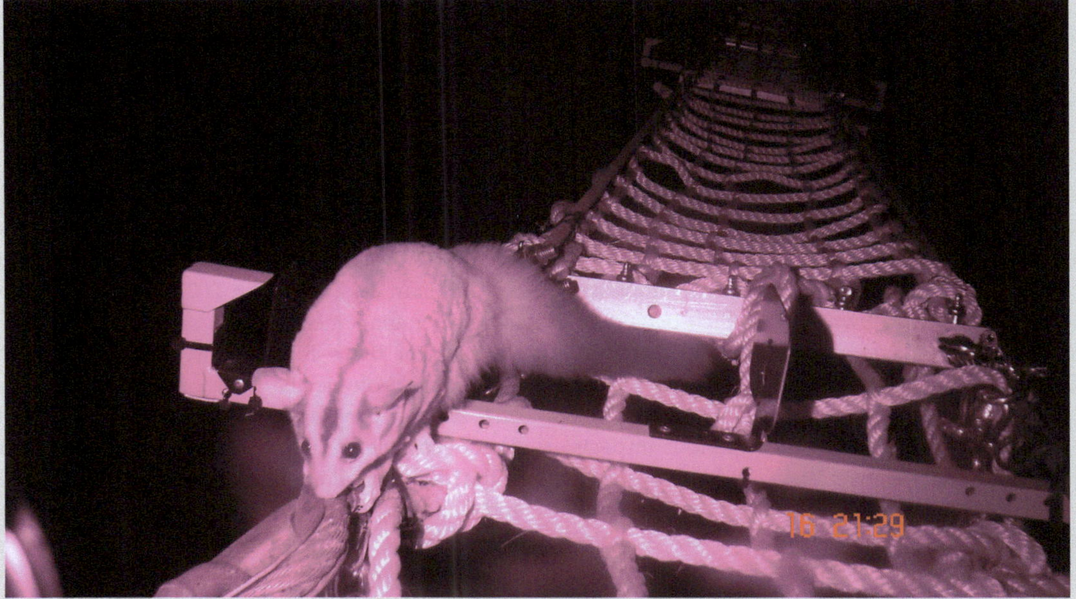

Figure 8.5 A squirrel glider *Petaurus norfolcensis* using a rope bridge to cross a freeway in northeast Victoria, Australia. These bridges were retrofitted to the highway and were found to increase the movement of squirrel gliders across a road that was previously a barrier to movement, with consequences for gene flow (Soanes et al. 2018).

Photo: used with permission from Kylie Soanes

as important practical considerations that need to be considered (see Box 8.1). First, we need a better understanding of differences in the susceptibility of different species to habitat loss and fragmentation. While some species, such as urban adapters, benefit under human-modified conditions (Chapter 9), others are negatively impacted (Sih 2013; Sih et al. 2016). Understanding how and why species vary in sensitivity to habitat change will be important for identifying those species that are most urgently in need of management action. Prioritization tools exist to help guide decisions about which species to focus on in conservation projects that consider a range of costs and benefits (both to the individual animal and the project more broadly), and likelihood of success (Joseph et al. 2009). Here, it would be beneficial to consider how behavioural information can be incorporated into the application of these tools, including the financial costs involved in collecting such information and how it might lead to better management actions. In some instances, management decisions about which species or locations to prioritize might change if information on behaviour and fitness is considered. For example, species distribution modelling based on occurrence data could prioritize areas for conservation that are flagged as potential ecological traps (Titeux et al. 2020). Therefore, behavioural information may help reduce the likelihood that assumptions about habitat suitability and preference could cause inadvertent poor management outcomes.

A second key area for future research is to better understand how habitat loss and fragmentation can affect the pathways via which behaviour affects different ecosystem functions (Wilson et al. 2020). This knowledge will help reveal which behavioural changes in response to habitat loss and fragmentation might have deleterious impacts on populations, communities, and ecosystems. As we have already argued, however, this will require taking an 'animal-centric' view of habitats (Van Dyck 2012). Without this understanding, we risk focusing management efforts on mitigating changes in behaviour that have limited ecological relevance or, alternatively, mismanage circumstances where behavioural change drives ecosystem change.

While habitat loss and fragmentation are significant threats to biodiversity, they do not occur in isolation from other threats (see Chapter 10), and thus interactive effects are possible that could result in worse outcomes for affected wildlife. An important example is how climate change can influence the quality and distribution of habitats—both directly and indirectly—by increasing the extent, frequency, and magnitude of a suite of disturbances (Segan et al. 2016; Sitters and Di Stefano 2020; see also Chapter 1). It will thus be important to consider how climate change alters the behavioural performance of animals, such as the impacts of changes in biomechanics or physiology on movement (Seebacher and Post 2015; Domenici and Seebacher 2020). Given increasing global trends in pressures and the inevitability of climate change impacts across ecosystems, there is a need to not only follow the traditional approach of managing ecosystems predominantly through resisting change, but to be more flexible and accept or direct change (Lynch et al. 2021). Directing change involves anticipating likely responses to current or future conditions, and then implementing strategies to promote preferred ecosystem characteristics under these altered conditions (e.g. composition, structure, processes, function; Thompson et al. 2021). Behavioural knowledge could be important in implementing such strategies. For instance, we know that ecologically central species (e.g. keystone species, ecosystem engineers, dominant species) influence species interaction networks and ultimately ecosystem structure, function, and stability (Rahman and Candolin 2022). By undertaking targeted management actions based on our understanding of these interactions, such as restoring habitat for ecologically central species that will in turn influence other species, we may be able to direct ecosystems in ways that ensure persistence in the face of habitat loss and degradation (Rahman and Candolin 2022).

8.7 Concluding remarks and moving forward

We have described how animal behaviour can be influenced by habitat loss and fragmentation, and outlined how these behavioural changes might affect individual fitness, populations, communities, and ecosystems. As with other forms of human-induced environmental change, some animals (species and individuals) will be able to adapt and adjust accordingly, and some may even thrive, whilst others will struggle. While restoration offers hope of reversing habitat loss and fragmentation, management and conservation approaches that do not give appropriate consideration to animal behaviour can have unintended biodiversity consequences. Ultimately, knowledge of animal behaviour can assist in mitigating the effects of habitat loss and fragmentation, but requires hypothesis-driven behavioural research tied to well-defined conservation problems, and simultaneously, a better understanding by behavioural ecologists of the decision-making processes used to guide management and conservation decisions (Greggor et al. 2016).

Acknowledgements

M.S. acknowledges support from the Global Wetlands Project, supported by a charitable organization that neither seeks nor permits publicity for its efforts. M.S. was funded by Australian Research Council (ARC) Discovery Early Career Researcher Award no. DE220100079. This is a publication of the National Centre for Coasts and Climate, funded by the Australian Government's National Environmental Science Program.

References

Abrahms, B., Aikens, E.O., Armstrong, J.B., et al. (2021). Emerging perspectives on resource tracking and animal movement ecology. *Trends in Ecology & Evolution*, 36, 308–320.

Aguilar, R., Ashworth, L., Galetto, L., and Aizen, M.A. (2006). Plant reproductive susceptibility to habitat fragmentation: review and synthesis through a meta-analysis. *Ecology Letters*, 9, 968–980.

Aguilar, R., Cristóbal-Pérez, E.J., Balvino-Olvera, F.J., et al. (2019). Habitat fragmentation reduces plant progeny quality: a global synthesis. *Ecology Letters*, 22, 1163–1173.

Ahtiainen, J.J., Alatalo, R.V., Mappes, J., and Vertainen, L. (2004). Decreased sexual signalling reveals reduced viability in small populations of the drumming wolf spider *Hygrolycosa rubrofasciata*. *Proceedings of the Royal Society B: Biological Sciences*, 271, 1839–1845.

Almond, R.E., Grooten, M., and Peterson, T. (2020). *Living Planet Report 2020: Bending the Curve of Biodiversity Loss*. World Wildlife Fund, Gland, Switzerland.

Andrieu, E., Dornier, A., Rouifed, S., et al. (2009). The town Crepis and the country Crepis: how does fragmentation affect a plant–pollinator interaction? *Acta Oecologica*, 35, 1–7.

Banks, S.C., Finlayson, G.R., Lawson, S., et al. (2005). The effects of habitat fragmentation due to forestry plantation establishment on the demography and genetic variation of a marsupial carnivore, *Antechinus agilis*. *Biological Conservation*, 122, 581–597.

Banks, S.C., Lindenmayer, D.B., McBurney, L., et al. (2011). Kin selection in den sharing develops under limited availability of tree hollows for a forest marsupial. *Proceedings of the Royal Society B: Biological Sciences*, 278, 2768–2776.

Banks, S.C., Piggott, M.P., Stow, A.J., and Taylor, A.C. (2007). Sex and sociality in a disconnected world: a review of the impacts of habitat fragmentation on animal social interactions. *Canadian Journal of Zoology*, 85, 1065–1079.

Banks-Leite, C., Ewers, R.M., Folkard-Tapp, H., and Fraser, A. (2020). Countering the effects of habitat loss, fragmentation, and degradation through habitat restoration. *One Earth*, 3, 672–676.

Barbarossa, V., Schmitt, R.J., Huijbregts, M.A., et al. (2020). Impacts of current and future large dams on the geographic range connectivity of freshwater fish worldwide. *Proceedings of the National Academy of Sciences*, 117, 3648–3655.

Battin, J. (2004). When good animals love bad habitats: ecological traps and the conservation of animal populations. *Conservation Biology*, 18, 1482–1491.

Belder, D.J., Pierson, J.C., Ikin, K., and Lindenmayer, D.B. (2018). Beyond pattern to process: current themes and future directions for the conservation of woodland birds through restoration plantings. *Wildlife Research*, 45, 473–489.

Bélisle, M., and Desrochers, A. (2002). Gap-crossing decisions by forest birds: an empirical basis for parameterizing spatially-explicit, individual-based models. *Landscape Ecology*, 17, 219–231.

Berger-Tal, O., Blumstein, D., and Swaisgood, R.R. (2020). Conservation translocations: a review of common difficulties and promising directions. *Animal Conservation*, 23, 121–131.

Berger-Tal, O., Polak, T., Oron, A., et al. (2011). Integrating animal behavior and conservation biology: a conceptual framework. *Behavioral Ecology*, 22, 236–239.

Beyer, R.M., and Manica, A. (2020). Historical and projected future range sizes of the world's mammals, birds, and amphibians. *Nature Communications*, 11, 5633.

Blumstein, D.T., Hayes, L.D., and Pinter-Wollman, N. (2022). Social consequences of rapid environmental change. *Trends in Ecology & Evolution*, 38, 337–345.

Bonte, D., and Van Dyck, H. (2009). Mate-locating behaviour, habitat-use, and flight morphology relative to rainforest disturbance in an Afrotropical butterfly. *Biological Journal of the Linnean Society*, 96, 830–839.

Bosco, L., Cushman, S., Wan, H., et al. (2021). Fragmentation effects on woodlark habitat selection depend on habitat amount and spatial scale. *Animal Conservation*, 24, 84–94.

Bowler, D.E., and Benton, T.G. (2005). Causes and consequences of animal dispersal strategies: relating individual behaviour to spatial dynamics. *Biological Reviews of the Cambridge Philosophical Society*, 80, 205–225.

Brooker, L., and Brooker, M.G. (2002). Dispersal and population dynamics of the blue-breasted fairy-wren, *Malurus pulcherrimus*, in fragmented habitat in the Western Australian wheatbelt. *Wildlife Research*, 29, 225–233.

Brudvig, L.A., Damschen, E.I., Haddad, N.M., et al. (2015). The influence of habitat fragmentation on multiple plant–animal interactions and plant reproduction. *Ecology*, 96, 2669–2678.

Bryan-Brown, D.N., Connolly, R.M., Richards, D.R., et al. (2020). Global trends in mangrove forest fragmentation. *Scientific Reports*, 10, 1–8.

Carstens, K.F., Kassanjee, R., Little, R.M., et al. (2019). Breeding success and population growth of Southern Ground Hornbills *Bucorvus leadbeateri* in an area supplemented with nest-boxes. *Bird Conservation International*, 29, 627–643.

Cattarino, L., McAlpine, C.A., and Rhodes, J.R. (2016). Spatial scale and movement behaviour traits control the impacts of habitat fragmentation on individual fitness. *Journal of Animal Ecology*, 85, 168–177.

Charter, M., Izhaki, I., Bouskila, A., and Leshem, Y. (2007). Breeding success of the Eurasian Kestrel (*Falco tinnunculus*) nesting on buildings in Israel. *Journal of Raptor Research*, 41, 139–143.

Cheptou, P.O., and Avendaño, L.G. (2006). Pollination processes and the Allee effect in highly fragmented populations: consequences for the mating system in urban environments. *New Phytologist*, 172, 774–783.

Clevenger, A.P., and Huijser, M.P. (2011). *Wildlife Crossing Structure Handbook: Design and Evaluation in North America*. Publication no. FHWA-CFL/TD-11-003. Federal Highway Administration, Central Federal Lands Highway Division, USA.

Cooke, S.J., and Hinch, S.G. (2013). Improving the reliability of fishway attraction and passage efficiency estimates to inform fishway engineering, science, and practice. *Ecological Engineering*, 58, 123–132.

Cornelius, C., Awade, M., Cândia-Gallardo, C., et al. (2017). Habitat fragmentation drives inter-population variation in dispersal behavior in a Neotropical rainforest bird. *Perspectives in Ecology and Conservation*, 15, 3–9.

Cote, J., Clobert, J., Brodin, T., et al. (2010). Personality-dependent dispersal: characterization, ontogeny and consequences for spatially structured populations. *Philosophical Transactions of the Royal Society B: Biological Sciences*, 365, 4065–4076.

Cousseau, L., Hammers, M., Van de Loock, D., et al. (2020). Habitat fragmentation shapes natal dispersal and sociality in an Afrotropical cooperative breeder. *Proceedings of the Royal Society B: Biological Sciences*, 287, 20202428.

Dahirel, M., Olivier, E., Guiller, A., et al. (2015). Movement propensity and ability correlate with ecological specialization in European land snails: comparative analysis of a dispersal syndrome. *Journal of Animal Ecology*, 84, 228–238.

Davidson, N.C. (2014). How much wetland has the world lost? Long-term and recent trends in global wetland area. *Marine and Freshwater Research*, 65, 934–941.

Davis, J.M., and Stamps, J.A. (2004). The effect of natal experience on habitat preferences. *Trends in Ecology & Evolution*, 19, 411–416.

Desrochers, A., and Hannon, S.J. (1997). Gap crossing decisions by forest songbirds during the post-fledging period: decisíones de cruce de claros por aves paserinas de bosques durante el periodo post-juvenil. *Conservation Biology*, 11, 1204–1210.

Doherty, T.S., and Driscoll, D.A. (2018). Coupling movement and landscape ecology for animal conservation in production landscapes. *Proceedings of the Royal Society B: Biological Sciences*, 285, 20172272.

Domenici, P., and Seebacher, F. (2020). The impacts of climate change on the biomechanics of animals. *Conservation Physiology*, 8, coz102.

Dunic, J.C., Brown, C.J., Connolly, R.M., et al. (2021). Long-term declines and recovery of meadow area across the world's seagrass bioregions. *Global Change Biology*, 27, 4096–4109.

Ewers, R.M., and Didham, R.K. (2006). Confounding factors in the detection of species responses to habitat fragmentation. *Biological Reviews*, 81, 117–142.

Fahrig, L. (2003). Effects of habitat fragmentation on biodiversity. *Annual Review of Ecology, Evolution, and Systematics*, 34, 487–515.

Fahrig, L. (2017). Ecological responses to habitat fragmentation per se. *Annual Review of Ecology, Evolution, and Systematics*, 48, 1–23.

Fahrig, L., Arroyo-Rodríguez, V., Bennett, J.R., et al. (2019). Is habitat fragmentation bad for biodiversity? *Biological Conservation*, 230, 179–186.

Farneda, F.Z., Rocha, R., López-Baucells, A., et al. (2015). Trait-related responses to habitat fragmentation in Amazonian bats. *Journal of Applied Ecology*, 52, 1381–1391.

Filgueiras, B.K., Peres, C.A., Melo, F.P., et al. (2021). Winner–loser species replacements in human-modified landscapes. *Trends in Ecology & Evolution*, 36, 545–555.

Fisher, D.N., Kilgour, R.J., Siracusa, E.R., et al. (2021). Anticipated effects of abiotic environmental change on intraspecific social interactions. *Biological Reviews*, 96, 2661–2693.

Fletcher Jr, R.J., Didham, R.K., Banks-Leite, C., et al. (2018). Is habitat fragmentation good for biodiversity? *Biological Conservation*, 226, 9–15.

Flood, C.E., and Wong, M.Y. (2017). Social stability in times of change: effects of group fusion and water depth on sociality in a globally invasive fish. *Animal Behaviour*, 129, 71–79.

Forman, R.T. (2012). *Safe Passages: Highways, Wildlife, and Habitat Connectivity*. Island Press, Washington, DC.

Friess, D.A., Rogers, K., Lovelock, C.E., et al. (2019). The state of the world's mangrove forests: past, present, and future. *Annual Review of Environment and Resources*, 44, 89–115.

Fuller, M.R., Doyle, M.W., and Strayer, D.L. (2015). Causes and consequences of habitat fragmentation in river networks. *Annals of the New York Academy of Sciences*, 1355, 31–51.

Gil, M.A., Emberts, Z., Jones, H., and St. Mary, C.M. (2016). Social information on fear and food drives animal grouping and fitness. *The American Naturalist*, 189, 227–241.

Gilby, B.L., Olds, A.D., Connolly, R.M., et al. (2018). Spatial restoration ecology: placing restoration in a landscape context. *Bioscience*, 68, 1007–1019.

Gillies, C.L., McLeod, I.M., Alleway, H.K., et al. (2018). Australian shellfish ecosystems: past distribution, current status and future direction. *PLOS ONE*, 13, e0190914.

Giuntini, S., and Pedruzzi, L. (2022). Sex and the patch: the influence of habitat fragmentation on terrestrial vertebrates' mating strategies. *Ethology, Ecology & Evolution*, 1–30.

Goerig, E., and Castro-Santos, T. (2017). Is motivation important to brook trout passage through culverts? *Canadian Journal of Fisheries and Aquatic Sciences*, 74, 885–893.

Gordon, T.A., Radford, A.N., Davidson, I.K., et al. (2019). Acoustic enrichment can enhance fish community development on degraded coral reef habitat. *Nature Communications*, 10, 1–7.

Greggor, A.L., Berger-Tal, O., and Blumstein, D.T. (2020). The rules of attraction: the necessary role of animal cognition in explaining conservation failures and successes. *Annual Review of Ecology Evolution and Systematics*, 51, 483–503.

Greggor, A.L., Berger-Tal, O., Blumstein, D.T., et al. (2016). Research priorities from animal behaviour for maximising conservation progress. *Trends in Ecology & Evolution*, 31, 953–964.

Grill, G., Lehner, B., Lumsdon, A.E., et al. (2015). An index-based framework for assessing patterns and trends in river fragmentation and flow regulation by global dams at multiple scales. *Environmental Research Letters*, 10, 015001.

Haddad, N.M., Brudvig, L.A., Clobert, J., et al. (2015). Habitat fragmentation and its lasting impact on Earth's ecosystems. *Science Advances*, 1, e1500052.

Hadley, A.S., and Betts, M.G. (2012). The effects of landscape fragmentation on pollination dynamics: absence of evidence not evidence of absence. *Biological Reviews*, 87, 526–544.

Hale, R., Blumstein, D.T., Mac Nally, R., and Swearer, S.E. (2020). Harnessing knowledge of animal behavior to improve habitat restoration outcomes. *Ecosphere*, 11, e03104.

Hale, R., Coleman, R., Pettigrove, V., and Swearer, S.E. (2015a). Identifying, preventing and mitigating ecological traps to improve the management of urban aquatic ecosystems. *Journal of Applied Ecology*, 52, 928–939.

Hale, R., Coleman, R., Sievers, M., et al. (2018). Using conservation behavior to manage ecological traps for a threatened freshwater fish. *Ecosphere*, 9, e02381.

Hale, R., Mac Nally, R., Blumstein, D.T., and Swearer, S.E. (2019). Evaluating where and how habitat restoration is undertaken for animals. *Restoration Ecology*, 27, 775–781.

Hale, R., and Swearer, S.E. (2016). Ecological traps: current evidence and future directions. *Proceedings of the Royal Society B: Biological Sciences*, 283, 20152647.

Hale, R., and Swearer, S.E. (2017). When good animals love bad restored habitats: how maladaptive habitat selection can constrain restoration. *Journal of Applied Ecology*, 54, 1478–1486.

Hale, R., Treml, E.A., and Swearer, S.E. (2015b). Evaluating the metapopulation consequences of ecological traps. *Proceedings of the Royal Society B: Biological Sciences*, 282, 20142930.

Hamilton, S.E., and Casey, D. (2016). Creation of a high spatio-temporal resolution global database of continuous mangrove forest cover for the 21st century

(CGMFC-21). *Global Ecology and Biogeography*, 25, 729–738.

Harrison, T., and Winfree, R. (2015). Urban drivers of plant-pollinator interactions. *Functional Ecology*, 29, 879–888.

Hawlena, D., Saltz, D., Abramsky, Z., and Bouskila, A. (2010). Ecological trap for desert lizards caused by anthropogenic changes in habitat structure that favor predator activity. *Conservation Biology*, 24, 803–809.

Hermansen, T.D., Minchinton, T.E., and Ayre, D.J. (2017). Habitat fragmentation leads to reduced pollinator visitation, fruit production and recruitment in urban mangrove forests. *Oecologia*, 185, 221–231.

Huijbers, C.M., Nagelkerken, I., Lössbroek, P.A.C., et al. (2012). A test of the senses: fish select novel habitats by responding to multiple cues. *Ecology*, 93, 46–55.

Jones, M.E., Bain, G.C., Hamer, R.P., et al. (2021). Research supporting restoration aiming to make a fragmented landscape 'functional' for native wildlife. *Ecological Management & Restoration*, 22, 65–74.

Jones, M.J., and Hale, R. (2020). Using knowledge of behaviour and optic physiology to improve fish passage through culverts. *Fish and Fisheries*, 21, 557–569.

Joseph, L.N., Maloney, R.F., and Possingham, H.P. (2009). Optimal allocation of resources among threatened species: a project prioritization protocol. *Conservation Biology*, 23, 328–338.

Karkarey, R., Alcoverro, T., Kumar, S., and Arthur, R. (2017). Coping with catastrophe: foraging plasticity enables a benthic predator to survive in rapidly degrading coral reefs. *Animal Behaviour*, 131, 13–22.

Keenan, R.J., Reams, G.A., Achard, F., et al. (2015). Dynamics of global forest area: results from the FAO Global Forest Resources Assessment 2015. *Forest Ecology and Management*, 352, 9–20.

Kentula, M.E., Gwin, S.E., and Pierson, S.M. (2004). Tracking changes in wetlands with urbanization: sixteen years of experience in Portland, Oregon, USA. *Wetlands*, 24, 734–743.

Kintsch, J., Jacobson, S., and Cramer, P. (2015). The Wildlife Crossing Guilds decision framework: a behavior-based approach to designing effective wildlife crossing structures. In: Proceedings of the 2015 International Conference on Ecology and Transportation, ICOET Conference 2015 Session (Vol. 201).

Komyakova, V., Chamberlain, D., and Swearer, S.E. (2021). A multi-species assessment of artificial reefs as ecological traps. *Ecological Engineering*, 171, 106394.

Kupfer, J.A., Malanson, G.P., and Franklin, S.B. (2006). Not seeing the ocean for the islands: the mediating influence of matrix-based processes on forest fragmentation effects. *Global Ecology and Biogeography*, 15, 8–20.

Laiolo, P., Vögeli, M., Serrano, D., and Tella, J.L. (2008). Song diversity predicts the viability of fragmented bird populations. *PLOS ONE*, 3, e1822.

Lane, J.E., Forrest, M.N., and Willis, C.K. (2011). Anthropogenic influences on natural animal mating systems. *Animal Behaviour*, 81, 909–917.

Lattanzio, M.S., and Miles, D.B. (2014). Ecological divergence among colour morphs mediated by changes in spatial network structure associated with disturbance. *Journal of Animal Ecology*, 83, 1490–1500.

Laurance, W.F. (2014). Contemporary drivers of habitat fragmentation. In: Chris J. Kettle and Lian Pin Koh (eds), *Global Forest Fragmentation*, pp. 20–27. CABI, Boston, MA.

Laurent, E., Schtickzelle, N., and Jacob, S. (2020). Fragmentation mediates thermal habitat choice in ciliate microcosms. *Proceedings of the Royal Society B: Biological Sciences*, 287, 20192818.

Lindell, C.A. (2008). The value of animal behavior in evaluations of restoration success. *Restoration Ecology*, 16, 197–203.

Lino, A., Fonseca, C., Rojas, D., et al. (2019). A meta-analysis of the effects of habitat loss and fragmentation on genetic diversity in mammals. *Mammalian Biology*, 94, 69–76.

Lister, N.-M., Brocki, M., and Ament, R. (2015). Integrated adaptive design for wildlife movement under climate change. *Frontiers in Ecology and the Environment*, 13, 493–502.

Londe, D.W., Elmore, R.D., Davis, C.A., et al. (2022). Why did the chicken not cross the road? Anthropogenic development influences the movement of a grassland bird. *Ecological Applications*, 32, e2543.

Lynch, A.J., Thompson, L.M., Beever, E.A., et al. (2021). Managing for RADical ecosystem change: applying the Resist-Accept-Direct (RAD) framework. *Frontiers in Ecology and the Environment*, 19, 461–469.

Macdonald, D., Harmsen, B., Johnson, P., and Newman, C. (2004). Increasing frequency of bite wounds with increasing population density in Eurasian badgers, *Meles meles*. *Animal Behaviour*, 67, 745–751.

Maciel, G.A., and Lutscher, F. (2013). How individual movement response to habitat edges affects population persistence and spatial spread. *The American Naturalist*, 182, 42–52.

Maggs, G., Norris, K., Zuël, N., et al. (2019). Quantifying drivers of supplementary food use by a reintroduced, critically endangered passerine to inform management and habitat restoration. *Biological Conservation*, 238, 108240.

Mathot, K.J., and Giraldeau, L.-A. (2010). Within-group relatedness can lead to higher levels of exploitation:

a model and empirical test. *Behavioral Ecology*, 21, 843–850.

Maurer, C., Bosco, L., Klaus, E., et al. (2020). Habitat amount mediates the effect of fragmentation on a pollinator's reproductive performance, but not on its foraging behaviour. *Oecologia*, 193, 523–534.

Mazza, V., Dammhahn, M., Lösche, E., and Eccard, J.A. (2020). Small mammals in the big city: behavioural adjustments of non-commensal rodents to urban environments. *Global Change Biology*, 26, 6326–6337.

McWilliams, C., Lurgi, M., Montoya, J.M., et al. (2019). The stability of multitrophic communities under habitat loss. *Nature Communications*, 10, 1–11.

Mekonnen, A., Fashing, P.J., Sargis, E.J., et al. (2018). Flexibility in positional behavior, strata use, and substrate utilization among Bale monkeys (*Chlorocebus djamdjamensis*) in response to habitat fragmentation and degradation. *American Journal of Primatology*, 80, e22760.

Méndez, M., Tella, J.L., and Godoy, J.A. (2011). Restricted gene flow and genetic drift in recently fragmented populations of an endangered steppe bird. *Biological Conservation*, 144, 2615–2622.

Morgado, R., Beja, P., Reino, L., et al. (2010). Calandra lark habitat selection: strong fragmentation effects in a grassland specialist. *Acta Oecologica*, 36, 63–73.

Nathan, R., Getz, W.M., Revilla, E., et al. (2008). A movement ecology paradigm for unifying organismal movement research. *Proceedings of the National Academy of Sciences*, 105, 19,052–19,059.

Nisi, A.C., Benson, J.F., and Wilmers, C.C. (2022). Puma responses to unreliable human cues suggest an ecological trap in a fragmented landscape. *Oikos*, 2022, e09051.

Parris, K.M. (2016). *Ecology of Urban Environments*. John Wiley & Sons, Chichester.

Pfeifer, M., Lefebvre, V., Peres, C., et al. (2017). Creation of forest edges has a global impact on forest vertebrates. *Nature*, 551, 187–191.

Püttker, T., Crouzeilles, R., Almeida-Gomes, M., et al. (2020). Indirect effects of habitat loss via habitat fragmentation: a cross-taxa analysis of forest-dependent species. *Biological Conservation*, 241, 108368.

Rahman, T., and Candolin, U. (2022). Linking animal behavior to ecosystem change in disturbed environments. *Frontiers in Ecology and Evolution*, 10.

Raynor, J.L., Grainger, C.A., and Parker, D.P. (2021). Wolves make roadways safer, generating large economic returns to predator conservation. *Proceedings of the National Academy of Sciences*, 118, e2023251118.

Resetarits, W.J., and Binckley, C.A. (2009). Spatial contagion of predation risk affects colonization dynamics in experimental aquatic landscapes. *Ecology*, 90, 869–876.

Resetarits, W.J., and Silberbush, A. (2016). Local contagion and regional compression: habitat selection drives spatially explicit, multiscale dynamics of colonisation in experimental metacommunities. *Ecology Letters*, 19, 191–200.

Robertson, B.A., and Hutto, R.L. (2006). A framework for understanding ecological traps and an evaluation of existing evidence. *Ecology*, 87, 1075–1085.

Rybicki, J., Abrego, N., and Ovaskainen, O. (2020). Habitat fragmentation and species diversity in competitive communities. *Ecology Letters*, 23, 506–517.

Schofield, L.N., Loffland, H.L., Siegel, R.B., et al. (2018). Using conspecific broadcast for Willow Flycatcher restoration. *Avian Conservation and Ecology*, 13.

Seebacher, F., and Post, E. (2015). Climate change impacts on animal migration. *Climate Change Responses*, 2, 1–2.

Segan, D.B., Murray, K.A., and Watson, J.E.M. (2016). A global assessment of current and future biodiversity vulnerability to habitat loss–climate change interactions. *Global Ecology and Conservation*, 5, 12–21.

Selwood, K., Mac Nally, R., and Thomson, J.R. (2009). Native bird breeding in a chronosequence of revegetated sites. *Oecologia*, 159, 435–446.

Severns, P.M. (2011). Habitat restoration facilitates an ecological trap for a locally rare, wetland-restricted butterfly. *Insect Conservation and Diversity*, 4, 184–191.

Shettleworth, S.J. (2009). *Cognition, Evolution, and Behavior*. Oxford University Press, New York, NY.

Sievers, M., Brown, C.J., Buelow, C., et al. (2022). Greater consideration of animals will enhance coastal restoration outcomes. *Bioscience*, 72, 1088–1098.

Sievers, M., Parris, K.M., Swearer, S.E., and Hale, R. (2018). Stormwater wetlands can function as ecological traps for urban frogs. *Ecological Applications*, 28, 1106–1115.

Sih, A. (2013). Understanding variation in behavioural responses to human-induced rapid environmental change: a conceptual overview. *Animal Behaviour*, 85, 1077–1088.

Sih, A., Montiglio, P.-O., Wey, T.W., and Fogarty, S. (2017). Altered physical and social conditions produce rapidly reversible mating systems in water striders. *Behavioral Ecology*, 28, 632–639.

Sih, A., Trimmer, P.C., and Ehlman, S.M. (2016). A conceptual framework for understanding behavioral responses to HIREC. *Current Opinion in Behavioral Sciences*, 12, 109–114.

Silva, A.T., Lucas, M.C., Castro-Santos, T., et al. (2018). The future of fish passage science, engineering, and practice. *Fish and Fisheries*, 19, 340–362.

Sitters, H., and Di Stefano, J. (2020). Integrating functional connectivity and fire management for better conservation outcomes. *Conservation Biology*, 34, 550–560.

Sloman, K.A., Taylor, A.C., Metcalfe, N.B., and Gilmour, K.M. (2001). Effects of an environmental perturbation on the social behaviour and physiological function of brown trout. *Animal Behaviour*, 61, 325–333.

Smith, D.J., Van Der Ree, R., and Rosell, C. (2015). Wildlife crossing structures: an effective strategy to restore or maintain wildlife connectivity across roads. In: R. Van Der Ree, D.J. Smith, and C. Grilo (eds), *Handbook of Road Ecology*, pp. 172–183. John Wiley & Sons, Chichester.

Soanes, K., Taylor, A.C., Sunnucks, P., et al. (2018). Evaluating the success of wildlife crossing structures using genetic approaches and an experimental design: lessons from a gliding mammal. *Journal of Applied Ecology*, 55, 129–138.

Sol, D., Lapiedra, O., and Ducatez, S. (2020). Cognition and adaptation to urban environments. In: M. Szulkin, J. Munshi-South, and A. Charmantier (eds), *Urban Evolutionary Biology*, pp. 253–265. Oxford University Press, New York, NY.

Spiegel, O., Leu, S.T., Bull, C.M., and Sih, A. (2017). What's your move? Movement as a link between personality and spatial dynamics in animal populations. *Ecology Letters*, 20, 3–18.

Stevens, V.M., Whitmee, S., Le Galliard, J.F., et al. (2014). A comparative analysis of dispersal syndromes in terrestrial and semi-terrestrial animals. *Ecology Letters*, 17, 1039–1052.

Sun, J., Tummers, J.S., Galib, S.M., and Lucas, M.C. (2022). Fish community and abundance response to improved connectivity and more natural hydromorphology in a post-industrial subcatchment. *Science of the Total Environment*, 802, 149720.

Szulkin, M., Munshi-South, J., and Charmantier, A. (2020). *Urban Evolutionary Biology*. Oxford University Press, New York, NY.

Szulkin, M., Stopher, K.V., Pemberton, J.M., and Reid, J.M. (2013). Inbreeding avoidance, tolerance, or preference in animals? *Trends in Ecology & Evolution*, 28, 205–211.

Taubert, F., Fischer, R., Groeneveld, J., et al. (2018). Global patterns of tropical forest fragmentation. *Nature*, 554, 519–522.

Teckentrup, L., Kramer-Schadt, S., and Jeltsch, F. (2019). The risk of ignoring fear: underestimating the effects of habitat loss and fragmentation on biodiversity. *Landscape Ecology*, 34, 2851–2868.

Thompson, L.M., Lynch, A.J., Beever, E.A., et al. (2021). Responding to ecosystem transformation: resist, accept, or direct? *Fisheries*, 46, 8–21.

Thomson, J.R., Moilanen, A.J., Vesk, P.A., et al. (2009). Where and when to revegetate: a quantitative method for scheduling landscape reconstruction. *Ecological Applications*, 19, 817–828.

Titeux, N., Aizpurua, O., Hollander, F.A., et al. (2020). Ecological traps and species distribution models: a challenge for prioritizing areas of conservation importance. *Ecography*, 43, 365–375.

Tonkin, J.D., Stoll, S., Sundermann, A., and Haase, P. (2014). Dispersal distance and the pool of taxa, but not barriers, determine the colonisation of restored river reaches by benthic invertebrates. *Freshwater Biology*, 59, 1843–1855.

Tucker, M.A., Böhning-Gaese, K., Fagan, W.F., et al. (2018). Moving in the Anthropocene: global reductions in terrestrial mammalian movements. *Science*, 359, 466–469.

Tuomainen, U., and Candolin, U. (2011). Behavioural responses to human-induced environmental change. *Biological Reviews*, 86, 640–657.

Valente, J.J., LeGrande-Rolls, C.L., Rivers, J.W., et al. (2021). Conspecific attraction for conservation and management of terrestrial breeding birds: current knowledge and future research directions. *The Condor*, 123, duab007.

Valiente-Banuet, A., Aizen, M.A., Alcántara, J.M., et al. (2015). Beyond species loss: the extinction of ecological interactions in a changing world. *Functional Ecology*, 29, 299–307.

van Baaren, J., and Candolin, U. (2018). Plasticity in a changing world: behavioural responses to human perturbations. *Current Opinion in Insect Science*, 27, 21–25.

Van Dyck, H. (2012). Changing organisms in rapidly changing anthropogenic landscapes: the significance of the Umwelt'-concept and functional habitat for animal conservation. *Evolutionary Applications*, 5, 144–153.

Van Moorter, B., Rolandsen, C.M., Basille, M., and Gaillard, J.M. (2016). Movement is the glue connecting home ranges and habitat selection. *Journal of Animal Ecology*, 85, 21–31.

Wang, M., and Schreiber, A. (2001). The impact of habitat fragmentation and social structure on the population genetics of roe deer (Capreolus capreolus L.) in Central Europe. *Heredity*, 86, 703–715.

Wilson, M.W., Ridlon, A.D., Gaynor, K.M., et al. (2020). Ecological impacts of human-induced animal behaviour change. *Ecology Letters*, 23, 1522–1536.

CHAPTER 9

Urbanization

Daniel Sol, Yolanda Melero, Lisieux Fuzessy, and César González-Lagos

Overview

Urbanization is a rapid and extreme form of environmental alteration that threatens biodiversity by causing a mismatch between an organism's phenotype and its environment. Despite the mismatch, some animals demonstrate a remarkable ability to thrive in urbanized environments. Here, we explore the role of behaviour in mediating the varying success of animals in an increasingly urbanized world. We contend that behaviour is of utmost importance because it reflects decision-making processes that influence the settlement of animals in urban environments and their ability to cope with the challenges posed by phenotype-environment mismatches. Such decisions range from simple, innate reactions to cognitively sophisticated responses based on learning, with ample evidence that they are often adaptive. Despite their adaptive significance, however, their demographic and evolutionary implications remain poorly understood, primarily due to a research focus on less cognitively demanding decisions and a tendency to neglect the importance of life history variation. By delving deeper into the intricacies of animal behaviour, we have the potential to fill these gaps and gain valuable insights into the varying degrees of success observed among animals in urban environments and thereby enhancing our understanding of urban biodiversity.

9.1 Introduction

Urbanization is widely recognized as a major driver of global change, representing one of the fastest and most extreme forms of human-induced environmental alteration (Grimm et al. 2008; Johnson and Munshi-South 2017; Marzluff et al. 2001). With urban land projected to increase by a factor of 1.8–5.9 over the 21st century (Gao and O'Neill 2020), concerns are growing about the global impact of such rapid expansion on biodiversity (Güneralp and Seto 2013). At the local scale, there is also concern that urbanization will alter the functioning and stability of ecosystems, endangering the services on which human societies depend. In India, for instance, the decline in vulture populations has resulted in increased availability of waste food for feral dogs, leading to their proliferation in cities; this has resulted in intensified pressures on wildlife through competition and predation, and a higher risk of rabies transmission to humans (Soulsbury and White 2015). There is thus a growing pressure to understand how urbanization will impact biodiversity.

Previous studies have uncovered several consistent biodiversity patterns in urban areas (McKinney 2002; Shochat et al. 2006; Figure 9.1). First, highly urbanized areas, such as city centres, tend to contain fewer species than the non-urban areas that surround them (Aronson et al. 2014). Second, while most of the species living in cities occur at low abundance, a few can become extremely abundant and dominate the entire community (Shochat et al. 2006; Sol et al. 2020a). We call these species urban exploiters (Blair 2001). Third, non-native species tend to be more frequent and abundant within cities than in the surrounding non-urban environments (Cadotte et al. 2017). Finally, both urban exploiters and non-native species are overrepresented in

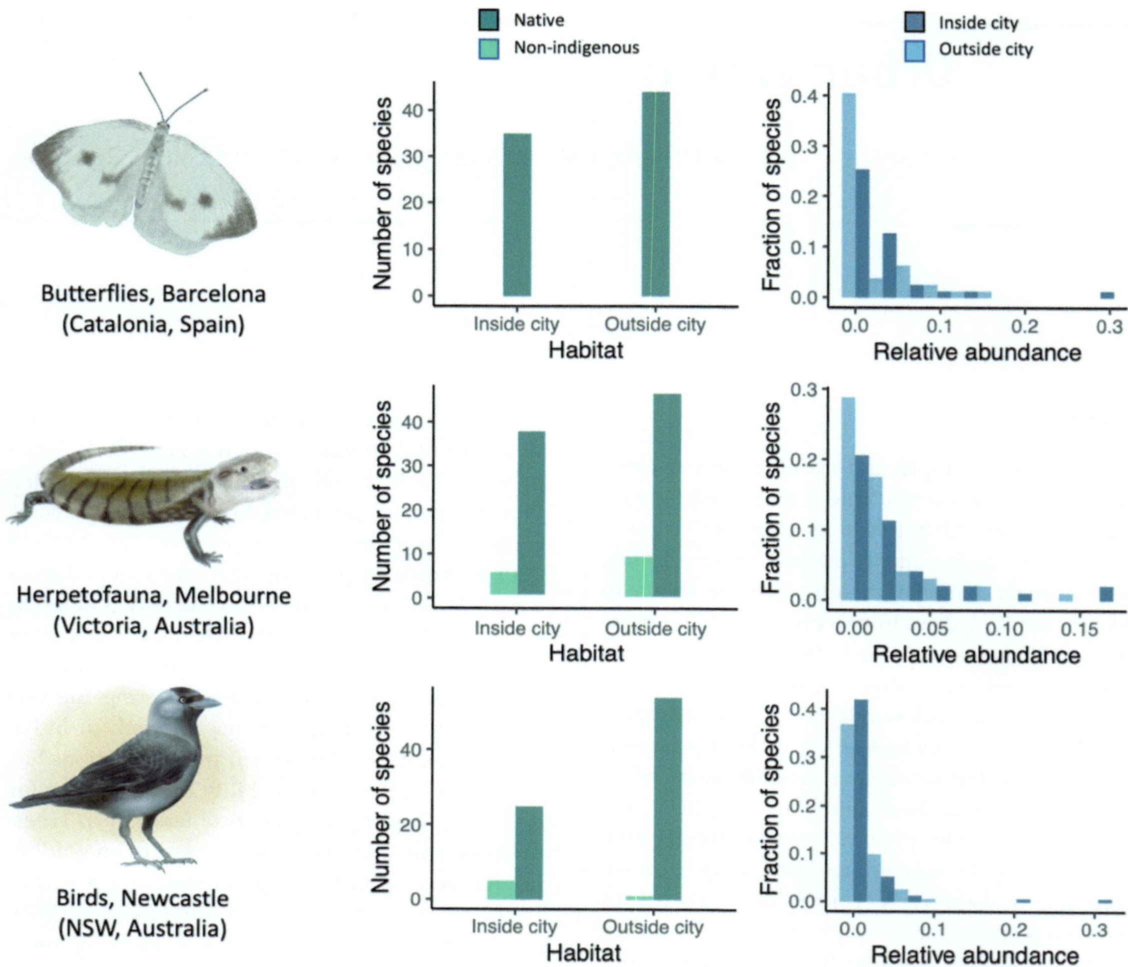

Figure 9.1 Biodiversity patterns in urban areas compared to surrounding non-urban areas, in butterflies from Barcelona (Y. Melero, unpublished), herpetofauna (amphibians and reptiles) from Melbourne (Hamer and Mcdonnell 2009), and birds from Newcastle (Sol et al. 2012). Compared to surrounding rural and wildland areas, cities tend to contain fewer species and a greater dominance of a few superabundant species, which are often non-native and belong to a few taxonomic clades.

All animal drawings by D.S.

a few taxonomic groups (Duncan et al. 2003; Sol et al. 2017a), contributing to biotic homogenization (McKinney 2006). The above patterns suggest that cities present both challenges and opportunities for organisms, and that to gauge the implications of urbanization for biodiversity, we must delve into the factors that determine why some species are more successful than others in colonizing and exploiting urban habitats.

In this chapter, we focus on the role of behaviour in the varying success of animals in urbanized environments. We contend that the importance of behaviour lies in the fact that it expresses the outcome of decision-making processes in two relevant contexts: first, the choice of whether to settle in urban environments and, second, how to navigate the novel challenges posed by such environments. Cities, with their bustling human activity, constant traffic, dense building infrastructure, and pervasive pollution, starkly differ from the surrounding natural habitats that have shaped the evolution of organisms over millennia.

Consequently, a profound mismatch often emerges between the current phenotype of organisms and the urban environment. This should diminish the likelihood of animals perceiving cities as ecological opportunities. In addition, for animals already present or opting to settle in urban areas, the mismatch could also increase the risk of extinction due to maladaptation. While these two processes may account for why most animals avoid cities or occur there in low numbers, the observation that certain species exhibit a remarkable ability to thrive and proliferate in urban areas warrants explanation. Our central thesis in this chapter is that decision-making processes can potentially help to bridge the gap between current phenotype and the new environment, thus increasing the probability of successfully colonizing and proliferating in cities (Figure 9.2). Throughout the chapter, we explore the complexities of animal decision-making, shed light on the unique challenges organisms face in urban contexts, assess the types of decisions most pertinent in these environments, and evaluate the implications of those decisions for population persistence and adaptive evolution. We also acknowledge the gaps in our current understanding and propose avenues for future research.

9.2 The hurdles to becoming an urban-dweller

As with any colonization process, urban biodiversity can be seen as the outcome of two main hurdles (Stamps 2001; Evans et al. 2010; Sol et al. 2013; Caspi et al. 2022). The first is the dispersal filter. Although some species may have been present in the area long before it was urbanized, or may have been introduced by humans after the area was developed, the most likely scenario is that many species from urban communities come from surrounding non-urbanized environments (Clergeau et al. 2001; Sol et al. 2014; Johnson and Munshi-South 2017). For example, in the city of Hamburg, Germany, northern goshawks *Accipiter gentilis* regularly visited the city centre decades before the first successful breeding attempts were ever recorded (Rutz 2008). Now the species is well-established in the city. Likewise, foxes *Vulpes vulpes* in the UK were initially confined to rural and natural habitats until the 1930s, when they were first reported to be resident in urban areas in southern England (Scott et al. 2014). If urban communities mainly result from dispersal, some species can be absent there simply because they have failed to disperse from their original habitats (Blair 2001; Clergeau et al. 2001; Sol et al. 2014).

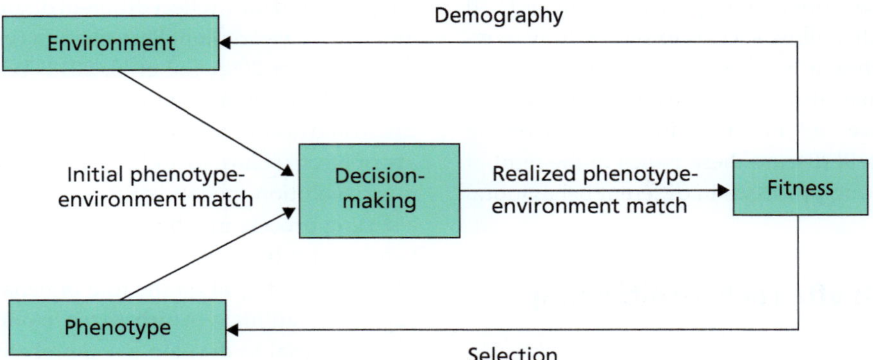

Figure 9.2 The role of decision-making in the success of animals in novel environments. The fitness of individuals relies on the level of match between their phenotype and the environment, diminishing when the match is poor. In the novel environment, the influence of decisions on fitness lies in establishing a general alignment between the phenotype and the environment (realized phenotype-environment match). In instances where the initial phenotype-environment match is appropriate, decisions based on matching habitat choice are optimal. However, in cases of phenotype-environment mismatch, learning becomes crucial in adjusting decision-making processes to mitigate the mismatch. If there is a weak initial phenotype-environment match due to the animal's generalist nature, the fitness of individuals may not rely heavily on matching habitat choices, though learning may still play a central role.

Dispersal, however, is only part of the process needed to become an urban-dweller. After arrival, a second hurdle comes from the need to successfully survive and reproduce in the urban environment. Many species are absent or rare in cities despite being abundant in surrounding natural and rural habitats, suggesting that they may not tolerate urbanization well. This is to be expected, considering that cities can pose myriad challenges for organisms that can lead to maladaptation, including the need to find new resources, deal with disturbances by people and pets (see also Chapter 13), avoid risks associated with infrastructure and transportation systems (Chapter 8), and cope with noise (Chapter 2), chemical (Chapter 3), and light (Chapter 4) pollution. Even non-native introduced species, or those that were already present in the area before it was urbanized, can suffer from the difficulties of having to deal with these challenges. Consequently, many species are absent or rare in cities because they simply failed to establish a self-sustaining population.

Thus, the combined hurdles associated with dispersal and population persistence are crucial to understanding biodiversity patterns in urban environments. If only certain animals recognize cities as ecological opportunities—and possess the appropriate phenotypes to effectively cope with, and adapt to, the new challenges—this may explain why cities have a reduced species diversity and a prevalence of a few dominant species (Figure 9.1). If the features that allow species to successfully overcome the different hurdles are evolutionarily conserved, this may also explain the high biotic homogenization observed in cities. In the forthcoming sections we will discuss these issues in the light of decision-making processes in different behavioural domains.

9.3 Animals often behave differently in cities

If individuals within species exhibit variation in phenotypic suitability to the new environmental conditions, the filtering that takes place during the dispersal and establishment stages should result in phenotypic differentiation from source populations (Price and Sol 2008). There is a wealth of evidence indicating that urban animals often behave differently from non-urban animals (Lowry et al. 2013; Sol et al. 2013; Alberti et al. 2017; Ritzel and Gallo 2020; Diamond and Martin 2021). One example is the establishment of dark-eyed juncos *Junco hyemalis* on the campus of the University of California San Diego, USA, which was accompanied by an increased tolerance towards people, a reduced aggressiveness among individuals, a preference for elevated nesting sites, and an extended reproductive season (Yeh and Price 2004; Yeh et al. 2007). Behavioural differences between urban and non-urban individuals have been documented in a plethora of behavioural domains, such as resource utilization, temporal organization, risk avoidance, and communication (Table 9.1), highlighting the pivotal role of behaviour in explaining the varying success of animals in urbanized environments.

The existence of behavioural differences between urban and non-urban animals implies that the decisions individuals make in their natural habitats may not always be optimal in urban environments. However, documenting behavioural differences is insufficient to understand the mechanisms through which decisions made by animals affect their likelihood of becoming an urban-dweller. Differences may come from decisions that animals make before and/or after settlement in urban environments; furthermore, such decisions may be adaptive or non-adaptive, and may reflect different degrees of plastic and heritable variation (Partecke and Gwinner 2007; Price and Sol 2008; Sol et al. 2013; Winchell et al. 2023). Therefore, behavioural differences between urban and non-urban animals may arise from a variety of mechanisms, including phenotypic plasticity, microevolution (genetic drift or selection), and/or a sorting process in which individuals with certain behavioural traits actively choose to inhabit urban areas. To shed light on the role of behaviour in the success of animals in urbanized environments, it is thus crucial to develop a comprehensive mechanistic understanding that establishes links between relevant decision-making processes and their demographic and evolutionary consequences. Examining the nature of the cognitive processes involved and their role in adjusting potential phenotype-environment mismatches is particularly relevant for such a purpose.

Table 9.1 Behavioural differences between urban and non-urban animals.

Behavioural domain	Behaviour change	Example	Reference
Foraging	Feeding items	Cooper's hawks *Accipiter cooperii* delivered higher rates of prey items to juveniles in urban than in rural areas.	Estes and Mannan 2003
		Urban bushbabies *Galago moholi* shifted food items from gum and arthropods to anthropogenic resources.	Scheun et al. 2019
	Movement patterns	Urban-dwelling hedgehogs *Erinaceus europaeus* changed their nocturnal foraging movements to avoid roads.	Dowding et al. 2010
	Food acquisition behaviour	Sulphur-crested cockatoos *Cacatua galerita* developed a set of behaviours to exploit human-generated resources, specifically different tactics to open garbage bins.	Klump et al. 2021
		Urban bonnet macaques *Macaca radiata* used hands more frequently than their mouths to acquire food when compared to rural individuals.	Dhananjaya et al. 2022
Activity patterns	Increased movement	Male rock agamas *Psammophilus dorsalis* showed higher movement rates in rural than in urban areas.	Batabyal et al. 2017
	Increased nocturnality	A meta-analysis showed a global increase of nocturnality by an average factor of 1.36 among wildlife in human-dominated areas.	Gaynor et al. 2018
	Increased sedentarism	In blackbirds *Turdus merula*, urban birds kept from early life under common garden conditions exhibited reduced migratoriness compared to rural birds.	Partecke and Gwinner 2007
Risk-taking	Response to predators	A set of 44 European bird species in urban areas exhibited shorter flight initiation distances than rural birds.	Møller 2008
		Fox squirrels *Sciurus niger* in more urbanized areas invested less time in vigilance and reacted less to predator vocalization than individuals in rural areas.	Mccleery 2009
		Flight initiation distance of rock agamas *Psammophilus dorsalis* after first attack was higher in males from rural areas than males from urban areas.	Batabyal et al. 2017
		Eastern cottontails *Sylvilagus floridanus* showed higher vigilance in the presence of coyotes in less urbanized areas than in more urbanized areas.	Gallo et al. 2019
Exploratory responses	Personality	Hand-raised blackbirds *Turdus merula* from the urban population were more neophobic and seasonally less neophilic than blackbirds from the nearby rural area.	Miranda et al. 2013
		Anolis sagrei lizards from urban areas were more tolerant of humans, less aggressive, bolder after a simulated predator attack, and spent more time exploring new environments than nearby forest lizards.	Lapiedra et al. 2017

continued

Table 9.1 *Continued*

Behavioural domain	Behaviour change	Example	Reference
	Use of new material	Megachilid bees *Megachile campanulae* and *M. rotundata* used different types of polyurethane and polyethylene plastics to construct and close brood cells in nests containing successfully emerging brood in urban areas.	MacIvor and Moore 2013
	Neophilic responses	In city centres, common mynas *Acridotheres tristis* were more responsive to approaching novel objects and adopting novel foods than in moderately urbanized areas.	Sol et al. 2011
		Neophobic corvids and a set of less fearful bird species were faster to approach objects made from human litter in urban than in rural areas.	Greggor et al. 2016
	Higher exploratory tendency	Barbados bullfinches *Loxigilla barbadensis* from urban areas were faster in accomplishing lid-drawer tasks than rural birds.	Audet et al. 2016
		Coyotes *Canis latrans* were bolder and more exploratory in urban areas than in rural areas.	Breck et al. 2019
		Urban blackcapped chickadees *Poecile atricapillus* were faster explorers in novel environments.	Thompson et al. 2018
Communication	Change in communication strategies	Great tits *Parus major* used higher-frequency singing types under louder urban noise.	Slabbekoorn and Peet 2003
		Urban eastern grey squirrels *Sciurus carolinensis* relied more on visual anti-predator signals in noisy environments.	Partan et al. 2010
		Noisy miners *Manorina melanocephala* at noisier locations (arterial roads) alarm-called significantly more loudly than those at quieter locations (residential streets).	Lowry et al. 2012
Sociability	Interspecific interactions	Feral pigeons *Columba livia* were able to recognize individual human faces and learned to use this information to modify their foraging behaviour in urban areas.	Belguermi et al. 2011
	Intraspecific interactions	Urban individuals of bushbabies *Galago moholi* showed a higher degree of sociality (over solitary behaviour) than rural individuals, and spent less time moving than communicating and/or in pair-grooming.	Scheun et al. 2019
		Striped field mice *Apodemus agrarius* were less likely to avoid close contact with each other and were more likely to show tolerant behaviour in urban areas.	Łopucki et al. 2021

9.4 Why is decision-making relevant in novel environments?

Decision-making involves the cognitive processes through which individuals assess different options, evaluate potential outcomes, and make choices (Shettleworth 2010). Decisions are primarily influenced by two factors. The first is the cognitive machinery of the animals involved in gathering, integrating, and responding to information. Cognitive processes that are fundamental for decision-making—and that exhibit substantial variation among animals—include perception, attention, learning, memory, reasoning, planning, and metacognition (Owen et al. 2017). The second factor refers to the accessibility and reliability of available information (Dall et al. 2005). Information can influence the perceived costs and benefits of different options. When it is unreliable, information can also lead to wrong decisions (Robertson et al. 2013). Besides these two factors, decisions may also be shaped, to varying degrees, by cognitive biases, emotional responses, and phenotypic constraints. Fear of novelty (neophobia), for example, has been proposed to promote conservative decisions by inducing a generalized aversion to new situations, thereby limiting individuals to utilizing resources with which they are most familiarized (Greenberg 1983).

To make decisions, animals often rely on environmental stimuli (i.e. 'cues') that provide information without having to invest the time and energy needed to directly evaluate all alternative options (Stamps 2001). These cues can elicit automatic responses, if they express genetic predispositions or have been acquired through imprinting during early life stages (Gilroy and Sutherland 2007; Stamps and Swaisgood 2007; Figure 9.3a). When these automatic responses result from the activation of a pre-existing neural network, they are referred to as activational plasticity (Snell-Rood et al. 2018). Animals that primarily rely on such a form of plasticity to make decisions will tend to ignore and/or avoid novel cues, and hence be highly conservative in their choices. While this will enhance the match of their phenotype with the environment, it may concurrently diminish their ability to recognize and capitalize on ecological opportunities.

Although all animals make decisions during their life, the process of making decisions in urban environments has two particularities. First, individuals often confront novel or infrequent challenges that require making decisions with limited previous experience. Second, the cues relied upon by the animal to inform its decision-making processes may have been altered in urban environments, causing them to be less reliable. Under such circumstances, choices based on activational plasticity become less useful and can even lead to maladaptive decisions that reduce fitness, a phenomenon known as an ecological trap (Kokko and Sutherland 2001; Robertson et al. 2013; see also Chapter 8). However, animals can modify decision-making rules through learning; that is, by creating new neural connections that store information acquired in previous experiences (Dukas 2004; Figure 9.2). For example, when exposed to acoustic cues simulating increased nest predation risk, Siberian jays *Perisoreus infaustus* can learn to select nest sites with greater protective covering, demonstrating a facultative response to the perceived risk of nest predation (Eggers et al. 2006). Animals that can plastically modify their decision-making processes in such ways have an advantage in coping with new conditions (Figure 9.3a).

When challenges are new and very different from those that animals encounter in their ancestral environments, decision-making transcends the domain of incomplete or unreliable information. In such instances, animals must modify their decision-making processes in innovative ways to navigate and find solutions to novel challenges. Some animals have an unusual ability to plastically adjust decisions in innovative ways. A compelling illustration is the utilization of tools by certain birds and mammals to access food that would otherwise remain out of reach (Lefebvre et al. 2004; Taylor et al. 2012). This means that behaviour has greater potential to change in the direction of new adaptive optima, improving fitness and population persistence in novel environments (Price et al. 2003; Sol et al. 2020b; see also Chapters 14 and 19).

Gathering information is time-consuming and can be risky, but many animals can copy decisions from more informed individuals, or acquire new behaviours through social learning (Dall et al. 2005; Reader 2016). This does not only spread the benefits

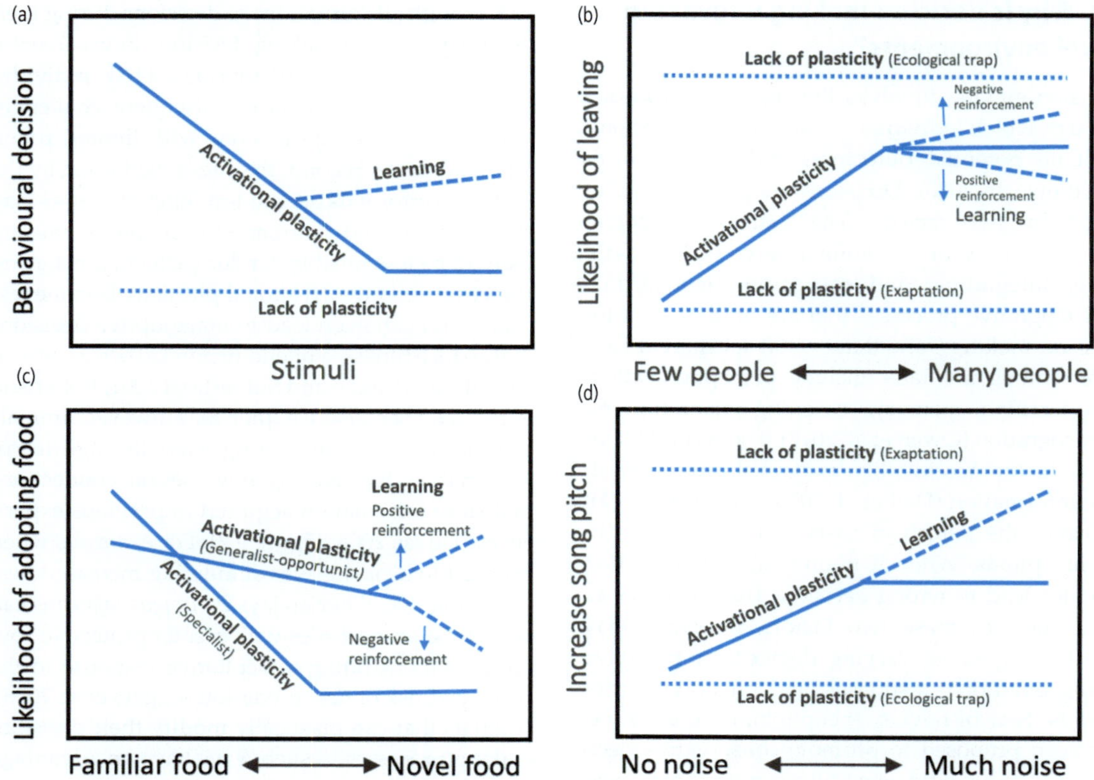

Figure 9.3 (a) Framework to understand behavioural decisions in novel environments. The decision of animals in response to stimuli can be described as a reaction norm, with activational plasticity corresponding to the slope value. The shape of the reaction norm (slope and intercept) can be modified through learning, providing an additional source of plasticity in decision-making. (b–d) Examples of how plasticity can influence responses to challenges in urban areas. The intercept and slope of the reaction norm can vary among individuals, implying that they will respond differently to the same stimulus and exhibit differences in plasticity. For example, some individuals may be more inclined than others to attack small prey while simultaneously displaying higher hesitation to engage with larger prey. Although decisions driven by activational plasticity can be adaptive, having evolved through natural selection, they typically function within the range of stimuli the species has previously encountered in the past (otherwise, the reaction norm cannot evolve). In a new environment, this may mean that individuals will only respond to familiar stimuli and disregard novel ones, favouring 'habitat matching choice'. However, innovation and learning enable individuals to modify the reaction norm beyond the range of conditions to which they have frequently been exposed; for example, allowing individuals to recognize, capture, and eat novel preys. These mechanisms may plastically rescue the population from extinction.

of knowledge more rapidly, but can also facilitate the emergence of innovative solutions through cooperation and skill pool effects (Giraldeau 1984; Morand-Ferron and Quinn 2011). Thus, the power of learning lies in its potential to modify decision-making, not only based on previous experiences of the individual, but also by using information generated by others.

Experiments have revealed that urban animals largely differ in the way they make decisions, with some primarily relying on activational plasticity while others depend more on decisions based on knowledge acquired through learning. Major factors that contribute to such variation include brain size and architecture, which determine cognitive performance, the species' life history, which influences the costs and benefits of learning, and the social organization of individuals, which facilitates knowledge acquisition and sharing (Lefebvre et al. 2004; Belguermi et al. 2011; Sol et al. 2022; Section 9.7). These features tend to be evolutionary conserved, with closely related species often exhibiting similar decision-making strategies. Corvids, parrots, and primates are notable examples of animals with complex decision-making processes. In particular, they are renowned for their ability to engage in prospective cognition, which involves planning, reasoning, and anticipating future events (e.g. Emery et al. 2007; Lee et al. 2011; Taylor et al. 2011). Corvids, parrots, and primates also have in common highly encephalized brains (i.e. brains that are larger than expected by body size), slow life histories, and complex social structures.

9.5 Staying home or embracing city life?

When given the choice between two habitats, we expect that most animals would choose the habitat to which they have had more opportunity to adapt. The reason for using such a 'habitat matching' choice strategy is that the likelihood that a species can survive and reproduce in a habitat depends on the existence of a good match between its phenotype and the environment (Nicolaus and Edelaar 2018). A preference for habitats that resemble ancestral habitats also reduces the time and energy needed to accurately assess different habitat alternatives (Stamps 2001). In the decision-making process, the implication is that choices based on environmental cues that inform such a match will be favoured by selection. Although preferences for ancestral habitats can have a genetic basis shaped by selection, they can also be environmentally induced. One mechanism is natal habitat preference induction, in which experience with a natal habitat shapes the habitat preferences of individuals (Stamps 2001). While a preference for habitats that are comparable to the ancestral habitat can provide important benefits, it also means that a disperser might bypass a habitat of higher quality than its original habitat (see Section 9.6).

If animals strictly follow a 'habitat matching' choice strategy, we would expect that they avoid establishing themselves in novel habitats, such as cities. Thus, the observation that some animals actively choose urban habitats raises questions as to what influences their decision-making and what prompts them to alter it. The decision of animals to abandon their original habitat and settle in cities has been little studied, but one would expect that such habitat shifts are more likely in generalist-opportunist species, which exhibit a more relaxed phenotype-environment match compared to species that primarily rely on matching habitat choice. Although this specific issue has yet to be directly addressed in any study, there is substantial evidence indicating that urban-dwellers exhibit habitat generalist tendencies (Bonier et al. 2007; Ducatez et al. 2018; Sol et al. 2014). For instance, Ducatez et al. (2018) employed habitat data encompassing all terrestrial vertebrate groups to show that urban-dwellers are not confined to urbanized environments as specialists but, rather, display ecological generalist habits, exploiting opportunities across a range of both human-altered and natural habitats.

If, when selecting a new habitat, individuals look for features that resemble those in their former habitats, the likelihood of settling in urbanized environments should also increase if the environment still preserves some of these features (Stamps 2001). In the city of Puerto Rico, for example, the lizard *Anolis stratulus* tends to restrict itself to the more 'natural' portions of the urban environment, such as trees and cultivated vegetation (Winchell et al. 2018). There is evidence that moderately urbanized areas, which still preserve important elements of natural habitats, can harbour a substantial fraction of the species occurring in the surrounding non-urbanized areas (Sol et al. 2020a; Sol et al. 2017a; Sol et al. 2014).

For some species, settling in cities may represent the best of a bad job. For instance, in the case of the goshawks from Hamburg (discussed in Section 9.2), the decision to settle in the city was apparently associated with the deterioration of conditions in nearby rural areas due to an increase in (legal) hunting pressure and a succession of severe winters (Rutz 2008). When conditions deteriorate—whether by natural or human-related disturbances, or because of density-dependent effects caused by population growth—it can pay to move to habitats that might otherwise have been considered suboptimal and, hence, been rarely used previously. In herring gulls *Larus argentatus*, the colonization of cities and subsequent increase in numbers coincided with the growth of traditional colonies associated with their protection and the increased availability of food discarded by humans (Coulson and Coulson 2009).

Classic models of habitat selection by dispersers often assume that, in a given place and time frame, a particular habitat is equally suitable for all members of the population (Stamps 2001). However, this is not necessarily true if the benefits of choosing a particular habitat vary with the phenotype. Likewise, the costs of searching for a habitat by dispersers can vary depending on their phenotype. Although making a priori predictions is challenging, we would expect that bolder and more exploratory individuals should generally be more likely to move into urban environments than those that are shyer and less curious (Rehage and Sih 2004; Duckworth and Badyaev 2007). The non-random sorting of individuals based on their phenotype is thought to be one of the factors behind the striking behavioural differences that are often observed between urban and non-urban animals (Table 9.1), although the general importance of this mechanism still awaits empirical verification.

9.6 Challenges to thriving in urban habitats

While the decision to move into a city can be challenging, thriving within that environment can prove to be even more demanding. Urbanization reduces and fragments natural habitats, increases the impervious surface in the form of buildings and roads, alters resource availability, and increases the frequency of human-related disturbances (Johnson and Munshi-South 2017; see also Chapter 8). As these challenges present situations for which most animals lack prior adaptation or experience, the risk of extinction due to maladaptation should usually be high. Yet, cities also provide resources and appropriate conditions for those species that can overcome these challenges. Disregarding these ecological opportunities can result in missed opportunities for individuals to maximize their fitness (Stamps 2001; Gilroy and Sutherland 2007).

A major challenge animals face in cities is the presence of people. It is estimated that predation pressure exerted by humans is up to 14 times higher than that exerted by other superpredators, such as lions or wolves (Darimont et al. 2015). It should come as no surprise, then, that many animals avoid human presence and, therefore, their landscape of

fear is heavily influenced by people (Blumstein 2006; Chapter 14). Indeed, even superpredators can be afraid of humans. In California, cougars are so scared of humans that they abandon their prey when they hear people nearby (Smith et al. 2017). The increased observation of animals that are rare in cities following the confinement of people during the COVID-19 pandemic in 2019 is a prime example of how humans impact the movement of animals along the natural-urban gradient (Bates et al. 2021; Tucker et al. 2023).

To minimize conflict with humans, a common decision of urban animals is to shift towards nocturnal behaviour (Gaynor et al. 2018). A study on GPS-collared coyotes *Canis latrans* found that road-killed individuals were more active during rush hour and crossed roads at dusk, while surviving coyotes still crossed roads frequently, but their crossings were more common near midnight (Murray and St. Clair 2015). It is worth noting that nocturnal shifts have also been documented in animals that exhibit a high tolerance for humans, such as feral pigeons and gulls. This observation suggests that the avoidance of human conflict may not be the sole purpose of these nocturnal behaviours. In some cases, nocturnality may reflect the need to extend foraging periods to meet daily energetic requirements.

If human disturbances are frequent—but pose no immediate threat—animals should develop a certain level of tolerance to human presence. This possibility is supported by ample evidence. For example, numerous studies have reported city-dwelling animals exhibiting shorter flight initiation distances (i.e. the distance at which an animal takes flight when a human is approaching) compared to their non-urban counterparts (e.g. Blumstein 2006; Møller 2008; Figure 9.3b). By tolerating the presence of people, animals can better capitalize on the abundant resources deliberately or accidentally provided by humans (Blumstein 2006; Carrete and Tella 2011).

The approach-avoidance conflict in foraging contexts is another crucial decision-making process animals face in novel environments (Sol et al. 2011). Adopting novel food resources may entail risks (e.g. poisons, diseases, etc.) and, hence, some animals consistently avoid them. In a novel environment, however, individuals are confronted more often with novel foods than familiar ones, so ignoring them may increase the risk of starvation. Such a constraint is relaxed in generalist-opportunist animals (Bonier et al. 2007; Sol et al. 2014; Ducatez et al. 2018), which show less attachment to specific food sources and are therefore more open to adopting new foods. The opportunistic nature of some urban exploiters is reflected in their reduced fear of exploring and adopting novel food opportunities compared to non-urban individuals, as shown in several field and common garden experiments (Liker and Bokony 2009; Sol et al. 2012, 2011; Gregor et al. 2016; Figure 9.3c).

Another important challenge for animals in cities is pollution, including noise (Chapter 2), chemical (Chapter 3), and light (Chapter 4). For instance, in highly vocal animals, noise associated with people, road traffic, and industrial activities can interfere with communication, affecting mating success and social interactions (Halfwerk and Slabbekoorn 2009; Chapter 2). Because human activities typically generate noise biased towards low frequencies, a strategy adopted by many urban-dwelling birds to partially overcome this problem is to increase the pitch of their songs in noisy areas (Slabbekoorn and Peet 2003). By contrast, evidence suggests that some birds, like grey flycatchers *Empidonax wrightii*, do not adjust their vocal frequency in response to noise but, instead, avoid highly noisy places altogether. In grey squirrels *Sciurus carolinensis*, the solution to noisy environments has been to rely more on visual signals, for example when they communicate the presence of a predator (Partan et al. 2010; Figure 9.3d).

A last important challenge animals face in cities is the need to cope with infrastructure and transportation systems, which can be major causes of mortality (see also Chapters 8 and 13). It is estimated that in the USA and Canada alone, nearly 600 million birds die each year from collisions with tall buildings (Horton et al. 2019). Attracted to the lights of the buildings, birds can become disorientated and crash into walls (Chapter 4). Collisions with vehicles have also steadily increased over the last few decades (Canal et al. 2018) and can be a major safety issue for animals (Martin et al. 2018; Kent et al. 2021). One solution is to avoid areas densely occupied by infrastructure and intense traffic, as observed in some mammals (McGregor et al.

2008). However, animals can still use these areas if they make decisions to reduce the risk of collisions (Husby and Husby 2014; Møller and Erritzøe 2017; see also Section 9.7 below).

9.7 The cognitive nature of decisions matters in urban habitats

To understand how a species thrives in a city, we must know not only what decisions animals make, but also how these are made. This is important to comprehend whether the behavioural adjustment will be necessary and sufficient—and occur fast enough—to avoid population extinction. For example, if the presence of people does not elicit any behavioural response (lack of plasticity), the behaviour can still help to facilitate life in the city if it aligns closely with an organism's optimal behaviour (i.e. if the behaviour is an exaptation; Sol et al. 2018; Winchell et al. 2023). However, such a lack of plasticity can be maladaptive in an environment where humans or their pets are dangerous, potentially leading to ecological traps (Kokko and Sutherland 2001).

Many decisions animals make do not seem to be cognitively complex, but probably reflect some form of activational plasticity (Figure 9.3a–d). For example, activational plasticity can be behind the decision to stay or leave when animals find themselves in the presence of humans, thereby allowing individuals to rapidly adjust their behaviour optimally in different contexts (Figure 9.3b). Likewise, in the case of birds, the plastic adjustment in song pitch in noisy urban environments (Section 9.6) probably reflects innate fixed action patterns (Halfwerk and Slabbekoorn 2009; Halfwerk et al. 2011). The degree to which behaviour requires modification is anticipated to hinge on the extent to which the existing behaviour already serves a function aligned with the challenges posed by the new environment (Figure 9.3). For instance, it is reasonable to assume that birds producing high-frequency sounds would experience minimal challenges in communication, since those calls would already be produced at a pitch that can be heard above the low-frequency din of urban noise (Moiron et al. 2015).

Although activational plasticity offers rapid responses to specific cues, as we have already discussed (Section 9.4), it is only effective within the range of stimuli the reaction norm has had the opportunity to evolve under. This means that responding to novel stimuli will require the evolution of a new reaction norm, a process that takes time and relies on existing heritable variation. However, learning offers a potential alternative for rapidly adjusting the decision-making process to novel contexts. Repeated exposures to people (i.e. a positive reinforcement to the stimuli) can, for instance, reduce the fear of humans through habituation (Stansell et al. 2020). Some animals can even recognize human faces, allowing them to distinguish people that can pose a risk from those who do not. Marzluff et al. (2010), for example, demonstrated experimentally that American crows *Corvus brachyrhynchos* rapidly learn to identify dangerous humans, a skill they retain for several years.

Devising sophisticated techniques to exploit new food opportunities is another benefit of learning (Lefebvre et al. 1997; Overington et al. 2009; Sol et al. 2022). Sulphur-crested cockatoos *Cacatua galerita* have learned a set of complex behaviours to exploit human-generated resources, including different tactics to open garbage bins (Klump et al. 2021). Herring gulls, on the other hand, use human behaviour as a cue to make foraging decisions. Experiments show, for instance, that human handling of food captures the attention of the gulls, making the food more appealing to the birds compared to food that has not been observed being handled (Goumas et al. 2020).

Cognitive performance ultimately depends on the number of neuronal connections in the associative areas of the brain (see Sol et al. 2022). Because the number of connections largely varies among species, it follows that not all animals may have the cognitive capacity to devise appropriate learned solutions to cope with the challenges of an urban life (see also Section 9.4). The importance of advanced cognition in urban contexts stems from multiple analyses highlighting the benefits of having an encephalized brain (a surrogate of neuron numbers; see Sol et al. 2022). Avoiding traffic, for example, requires assessing speed and directionality of oncoming cars, which implies some understanding of the behaviour of vehicles (Husby and Husby 2014; Møller and Erritzøe 2017). There is evidence in birds showing that species with a more encephalized brain tend to fly away from the road more often than species with a small brain size (Husby and Husby 2014; Møller and Erritzøe 2017). A highly encephalized brain and enhanced innovative ability have also been found to be distinctive features of animals that have successfully established in novel environments in general (Sol et al. 2005), and urban environments in particular (Carrete and Tella 2011; Maklakov et al. 2011). Indeed, this may be one of the reasons why taxa such as corvids, parrots, and primates, are among the animals that have achieved greatest success in urban environments (see also Section 9.4).

It is important to note, however, that enhanced cognition is not the only pathway to becoming an urban exploiter (Dale et al. 2015; Santini et al. 2019; Sayol et al. 2020). Diquelou et al. (2016) examined problem solving in seven bird species that were commonly found in urban areas and that were characterized by different degrees of encephalization. The researchers found that birds largely differed in their ability to solve a motor innovation task, with species ranging from being quick and reliable at solving problems to never solving them at all. Thus, there are a variety of strategies that can enable animals to thrive in urban areas, of which being innovative in embracing new resource opportunities is just one (Santini et al. 2019; Sol et al. 2014, 2012).

9.8 Demographic consequences of decision-making

If decisions made by animals affect their fitness, these may have important population-level consequences. Some of the consequences may be negative for the animal. For example, animals that primarily rely on activational plasticity to make decisions will tend to ignore or avoid novel cues, and hence be highly conservative in their choices (discussed in Section 9.4). This will generally result in such animals avoiding novel habitats, despite them being favourable for population growth (Gilroy and Sutherland 2007). For instance, woodlarks *Lullula arborea* actively avoid nesting in areas adjacent to busy footpaths, despite minimal risks posed by humans and the enhanced breeding success resulting from low nest densities in such locations (Mallord et al. 2007). The proclivity for animals to settle in their natural habitats (rather than novel ones), even when the latter offer superior conditions and resource opportunities, can result in colonization delays (the so-called 'colonization credit') and underutilized resources in urban areas (Gilroy and Sutherland 2007). Hence, the observed biodiversity within urban areas may be significantly lower than what the city is actually capable of supporting, as exemplified by the success and prevalence of a select number of non-native introduced species in many cities around the world (Cadotte et al. 2017; Sol et al. 2017b).

Ecological traps are another manifestation of wrong decisions that may potentially lead to declines and extinctions of populations in human-altered environments (Kokko and Sutherland 2001; Schlaepfer et al. 2002; Battin 2004; Robertson et al. 2013; see Figure 9.3). This phenomenon is of particular concern in urban environments, where the cues used in decision-making are often disrupted by human activities, resulting in varying and inconsistent information across time and space (Section 9.4). A well-known example is the disorientation of sea turtle hatchlings caused by light pollution originating from beachfront structures, which prevents them from successfully reaching the ocean after emerging from the nest at night (Schlaepfer et al. 2002; Chapters 4 and 11). Another example is the well-intentioned attempt to mitigate the decline of monarch butterflies *Danaus plexippus* in North America by encouraging citizens to plant gardens with milkweed *Asclepias* spp., the obligate larval host plants of the butterfly. Despite being effective at attracting ovipositing adults, these gardens can actually act as an ecological trap by exposing the butterflies to an elevated risk of predation by invasive paper wasps *Polistes dominula* (Baker and Potter 2020).

Many decisions animals make in urban contexts seems nonetheless to be adaptive (i.e. they have positive fitness effects), as we have described in Section 9.6. Thus, they may potentially contribute to population growth and persistence. The process

to avoid extinction by decision-making is part of a more general process known as plastic rescue, and represents an important mechanism through which individuals can thrive in novel environments (Snell-Rood et al. 2018; Maspons et al. 2019). Studies investigating the effects of human food provisioning on the demography of opportunistic species provide some of the most compelling evidence for the positive impact of decision-making on urban population dynamics. For instance, the rise in some bird populations within urban areas can be in part attributed to the efficient use of food resources stemming from human activities (Box 9.1). Herring gulls are one of the species that have benefited the most; they exploit food at refuse tips and will travel considerable distances to access those food sources. To efficiently exploit refuse tips, individuals adjust the timing of their visits, and have learnt to avoid days when there is no uncovered refuse available (Coulson et al. 1987). The regular provisioning of food, together with low levels of interspecific competition and predation, could explain why species such as gulls are extremely abundant in cities, even when they have had limited opportunity to adapt to the urban environment (Box 9.1).

Changes in time schedules to more efficiently track resources are other decision-making processes that have demonstrable demographic consequences. A recent meta-analysis found that in urban environments, birds tend to start breeding earlier and extend the reproductive season longer than in the surrounding rural and natural habitats (Capilla-Lasheras et al. 2022). In San Diego, for example, dark-eyed juncos experience a milder climate than the nearby mountain population, allowing them to have a breeding season that is more than twice as long. This extended breeding season has proven to be critical for population maintenance as it enables the rearing of approximately twice as many young per individual, partially compensating for the high mortality observed among juveniles during their first year of life (Yeh and Price 2004).

In most cases, however, determining the demographic impacts of decisions is less straightforward. This is primarily due to the fact that animals make numerous decisions throughout their lives, not all of which will necessarily have demographic relevance. Furthermore, decisions that influence

Box 9.1 How behaviour can contribute to fostering biodiversity in cities

Urban biodiversity provides a wide range of environmental and human health benefits. However, urban expansion also brings humans into closer proximity with wildlife, which may escalate conflicts. There is thus a critical need to comprehensively understand human-wildlife interactions in order to develop strategies that can promote biodiversity while effectively mitigating conflicts in urban environments (Soulsbury et al. 2015; see also Chapter 13).

As we have discussed throughout the chapter, animal decision-making processes are likely to play a crucial role in mediating their success in urban environments. This means that there is considerable opportunity to leverage behavioural manipulations as a means to promote biodiversity in cities through enhanced conservation and management outcomes. For example, research on animal behaviour has already led to the development of building design guidelines that reduce bird collisions with windows, such as by incorporating window films or decals that break up reflections (Martin 2011). Learning principles have also been applied to mitigate human-wildlife conflicts. To reduce the incidence of bear-train collisions, for instance, researchers have developed an electronic system that can detect passing trains via vibrations and that, when activated, triggers sounds and flashing lights to ward off bears (St. Clair et al. 2019; Figure 9.4a).

Many interventions aimed at promoting the coexistence of humans and wildlife focus on modifying the behaviour of the animals (Chapter 13). However, modifying human behaviour can also be a powerful strategy to support these interventions (Dobson et al. 2019). For example, Haag-Wackernagel (1993) devised a control programme in Basel, Switzerland, to reduce the overpopulation of feral pigeons *Columba livia* (Figure 9.4b), which was causing environmental and human health issues. Instead of resorting to measures such as lethal control or sterilization, which had proven to be ineffective in the past, the programme targeted humans through an education programme aimed at convincing people to refrain from feeding pigeons. By emphasizing the negative consequences of overcrowding on pigeon health and the risk of disease, the programme successfully reduced the population of pigeons by 50% within 50 months.

Figure 9.4 Behavioural interventions for promoting urban biodiversity. (a) Capitalizing on animal learning as a tool to mitigate collisions between bears and trains; and (b) educating humans to reduce feeding of feral pigeons as a management strategy.

Photos: (a) Stock Photo ID: 347580623 (https://www.shutterstock.com/image-photo/grizzly-bear-on-train-tracks-banff-347580623); (b) Daniel Sol

survival or reproduction do not have the same demographic impact in species with different life-history strategies. Species that prioritize fecundity over longevity, for instance, experience higher fitness costs when a breeding attempt fails, as they have fewer future reproductive opportunities compared to longer-lived species. Consequently, the decision to choose an unsuitable breeding location will have more pronounced negative repercussions on the demography of species that prioritize fecundity (Maspons et al. 2019).

The demographic consequences of decision-making will also depend on whether a significant fraction of the population makes similar choices. This will, in turn, depend on how rapidly the behaviour can change and spread through the population. Decisions that are made collectively or transmitted among individuals by social learning have a greater potential to impact on population dynamics. One possible example of cultural transmission of knowledge in urban contexts is the milk-bottle opening developed by great tits to access milk-cream as a food resource during the winter. The behaviour was first observed in England in the early 1920s, and 30 years later, it had become widespread in England and parts of Wales, Scotland, and Ireland (Fisher and Hinde 1949; Aplin et al. 2015). Natal habitat preference induction is another transmission mechanism that may facilitate population maintenance in urban environments. This is because individuals born in urban environments will then be more inclined to settle in similar environments when they choose a habitat to breed (Stamps 2001). The rapid spread of urban blackbirds *Turdus merula* in the western Palearctic has in part been attributed to this type of environmental induction (Evans et al. 2009).

Despite the obvious limitations in our comprehension of how decision-making influences the demography of urban populations, the available evidence underscores the pivotal role of behavioural decisions in enabling animals to endure and proliferate in urban environments. This acknowledgment prompts a call for further research to unravel the intricate dynamics of decision-making and its cascading effects on population abundance and expansion within urban settings.

9.9 The evolutionary side of behavioural decisions

Given that selection is ultimately responsible for the existence of a phenotype-environment match, it also holds the potential to rescue a population from extinction by facilitating a new evolutionary equilibrium (Figure 9.2; Bell 2017). However, evidence that evolutionary rescue can play a major role in urban environments is lacking, even though some of the highest rates of phenotypic change have been observed in populations exposed to human-induced changes (Alberti et al. 2017; Johnson and Munshi-South 2017). While this lack of evidence can, in part, reflect the scarcity of studies (Caspi et al. 2022), it could also indicate that evolutionary adaptation to urban contexts is particularly difficult. One factor is the lack of standing genetic variation. Insufficient genetic variation for selection to act upon might be expected due to bottlenecks (if individuals are relics that pre-dated urbanization) or founder effects (if they established from a small number of colonizers after the city was formed; Evans et al. 2009; Johnson and Munshi-South 2017), and also because the challenges animals face in cities are drastically different from those found in their ancestral environments (see Section 9.6). Given that these challenges often act simultaneously—and can rapidly vary over time and space—they can also cause selection to fluctuate in ways that make local adaptation more difficult (Sol et al. 2020b).

Behavioural decisions may also contribute to a slowing down of the process needed to reach a new phenotype-environment equilibrium. If, by making decisions, individuals minimize exposure to new selective pressures, this should reduce the strength of natural selection and, hence, inhibit evolutionary change (Huey et al. 2003; Price et al. 2003). In novel environments, however, behavioural decisions can also facilitate adaptive responses by increasing the chances that the population reaches a new fitness optimum before it goes extinct (Price et al. 2003). For example, developing a foraging technique to exploit a novel food can enhance fitness while exposing individuals to divergent selection on morphology, physiology, and behaviour, moving the population toward a new phenotype-environment equilibrium. As suggested by the behavioural drive hypothesis

(Mayr 1963), this opens the possibility for phenotypic evolution to move towards new directions.

Additional behavioural decisions that can facilitate adaptation to urban environments include assortative mating and learned habitat preferences (Caspi et al. 2022). Assortative mating may reduce gene flow with the ancestral population, and it is more likely to occur in small populations colonizing novel environments (Berner and Thibert-Plante 2015; Caspi et al. 2022). In cane toads *Rhinella marina* introduced to Australia, for example, the rapid expansion of the species has in part been attributed to the fact that individuals in the invasion front show higher propensity to disperse and, at the same time, are also more likely to mate with individuals showing a similar propensity, further selecting them for increased dispersal ability (Phillips 2016). Preferences for certain habitats can also drive assortative mating by reducing reproductive interactions with individuals from other habitats (Berner and Thibert-Plante 2015), as well as reinforcing the exposure of populations to specific selective pressures. In European blackbirds *Turdus merula*, genetic isolation in large cities might have been achieved, to some degree, from both geographic isolation and assortative mating of sedentary urban individuals (Partecke and Gwinner 2007; Evans et al. 2009).

Apart from influencing the evolution of the overall phenotype, the ability to make decisions may also undergo evolutionary changes when the underlying cognitive and motor functions display heritable variation and have impacts on fitness. Selection for enhanced learning is expected in animals that are frequently exposed to novel challenges or in those species that invade (or build) niches that are cognitively demanding (see references in Kaufman et al. 2021). If a particular behaviour affects fitness and is heritable, selection can facilitate the emergence of new reaction norms to allow animals to rapidly respond to novel cues generated by human activities. Complex plastically induced decisions can also end up being genetically encoded through genetic assimilation (West-Eberhard 2003).

Therefore, behavioural decisions have a great potential to shape evolutionary trajectories of the overall phenotype, either by facilitating or inhibiting adaptive responses (Mayr 1963; Price et al. 2003; Duckworth 2009; Sol et al. 2020b; Caspi et al. 2022).

Because evolutionary changes are expected to be crucial for the long-term persistence of populations in urban environments, more research is needed to understand the role of decision-making in the context of urban evolution.

9.10 Conclusions

Growing evidence suggests that understanding the decision-making processes of animals is crucial for unravelling why some species thrive while others avoid or struggle in urban environments. Making inappropriate decisions can hamper the ability of animals to successfully establish and persist in cities. Conversely, adaptive decisions, together with the ability to plastically modify them through learning and microevolution, may facilitate animals to settle in urban environments and better cope with the novel challenges and opportunities that such environments offer. By studying the choices animals make, we can thus gain valuable insights into the factors contributing to the impoverishment, unevenness, and homogenization of urban biotas.

The notion that enhanced cognition for making decisions improves the success of animals in urban environments remains insufficiently understood. While some species demonstrate a reliance on their innovative capacity to thrive in cities, we do not yet understand well how this affects their fitness components. It is evident, however, that thriving in cities is not solely dependent on animals having cognitively advanced forms of decision-making.

Similarly, our understanding of the demographic consequences of decision-making processes in an urban context remains rudimentary. While it is often assumed that behaviours that directly impact survival or reproduction should ultimately affect demography and population persistence, this assumption does not universally hold true. Demographic responses can be shaped by myriad factors, including an organism's life history. The notion that plasticity is universally beneficial in novel environments is also debatable. While a lack of plasticity in certain decision-making processes can result in ecological traps, it can also be adaptive when the ancestral behaviour is already well-suited to the new environment. Conversely, plastic decisions have the potential to lead individuals into ecological traps when the available information is unreliable.

Further efforts are also warranted in determining the extent to which behavioural decisions help to reduce phenotype-environment mismatches through evolutionary mechanisms, which is crucial for the long-term persistence of populations. Behavioural decisions have a great potential to shape the evolutionary trajectories of the overall phenotype, yet whether they may primarily facilitate or inhibit adaptive responses is still unclear. Evidence for evolution of decision-making mechanisms, like reaction norms or improved learning skills, is also notoriously absent for urban environments.

There are thus important gaps in our understanding of how behaviour affects the success of animals in urban environments. Addressing the above gaps is challenging. First, it requires long-term data to identify decisions with relevant fitness consequences across the lifetime of an animal. Next, the cognitive basis of these decisions, and their effect on fitness, need to be assessed. Finally, population and genetic approaches are needed to assess how decisions affect demography and evolutionary responses (Maspons et al. 2019; Acker et al. 2021; Paniw et al. 2021). Studies that comprehensively address all these issues are still uncommon in general, and notably absent in the case of urban populations. To make further progress, we need to consider behaviour as part of a life history strategy inherited from ancestors and shaped by selection to deal with problems that have been relevant for a substantial part of an animal's evolutionary history.

By deepening our understanding of the decision-making processes in urban animals, we can strive to enhance our comprehension of how animals respond and adapt to rapid environmental changes caused by human activities. The implications for understanding urban biodiversity are important. Because success in cities depends not only on animals having the capacity to withstand urban challenges, but also on their ability to recognize the ecological opportunities that urban environments may provide, it is probable that more species are capable of tolerating urban conditions than those we currently observe. The existence of such a colonization credit suggests that we might be overestimating the overall impact of urbanization on biodiversity. However, this does not negate the fact that the impact is significant, especially in intensively urbanized regions, as there is compelling evidence that many species struggle to cope with the abrupt changes associated with urbanization. It is even plausible that there is an extinction debt, where certain maladapted populations persist in cities only through the influx of individuals from the surrounding non-urbanized environments. Since the decision-making processes in animals are often evolutionarily conserved, investigating the importance of colonization credits and extinction debts may shed light on other aspects of urban biodiversity, like the loss of functional diversity and the increase of biotic homogenization. Such a research programme holds the potential to inform and guide the development of effective conservation and management strategies aimed at promoting and preserving biodiversity in urban environments (Box 9.1), as we attempt to foster a more harmonious coexistence between humans and wildlife.

Acknowledgements

We thank Julie Morand-Ferron, Simon Ducatez, Andrea Griffin, Oriol Lapiedra, Louis Lefebvre, and Ferran Sayol for helping us to structure our ideas over the years. This contribution was funded by the grant PID2020-119514GB-I00 (MCIN/AEI/ 10.13039/501100011033) to D.S., MEDYCI PID2020-113133RB-I00 (MCIN/AEI/ 10.13039/501100011033) and Severo Ochoa (CEX-2018-000828-S) to Y.M., ANID (PIA BASAL FB0002 and FONDECYT n°1231191) to C.G.L., and EU's Horizon 2020 MSCA n° 101030199 to L.F.

References

Acker, P., Burthe, S.J., Newell, M.A., et al. (2021). Episodes of opposing survival and reproductive selection cause strong fluctuating selection on seasonal migration versus residence. *Proceedings of the Royal Society B: Biological Sciences*, 288, 20210404.

Alberti, M., Correa, C., Marzluff, J.M., et al. (2017). Global urban signatures of phenotypic change in animal and plant populations. *Proceedings of the National Academy of Sciences*, 114, 8951–8956.

Aplin, L.M., Farine, D.R., Morand-Ferron, J., et al. (2015). Experimentally induced innovations lead to persistent culture via conformity in wild birds. *Nature*, 518, 538–541.

Aronson, M.F.J., Sorte, F.A.L., Nilon, C.H., et al. (2014). A global analysis of the impacts of urbanization on bird and plant diversity reveals key anthropogenic drivers. *Proceedings of the Royal Society B: Biological Sciences*, 281, 20133330.

Audet, J.-N., Ducatez, S., and Lefebvre, L. (2016). The town bird and the country bird: problem solving and immunocompetence vary with urbanization. *Behavioral Ecology*, 27, 637–644.

Baker, A.M., and Potter, D.A. (2020). Invasive paper wasp turns urban pollinator gardens into ecological traps for monarch butterfly larvae. *Scientific Reports*, 10, 9553.

Batabyal, A., Balakrishna, S., and Thaker, M. (2017). A multivariate approach to understanding shifts in escape strategies of urban lizards. *Behavioral Ecology and Sociobiology*, 71, 83.

Bates, A.E., Primack, R.B., Biggar, B.S., et al. (2021). Global COVID-19 lockdown highlights humans as both threats and custodians of the environment. *Biological Conservation*, 263, 109175.

Battin, J. (2004). When good animals love bad habitats: ecological traps and the conservation of animal populations. *Conservation Biology*, 18, 1482–1491.

Belguermi, A., Bovet, D., Pascal, A., et al. (2011). Pigeons discriminate between human feeders. *Animal Cognition*, 14, 909–914.

Bell, G. (2017). Evolutionary rescue. *Annual Review of Ecology, Evolution, and Systematics*, 48, 605–627.

Berner, D., and Thibert-Plante, X. (2015). How mechanisms of habitat preference evolve and promote divergence with gene flow. *Journal of Evolutionary Biology*, 28, 1641–1655.

Blair, R.B. (2001). Birds and butterflies along urban gradients in two ecoregions of the U.S. In: J.L. Lockwood and M.L. McKinney (eds), *Biotic Homogenisation*, pp. 33–56. Kluwer, Norwell, MA.

Blumstein, D.T. (2006). Developing an evolutionary ecology of fear: how life history and natural history traits affect disturbance tolerance in birds. *Animal Behaviour*, 71, 389–399.

Bonier, F., Martin, P.R., and Wingfield, J.C. (2007). Urban birds have broader environmental tolerance. *Biology Letters*, 3, 670–673.

Breck, S.W., Poessel, S.A., Mahoney, P., and Young, J.K. (2019). The intrepid urban coyote: a comparison of bold and exploratory behavior in coyotes from urban and rural environments. *Scientific Reports*, 9, 2104.

Cadotte, M.W., Yasui, S.L.E., Livingstone, S., and MacIvor, J.S. (2017). Are urban systems beneficial, detrimental, or indifferent for biological invasion? *Biological Invasions*, 19, 3489–3503.

Canal, D., Martín, B., de Lucas, M., and Ferrer, M. (2018). Dogs are the main species involved in animal-vehicle collisions in southern Spain: daily, seasonal and spatial analyses of collisions. *PLOS ONE*, 13, e0203693.

Capilla-Lasheras, P., Thompson, M.J., Sánchez-Tójar, A., et al. (2022). A global meta-analysis reveals higher variation in breeding phenology in urban birds than in their non-urban neighbours. *Ecological Letters*, 25, 2552–2570.

Carrete, M., and Tella, J.L. (2011). Inter-individual variability in fear of humans and relative brain size of the species are related to contemporary urban invasion in birds. *PLOS ONE*, 6, e18859.

Caspi, T., Johnson, J.R., Lambert, M.R., et al. (2022). Behavioral plasticity can facilitate evolution in urban environments. *Trends Ecology & Evolution*, 37, 1092–1103.

Clergeau, P., Jokimäki, J., and Savard, J.P.L. (2001). Are urban bird communities influenced by the bird diversity of adjacent landscapes? *Journal of Applied Ecology*, 38, 1122–1134.

Coulson, J.C., Butterfield, J., Duncan, N., and Thomas, C. (1987). Use of refuse tips by adult British herring gulls *Larus argentatus* during the week. *Journal of Applied Ecology*, 24, 789.

Coulson, J.C., and Coulson, B.A. (2009). Ecology and colonial structure of large gulls in an urban colony: investigations and management at Dumfries, SW Scotland. *Waterbirds*, 32, 1–15.

Dale, S., Lifjeld, J.T., and Rowe, M. (2015). Commonness and ecology, but not bigger brains, predict urban living in birds. *BMC Ecology*, 15, 12.

Dall, S.R.X., Giraldeau, L.A., Olsson, O., et al. (2005). Information and its use by animals in evolutionary ecology. *Trends in Ecology & Evolution*, 20, 187–193.

Darimont, C.T., Fox, C.H., Bryan, H.M., and Reimchen, T.E. (2015). The unique ecology of human predators. *Science*, 349, 858–860.

Dhananjaya, T., Das, S., Harpalani, M., et al. (2022). Can urbanization accentuate hand use in the foraging activities of primates? *Biological Anthopology*, 178, 667–677.

Diamond, S.E., and Martin, R.A. (2021). Evolution in cities. *Annual Review of Ecology, Evolution, and Systematics*, 52, 519–540.

Diquelou, M.C., Griffin, A.S., and Sol, D. (2016). The role of motor diversity in foraging innovations: a cross-species comparison in urban birds. *Behavioral Ecology*, 27, 584–591.

Dobson, A.D.M., De Lange, E., Keane, A., et al. (2019). Integrating models of human behaviour between the individual and population levels to inform conservation interventions. *Philosophical Transactions of the Royal Society B: Biological Sciences*, 374, 20180053.

Dowding, C.V., Harris, S., Poulton, S., and Baker, P.J. (2010). Nocturnal ranging behaviour of urban hedgehogs, *Erinaceus europaeus*, in relation to risk and reward. *Animal Behaviour*, 80, 13–21.

Ducatez, S., Sayol, F., Sol, D., and Lefebvre, L. (2018). Are urban vertebrates city specialists, artificial habitat exploiters, or environmental generalists? *Integrative and Comparative Biology*, 58, 929–938.

Duckworth, R.A. (2009). The role of behavior in evolution: a search for mechanism. *Evolutionary Ecology*, 23, 513–531.

Duckworth, R.A., and Badyaev, A.V. (2007). Coupling of dispersal and aggression facilitates the rapid range expansion of a passerine bird. *Proceedings of the National Academy of Sciences*, 104, 15,017–15,022.

Dukas, R. (2004). Evolutionary biology of animal cognition. *Annual Review of Ecology, Evolution, and Systematics*, 35, 347–374.

Duncan, R.P., Blackburn, T.M., and Sol, D. (2003). The ecology of bird introductions. *Annual Review of Ecology, Evolution, and Systematics*, 34, 71–98.

Eggers, S., Griesser, M., Nystrand, M., and Ekman, J. (2006). Predation risk induces changes in nest-site selection and clutch size in the Siberian jay. *Proceedings of the Royal Society B: Biological Sciences*, 273, 701–706.

Emery, N.J., Seed, A.M., Bayern, A.M.P., and Von Clayton, N.S. (2007). Cognitive adaptations of social bonding in birds. *Philosophical Transactions of the Royal Society B: Biological Sciences*, 362, 489–505.

Estes, W.A., and Mannan, R.W. (2003). Feeding behavior of cooper's hawks at urban and rural nests in Southeastern Arizona. *The Condor*, 105, 107–116.

Evans, K.L., Gaston, K.J., Frantz, A.C., et al. (2009). Independent colonization of multiple urban centres by a formerly forest specialist bird species. *Proceedings of the Royal Society B: Biological Sciences*, 276, 2403–2410.

Evans, K.L., Hatchwell, B.J., Parnell, M., and Gaston, K.J. (2010). A conceptual framework for the colonisation of urban areas: the blackbird *Turdus merula* as a case study. *Biological Reviews*, 643–667.

Fisher, J., and Hinde, R.A. (1949). The opening of milk bottles by birds. *British Birds*, 42, 347–357.

Gallo, T., Fidino, M., Gerber, B., et al. (2019). Mammals adjust diel activity across gradients of urbanization. *Journal of Animal Ecology*, 88, 793–803.

Gao, J., and O'Neill, B.C. (2020). Mapping global urban land for the 21st century with data-driven simulations and shared socioeconomic pathways. *Nature Communications*, 11, 2302.

Gaynor, K.M., Hojnowski, C.E., Carter, N.H., and Brashares, J.S. (2018). The influence of human disturbance on wildlife nocturnality. *Science*, 360, 1232–1235.

Gilroy, J.J., and Sutherland, W.J. (2007). Beyond ecological traps: perceptual errors and undervalued resources. *Trends in Ecology & Evolution*, 22, 351–356.

Giraldeau, L.A: (1984). Group foraging: the skill pool effect and frequency dependent effect. *The American Naturalist*, 124, 72–79.

Goumas, M., Boogert, N.J., and Kelley, L.A. (2020). Urban herring gulls use human behavioural cues to locate food. *Royal Society of Open Science*, 7, 191959.

Greenberg, R. (1983). The role of neophobia in determining the degree of foraging specialization in some migrant warblers. *The American Naturalist*, 122, 444–453.

Greggor, A.L., Clayton, N.S., Fulford, A.J.C., and Thornton, A. (2016). Street smart: faster approach towards litter in urban areas by highly neophobic corvids and less fearful birds. *Animal Behaviour*, 117, 123–133.

Grimm, N.B., Faeth, S.H., Golubiewski, N.E., et al. (2008). Global change and the ecology of cities. *Science*, 319, 756–760.

Güneralp, B., and Seto, K.C. (2013). Futures of global urban expansion: uncertainties and implications for biodiversity conservation. *Environmental Research Letters*, 8, 014025.

Haag-Wackernagel, D. (1993). Street pigeons in Basel. *Nature*, 361, 200.

Halfwerk, W., Bot, S., Buikx, J., et al. (2011). Low-frequency songs lose their potency in noisy urban conditions. *Proceedings of the National Academy of Sciences*, 108, 14,549–14,554.

Halfwerk, W., and Slabbekoorn, H. (2009). A behavioural mechanism explaining noise-dependent frequency use in urban birdsong. *Animal Behaviour*, 78, 1301–1307.

Hamer, A.J., and Mcdonnell, M.J. (2009). The response of herpetofauna to urbanization: inferring patterns of persistence from wildlife databases: Persistence of herpetofauna under urbanization. *Austral Ecology*, 35, 568–580.

Horton, K.G., Nilsson, C., Van Doren, B.M., et al. (2019). Bright lights in the big cities: migratory birds' exposure to artificial light. *Frontiers in Ecology and the Environment*, 17, 209–214.

Husby, A., and Husby, M. (2014). Interspecific analysis of vehicle avoidance behavior in birds. *Behavioral Ecology*, 25, 504–508.

Huey, R.B., Hertz, P.E. and Sinervo, B. (2003). Behavioral drive versus behavioral inertia in evolution: A null model approach. *The American Naturalist*, 161, 357–366.

Johnson, M.T.J., and Munshi-South, J. (2017). Evolution of life in urban environments. *Science*, 358, eaam8327.

Kaufman, A.B., Kaufman, J.C., and Call, J. (eds) (2021). *The Cambridge Handbook of Animal Cognition, Cambridge Handbooks in Psychology*, pp. 637–791. Cambridge University Press, Cambridge.

Kent, E., Schwartz, A.L.W., and Perkins, S.E. (2021). Life in the fast lane: roadkill risk along an urban–rural gradient. *Journal of Urban Ecology*, 7, juaa039.

Klump, B.C., Martin, J.M., Wild, S., et al. (2021). Innovation and geographic spread of a complex foraging culture in an urban parrot. *Science*, 373, 456–460.

Kokko, H., and Sutherland, W.J. (2001). Ecological traps in changing environments: ecological and evolutionary consequences of a behaviourally mediated Allee effect. *Evolutionary Ecology Research*, 3, 537–551.

Lapiedra, O., Chejanovski, Z., and Kolbe, J.J. (2017). Urbanization and biological invasion shape animal personalities. *Global Change Biology*, 23, 592–603.

Lee, W.Y., Lee, S., Choe, J.C., and Jablonski, P.G. (2011). Wild birds recognize individual humans: experiments on magpies, *Pica pica*. *Animal Cognition*, 14, 817–825.

Lefebvre, L., Reader, S.M., and Sol, D. (2004). Brains, innovations and evolution in birds and primates. *Brain, Behavior and Evolution*, 63, 233–246.

Lefebvre, L., Whitle, P., Lascaris, E., and Finkelstein, A. (1997). Feeding innovations and forebrain size in birds. *Animal Behaviour*, 53, 549–560.

Liker, A., and Bokony, V. (2009). Larger groups are more successful in innovative problem solving in house sparrows. *Proceedings of the National Academy of Sciences*, 106, 7893–7898.

Łopucki, R., Klich, D., and Kiersztyn, A. (2021). Changes in the social behavior of urban animals: more aggression or tolerance? *Mammalian Biology*, 101, 1–10.

Lowry, H., Lill, A., and Wong, B.B.M. (2012). How noisy does a noisy miner have to be? Amplitude adjustments of alarm calls in an avian urban 'adapter'. *PLOS ONE*, 7, 1–5.

Lowry, H., Lill, A., and Wong, B.B.M. (2013). Behavioural responses of wildlife to urban environments. *Biological Reviews*, 88, 537–549.

MacIvor, J.S., and Moore, A.E. (2013). Bees collect polyurethane and polyethylene plastics as novel nest materials. *Ecosphere*, 4, 155.

Maklakov, A.A., Immler, S., Gonzalez-Voyer, A., et al. (2011). Brains and the city: big-brained passerine birds succeed in urban environments. *Biological Letters*, 7, 730–732.

Mallord, J.W., Dolman, P.M., Brown, A.F., and Sutherland, W.J. (2007). Linking recreational disturbance to population size in a ground-nesting passerine. *Journal Applied Ecology*, 44, 185–195.

Martin, A.E., Graham, S.L., Henry, M., et al. (2018). Flying insect abundance declines with increasing road traffic. *Insect Conservation and Diversity*, 11, 608–613.

Martin, G.R. (2011). Understanding bird collisions with man-made objects: a sensory ecology. *Ibis*, 153, 239–254.

Marzluff, J.M., Bowman, R., and Donnelly, R. (2001). A historical perspective on urban bird research: trends, terms, and approaches. *Avian Ecology and Conservation in an Urbanizing World*, 1–17.

Marzluff, J.M., Walls, J., Cornell, H.N., et al. (2010). Lasting recognition of threatening people by wild American crows. *Animal Behaviour*, 79, 699–707.

Maspons, J., Molowny-Horas, R., and Sol, D. (2019). Behaviour, life history and persistence in novel environments. *Philosophical Transactions of the Royal Society B: Biological Sciences*, 374, 20180056.

Mayr, E. (1963). *Animal Species and Evolution*. Harvard University Press, London.

Mccleery, R.A. (2009). Changes in fox squirrel antipredator behaviors across the urban-rural gradient. *Landscape Ecology*, 24, 483–493.

McGregor, R.L., Bender, D.J., and Fahrig, L. (2008). Do small mammals avoid roads because of the traffic? *Journal of Applied Ecology*, 45, 117–123.

McKinney, M.L. (2002). Urbanization, biodiversity, and conservation. *BioScience*, 52, 883–890.

McKinney, M.L. (2006). Urbanization as a major cause of biotic homogenisation. *Biological Conservation*, 127, 247–260.

Miranda, A.C., Schielzeth, H., Sonntag, T., and Partecke, J. (2013). Urbanization and its effects on personality traits: a result of microevolution or phenotypic plasticity? *Global Change Biology*, 19, 2634–2644.

Moiron, M., González-Lagos, C., Slabbekoorn, H., and Sol, D. (2015). Singing in the city: high song frequencies are no guarantee for urban success in birds. *Behavioral Ecology*, 26, 843–850.

Møller, A.P. (2008). Flight distance of urban birds, predation, and selection for urban life. *Behavioral Ecology and Sociobiology*, 63, 63–75.

Møller, A.P., and Erritzøe, J. (2017). Brain size in birds is related to traffic accidents. *Royal Society of Open Science*, 4, 161040.

Morand-Ferron, J., and Quinn, J.L. (2011). Larger groups of passerines are more efficient problem solvers in the wild. *Proceedings of the National Academy of Sciences*, 108, 15,898–15,903.

Murray, M.H., and St. Clair, C.C. (2015). Individual flexibility in nocturnal activity reduces risk of road mortality for an urban carnivore. *Behavioral Ecology*, 26, 1520–1527.

Nicolaus, M., and Edelaar, P. (2018). Comparing the consequences of natural selection, adaptive phenotypic plasticity, and matching habitat choice for phenotype-environment matching, population genetic structure, and reproductive isolation in meta-populations. *Ecology and Evolution*, 8, 3815–3827.

Overington, S.E., Morand-Ferron, J., Boogert, N.J., and Lefebvre, L. (2009). Technical innovations drive the relationship between innovativeness and residual brain size in birds. *Animal Behaviour*, 78, 1001–1010.

Owen, M.A., Swaisgood, R.R., and Blumstein, D.T. (2017). Contextual influences on animal decision-making: significance for behavior-based wildlife conservation and management. *Integrative Zoology*, 12, 32–48.

Paniw, M., Cozzi, G., Sommer, S., and Ozgul, A. (2021). Demographic processes in socially structured populations. In: *Demographic Methods across the Tree of Life*, pp. 341–350. Oxford University Press. New York, NY.

Partan, S.R., Fulmer, A.G., Gounard, M.A.M., and Redmond, J.E. (2010). Multimodal alarm behavior in urban and rural gray squirrels studied by means of observation and a mechanical robot. *Current Zoology*, 56, 313–326.

Partecke, J., and Gwinner, E. (2007). Increased sedentariness in European blackbirds following urbanization: a consequence of local adaptation? *Ecology*, 88, 882–890.

Phillips, B.L. (2016). Behaviour on invasion fronts, and the behaviour of invasion fronts. In: D. Sol and J.S. Weis (eds), *Biological Invasions and Animal Behaviour*, pp. 82–95. Cambridge University Press, Cambridge.

Price, T.D., Qvarnström, A., Irwin, D.E., et al. (2003). The role of phenotypic plasticity in driving genetic evolution. *Proceedings of the Royal Society B: Biological Sciences*, 270, 1433–1440.

Price, T.D., and Sol, D. (2008). Introduction: genetics of colonising species. *The American Naturalist*, 172, S1–S3.

Reader, S.M. (2016). Animal social learning: associations and adaptations. *F1000Research*, 5, 2120.

Rehage, J.S., and Sih, A. (2004). Dispersal behavior, boldness, and the link to invasiveness: a comparison of four gambusia species. *Biological Invasions*, 6, 379–391.

Ritzel, K., and Gallo, T. (2020). Behavior change in urban mammals: a systematic review. *Frontiers in Ecology and Evolution*, 8, 576665.

Robertson, B.A., Rehage, J.S., and Sih, A. (2013). Ecological novelty and the emergence of evolutionary traps. *Trends in Ecology & Evolution*, 28, 552–560.

Rutz, C. (2008). The establishment of an urban bird population. *Journal of Animal Ecology*, 77, 1008–1019.

Santini, L., González-Suárez, M., Russo, D., et al. (2019). One strategy does not fit all: determinants of urban adaptation in mammals. *Ecology Letters*, 22, 365–376.

Sayol, F., Sol, D., and Pigot, A.L. (2020). Brain size and life history interact to predict urban tolerance in birds. *Frontiers in Ecology and Evolution*, 8, 58.

Scheun, J., Greeff, D., and Nowack, J. (2019). Urbanization as an important driver of nocturnal primate sociality. *Primates*, 60, 375–381.

Schlaepfer, M.A., Runge, M.C., and Sherman, P.W. (2002). Ecological and evolutionary traps. *Trends in Ecology & Evolution*, 17, 474–480.

Scott, D.M., Berg, M.J., Tolhurst, B.A., et al. (2014). Changes in the distribution of red foxes (*Vulpes vulpes*) in urban areas in Great Britain: findings and limitations of a media-driven nationwide survey. *PLOS ONE*, 9, e99059.

Shettleworth, S.J. (2010). *Cognition, Evolution, and Behavior*. Oxford University Press. New York, NY.

Shochat, E., Warren, P., Faeth, S., et al. (2006). From patterns to emerging processes in mechanistic urban ecology. *Trends in Ecology & Evolution*, 21, 186–191.

Slabbekoorn, H., and Peet, M. (2003). Ecology: birds sing at a higher pitch in urban noise. *Nature*, 424, 267.

Smith, J.A., Suraci, J.P., Clinchy, M., et al. (2017). Fear of the human 'super predator' reduces feeding time in large carnivores. *Proceedings of the Royal Society B: Biological Sciences*, 284, 20170433.

Snell-Rood, E.C., Kobiela, M.E., Sikkink, K.L., and Shephard, A.M. (2018). Mechanisms of plastic rescue in novel environments. *Annual Review of Ecology, Evolution, and Systematics*, 49, 331–354.

Sol, D., Bartomeus, I., González-Lagos, C., and Pavoine, S. (2017a). Urbanization and the loss of phylogenetic diversity in birds. *Ecology Letters*, 20, 721–729.

Sol, D., Bartomeus, I., and Griffin, A.S. (2012). The paradox of invasion in birds: competitive superiority or ecological opportunism? *Oecologia*, 169, 553–564.

Sol, D., Duncan, R.P., Blackburn, T.M., et al. (2005). Big brains, enhanced cognition, and response of birds to novel environments. *Proceedings of the National Academy of Sciences*, 102, 5460–5465.

Sol, D., González-Lagos, C., Lapiedra, O., and Díaz, M. (2017b). Why are exotic birds so successful in urbanized environments? In: E. Murgui and M. Hedblom (eds), *Ecology and Conservation of Birds in Urban Environments*, pp. 75–89. Springer International Publishing, Cham.

Sol, D., González-Lagos, C., Moreira, D., et al. (2014). Urbanization tolerance and the loss of avian diversity. *Ecology Letters*, 17, 942–950.

Sol, D., Griffin, A.S., Bartomeus, I., and Boyce, H. (2011). Exploring or avoiding novel food resources? The novelty conflict in an invasive bird. *PLOS ONE*, 6, e19535.

Sol, D., Lapiedra, O., and Ducatez, S. (2020b). Cognition and adaptation to urban environments. In: *Urban Evolutionary Biology*, pp. 253–267. Oxford University Press. Oxford.

Sol, D., Lapiedra, O., and González-Lagos, C. (2013). Behavioural adjustments for a life in the city. *Animal Behaviour*, 85, 1101–1112.

Sol, D., Maspons, J., Gonzalez-Voyer, A., et al. (2018). Risk-taking behavior, urbanization and the pace of life in birds. *Behavioral Ecology and Sociobiology*, 72, 59.

Sol, D., Olkowicz, S., Sayol, F., et al. (2022). Neuron numbers link innovativeness with both absolute and relative brain size in birds. *Nature Ecology & Evolution*, 6, 1381–1389.

Sol, D., Trisos, C., Múrria, C., et al. (2020a). The worldwide impact of urbanization on avian functional diversity. *Ecology Letters*, 23, 962–972.

Soulsbury, C.D., and White, P.C.L. (2015). Human–wildlife interactions in urban areas: a review of conflicts, benefits and opportunities. *Wildlife Research*, 42, 541–553.

Stamps, J. (2001). Habitat selection by dispersers: integrating proximate and ultimate approaches. In: J. Clobert, E. Danchin, A. Donth, and J. Nichols (eds), *Dispersal*, pp. 230–242. Oxford University Press, Croydon, USA.

Stamps, J.A., and Swaisgood, R.R. (2007). Someplace like home: experience, habitat selection and conservation biology. *Applied Animal Behaviour Science*, 102, 392–409.

Stansell, H.M., Blumstein, D.T., Yeh, P.J., and Nonacs, P. (2020). Individual variation in tolerance of human activity by urban dark-eyed juncos (*Junco hyemalis*). *The Wilson Journal of Ornithology*, 134, 43–51.

St. Clair, C.C., Backs, J., Friesen, A., et al. (2019). Animal learning may contribute to both problems and solutions for wildlife–train collisions. *Philosophical Transactions of the Royal Society B: Biological Sciences*, 374, 20180050.

Taylor, A.H., Elliffe, D.M., Hunt, G.R., et al. (2011). New Caledonian crows learn the functional properties of novel tool types. *PLOS ONE*, 6, 26887.

Taylor, A.H., Miller, R., and Gray, R.D. (2012). New Caledonian crows reason about hidden causal agents. *Proceedings of the National Academy of Sciences*, 109, 16,389–16,391.

Thompson, M.J., Evans, J.C., Parsons, S., and Morand-Ferron, J. (2018). Urbanization and individual differences in exploration and plasticity. *Behavioral Ecology*, 29, 1415–1425.

Tucker, M.A., Schipper, A.M., Adams, T.F., et al. (2023). Behavioral responses of terrestrial mammals to COVID-19 lockdowns. *Science*, 380, 1059–1064.

West-Eberhard, M.J. (2003). *Developmental Plasticity and Evolution*. Oxford University Press, New York.

Winchell, K.M., Carlen, E.J., Puente-Rolón, A.R., and Revell, L.J. (2018). Divergent habitat use of two urban lizard species. *Ecology and Evolution*, 8, 25–35.

Winchell, K.M., Losos, J.B., and Verrelli, B.C. (2023). Urban evolutionary ecology brings exaptation back into focus. *Trends in Ecology & Evolution*, S0169534723000605.

Yeh, P.J., Hauber, M.E., and Price, T.D. (2007). Alternative nesting behaviours following colonisation of a novel environment by a passerine bird. *Oikos*, 116, 1473–1480.

Yeh, P.J., and Price, T.D. (2004). Adaptive phenotypic plasticity and the successful colonisation of a novel environment. *The American Naturalist*, 164, 531–542.

CHAPTER 10

Multiple stressors

Amelia Munson, Daphne Cortese, and Shaun S. Killen

Overview

In a rapidly changing world, animals must respond to multiple stressors that occur either simultaneously or in rapid succession. While more challenging than studying responses to single stressors, directly studying multiple stressors is important because not all effects are additive; multiple stressors can interact in synergistic ways where exposure to two or more stressors is greater than the sum of exposure to either stressor in isolation. It is important to realize that exposure to multiple stressors not only can impact behavioural responses, but behaviour can also alter the effects of a stressor by determining the extent to which animals are exposed. A variety of factors, including timing of stressor exposure and the potential for acclimation, can also influence the impact of multiple stressors. Here, we describe the effects of multiple stressors on different aspects of animal behaviour, including activity, hiding, foraging, and sociality. Ultimately, changes in behaviour due to stressor exposure can have varying effects at different levels of biological organization. For example, individual responses can alter group dynamics, which may, in turn, influence community dynamics and affect ecosystem function. A better understanding of behavioural responses to multiple stressors can help to direct conservation efforts in a changing world.

10.1 Behaviour and multiple stressors

The world is rapidly changing, introducing animals to conditions not previously encountered during their evolutionary history. In nature, animals typically do not experience stressors in isolation and are, instead, often exposed to multiple stressors—either simultaneously or in close succession—with conflicting demands on their behaviour or energy allocation. Anthropogenic change can bring about novel combinations of challenges or exacerbate existing stressors, thus leading to unexpected interactions. Despite the ubiquity of multiple stressors in nature, most empirical research has focused on wildlife responses to individual stressors. While this focus has led to important insights regarding animals' responses to a variety of challenges, it likely misses the real ecological consequences of human-induced rapid environmental change. For these reasons, research has increasingly emphasized the importance of understanding interactions among multiple stressors (Todgham and Stillman 2013). Yet, we still lack detailed knowledge of the effects of multiple stressors on wild animals. Most of the work done in this area so far has been focused on how interacting stressors affect physiological mechanisms. Crucially, however, multiple stressors do not just impact the physiology of animals; they also impact a variety of behaviours, including hiding, foraging, sociality, and activity. Indeed, recent studies have suggested that the immediate responses of animals to stressors are primarily behavioural, with morphological and life-history changes occurring when optimal fitness conditions cannot be maintained by behavioural shifts alone (Cerini et al. 2023).

In this chapter, we discuss what is known—and what still needs to be learned—about the effects of interacting stressors on animal behaviour. We consider situations where: (1) there are two or more

Amelia Munson, Daphne Cortese, and Shaun S. Killen, *Multiple stressors*. In: *Behavioural Responses to a Changing World*. Edited by: Bob B. M. Wong and Ulrika Candolin, Oxford University Press. © Oxford University Press (2024). DOI: 10.1093/oso/9780192858979.003.0010

environmental factors changing simultaneously or in succession; and (2) at least one of these factors causes some degree of behavioural or physiological adjustment. We discuss the types of interactions between stressors and ways to classify their effects, the factors that influence the magnitude and direction of the interactive effects, and approaches for studying these interactions. Throughout, we highlight how multiple stressors may influence behaviour, as well as how behaviour may generate feedbacks that influence how the stressors themselves interact. Finally, we outline the importance of interacting stressors for community dynamics and conservation efforts.

10.2 Stressor interactions in a changing world

Interactions among stressors can take several forms (Figure 10.1). One way of classifying interacting stressors, for example, is whether they have commonly co-occurred throughout the evolutionary history of a given species, or if their co-occurrence is relatively novel. Increasing water temperatures, for instance, can also mean reductions in the oxygen available to aquatic organisms, either directly via reduced oxygen solubility or indirectly by increasing the effects of eutrophication on dissolved oxygen (Saari et al. 2018). While changes in temperature and oxygen availability are becoming more common or severe due to anthropogenic effects, these factors will have also exhibited natural correlated fluctuations over various timescales during species' evolution. In these cases, animals may have evolved mechanisms and behaviours for dealing with the combined effects of commonly co-occurring stressors, which may be useful for coping with anthropogenic change, even if the magnitude or frequency of perturbation is altered. For example, many freshwater tropical fishes have not only evolved a higher tolerance for warm water, but also possess behavioural and physiological adaptations for air-breathing that allow them to tolerate extremely hypoxic conditions (Damsgaard et al. 2020; Figure 10.1a). In contrast, other human-induced changes can bring about multiple challenges that are not mechanistically linked, but that have begun to co-occur much more recently in

evolutionary history. For example, animals living in urban areas must deal with a host of novel stressors, including increasing temperature due to the urban heat island effect (Rizwan et al. 2008), novel predators (Baker et al. 2005), and increased exposure to toxicants (Pollack et al. 2017; see also Chapter 9). The relative novelty of these stressor combinations may challenge the ability of animals to adapt, thus making behavioural adjustments especially important for survival (Figure 10.1a).

Even singular novel stressors are likely to be more complicated than empirical research may suggest, because animals must respond to them while still maintaining responses to historic challenges. For example, anthropogenic noise, a common novel stressor in a human-altered environment (Chapter 2), reduces foraging behaviour (Shannon et al. 2014; Luo et al. 2015), alters anti-predator behaviour (Spiga et al. 2017), and can affect susceptibility to disease (Berkhout et al. 2022) and conspecific aggression (Bruintjes and Radford 2013). In other words, the addition of a novel challenge alters the response to historic challenges (Figure 10.1b).

What constitutes a stressor may also shift in a changing world depending on other environmental factors. For example, under normal seasonal temperatures, drier conditions are more detrimental to the survival of streamside salamanders *Ambystoma barbourin*; however, when temperatures are elevated, mortality occurs more frequently when conditions are wet (Rohr and Palmer 2013). While the exact reasons for this are unclear, it is potentially because warmer conditions cause a direct increase in maintenance metabolism, while wetter conditions lead to increased opportunity for spontaneous activity with associated locomotor costs. Recent increases in the frequency or severity of warming may cause activity costs generated during foraging to outpace energy intake, eventually leading to increased mortality risk due to an inability to meet overall energy demand. These effects may be especially likely if food becomes more limited or requires more effort to find. Altered precipitation patterns that occur out of season may thus generate behaviours that are maladaptive for the temperature to which animals are exposed (Figure 10.1c). This can lead to conditions that would not previously have been considered

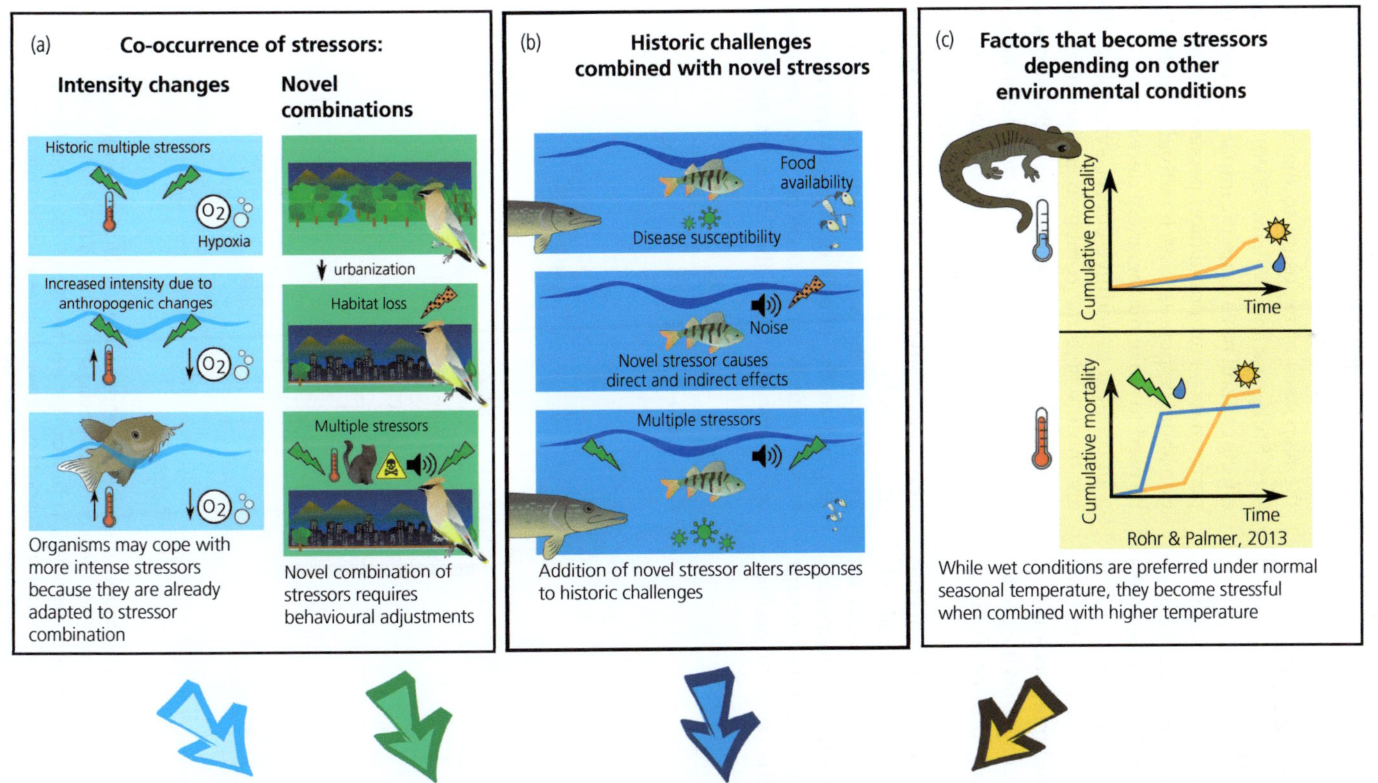

Figure 10.1 Summary of potential stressor interactions and their effect on animal behaviour following human-induced changes. Each panel represents a different type of multiple stressor combination: (a) stressor co-occurrence, (b) interaction between novel stressors and historic challenges, and (c) factors that become stressors depending on other environmental conditions. Green plain thunderbolt shapes indicate multiple stressors and orange dotted thunderbolts stand for a single stressor. In panel c) the sun indicates dry conditions, and the water droplet indicates wet conditions.

Graph adapted from data in Rohr and Palmer 2013.

stressful to become potentially lethal because the environmental cues no longer elicit appropriate behavioural responses.

10.3 Effects of interactions: additive, synergistic, and antagonistic

Responses to multiple stressors are often classified into three main types of interaction: additive, synergistic, and antagonistic (Figure 10.2). These interactions have been used to specify how multiple stressors affect overall fitness, as well as to classify behavioural responses to multiple stressors (e.g. Dominoni et al. 2020). Behaviour also has the potential to shape additive versus non-additive effects on fitness.

Additive effects are those where the effect of both stressors is the same as the sum of the individual stressors. For example, when experienced on their own, both resource limitation and exposure to the pesticide imidacloprid caused blue orchard bees *Osmia lignaria* to delay the onset of nesting,

stop nesting sooner, and lower their rate of offspring provisioning. When exposed to both stressors in combination there were additive decreases in reproduction (i.e. the decrease was equal to the sum of the responses of exposure to the individual stressors; Stuligross and Williams 2020). While additive effects can still have dramatic impacts on individual performance and population persistence, information about responses to each individual stressor can be used to make management decisions.

In contrast to additive effects, synergistic effects are those where the combined impact of multiple stressors is greater than the sum of the stressors in isolation. This can occur when one factor amplifies the intensity of another. For example, both temperature change and low food availability may be only mildly stressful on their own but when they occur together, they may cause high mortality if, for example, the animal needs increased food to respond to the effects of high temperature. Synergistic effects can also occur when the effect of one factor constrains the ability of an animal to

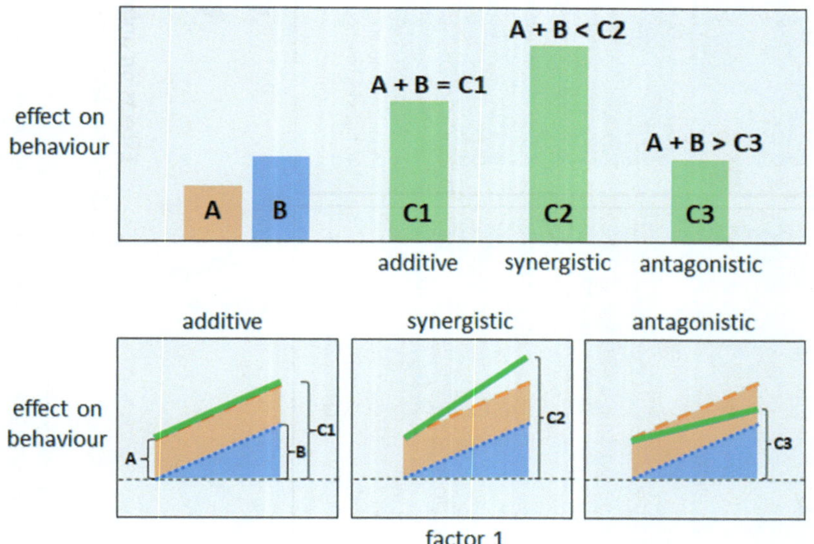

Figure 10.2 Conceptual models for illustrating interactions between stressors and the effect on behaviour. On the top panel there is an illustration for interaction between two discrete stressors, with their effects on behaviour being shown by A and B. The combined effect may be additive (C1), synergistic (C2), or antagonistic (C3). On the bottom panel is an illustration between two levels of one discrete factor ('factor 2', illustrated by the orange dashed lines and blue dotted lines), and one continuous stressor, shown along the x-axis (here referred to as 'factor 1'). In the bottom panel, the effects of each factor on behaviour are also shown as either A or B and are analogous to the effects shown in the top panel when the effect of each is measured when the other is held at a baseline minimal level. The solid green lines show the potential additive, synergistic, or antagonistic effect of the combined factors across a gradient of factor 1 on animal behaviour.

adjust behaviourally or physiologically to cope with other additional stressors, or causes increased likelihood of the animal being exposed to an additional stressor. For example, hypoxia can increase spontaneous activity and aquatic-surface respiration in fish, thereby increasing their exposure to predation risk (Killen et al. 2012).

Antagonistic effects are those where the effect of both stressors is less than the sum of the response to either stressor individually. This can occur if, for instance, one stressor induces defences that can also be used against the other stressor. For example, eastern mosquitofish *Gambusia holbrooki* kept in a high-flow environment that required elevated levels of activity were less affected by UV-B radiation compared to individuals kept in a less stressful low-flow environment (Ghanizadeh-Kazerouni et al. 2017). This is presumably because prolonged exercise increases antioxidant activity, which is effective at reducing the impacts of UV-B radiation. However, effects of multiple stressor combinations are challenging to predict, and whether stressor combinations are additive, synergistic, or antagonistic can be species-dependent (Mahon et al. 2019) and requires specific testing.

Behaviour can also change the type of interaction between stressors. Altered behaviour in response to one stressor may exacerbate the effects of a second stressor, thus changing something that, from a mechanistic perspective, might have additive effects to an interaction that is synergistic. For example, Van Dievel et al. (2019) found additive negative effects on growth rate via changes to assimilation and conversion efficiency when damselfly *Enallagma cyathigerum* larvae were exposed to both chlorpyrifos (a pesticide) and predation risk in the laboratory. However, all treatments had easy access to an ecologically relevant amount of food, and there was no difference in consumption among treatment groups. By contrast, in the wild, predation risk often leads to reductions in consumption via reduced time spent foraging (Brown and Kotler 2004). In relatively small experimental chambers, different behavioural outcomes from stressor exposure, like differences in time spent foraging, may not be present because food is readily available and requires limited effort to locate. If damselfly larvae spent less time foraging in response to predation risk, then the bioenergetic effects of pesticides would be exacerbated because they are obtaining less food, leading to a potential synergistic interaction between stressors (Van Dievel et al. 2019). It is therefore important in multiple stressor studies to allow animals to engage in species-typical behaviours to better predict the effects of stressor interactions.

In recent years the number of mentions of 'synergy' in the titles and abstracts of ecological journal articles has rapidly increased (Côté et al. 2016). Although mentions of additive and antagonistic interactions are also on the rise, neither rival that of synergies. However, the increased use of the word synergy may not reflect the actual prevalence of synergistic interactions (Côté et al. 2016). More rigorous meta-analyses found that while the type of interaction depended on the stressors and level of organization, synergies only account for about a third of all stressor responses (Crain et al. 2008; Darling and Côté 2008). This mismatch in reporting of interactions and prevalence of actual synergies is particularly important to consider for conservation-directed research because, if the assumption is that synergies are the likely outcome of multiple stressors, conservation practitioners and the public may become pessimistic about the potential to combat anthropogenic stressors, as additive interactions are much easier to predict and respond to (Côté et al. 2016; see Box 10.1). However, it is important to be realistic about the threats that multiple stressors pose. While it was found that synergies were not the most common outcome of stressor interactions, the inclusion of a third stressor in an interaction did increase the likelihood of a synergism (Crain et al. 2008).

10.4 Timing and acclimation

The consequences of being exposed to multiple stressors can also differ depending on the timing of those stressors relative to the organism's generation time. Stressors can occur as discrete isolated events, oscillatory repeated events, or continuous events (Galsby and Underwood 1996, Jackson et al. 2021). Moreover, stressors can either occur at the same time or in rapid succession, which can change their impacts, as time between exposures may reduce

Box 10.1 Multiple stressors and conservation

Conservation managers and practitioners are already making decisions in systems where multiple stressors are the rule, rather than the exception. Animals must regularly contend with multiple anthropogenic changes against a backdrop of historical challenges. Because multiple stressors are ubiquitous in nature, understanding behavioural responses to direct conservation efforts is of crucial importance.

When making management decisions, not all stressors are created equal. 'Global drivers' like climate change are typically much harder to address compared to more local stressors. However, targeting and alleviating local stressors can affect the magnitude of larger, harder to control challenges (Strain et al. 2015). However, the type of interaction is always important; reducing local stressors has the greatest impact on the effects of global stressors when there is a synergistic interaction and can actually be detrimental overall if the stressors are interacting antagonistically (Brown et al. 2013). In an antagonistic interaction, the sum of two stressors is less than the sum of the stressors in isolation, so mediating or changing one stressor may actually increase the effect of the other stressor. For example, while both high temperature and high turbidity affected coral survival, high turbidity reduced the risk of mortality under high temperature, potentially because turbidity provided protection from light and served as an alternative food source for bleached corals (Anthony et al. 2007). Conservation activities that aim at reducing turbidity may thus have the unintended consequence of increasing the risk of death from other sources.

When making community-level conservation management decisions, potential correlations between species' sensitivities to different stressors should also be considered. If species tolerances are positively correlated ('co-tolerance'), then the impact of multiple stressors is predicted to be less than if there are negative relationships between stressor tolerances (Vinebrooke et al. 2004). This is because, in the case of positive co-tolerance, one stressor eliminates individuals or species that are also susceptible to a second stressor, so when the community must face the second stressor, there are fewer vulnerable species to be negatively impacted. Conversely, if tolerances to stressors are negatively related, then one stressor will eliminate the species or individuals that are less affected by a second stressor, leaving vulnerable individuals in the community. Identifying potential positive co-tolerances can help to direct management decisions. For example, while both fishing and high temperature negatively impacted coral, following a high temperature event, coral cover declined less on fished versus unfished reefs (Darling et al. 2010). It was hypothesized that this was because fishing had removed species that were not only sensitive to fishing stress but temperature stress as well; this was supported by the similarity in species assemblages between fished and unfished sites after the high temperature event (Darling et al. 2010). This is not to suggest that fishing should be used as a conservation practice necessarily, but understanding of these dynamics can help to direct management practices and identify regions most vulnerable to stressors. Alternatively, if no one stressor entirely removes a species from an area, then negatively correlated tolerances should result in higher biodiversity because all species are able to persist. In this case, positive co-tolerance would then lead to reduced biodiversity when stressors co-occur because vulnerable species will be removed, leaving only species unaffected by either stressor.

Identifying and avoiding tipping points in population dynamics is particularly important from a conservation perspective. An 'extinction vortex' occurs when positive feedback between biotic and abiotic forces leads to declines in population size and eventual extinction (Fagan and Holmes 2006). Multiple stressors can lead to this if, for example, animals alter their behaviour in response to stressor exposure to invest less in reproduction in favour of diverting energy to maintenance processes (Sokolova et al. 2012; Lopez et al. 2023).

The role of understanding animal behaviour in conservation decisions has, historically, been a contentious one (Caro and Sherman 2013). However, integrating behaviour—specifically movement, foraging decisions, and mating and social behaviours—into conservation biology has been proposed as a way to further develop both disciplines (Berger-Tal et al. 2011; see also Chapter 11). Considering the ways that behaviour (including altered foraging and moving to avoid stressors) can expose animals to additional stressors further supports the need for including behaviour into management and conservation decisions.

Conservation practitioners must make decisions about managing ecosystems that scientists do not always have answers for. Some decisions involve judgement calls; for example, many effects of multiple stressors are species-dependent and, in this regard, what may be harmful to one species may be of less concern or even beneficial to another. Science can only provide information on these effects, but cultural and practical knowledge are critical in making decisions about what to do with that information. However, experiments can be designed in ways that better inform management and conservation decisions and to improve the

Box 10.1 *Continued*

outcomes of these decisions (Côté et al. 2016). Experiments can be designed with ecologically relevant levels of stressors with durations to match real-world events. Additionally, most ecological literature on multiple stressors focuses on comparing the addition of stressors to a control 'pristine' environment; however, conservation managers are typically concerned with the outcome of removals of stressors. Because mitigating and removing a stressor may not have the reciprocal effect of adding a stressor, more research is needed on the effects of and strategies for effective stressor removal (Côté et al. 2016). Ultimately, balancing prolonged monitoring to facilitate adaptive responses with actions that are flexible enough to allow for different types of interactions between stressors is the most effective way to combat the myriad multiple stressors associated with anthropogenic change.

any potential interactive effects. If individuals can respond to each stressor separately, they may be able to rapidly recover before needing to respond to additional stressors. For example, in Olympic oysters *Ostrea lurida*, periods of co-occurring low salinity and high temperature had synergistic effects on survival; however, when periods of low salinity and high temperature were separated by at least two weeks of recovery time, the synergistic effects were eliminated (Bible et al. 2017).

Even with time between exposure to two stressors, exposure to the first stressor may alter the response to the second stressor. Carryover effects are responses that occur because exposure to one stressor can have consequences for an individual across time or developmental stages. Typically, if there is less time between exposures, there are stronger carryover effects, but even exposures across relatively prolonged time periods can have effects on responses to future stressors. This is especially true when exposure to the first stressor occurs during early development. For example, exposure to the herbicide atrazine during the embryo and larval stage influenced water-conserving behaviours, foraging efficiency, mass, and time to death of postmetamorphic streamside salamanders exposed to altered temperature and decreased water availability (Rohr and Palmer 2013). Interestingly, research looking at the interactive effect of increased temperature and pollutant exposure on invertebrates suggests that the timing of exposures, including the order of exposure and ontogenetic and transgenerational effects, can change whether the interaction is synergistic or antagonistic (Dinh et al. 2023).

Thus, research on the effects of anthropogenic stressors should carefully consider the timing and rate of exposure to closely match ecologically relevant scenarios.

When different stressors require similar behavioural or physiological responses, timing of exposure to the stressors may be particularly important. Shorter delays between stressors may reduce the impact of the second stressor if the individual is already displaying appropriate defences due to their exposure to the initial stressor. For example, if individuals are more vigilant immediately after a predator attack, they may be less susceptible to a second predator. In the context of anthropogenic change, this could be particularly beneficial in the event of an invasive novel predator. Experience with native predators may make individuals generally more vigilant for predators they have not experienced before (West et al. 2017; Pollack et al. 2022; see also Chapter 6). Alternatively, if different stressors require an energetically costly response, then challenges that occur in rapid succession may be more detrimental because they may require more of a response than an individual is able to mount. This can be due to physiological capacity or behavioural constraints. For example, while theory predicts that experience with native predators can decrease the impact of novel predators, if predators are always present, animals may, counterintuitively, resume foraging sooner following a threat or show a reduced response to threat (Ehlman et al. 2019b; see also Chapter 19). This is because constant exposure to predatory threats can actually make individuals less wary of predators;

if predators are often present, prey are forced to forage under potentially dangerous conditions in order to obtain enough food resources.

Acclimation should also be considered when predicting responses to multiple stressors. If the onset of one stressor occurs before another stressor, then, even if the first stressor is still present at the onset of the second stressor, acclimation to the first stressor may reduce the potential for interactive effects compared to if both stressors occurred simultaneously. In ectotherms, for example, the behavioural response to and interaction between temperature and other stressors may initially be strong but dampen over time as individuals physiologically acclimate to the new temperature. The effect of multiple anthropogenic stressors may also be different for populations that have had experience with those changes compared to populations experiencing them for the first time, even over relatively short, within-generational timescales (see also Chapter 14 and Chapter 19). For example, artificial light at night and noise pollution affected the activity of great tits *Parus major* in different ways depending on whether the birds came from an urban or rural population (Dominoni et al. 2020).

10.5 Interactions across different levels of biological organization

An important consideration is that the nature or magnitude of an interaction between stressors may depend on the trait being measured and the biological scale at which the traits are observed. Interactive effects on specific behaviours may be different from those observed on life-history traits, such as growth or survival, in response to the same combination of stressors. As an example, consider the potential interactive effects of changes in temperature and predation risk for an ectothermic species (Figure 10.3). Warming may require an increase in foraging activity to support an increased metabolic rate, but increased predator density may suppress the ability to forage enough to satisfy these requirements (Killen and Brown 2006; Killen et al. 2007). As a result, although we might expect an additive or mildly synergistic interaction between warming and increased predation risk on foraging activity, this stressor combination may have a strongly

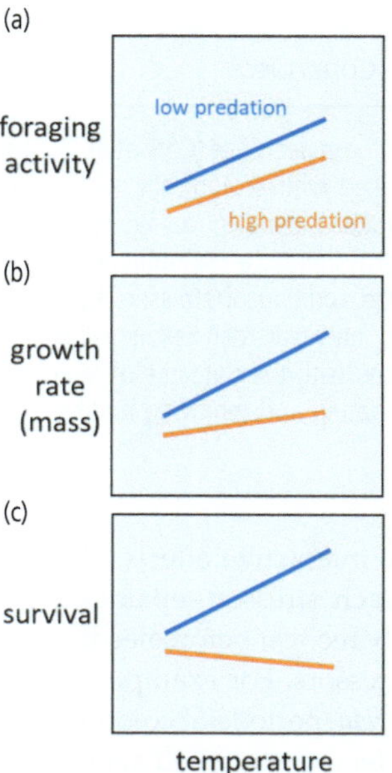

Figure 10.3 Combined effects of temperature and predation on a theoretical ectothermic species across multiple levels of study. (a) Overall, foraging increases with temperature, but this effect is decreased at high predator density because foraging activity is suppressed, generating an approximately additive negative effect on foraging. (b) Reductions in foraging due to predator presence cause mass loss, especially at higher temperatures where metabolic rates are highest, leading to a synergistic interaction on growth rate. (c) Direct and indirect effects of predators on survival, including increased likelihood of starvation on predator-associated stress, can cause strongly synergistic effects on survival.

synergistic effect on growth rate. Furthermore, in addition to any mortality stemming from starvation or predator-associated stress, any foraging that does occur may result in individuals incurring an increased risk of mortality by predation, leading to synergistic effects of these factors on survival that may be even stronger than that observed for growth.

As a consequence, stressors may have very different effects depending on the level of organization being observed and measured. Researchers must recognize that measuring interactive effects on one response variable may not give a complete picture of the ecological consequences of the stressors of interest—or the true nature of their interaction. Furthermore, because the effects of interactions may be hierarchical, with the energetic costs of behavioural changes cascading to affect growth rate or survival, experiments in which animals are not permitted to exercise their full behavioural repertoire may not provide the same results as would be observed in nature.

10.6 Studying the effects of multiple stressors

10.6.1 Recognizing and overcoming logistical constraints

Given the critical importance of multiple stressors for understanding the ecological effects of anthropogenic change, it is reasonable to ask why there has been relatively little research studying these interactions, as compared to isolated effects. From a purely logistical standpoint, the answer is that studying interactive effects is extremely complex, and generally requires more time, space, and resources to perform studies with sufficient statistical power to confidently identify the effects of interest. Historically, it has made sense to begin with the relatively straightforward task of understanding the main effects of isolated stressors on behaviour before delving into the myriad ways that these various effects might interact to alter trait expression.

Laboratory studies of multiple stressors commonly involve factorial designs with various levels of two (or more) stressors being measured in combination to test their effects on behaviour. While extremely useful, these studies have important limitations that must be recognized. Firstly, the use of categorical levels of factors for potentially continuous variables, like temperature, can be problematic because the effects of these variables on behaviour may not be linear and, as a result, may not be accurately modelled with only a few factor levels (e.g. Hewitt et al. 2016). This is a limitation in the study of isolated effects and becomes even more of a concern when studying interactions. In a similar vein, lab experiments can typically only examine combinations of two or three stressors at a time; additional factors may be ecologically relevant but would also require more time, space, and resources to study.

Another approach for studying multiple stressors is to use field studies, taking advantage of existing spatial or temporal variation in environmental conditions to examine statistical interactions among variables (e.g. Toft et al. 2018). Field studies can often capture ecologically relevant ranges of the factors of interest along a continuous scale and, ideally, provide measurements of behaviour and other traits of animals in natural or semi-natural conditions. During field studies, researchers can also take measurements of numerous environmental variables to examine complex interactions among many stressors using multivariate analyses. That said, there are significant challenges with wild studies. For instance, similar to laboratory studies, as more factors are measured, a higher sample size and sampling effort are required for adequate statistical analysis. Uncontrolled correlations between factors can also be an issue in field studies. Furthermore, the sampling effort itself is not trivial, and consideration must be given to the spatial and temporal scales that are relevant for measuring environmental variables and the questions of interest. Notably, measuring the behaviour of wild animals can also be extremely challenging. Fortunately, while such studies have traditionally been difficult, the monitoring of environmental variables and behaviour of wild animals is increasingly being aided by rapidly developing technologies for data logging and transmission, including environmental sensors and tags for tracking animal movements (Nathan et al. 2022). Study systems where effects of stressors can be tested in both lab and field conditions (see e.g. work on tungara frogs *Engystomops pustulosus* and noise and light pollution that have been studied in both the lab (Smit et al. 2022) and the field (Cronin et al. 2022)) may be used to bridge the gap and further our understanding of the effects of multiple stressors on behaviour.

10.6.2 Conceptual frameworks

With so many factors being potentially important when measuring the responses of wildlife to multiple stressors, it is perhaps not surprising that meta-analyses have failed to find consistent general patterns predicting the effects of multi-stressor interactions (Orr et al. 2020). However, several conceptual frameworks have been proposed to better understand how animals respond to interacting environmental perturbations and to potentially predict responses in novel conditions. Although these models predominantly focus on physiology, the reasoning can be useful for understanding the mechanisms underlying behavioural responses to multiple stressors.

Drawing on biomedical and ecological research, the Allostasis Model describes the process by which animals maintain stability despite changes in the environment (McEwen and Wingfield 2003). This framework primarily focuses on the balance between energy inputs and expenditures and considers both daily and seasonal variation and more unpredictable changes. 'Allostatic load' refers to the overall energy that an animal must expend to maintain homeostasis. An animal is considered to be experiencing 'allostatic overload' when the energy required to maintain homeostasis exceeds the stored energy that is available to meet this demand. Building on the Allostasis Model, the Reactive Scope Model considers the effects of physiological mediators when they fall in different ranges that correspond to predictable and unpredictable environmental changes, and levels above and below this range (Romero et al. 2009). These models are useful for classifying how 'stressed' an individual is—which may correspond to the level of behavioural change needed to mitigate environmental effects—but are less useful for predicting the effect of specific stressors. In addition, the estimates of energy expenditure used in these frameworks are often proxies, such as hormonal changes (e.g. cortisol) or changes in circulating glucose, which may not correlate with behavioural responses, thus rendering the use of allostatic load inappropriate for gauging the behavioural response to a stressor or set of stressors. For example, the behavioural response of animals to many stressors may be to reduce activity, which, everything else being equal, would be expected to reduce overall energy expenditure.

Even if an animal has sufficient energy intake or stores to meet their overall demand, behavioural constraints can arise when there are allocation conflicts among competing processes. An animal can only perform so many tasks at once, largely due to limitations in how fast blood oxygen can be delivered and utilized by various tissues. An animal's aerobic scope is the absolute amount or factor by which it can increase oxygen delivery beyond that required for maintenance functions alone (Figure 10.4). Even under 'normal', nonstressful conditions, delivery of blood oxygen to tissue for digestion or other energetically demanding functions will limit the capacity to deliver blood oxygen to fuel the performance of behaviours. Under conditions of stress, these trade-offs can be more pronounced, if the response to a stressor effectively increases maintenance costs (e.g. as generally occurs during warming in ectotherms) or lowers the maximum capacity for oxygen delivery and use (e.g. as occurs during systemic hypoxia). This can occur in response to isolated stressors; the cumulative effects of multiple stressors would be expected to strongly constrain an animal's aerobic scope and the behaviours that they are able to perform (Claireaux et al. 2000). In addition, due to capacity-limited trade-offs between behaviours and traits, such as growth, allocation conflicts within an animal's aerobic scope may also contribute to differences in the strength of effects at different levels (i.e. behaviour-specific versus whole-animal growth or survival), as previously discussed (see Section 10.5). The concept of physiological trade-offs within an animal's aerobic scope has been combined with other energy allocation frameworks, including dynamic energy budgets, in an effort to predict responses to a variety of multiple stressor combinations (Sokolova et al. 2012). While these models have been useful for understanding immediate physiological reactions to stressors, and they acknowledge the importance of behaviour for predicting reactions to multiple stressors, the models often do not incorporate the nuanced ways that behaviour can affect responses to multiple stressors (Lopez et al. 2023). Many behaviours may fall into the category of 'activity' and thus, depending on the level of environmental challenge, the aerobic scope available to these activities is altered; however, behaviour can also alter the magnitude of the stressor (via movement) or the aerobic scope itself.

Understanding changes in perceptual abilities may also help to predict behavioural responses to multiple stressors (Halfwerk and Slabbekoorn 2015). One stressor may mask the detection of a secondary stressor, thus altering the behavioural response. For example, Caribbean hermit crabs *Coenobita clypeatus* allowed simulated predators to

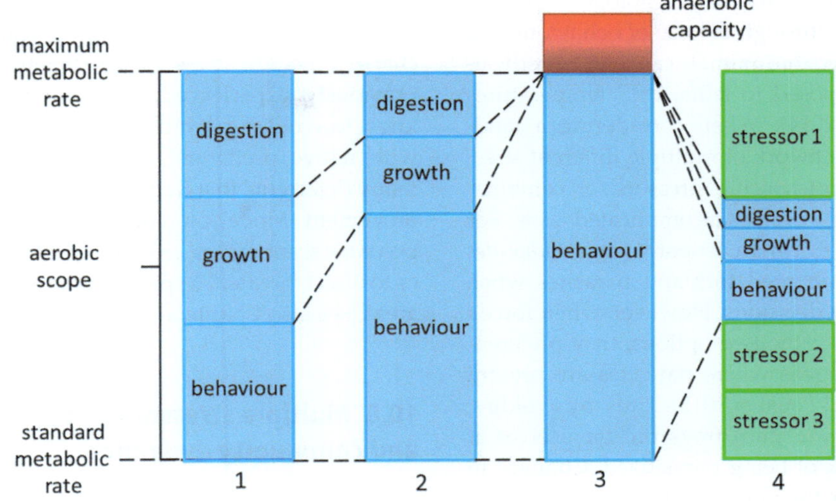

Figure 10.4 Schematic of allocation of capacity to use energy within an animal's available aerobic scope. Aerobic scope can be defined as the difference or ratio between an animal's minimum metabolic rate needed to sustain life (standard metabolic rate) and their maximum rate of aerobic metabolism (MMR; 1). Therefore, aerobic scope represents the capacity for the animal to deliver oxygen to tissues for physiological processes above those required for basic maintenance, including digestion, growth, and various behaviours (2). Allocation of aerobic scope to any of these processes can cause a trade-off, whereby there will be a reduced capacity to perform all others. During energetically demanding forms of behaviour (3), for example, the aerobic scope available for growth and digestion may be negligible, and the animal may temporarily engage in anaerobic metabolism. During exposure to a stressor or stressors (4), an animal may need to devote a portion of its aerobic capacity to the physiological processes associated with homeostasis (e.g. stressors 2 and 3), which effectively elevates the 'floor' of an animal's aerobic scope. Alternatively, stressors may lower the 'ceiling' by limiting the ability to obtain oxygen from the environment or distribute it among tissues. Multiple stressors may cause additive reductions in an animal's aerobic scope and intensify the trade-offs among oxygen-demanding physiological processes, potentially constraining behaviour.

approach more closely during boat motor noise playbacks (Chan et al. 2010). However, animals often rely on multiple cues to gain information about their environment (Parten and Marler 1999). So, a stressor may mask one type of cue associated with a secondary stressor, but animals may increase vigilance behaviour for other cues. For example, chaffinches *Fringilla coelebs* increased vigilance behaviour for visual cues associated with predators during periods of high background noise (Quinn et al. 2006). Increased vigilance, however, led to reductions in time spent foraging, suggesting that behavioural responses to alleviate one stressor can lead to increased potential costs for other tasks. Cues that alter perception may also generally cause stress to animals, which can interfere with an individual's ability to respond to cues associated with other stressors. For instance, moths were less responsive to ultrasonic calls of Cape serotine bats *Neromicia capensis* during periods of

light pollution (Minnaar et al. 2014). There are likely non-linear responses to multiple stressors on perception though, so more work is needed to specifically examine the relationship between perception and the resulting behavioural responses (Halfwerk and Slabbekoorn 2015).

10.7 The role of behaviour in mitigating and amplifying responses to multiple stressors

While multiple stressors can have various effects on animal behaviours, it is important to realize that behaviour may also feed back to alter an organism's response to stressors. Animals regularly respond to stressor exposure by altering movement patterns. For example, across a range of latitudes and elevations, the potential operative body temperature of many small terrestrial ectotherms in exposed habitats matches or exceeds physiological thermal limits. However, through the use of behaviour and microhabitat selection, animals can avoid overheating or being exposed to lethal cold temperatures (Sunday et al. 2014). When considering a landscape with a patchwork of multiple different stressors, avoidance of specific stressors, or combinations of stressors, becomes a complicated issue. For example, larval northern leopard frogs *Lithobates pipiens* avoid both predators and parasites when making foraging decisions. However, when forced to choose between the two options, they preferentially forage in areas where parasites are present (Koprivnikar and Penalva 2015). Thus, by avoiding a lethal predator, leopard frogs put themselves at an increased risk of being parasitized. Changes in activity can alter the effect of other anthropogenic changes, either because the change alters the likelihood of interacting with the disturbance or because it exacerbates the effect of the disturbance due to physiological changes associated with movement. For example, reduced swimming capacity during exposure to hypoxia can also increase susceptibility to capture in a scaled-down model of trawl fishing in zebrafish *Danio rerio* (Thambithurai et al. 2019). It is thus important to consider the network of different challenges that animals must respond to when predicting responses to a challenge (Lopez et al. 2023).

If a stressor increases energy requirements, this can lead to increased foraging motivation, which can then expose an individual to other additional stressors. For example, under warming conditions, shrimp *Palaemon* spp. foraged longer even in the presence of predators (Marangon et al. 2019). Alternatively, exposure to a stressor may lead to decreased foraging, which could decrease energy stores and body condition leading to an increased risk of being negatively affected by a secondary stressor (Lopez et al. 2023). For example, exposure to both predators and increased temperature decreased foraging in *Nucella lapillus* snails, reducing growth efficiency (Miller et al. 2014).

Behaviour can also mitigate the effects of multiple stressors by reducing exposure to one or more additional stressors. Avoiding one stressor may also reduce exposure to another stressor—either incidentally or because those stressors directly co-occur. Incidental avoidance may be particularly useful following human-induced changes if the changes include stressors that the animal has not previously experienced, and therefore does not know to avoid (e.g. if introduced predators co-occur with native predators). This is the logic behind many deterrents that humans use to control animal movement. Model predators (e.g. plastic owls) or predator scent can be used to control pest presence or to direct threatened species away from dangerous areas (see also Chapters 11 and 12).

10.8 Multiple stressors and population and community dynamics

Anthropogenic disturbances do not just affect the individual, but have the potential to influence ecological processes via changes to social behaviour between conspecifics and altered relationships between heterospecifics, including mutualisms and predator-prey dynamics (Figure 10.5). Interactions among species can be altered by the negative effects of one or more stressors on one species that indirectly affects another species (e.g. McInturf et al. 2022). Even if a species is resilient to a novel stressor, if other species it interacts with are affected by that stressor, it may face indirect effects. For example, if stressor A only impacts species A and stressor B only impacts species B, when stressors A and B are combined, their effect on each species will be the same as when each stressor is encountered in isolation. As a result, the combined stressors in this scenario may not be predicted to be more damaging at the species level than either stressor alone. However, because of interactions between species A and B, for example, if species A is the prey of species B, community level dynamics may be altered in ways that are difficult to predict.

Group behaviour is an important aspect of the lives of many species that can have substantial benefits for foraging, anti-predator defences, and reproduction (Krause and Ruxton 2002). Population level responses to stressors may be altered by changes in social behaviour, such that all individuals are interacting differently from before a disturbance. For example, temperature has been shown to influence fish social behaviour, with individuals

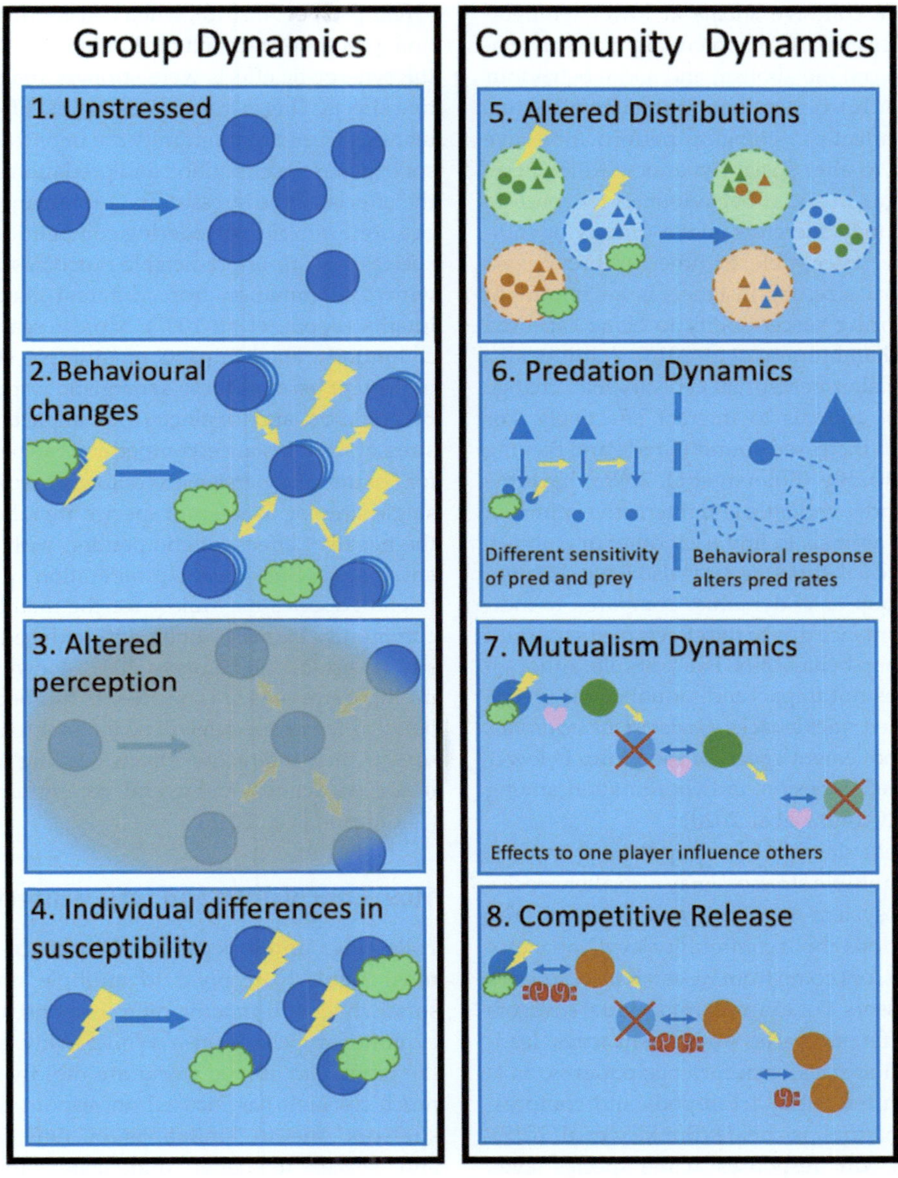

Figure 10.5 Potential effects of multiple stressors on groups and communities. Green clouds and yellow lightning bolts represent potential stressors. Coloured circles represent different species. Triangles represent potential predators. (1) In unstressed conditions, many animals engage in a variety of different group dynamics. (2) Following exposure to multiple stressors, animals may engage in behavioural changes that alter group dynamics. For example, increased activity at the individual level may lead to increased disorder at the group level. (3) Some stressors (e.g. turbidity) may also decrease cohesion by altering perceptual capacity of individuals. (4) Individual differences in susceptibility may also increase group level effects if some individuals are resistant to stressors that others are susceptible to, but susceptible to stressors that others are resistant to. (5) Exposure to multiple stressors across a landscape may also alter distributions of species as different species respond to individual tolerances, leading to different species interacting after onset of stressors. (6) Multiple stressors may also alter predator-prey dynamics. This could be because (left) prey (blue circles) are more susceptible to stressors than predators (blue triangles), leading to decreased numbers, which may have indirect effects on predators via increased difficulty of finding food. Alternatively (right), prey behaviour is altered in response to stressor exposure, making them more susceptible to predation. (7) Mutualisms may also be affected if one species (blue circles) in the interaction is negatively impacted by exposure to multiple stressors, leading to indirect negative effects on the other species (green circles). (8) Alternatively, multiple stressors may lead to competitive release if one species (blue circles) is negatively impacted by stressor exposure, leading to an increased abundance of species (orange circles) that are less affected by stressor exposure.

forming more cohesive shoals at lower temperatures (Bartolini et al. 2015), potentially due to a relationship between metabolism and social behaviour (Killen et al. 2016). Generally, environmental perturbations that affect physiological performance have the potential to alter social dynamics (Killen et al. 2021). Changes in social behaviour in response to one stressor may therefore make groups or individuals more susceptible to other anthropogenic changes. Shoal size, for instance, is an important factor influencing susceptibility to being captured by fishing (Thambithurai et al. 2018). Additionally, stressors that alter perceptual capacity may change the ability of animals to interact effectively. For example, both increased sound (Grade and Sieving 2016) and turbidity (Ehlman et al. 2019a) have the potential to independently and interactively impact the ability of animals to find each other or communicate. Individual stressors may also have different effects on group level dynamics compared to individual performance due to the effects of stressors on emergent group behaviours. For example, although darkness does not impact individual swim speeds of three-spined sticklebacks *Gasterosteus aculeatus*, the overall movement speed of the group is lower due to reduced cohesion and coordination among groupmates (Ginnaw et al. 2020).

Altered behaviour at the population level in response to multiple stressors may also alter species richness, ecosystem function, and biomass. Yet, many predictions about community level impacts of multiple stressors come from research on responses to single stressors. For example, individual stressors can affect different species in different ways, leading to altered species co-occurrence patterns, as is the case with terrestrial arthropods and compressor noise in natural gas fields (Bunkley et al. 2017). However, as with responses at the species level, the impacts of multiple stressors on the community as a whole are not always additive, and studying single stressors in isolation may not always allow for accurate predictions for community level responses.

Theoretical models that consider competition and predator-prey dynamics tend to predict synergistic effects of multiple stressors, whereas those considering facilitative interactions (including mutualisms and commensalisms) tend to lead to additive effects (Thompson et al. 2018); when considering competition, predator-prey, and facilitation simultaneously, the synergistic effects were stronger than the additive effects. This is in part because, as different populations within a community are impacted by stressor exposure, competition and predation exacerbate the effects of the stressor. In amphibian communities, for example, considering competition between species lead to unpredictable outcomes associated with contamination from different chemical pollutants (Boone et al. 2007). More research on the community level impacts of multiple stressors is important because these synergistic responses may be common and distinct from impacts of single stressors. For instance, competition can be beneficial for community persistence following exposure to a single stressor, as tolerant species are released from the negative effects of competition with less tolerant species. Subsequent proliferation of the more tolerant species may allow for the maintenance of community biomass, albeit with reduced biodiversity (Gonzalez and Loreau 2009). However, when multiple stressors are considered, this becomes less likely as fewer species will be tolerant to both stressors or their interactive effects, leading to declines in species richness and overall community biomass (Thompson et al. 2018).

10.9 Conclusions and future directions

Following anthropogenic disturbance, animals must regularly respond to multiple novel stressors while still maintaining responses to prior evolutionary challenges. While many aspects of the biotic and abiotic world are rapidly changing, much research has focused on responses to single stressors due to limitations of time, resources, and space. To better understand and predict how animals are responding to anthropogenic change, it is important to prioritize research that is designed to address pressing questions related to responses to multiple stressors, including studies that: (1) test three or more intensity levels for each stressor and not just relatively high and low levels; (2) observe responses across multiple levels of organization (including physiological, behavioural, growth/survival, population, and community) to understand how responses at one level differ from and affect responses at other levels; and (3) are specifically designed to address conservation questions, including ecologically relevant timing and magnitudes and 'removal of stressor' studies. Research with multiple magnitudes of each stressor is important because stressor effects are likely non-linear and may contain breakpoints where anthropogenic change becomes untenable. Identifying non-linear effects or tolerance thresholds is particularly important when considering multiple stressors, where effects may be additive for some values of a stressor but become synergistic at more intense levels. Measuring effects across different levels of organization (i.e. physiology, behaviour, growth, population, and community) can also help to better predict where the effects of multiple stressors are most severe. In addition to considering physiology and effects on growth, survival, and reproduction, behaviour is a key response to multiple stressors that can alter effects at other levels of organization. Often, if the impacts of stressor exposure on behaviour are measured, research focuses on changes to activity. However, to get a more complete picture of the impacts of stressor exposure, more complex suites of behaviours should be considered. Ultimately, studying the effect of a range of magnitudes of stressors across multiple behaviours will further our understanding of the influence of multiple stressors in a changing world.

References

Anthony, K.R.N., Connolly, S.R., and Hoegh-Guldberg, O. (2007). Bleaching, energetics, and coral mortality risk: effects of temperature, light, and sediment regime. *Limnology and Oceanoprography*, 52, 716–726.

Baker, P.J., Bentley, A.J., Ansell, R.J., and Harris, S. (2005). Impact of predation by domestic cats *Felis catus* in an urban area. *Mammal Review*, 35, 302–312.

Bartolini, T., Butail, S., and Porfiri, M. (2015). Temperature influences sociality and activity of freshwater fish. *Environmental Biology of Fishes*, 98, 825–832.

Berger-Tal, O., Polak, T., Oron, A., et al. (2011). Integrating animal behavior and conservation biology: a conceptual framework. *Behavioral Ecology*, 22, 236–239.

Berkhout, B.W., Budria, A., Thieltges, D.W., and Slabbekoorn, H. (2022). Trends in anthropogenic noise pollution and wildlife diseases. *Trends in Parasitology*, 1–10.

Bible, J.M., Cheng, B.S., Chang, A.L., et al. (2017). Timing of stressors alters interactive effects on a coastal foundation species. *Ecology*.

Boone, M.D., Semlitsch, R.D., Little, E.E., and Doyle, M.C. (2007). Multiple stressors in amphibian communities: effects of chemical contamination, bullfrogs, and fish. *Ecological Applications*, 17, 291–301.

Brown, C.J., Saunders, M.I., Possingham, H.P., and Richardson, A.J. (2013). Managing for interactions between local and global stressors of ecosystems. *PLOS ONE*, 8.

Brown, J.S., and Kotler, B.P. (2004). Hazardous duty pay and the foraging cost of predation. *Ecology Letters*, 7, 999–1014.

Bruintjes, R., and Radford, A.N. (2013). Context-dependent impacts of anthropogenic noise on individual and social behaviour in a cooperatively breeding fish. *Animal Behaviour*, 85, 1343–1349.

Bunkley, J.P., Christopher, J.W., McClure, A.Y., et al. (2017). Anthropogenic noise changes arthropod abundances. *Ecology and Evolution*, 7, 2977–2985.

Caro, T., and Sherman, P.W. (2013). Eighteen reasons animal behaviourists avoid involvement in conservation. *Animal Behaviour*, 85, 305–312.

Cerini, F., Childs, D.Z., and Clements, C.F. (2023). A predictive timeline of wildlife population collapse. *Nature Ecology & Evolution*.

Chan, A.A.Y.H., Giraldo-Perez, P., Smith, S., and Blumstein, D.T. (2010). Anthropogenic noise affects risk assessment and attention: the distracted prey hypothesis. *Biology Letters*, 6, 458–461.

Claireaux, G., Webber, D.M., Lagarde, J., and Kerr, S.R. (2000). Influence of water temperature and oxygenation on the aerobic metabolic scope of Atlantic cod (*Gadus Morhua*). *Journal of Sea Research*, 44, 257–265.

Côté, I.M., Darling, E.S., and Brown, C.J. (2016). Interactions among ecosystem stressors and their importance in conservation. *Proceedings of the Royal Society B: Biological Sciences*, 283, 20152592.

Crain, C.M., Kroeker, K., and Halpern, B.S. (2008). Interactive and cumulative effects of multiple human stressors in marine systems. *Ecology Letters*, 11, 1304–1315.

Cronin, A.D., Smit, J.A.H., and Halfwerk, W. (2022). Anthropogenic noise and light alter temporal but not spatial breeding behavior in a wild frog. *Behavioral Ecology*, 33, 1115–1122.

Damsgaard, C., Baliga, V.B., Bates, E., et al. (2020). Evolutionary and cardio-respiratory physiology of air-breathing and amphibious fishes. *Acta Physiologica*, 228, 1–22.

Darling, E.S., and Côté, I.M. (2008). Quantifying the evidence for ecological synergies. *Ecology Letters*, 11, 1278–1286.

Darling, E.S., Mcclanahan, T.R., and Côté, I.M. (2010). Combined effects of two stressors on Kenyan coral reefs are additive or antagonistic, not synergistic. *Conservation Letters*, 3, 122–130.

Dinh, K.V., Konestabo, H.S., Borgå, K., et al. (2023). Interactive effects of warming and pollutants on marine and freshwater invertebrates. *Current Pollution Reports*, 8, 341–359.

Dominoni, D., Smit, J.A.H., Visser, M.E., and Halfwerk, W. (2020). Multisensory pollution: artificial light at night and anthropogenic noise have interactive effects on activity patterns of great tits (*Parus Major*). *Environmental Pollution*, 256, 113314.

Ehlman, S.M., Halpin, R., Jones, C., et al. (2019a). Intermediate turbidity elicits the greatest antipredator response and generates repeatable behaviour in mosquitofish. *Animal Behaviour*, 158, 101–108.

Ehlman, S.M., Trimmer, P.C., and Sih, A. (2019b). Prey responses to exotic predators: effects of old risks and new cues. *The American Naturalist*, 193, 575–587.

Fagan, W.F., and Holmes, E.E. (2006). Quantifying the extinction vortex. *Ecology Letters*, 9, 51–60.

Ghanizadeh-Kazerouni, E., Franklin, C.E., and Seebacher, F. (2017). Living in flowing water increases resistance to ultraviolet B radiation. *Journal of Experimental Biology*, 220, 582–587.

Ginnaw, G.M., Davidson, I.K., Harding, H.R., et al. (2020). Effects of multiple stressors on fish shoal collective motion are independent and vary with shoaling metric. *Animal Behaviour*, 168, 7–17.

Glasby, T.M., and Underwood, A.J. (1996). Sampling to differentiate between pulse and press perturbations. *Environmental Monitoring and Assessment*, 42, 241–252.

Gonzalez, A., and Loreau, M. (2009). The causes and consequences of compensatory dynamics in ecological communities. *Annual Review of Ecology, Evolution, and Systematics*, 40, 393–414.

Grade, A.M., and Sieving, K.E. (2016). When the birds go unheard: highway noise disrupts information transfer between bird species. *Biology Letters*, 12, 7–10.

Halfwerk, W., and Slabbekoorn, H. (2015). Pollution going multimodal: the complex impact of the human-altered sensory environment on animal perception and performance. *Biology Letters*, 11, 20141051.

Hewitt, J.E., Ellis, J.I., and Thrush, S.F. (2016). Multiple stressors, nonlinear effects and the implications of climate change impacts on marine coastal ecosystems. *Global Change Biology*, 22, 2665–2675.

Jackson, M.C., Pawar, S., and Woodward, G. (2021). The temporal dynamics of multiple stressor effects: from individuals to ecosystems. *Trends in Ecology & Evolution*, 36, 402–410.

Killen, S.S., and Brown, J.A. (2006). Energetic cost of reduced foraging under predation threat in newly hatched ocean pout. *Marine Ecology Progress Series*, 321, 255–266.

Killen, S.S., Cortese, D., Cotgrove, L., et al. (2021). The potential for physiological performance curves to shape environmental effects on social behavior. *Frontiers in Physiology*, 12.

Killen, S.S, Fu, C., Wu, Q., et al. (2016). The relationship between metabolic rate and sociability is altered by food deprivation. *Functional Ecology*, 30, 1358–1365.

Killen, S.S., Gamperl, A.K., and Brown, J.A. (2007). Ontogeny of predator-sensitive foraging and routine metabolism in larval shorthorn sculpin, *Myoxocephalus Scorpius*. *Marine Biology*, 152, 1249–1261.

Killen, S.S., Marras, S., Ryan, M.R. et al. (2012). A relationship between metabolic rate and risk-taking behaviour is revealed during hypoxia in juvenile European sea bass, *Functional Ecology*, 26, 134–143.

Koprivnikar, J., and Penalva, L. (2015). Lesser of two evils? Foraging choices in response to threats of predation and parasitism. *PLOS ONE*, 10, 1–11.

Krause, J., and Ruxton, G. (2002). *Living in Groups*. Oxford University Press, Oxford.

Lopez, L.K., Gil, M.A., Crowley, P.H., et al. (2023). Integrating animal behaviour into research on multiple environmental stressors: a conceptual framework. *Biological Reviews*, 98, 1345–1364.

Luo, J., Siemers, B.M., and Koselj, K. (2015). How anthropogenic noise affects foraging. *Global Change Biology*, 21, 3278–3289.

Mahon, C.L., Holloway, G.L., Bayne, E.M., and Toms, J.D. (2019). Additive and interactive cumulative effects on boreal landbirds: winners and losers in a multi-stressor landscape. *Ecological Applications*, 29, 1–18.

Marangon, E., Goldenberg, S.U., and Nagelkerken, I. (2019). Ocean warming increases availability of crustacean prey via riskier behavior. *Behavioral Ecology*, 1–5.

McEwen, B.S., and Wingfield, J.C. (2003). The concept of allostasis in biology and biomedicine. *Hormones and Behavior*, 43, 2–15.

McInturf, A.G., Zillig, K.W., Cook, K., et al. (2022). In hot water? Assessing the link between fundamental thermal physiology and predation of juvenile chinook salmon. *Ecosphere*, 13, 1–19.

Miller, L.P., Matassa, C.M., and Trussell, G.C. (2014). Climate change enhances the negative effects of predation risk on an intermediate consumer. *Global Change Biology*, 20, 3834–3844.

Minnaar, C., Boyles, J.G., Minnaar, I.A., et al. (2014). Stacking the odds: light pollution may shift the balance in an ancient predator–prey arms race. *Journal of Applied Ecology*, 52, 522–531.

Nathan, R., Monk, C., Arlinghaus, R., et al. (2022). Big-data approaches enable increased understanding of animal movement ecology. *Science*, 375, eabg1780.

Orr, J.A., Vinebrooke, R.D., Jackson, M.C., et al. (2020). Towards a unified study of multiple stressors: divisions and common goals across research disciplines. *Proceedings of the Royal Society B: Biological Sciences*, 287.

Partan, S., and Marler, P. (1999). Communication goes multimodal. *Science*, 283, 1272–1273.

Pollack, L., Munson, A., Savoca, M.S., et al. (2022). Enhancing the ecological realism of evolutionary mismatch theory. *Trends in Ecology & Evolution*, 37, 233–245.

Pollack, L., Ondrasek, N.R., and Calisi, R. (2017). Urban health and ecology: the promise of an avian biomonitoring tool. *Current Zoology*, 63, 1–8.

Quinn, J.L., Whittingham, M.J., Butler, S.J., and Cresswell, W. (2006). Noise, predation risk compensation and vigilance in the chaffinch *Fringilla coelebs*. *Journal of Avian Biology*, 37, 601–608.

Rizwan, A.M., Leung, Y., Dennis, C., and Liu, C. (2008). A review on the generation, determination and mitigation of urban heat island. *Journal of Environmental Sciences*, 20, 120–128.

Rohr, J.R., and Palmer, B.D. (2013). Climate change, multiple stressors, and the decline of Ectotherms. *Conservation Biology*, 27, 741–751.

Romero, L.M., Dickens, M.J., and Cyr, N.E. (2009). The Reactive Scope Model—a new model integrating homeostasis, allostasis, and stress. *Hormones and Behavior*, 55, 375–389.

Saari, G.N., Wang, Z., and Brooks, B.W. (2018). Revisiting inland hypoxia: diverse exceedances of dissolved oxygen thresholds for freshwater aquatic life. *Environmental Science and Pollution Research*, 25, 3139–3150.

Sabet, S.S., Neo, Y.Y., and Slabbekoorn, H. (2016). Impact of anthropogenic noise on aquatic animals: from single species to community-level effects. *The Effects of Noise on Aquatic Life II*, 875, 957–961.

Shannon, G., Angeloni, L.M., Wittemyer, G., et al. (2014). Road traffic noise modifies behaviour of a keystone species. *Animal Behaviour*, 94, 135–141.

Smit, J.A.H., Cronin, A.D., van der Wiel, I., et al. (2022). Interactive and independent effects of light and noise pollution on sexual signaling in frogs. *Frontiers in Ecology and Evolution*, 757.

Sokolova, I.M., Frederich, M., Bagwe, R., et al. (2012). Energy homeostasis as an integrative tool for assessing limits of environmental stress tolerance in aquatic invertebrates. *Marine Environmental Research*, 79, 1–15.

Spiga, I., Aldred, N., and Caldwell, G.S. (2017). Anthropogenic noise compromises the anti-predator behaviour of the European seabass, *Dicentrarchus Labrax* (L.). *Marine Pollution Bulletin*, 122, 297–305.

Strain, E.M.A., Van Belzen, J., Van Dalen, J., et al. (2015). Management of local stressors can improve the resilience of marine canopy algae to global stressors. *PLOS ONE*, 10, 1–15.

Stuligross, C., and Williams, N.M. (2020). Pesticide and resource stressors additively impair wild bee reproduction. *Proceedings of the Royal Society B: Biological Sciences*, 287, 1–7.

Sunday, J.M., Bates, A.E., Kearney, M.R., et al. (2014). Thermal-safety margins and the necessity of thermoregulatory behavior across latitude and elevation. *Proceedings of the National Academy of Sciences*, 111, 5610–5615.

Thambithurai, D., Crespel, A.T., Norin, T., et al. (2019). Hypoxia alters vulnerability to capture and the potential for trait-based selection in a scaled-down trawl fishery. *Conservation Physiology*, 7, 1–12.

Thambithurai, D., Hollins, J., Van Leeuwen, T., et al. (2018). Shoal size as a key determinant of vulnerability to capture under a simulated fishery scenario. *Ecology and Evolution*, 8, 6505–6514.

Thompson, P.L., MacLennan, M.M., and Vinebrooke, R.D. (2018). Species interactions cause non-additive effects of multiple environmental stressors on communities. *Ecosphere*, 9, 11.

Todgham, A.E., and Stillman, J.H. (2013). Physiological responses to shifts in multiple environmental stressors: relevance in a changing world. *Integrative and Comparative Biology*, 53, 539–544.

Toft, J.D., Munsch, S.H., Cordell, J.R., et al. (2018). Impact of multiple stressors on juvenile fish in estuaries of the northeast Pacific. *Global Change Biology*, 24, 2008–2020.

Van Dievel, M., Janssens, L., and Stoks, R. (2019). Additive bioenergetic responses to a pesticide and predation risk in an aquatic insect. *Aquatic Toxicology*, 212, 205–213.

Vinebrooke, R.D., Cottingham, K.L., Norberg, J., et al. (2004). Impacts of multiple stressors on biodiversity and ecosystem functioning: the role of species co-tolerance. *Oikos*, 104, 451–457.

West, R., Letnic, M., Blumstein, D.T., and Moseby, K.E. (2017). Predator exposure improves anti-predator responses in a threatened mammal. *Journal of Applied Ecology*, 55, 147–156.

PART II

Applications

CHAPTER 11

Conservation behaviour

Oded Berger-Tal and Daniel T. Blumstein

Overview

In a rapidly changing world, an individual's behaviour is a key response to the changing environment, and it may permit individuals, populations, and species to survive, and sometimes even thrive, in human-dominated landscapes. Conservation behaviour is an emerging field focused on applying insights from animal behaviour research to conservation and management. In this chapter we provide an overview of how an understanding of animal behaviour can be used to predict the impacts of human activities on wildlife, and how it can be harnessed as a powerful tool in conservation and management interventions. We illustrate our points by describing a cognitive framework for conservation. We also include practical advice for behavioural ecologists seeking to have a greater impact on the conservation of species and habitats.

11.1 Introduction

For a newly hatched turtle, deciding where to go is straightforward. As it emerges from the beach burrow in which it hatched with its many siblings, it will start to crawl towards the light, because for millions of years, the reflections of the Moon and stars on the water reliably represented the hatchling's destination—the sea (Salmon 2005). Unfortunately for the turtles, human development along seashores, as well as inland, results in elevated artificial light levels that 'drown' the natural light on the shore (see Chapter 4 to read more about light pollution), causing the turtles to crawl away from the sea and towards the land (Tuxbury and

Salmon 2005). These hatchlings then either get captured by predators, get run over by vehicles, or just die of exhaustion and dehydration away from the shore (Witherington 1997). Light, a previously reliable cue signalling where the ocean is, now leads countless turtle hatchlings to their doom in a process that has become known as an evolutionary trap (Schlaepfer et al. 2002; Schlaepfer et al. 2005; see also Chapter 19). By understanding the mechanisms leading to the hatchlings' detrimental behaviour, researchers could devise effective mitigation strategies (Robertson and Blumstein 2019). These include turning off unnecessary lights, redirecting light sources away from the sea, altering the spectral properties of the lights to reduce the turtles' attraction to them (for example, by creating pulsing lights at frequencies in which the pulses are not visible to the human eye), and producing additional orientation cues for the turtles, such as restoring and vegetating dunes between the beaches and the land (Witherington 1997; Salmon 2005; see also Chapter 8).

Turtles are not the only species using light as a cue to guide their behaviour. Thousands of species of birds, insects, mammals, and amphibians rely on natural illumination (such as the lights of the Moon or the stars) for navigation, making them extremely sensitive to light pollution (Longcore and Rich 2004; see also Chapter 4). In the USA alone, between 100 million to 1 billion birds are killed every year by colliding with windows, and many of these birds are migrating birds attracted to artificial lights (Loss et al. 2014). Recent evidence suggests that social behaviour might increase the vulnerability of birds to collisions, with species that produce

Oded Berger-Tal and Daniel T. Blumstein, *Conservation behaviour*. In: *Behavioural Responses to a Changing World*. Edited by: Bob B. M. Wong and Ulrika Candolin, Oxford University Press. © Oxford University Press (2024). DOI: 10.1093/oso/9780192858979.003.0011

flight calls during nocturnal migration tending to be especially prone to collisions (Winger et al. 2019). These calls have probably evolved to facilitate collective decision-making during migration at night, but nowadays such collective decision-making may lead to mass collisions instead.

In an attempt to alleviate this huge problem, in many US cities, like New York, Philadelphia, and San Francisco, some skyscrapers and other landmarks have started a 'lights out' programme during bird migration seasons (Beatley 2020). Similar programmes are becoming more common around the world. For example, in Phillip Island, located just off mainland Australia, businesses turn off their lights at night to prevent shearwater *Ardenna tenuirostris* fledglings from becoming disoriented as they embark on their annual migration to Alaska (Rodriguez et al. 2014; see also Box 4.1 in Chapter 4), and Canada's Fatal Light Awareness Program (FLAP) has been operating for over 30 years to keep birds safe from deadly collisions with buildings. While all of these campaigns are crucial in reducing the number of bird collisions, understanding the way animals perceive light can help us do an even better job at creating sensory-attuned solutions (Adams et al. 2021).

These examples illustrate the power of understanding the cues animals use to make decisions and the promise of creating behaviourally informed mitigations. Such mitigations, in the case of turning off skyscraper lights, have the potential to save the lives of millions of individual animals globally at a relatively limited cost.

By the time this book is published, the global human population will have crossed the 8 billion mark. This growth comes with a heavy price tag for the planet's natural systems—there is virtually no ecosystem on our planet that has not been modified, at least to some extent, by anthropogenic disturbances (Bradshaw et al. 2021). Natural areas are converted to residential areas and agricultural fields (Chapter 8), invasive species wreak havoc on native communities (Chapter 6), noise pollution seeps into every corner of the Earth, including protected areas (Buxton et al. 2017; Chapter 2), light pollution completely transforms the nocturnal environment (Chapter 4), and climate change is fundamentally changing the conditions in both terrestrial and aquatic systems (Chapter 1;

Chapter 5). Individuals may survive these environmental changes (sometimes referred to as human-induced rapid environmental changes or HIREC; Sih 2013) if they are sufficiently plastic, but ultimately, a population will persist by adaptation (see also Chapter 14). However, adaptation is a relatively slow process, creating a mismatch between the rate of environmental change and the rate of the evolutionary response to it (Ehrlich and Blumstein 2018). In other words, we alter habitats at such a fast rate that wild populations may die off before selection has a chance to save them. This is where animal behavioural responses come into play, enabling animals to better confront a rapidly changing environment, and allowing populations and species to survive, and sometimes even thrive, in anthropogenically modified habitats (Berger-Tal and Saltz 2016a). In this chapter we provide an overview of conservation behaviour, a research field aimed at applying animal behaviour research to improve conservation and management. We will provide a cognitive-based framework for conservation, and demonstrate how understanding the mechanisms, consequences, challenges, and applications of how animals behaviourally respond to a rapidly changing environment is a powerful tool in the hands of people who want to ensure the continued survival of wildlife in the Anthropocene.

11.2 Using animal behaviour to improve conservation success

11.2.1 The links between conservation and animal behaviour

The field of conservation behaviour focuses on using insights from the field of animal behaviour (including studies into the mechanisms, development, function, and phylogeny of behavioural variation; sensu Tinbergen 1963) to aid the conservation and management of species and habitats. This can be achieved in three main ways (Berger-Tal et al. 2011; Figure 11.1). First of all, by understanding how animals behaviourally react to changing environments, we can better predict the outcomes of anthropogenic disturbances and the way wild populations are expected to be impacted by them. This

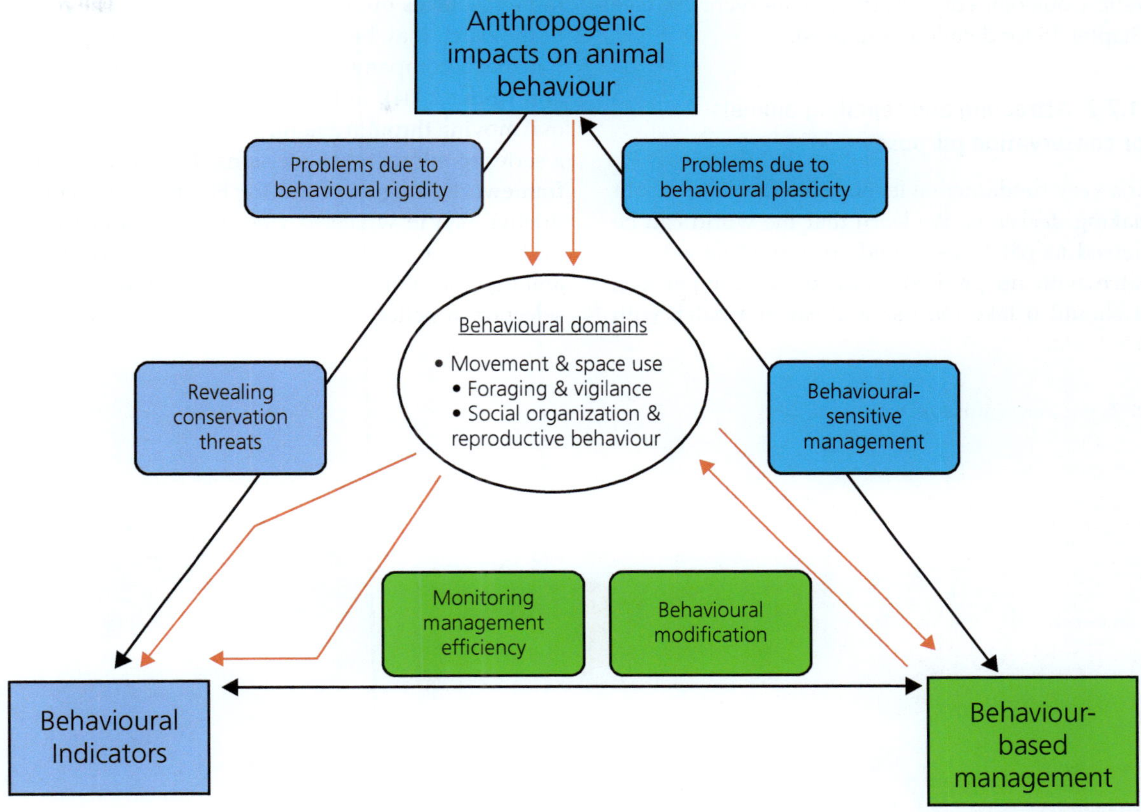

Figure 11.1 The conservation behaviour framework is composed of three basic interrelated conservation themes: (1) anthropogenic impacts on animal behaviour; (2) behaviour-based management; (3) behavioural indicators. The black arrows represent interactions between the conservation themes. Red arrows represent the pathways that connect each theme to the behavioural domains.

Reprinted, with permission of Oxford University Press, from Berger-Tal et al. (2011)

information can be instrumental in designing ways to reduce the impacts of these disturbances, either by targeting and modifying the disturbance itself (if eliminating it is not a feasible option), or by manipulating the behaviour of individuals of the species in question (Berger-Tal and Saltz 2016b). Second, knowledge of animal behaviour may be directly used in planning and executing conservation interventions. Understanding why animals behave the way they do (e.g. what attracts them to certain places and repels them from others) can be the difference between success and failure in conservation interventions (Greggor et al. 2020). Since resources in conservation and management are almost always scarce (Bottrill et al. 2008), and any resources invested in a failed intervention are resources that are denied from other pressing conservation issues that may be just as important, it is crucial that we maximize the chances of such interventions to succeed. Lastly, we can use the behaviour of animals to gain insights into the state and the welfare of animal populations. In some cases, behavioural indicators can reveal wildlife population changes long before demographic trends are evident, thus buying us precious time that can be used to try to reverse negative trends while this is still feasible (Kotler et al. 2016). Behavioural indicators also have an important role in the field of animal welfare, where behavioural knowledge can be harnessed for effective welfare interventions and to improve welfare outcomes of conservation interventions (see Chapter 15 for detailed examples).

11.2.2 Attracting and repelling animals for conservation purposes

At a very fundamental level, behaviour is all about making decisions. We learn that the world can be viewed as patches—should an individual eat in a patch with no predators but lots of competitors, or should it take the risk and eat in a patch with more resources but also more predators? The food in the patch may be very attractive, but the predators that accompany the food may be less so (Brown and Kotler 2004). Indeed, we can envision an animal moving through a natural landscape as making a series of attract-repel decisions. This attract-repel framework (Greggor et al. 2020; Figure 11.2) is a productive one in which to view many conservation problems. It provides insights into preferences that, among other things, can be formalized in resource selection function models (Boyce et al. 2002), or

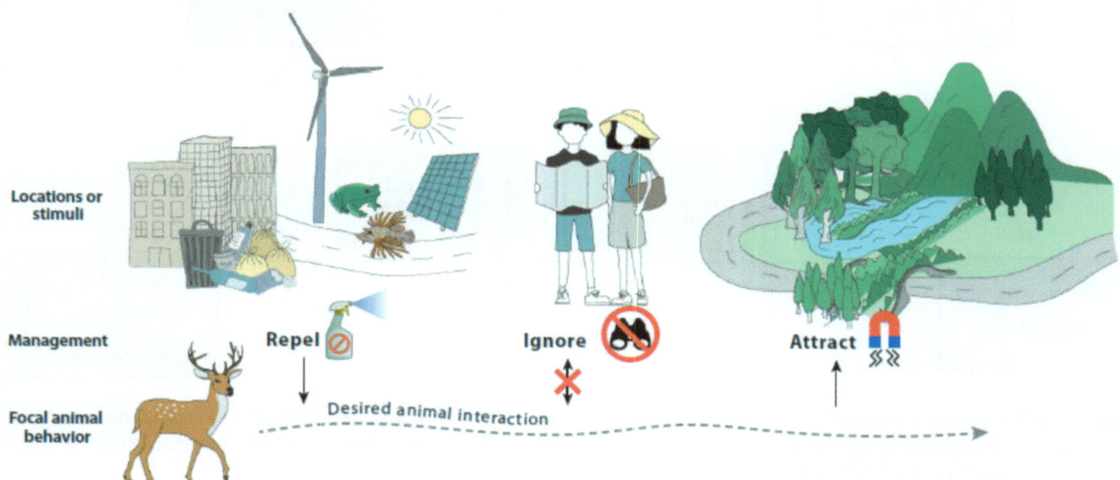

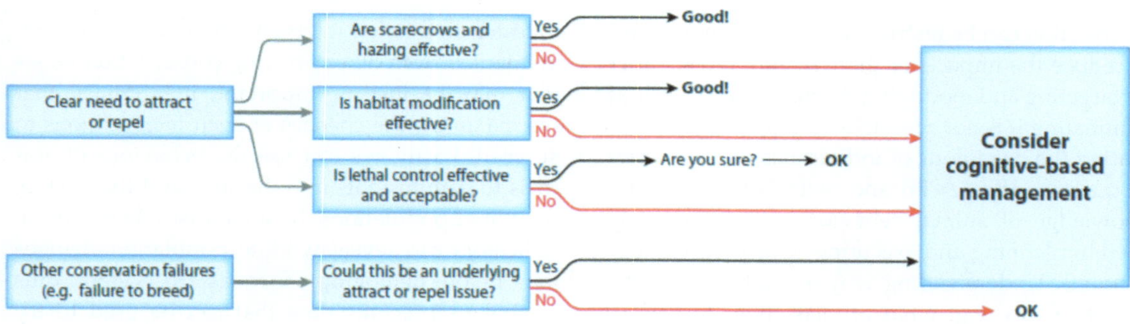

Figure 11.2 Attract-repel issues and the role of cognitive-based management. (a) Many conservation challenges can be defined along an attract-repel continuum in which we want to repel animals from certain locations or stimuli (left, e.g. human settlements, garbage, roads, wind turbines, invasive predators and prey, and ecological traps such as solar panels). At the same time, we want animals to ignore other stimuli (middle, e.g. ecotourists) and be attracted to specific locations and stimuli (right, e.g. overpasses and high-quality habitats). (b) With these cases in mind, we can then decide if cognitive-based management is appropriate for managing the issue.

used to understand wildlife responses to humans (Hebblewhite and Merrill 2008), and thus provides insights that can be mined to create attractants or deterrents.

We may wish to repel animals from sustainable power facilities (e.g. concentrated solar facilities incinerate many birds annually; Chock et al. 2021) and from roads (one review noted that in the USA alone, 1 million vertebrates are killed each day by cars; Forman and Alexander 1998). Or, we may wish to attract animals to road crossings that are installed to reduce vehicular collisions (Glista et al. 2009). Novel predators (i.e. predators introduced to the system by humans) are often something that prey animals must also avoid in order to survive; Stobo-Wilson et al. (2022: 984) noted that an estimated '697 million reptiles, 510 million birds and 1,435 million mammals are killed by foxes and cats across Australia each year'. This toll occurs despite substantial lethal predator control in Australia and further advances may require coming up with new ways to have individuals avoid predators, such as by anti-predator training (Griffin et al. 2000; Moseby et al. 2016; Blumstein et al. 2019). We may hope that animals ignore ecotourists, a process that can happen via habituation-like processes (Blumstein 2016), but sometimes they sensitize or otherwise learn to avoid humans (Blumstein 2014).

Understanding what attracts and repels animals becomes especially critical when attempting to estimate the consequences of habitat loss for wild populations. Habitat loss and habitat fragmentation are a major threat to biodiversity (Haddad et al. 2015; Chapter 8) as they limit dispersal (Crooks and Sanjayan 2006), reduce access to resources such as food, mates, and shelter (e.g. Bain et al. 2014; Lumsden et al. 2021), and create small populations that are more susceptible to Allee effects and loss of genetic variability (Crooks et al. 2017), and hence more likely to go extinct. The severity of the impact that habitat fragmentation will have on a given species or population will be determined by how a species moves around (i.e. does it fly? Can it walk?), its navigation capacity (i.e. how can it find its way across natural and altered habitat?), and its motivation to move (Nathan et al. 2008). An animal's motivation to move across an altered anthropogenic landscape will be directly determined by how repelled

it is by this environment (or more precisely, by the interactions between the various repelling and attracting forces within this matrix). Understanding these forces might allow us to discover that we can facilitate movement across such anthropogenically disturbed habitats, thus alleviating the detrimental impact of habitat loss and fragmentation (Chapter 8). However, it is also very important to remember that barriers to animal movement are not always visible to the human eye, and that fragmentation can occur even when no physical barrier is found (Berger-Tal and Saltz 2019; Chapter 8). Such invisible barriers may include (among other things) areas with altered acoustic conditions (Francis and Barber 2013), altered visual conditions (Jones and Hale 2020), or increased perceived risk of predation (Shamoon et al. 2018). Moreover, increased risk of predation may also be facilitated by altered acoustic, visual, or other sensory conditions, as well as by the presence of human hunters (Chapter 7). Many species, in particular species that rely on acoustic information for communication or for evaluating their environment (e.g. songbirds), are repelled by the noise emitted by human-made infrastructures such as roads (McClure et al. 2013). Thus, noise may create habitat loss on a vast scale, but this will be closely tied to the spectral properties of the noise, and the hearing thresholds of the species in question (Chapter 2).

Another important but non-intuitive driver of fragmentation is behavioural change in animals within human-dominated landscapes (see also Chapters 9 and 14). The mechanisms leading to such behavioural changes may vary from behavioural flexibility to anthropogenic selection on behaviour (Swaddle 2016), but regardless of the mechanism, the utilization of anthropogenic resources and the ensuing higher exposure of wildlife to humans may lead to various behavioural modifications, such as reduced anti-predatory behaviours (Saltz et al. 2018; Geffroy et al. 2020), increased dependence on anthropogenic resources (Oro et al. 2013; Hulme-Beaman et al. 2016), and reduced dispersal (Evans et al. 2012; Berger-Tal and Saltz 2019). Such local adaptations may create behavioural barriers between urban-adapted populations and rural populations of the same species, even when these populations are geographically adjacent. It is important to note, though, that the urban environment is not uniform, and different urban habitats (sometimes within the same city) may select for different behavioural adaptations (Uchida et al. 2021; Vardi and Berger-Tal 2022).

Just as it is imperative to have a good understanding of the drivers of animals' movement to be able to estimate the impacts of habitat fragmentation, we have to understand what attracts animals to certain areas and not to others when attempting to enhance connectivity among populations by means of habitat restoration and the creation of movement corridors (Chetkiewicz et al. 2006). Too many times humans have built structures for animals to use, only to discover that the animals, in their stubborn insistence on perceiving and experiencing the world in their own unique, non-human ways (defined as their *umwelt*; von Uexküll 1909), chose not to use these structures (van Dyck 2012; Hale and Swearer 2017; Chapter 8). Successful restoration (of habitats and of connectivity) requires an innovative approach that considers the way animals perceive their world, enhancing cues and habitats that the focal animals are attracted to, and removing cues and habitats that the focal animals are repelled from (Jones et al. 2021). In the same way, habitat restoration may include attempts to reduce the impacts of invasive species by tuning into the cues they respond to the most (Robertson et al. 2017; Jones et al. 2021).

As discussed in Chapter 13, a considerable amount of effort goes into managing wildlife-human interactions. For instance, the city of Toronto spent millions of dollars designing and deploying racoon *Procyon lotor*-proof trash cans that racoons quickly figured out how to open (Doubek 2018). Crop foraging by non-human primates and elephants is widely perceived as an expensive behavioural problem (Hill 2018). Predation by the recovered California sea lions *Zalophus californianus* and the endangered Steller sea lion *Eumetopias jubatus* on 13 species of endangered salmonids concentrated at the Bonneville dam tailrace creates a serious conservation challenge (Schakner and Blumstein 2021; Tidwell et al. 2021). While lethal control is often politically acceptable to prevent birds from eating crops (even if it may be only marginally cost-effective; Blackwell et al. 2003), killing elephants,

non-human primates, and marine mammals is not typically considered an acceptable intervention. Thus, managers must come up with non-lethal ways to repel animals from key resources.

Non-lethal control can be based on fear conditioning, a type of rapid associative learning where animals quickly learn to avoid an alarming stimulus. For instance, Götz and Janik (2013) and Schakner et al. (2016) fear conditioned harbour seals *Phoca vitulina* and sea lions (respectfully) using an acoustic startle device. Harbour seals were deterred from fish pens (Götz and Janik 2013), while sea lions rapidly learned to expect an aversive sound and modified their behaviour to avoid it, but it did not successfully prevent them from eating fish from bait barges or from the lines of commercial sport fishing boats (Schakner et al. 2017). Understanding the evolutionary basis of fear responses can lead to innovative management interventions. For example, painting artificial eyespots on cattle significantly reduced predation by large predators (Radford et al. 2020).

Conservation translocations are another important management strategy that may benefit from adopting an explicit attract-repel framework. Conservation translocations, as discussed in Chapter 12, are used to recover populations or to supplement extant populations' size. Many translocations fail in the sense that animals are not anchored to the release site, which may have been specifically selected because it had sufficient resources. This may lead to dispersal away from the release site followed by increased mortality and, consequently, failure of the translocation project (Berger-Tal et al. 2020). Anchoring can be framed as the need to make a location attractive to released individuals. While ecologists may initially view conspecifics as potential competitors, behavioural ecologists recognize that living with conspecifics may be beneficial. For an individual that must make a decision about a suitable habitat, the presence of conspecifics may provide information that others have found a given location to be suitable (Stamps 1988; Muller et al. 1997).

Conspecific attraction explicitly falls within the attract-repel framework. By providing cues of others, habitat may be viewed as an appropriate location to settle. While living animals may be the best evidence of suitable habitat, cues associated with animals may be sufficient and these cues may come from any modality; olfactory, visual, or acoustic (e.g. Kress 1977; Linklater et al. 2006). A recent example illustrates anchoring using conspecific cues nicely (Hennessy et al. 2022). Burrowing owls *Athene cunicularia hypugaea* were 20 times less likely to leave a release site when living owls were present at the release site or when artificial visual cues (simulated 'whitewash' placed outside release burrows) or artificial acoustic cues (broadcasting the sound of burrowing owls) were deployed. Conservation Evidence (https://www.conservationevidence.com) suggests that for birds, visual or acoustic cues are likely to be beneficial in keeping animals from leaving their release site, but conspecific attraction does not always work and more work is needed to understand when and why the application of such cues works (Putman and Blumstein 2019).

11.3 A cognitive framework for species conservation and management

To better understand the rules of attraction, we must understand cognition. Cognition, as Sara Shettleworth (2010: 278) defined it, 'includes perception, learning, memory and decision making, in short all ways in which animals take in information about the world through the senses, process, retain and decide to act on it'. Cognitive-based management focuses on decisions that animals make about approaching or avoiding specific features or resources. It provides a mechanistic understanding of conservation failures. For instance, culling badgers *Meles meles* in the UK (Woodroffe et al. 2006) failed to eliminate bovine tuberculosis because culling individuals attracted badgers from neighbouring territories, and facilitated the continued spread of the disease (see also Box 17.1 in Chapter 17). Cognitive-based management may explain why certain habitat manipulations create ecological and evolutionary traps (Robertson and Hutto 2006; Hale and Swearer 2016). For instance, planting trees in the Negev Desert created perches for predatory birds that then targeted the critically endangered Be'er Sheva fringe-fingered lizard *Acanthodactylus beershebensis*, which were unaware that the planted area was more dangerous for them (Hawlena et al.

2010). In another example, roadside signs with hollow poles created an ecological trap for mourning wheatears *Oenanthe lugens* looking for nesting cavities in what they perceived to be high-quality territories (Ben-Aharon et al. 2020). Greggor et al. (2020) proposed a set of rules that incorporates a cognitive view of why species are attracted by or repelled from various stimuli. Understanding these rules can help us decipher why animals behave the way they do, and consequently to design more successful conservation interventions.

The first two rules focus on perception and attention. The umwelt, a species' perceptual world, matters. To attract or repel an animal, we must understand what it can perceive, and realize that the sensory world of species may be vastly different from our own (Lim et al. 2008). This includes both the range of perception (e.g. bats using ultrasounds, elephants using infrasounds, snakes using infrared vision, and countless more examples), as well as the sensory modality such as magnetoreception (i.e. the detection of magnetic fields) and electroreception (i.e. sensitivity to electric fields). Dominoni et al. (2020) proposed three different mechanisms that underlie the ecological effects of sensory pollutants on wildlife: masking, misleading, and distracting. Sensory pollutants may mask stimuli. For example, the advertisement calls of many animal species—including frogs, birds, and insects—are often used by the females to assess male quality and choose the best male. Noise pollution (Chapter 2) may mask the males' calls and lead to females choosing lower-quality males, leading to reduced reproductive success (Halfwerk et al. 2011; Candolin and Wong 2019). Sensory pollutants may also mislead animals, leading them to lesser-quality areas and even into ecological traps. Misleading cues may also expedite unnecessary behaviours. For example, chemical compounds (Chapter 3) may be misidentified as predator cues, leading to costly behavioural reactions and sometimes even inhibiting growth (Lürling 2006). Lastly, for many species, including humans, the environment is filled with multiple sensory distractors. Attention, the ability to focus on a task, or acquire information through a sense organ, is always limited and animals are distractible (Chan and Blumstein 2011). Distractions may interfere with natural risk assessment or with the localization of preferred food sources (Riffell et al. 2014), inadvertently making a location undesirable, even when it's a high-quality site (a perceptual trap; Patten and Kelly 2010).

Once we have a deeper understanding of organisms' umwelt, we can use that knowledge to strategically make some locations attractive, and other locations undesirable. For example, by illuminating gillnets along Mexico's Baja California peninsula with green LED lights, researchers have reduced fisheries' bycatch by 63%, including a 95% reduction in shark, skates, and rays bycatch, while target fish catch and value were not affected. Moreover, illuminated nets significantly reduced the mean time required to retrieve and disentangle nets (Senko et al. 2022).

The next two rules of attraction (Greggor et al. 2020) focus on decision-making. All decisions are economic and thus involve trade-offs. Repelling an animal from a patch must be viewed explicitly in the context of other options it has. If there are no other desirable patches, then it will be much more difficult to successfully repel them. And, importantly, not all cues are equally valuable for decisions. This means that the cues we use as attractants or repellents must not only be perceivable, they must be valuable.

Much of cognitive-based management requires a foundational understanding of learning. Learning is Bayesian, which means that prior experience is important. Learning may be biased to focus on certain information that, over evolutionary time, has proved beneficial. Some associations, in particular those without any evolutionary context, are impossible to teach, as seen with tammar wallabies *Notamacropus eugenii* that can be easily conditioned to avoid foxes *Vulpes vulpes* (Griffin et al. 2001) but not goats *Capra hircus* (Griffin et al. 2002). The order of cues and experiences matter, and therefore we must understand how animals naturally learn if we are to modify their behaviour (for pre-release training examples, see Chapter 12). Many species have sensitive periods—a time interval that is particularly important in learning, after which it may be difficult to change specific preferences or learn a particular skill (e.g. bird song; Marler and Peters 1987). We must be explicit about what we are teaching animals; animals may learn other things associated with our desired target and this may

lead to ineffective interventions. And finally, social learning can be an accelerant, as seen in California sea lions where information about the location of endangered salmonids travels across social networks at sea lion haul-out sites and spreads like a pathogen through these networks (Schakner et al. 2016).

But it's not just how things are learned; it's how they are maintained. Thus, memory is important as well. Animals are more likely to remember survival-related information than 'random' information, and certain evolved contingencies than other non-evolved contingencies. Once memorized, memory may replace perception, to some extent, in directing the movement of animals (Fagan et al. 2013). This may be beneficial in some cases, since animals may be able to find memorized patches even when the signals leading to these patches are masked by sensory pollution. However, in other cases, animals relying solely on their memory may find themselves in trouble if this habitat has been modified since the animal last visited it.

11.4 Leveraging behavioural knowledge to make an impact

Behavioural ecologists, by their very nature, recognize that behaviour is an adaptation to the environment and thus, because environmental change is rapid in the Anthropocene, they are uniquely positioned to provide mechanistic insights that may improve wildlife conservation (as is detailed above). Additionally, many field behavioural ecologists recognize and mourn the loss of species—the species they study, the species that inspire them, the species that provide vital ecosystem services. In the Anthropocene, all field behavioural ecologists, whether they self-identify or not, are conservation scientists with much to share. Indeed, many behavioural ecologists have a visceral desire to do impactful work. But how to contribute?

One vital way that behavioural ecologists have contributed is by creating frameworks to identify the mechanistic basis of response to environmental changes (e.g. Francis and Barber 2013; Todgham and Stillman 2013; Swaddle et al. 2015; Sih et al. 2016; Razgour et al. 2018). Animals may respond to changes behaviourally, genetically, or physiologically. Gene-environment interactions are expected. Causal pathways are not always direct. Feedbacks may be important. Unpacking these causal models of phenotypic change are important advances that give us the tools to understand the response to rapid environmental change. This permits us to both identify the limits of plasticity, and to identify the precise way that animals do respond to changes, because these may offer concrete ways to manage behaviour in a rapidly changing world. Indeed, the attract-repel framework that guides much of this chapter illustrates the importance of considering behavioural mechanisms.

Because mechanisms are levers of change, mechanistic insights complement the already established genetic and ecological toolkits. But creating tools is not the same as applying tools, and there is a recognized evidence-knowledge gap whereby good ideas do not get applied because of a lack of appropriate knowledge in the hands of those who wish to use it (Dubois et al. 2020). This is not through the lack of outreach or the lack of communication. With a number of books (including this one!), and many review articles already written, behavioural ecologists have been doing an excellent job of making knowledge accessible to those who may need it. But making knowledge accessible is not the same as making knowledge useful.

While we may learn Hamilton's Rule and the Optimal Foraging theory in class, we're not often taught to apply knowledge. Applying knowledge has its own textbook, its own tricks, its own art. The application of knowledge is no less exciting and intellectually interesting than the creation of knowledge. For example, behavioural ecologists study communication, cooperation, and social networks of animals and know that information travels through social networks and that social bonds have many benefits. It's time to apply these same principles to ourselves in order to improve the usefulness of scientific knowledge (as conservation psychologists and other conservation social scientists have been doing for years; Clayton and Myers 2015).

To have an applied impact, behavioural ecologists must embed themselves into conservation networks and create opportunities to co-create projects with wildlife managers. Co-creation is the secret sauce of effective cooperation (Jones 2018).

Co-creation requires behavioural ecologists to seek out managers with a well-defined conservation problem and work with them to come up with hypotheses to solve it (sensu Caro and Sherman 2013). This contrasts an alternative model of advertising one's favourite tool. It's not the tool, it's the problem that matters.

Many solutions are likely to not be behavioural in any meaningful sense. If a species is being poached to extinction (e.g. there were fewer than ten remaining vaquita *Phocoena sinus* when we wrote the chapter), immediate cessation of illegal killing is essential. Yet many problems intersect in some meaningful way with behaviour. And this is where the value of creating a rich behavioural toolkit becomes important. Behavioural ecologists are in an excellent position to suggest possible tools for a variety of problems but they must do so in a collaborative way.

Developing mechanistic frameworks can be important, but empowering managers with decision support tools is *extremely* useful (Box 11.1). Decision support tools are algorithms that improve decision-making. Many can be structured as decision trees whereby if a specific situation is encountered, then there are options. Once those options are explored and a problem solved, there may be another question or challenge to be addressed. Greggor et al. (2020) structured the attract-repel framework as two decision trees; the first to ask whether cognition-based management should be considered (Figure 11.2) and the second to walk users through a series of steps to develop effective attractants or repellents. To make an impact, we encourage others to develop behaviourally informed decision support tools that can be used to address specific management problems.

We view the selection and application of a behavioural tool to address a conservation problem as a hypothesis that must be evaluated. Contemporary conservation science understands the value of adaptive management. Adaptive management either actively designs experiments to evaluate the efficacy of a management intervention or designs analyses to passively infer its efficacy. We assert that all behavioural interventions should be evaluated in the context of adaptive

---

Box 11.1 Using conservation behaviour to improve conservation decision-making

Many readers of this book are likely academics. As academics, we measure our impact by publications and citations, invited talks and awards. Conservation scientists and practitioners measure their impact by achieving their management objectives (e.g. keeping bears and racoons out of dumpsters and recovering a population or species from the verge of extinction). While academic behavioural biologists have much to contribute to conservation science, we can increase the impact of our research by understanding more about the context under which conservation and management interventions are implemented by conservation practitioners, and by developing tools to assist them.

Decisions regarding conservation behaviour interventions are just part of a larger series of decisions involved in planning and implementing any conservation project. To be effective, integrative conservation cannot focus strictly on the biological aspects of the intervention, and must consider the needs, wants, and desires of all human stakeholders. There is a suite of social science tools that can be used to engage stakeholders and genuinely understand their perspectives. Since most conservation actions are in some way related to humans, and since most problems can be framed in the context of a socioecological system (e.g. Berkes et al. 2001), this understanding is essential.

Most decisions also involve trade-offs, and options that may have different probabilities of success as well as associated risks. All decisions are made with some degree of uncertainty. Decision science (Hemming et al. 2022) provides a series of theories, frameworks, and tools to help make informed decisions (e.g. Kleindorfer et al. 1993; Burgman 2005). Structured decision-making breaks down a decision problem into a series of steps. The problem must be specified, and its objectives and performance measures defined. Alternatives must be developed and the consequences of these alternative actions must be evaluated. This permits trade-offs to be clearly understood. Based on this, an action is selected, but this action must be implemented in such a way that it can be monitored. Adaptive management (Williams 2011) is therefore a key attribute of application, and informed conservation behaviourists who wish to have

impactful work will work with managers to ensure proper monitoring and evaluation.

There are several decision-support frameworks that are used in natural resource management that facilitate employing these steps. For instance, the Open Standards for Conservation (CMP 2020), priority threat management (Carwardine et al. 2019), and structured decision-making (Gregory et al. 2012) all provide processes to break down complex problems into a series of discrete and assessable objectives.

What, then, is the role of an academic-based conservation behaviourist? One thing that academics can do is to make it easier for managers to make decisions. And this is where decision support tools come into play. Decision support tools can be quantitative or qualitative but their goal is to help a decision-maker address a specific problem. In this chapter we highlighted the Greggor et al. (2020) paper, which contained explicit decision support tools. Referring back to Figure 11.2, we note that this series of questions was structured in such a way as to help a manager determine whether cognitive-based management was potentially useful. Greggor et al. (2020) also contains a detailed figure, developed as a bifurcating tree, where a manager seeking to attract or repel animals in a particular situation has a series of questions to address. Only when a question is answered positively does one move to the next question. By troubleshooting management actions using this tool, Greggor et al. (2020) provides a process to achieve effective attract-repel outcomes.

We suggest that a valuable role of academic-based conservation behaviourists is to develop similar decision support tools for a variety of problems for which behavioural insights might offer solutions. Developing these is not easy and requires thought. Indeed, the successful translation of ideas to successful actions is in itself an intellectual challenge and can be highly rewarding. In the future we hope that managers will have access to a variety of decision support tools that explicitly involve conservation behaviour management actions, so that the right tools will be used to improve conservation and management outcomes.

---

management. But more importantly, because conservation funds are limited, it's also essential to conduct comparative effectiveness evaluation, whereby alternative interventions are evaluated not only with respect to their success, but with respect to their cost (Blumstein and Berger-Tal 2015, White

et al. 2022). In some cases, it may be relatively simple to use a behaviourally informed intervention, such as turning off lights to reduce bird strikes, and to have a huge impact. By critically evaluating behavioural interventions and selecting those that are cost-effective, we will develop a culture of evidence-based decision-making and, over time, will create improved conservation outcomes (Sutherland et al. 2004; Walsh et al. 2015).

References

Adams, C.A., Fernandez-Juricic, E., Bayne, E.M., and St. Clair, C.C. (2021). Effects of artificial light on bird movement and distribution: a systematic map. *Environmental Evidence*, 10, 37.

Bain, G.C., Hall, M.L., and Mulder, R.A. (2014). Territory configuration moderates the frequency of extra-group mating in superb fairy-wrens. *Molecular Ecology*, 23, 5619–5627.

Beatley, T. (2020). *The Bird-friendly City. Creating Safe Urban Habitats*. Island Press, Washington DC.

Ben-Aharon, N., Kapota, D., and Saltz, D. (2020). Roads and road-posts as an ecological trap for cavity nesting birds. *Frontiers in Conservation Science*, 1, 614899.

Berger-Tal, O., Blumstein, D.T., and Swaisgood, R.R. (2020). Conservation translocations: a review of common difficulties and promising directions. *Animal Conservation*, 23, 121–131.

Berger-Tal, O., Polak, T., Oron, A., et al. (2011). Integrating animal behavior and conservation biology: a conceptual framework. *Behavioral Ecology*, 22, 236–239.

Berger-Tal, O., and Saltz, D. (2016a). *Conservation Behavior: Applying Behavioral Ecology to Wildlife Conservation and Management*. Cambridge University Press, Cambridge.

Berger-Tal, O., and Saltz, D. (2016b). Behavioral rigidity in the face of rapid anthropogenic changes. In: O. Berger-Tal and D. Saltz (eds), *Conservation Behavior: Applying Behavioral Ecology to Wildlife Conservation and Management*, pp. 95–120. Cambridge University Press, Cambridge.

Berger-Tal, O., and Saltz, D. (2019). Invisible barriers: anthropogenic impacts on inter- and intra-specific interactions as drivers of landscape-independent fragmentation. *Philosophical Transactions of the Royal Society B: Biological Sciences*, 374, 20180049.

Berkes, F., Colding, J., and Folke, C. (2001). *Linking Social-ecological Systems*. Cambridge University Press, Cambridge.

Burgman, M. (2005). *Risks and Decisions for Conservation and Environmental Management*. Cambridge University Press, Cambridge.

Blackwell, B.F., Huszar, E., Linz, G.M., and Dolbeer, R.A. (2003). Lethal control of red-winged blackbirds to manage damage to sunflower: an economic evaluation. *The Journal of Wildlife Management*, 67, 818–828.

Blumstein, D.T. (2014). Attention, habituation, and antipredator behaviour: implications for urban birds. In: D. Gill and H. Brumm (eds), *Avian Urban Ecology: Behavioural and Physiological Adaptations*, pp. 41–53. Oxford University Press, Oxford.

Blumstein, D.T. (2016). Habituation and sensitization: new thoughts about old ideas. *Animal Behaviour*, 120, 255–262.

Blumstein, D.T., and Berger-Tal, O. (2015). Understanding sensory mechanisms to develop effective conservation and management tools. *Current Opinion in Behavioral Sciences*, 6, 13–18.

Blumstein, D.T., Letnic, M., and Moseby, K.E. (2019). *In situ* predator conditioning of naïve prey prior to reintroduction. *Philosophical Transactions of the Royal Society B: Biological Sciences*, 374, 20180058.

Bottrill, M.C., Joseph, L.N., Carwardine, J., et al. (2008). Is conservation triage just smart decision making? *Trends in Ecology & Evolution*, 23, 649–654.

Boyce, M.S., Vernier, P.R., Nielsen, S.E., and Schmiegelow, F.K.A. (2002). Evaluating resource selection functions. *Ecological Modelling*, 157, 281–300.

Bradshaw, C.J.A., Ehrlich, P.R., Beattie, A., et al. (2021). Underestimating the challenges of avoiding a ghastly future. *Frontiers in Conservation Science*, 1, 615419.

Brown, J.S., and Kotler, B.P. (2004). Hazardous duty pay and the foraging cost of predation. *Ecology Letters*, 7, 999–1014.

Buxton, R.T., McKenna, M.F., Mennitt, D., et al. (2017). Noise pollution is pervasive in U.S. protected areas. *Science*, 356, 531–533.

Candolin, U., and Wong, B.B.M. (2019). Mate choice in a polluted world: consequences for individuals, populations and communities. *Philosophical Transactions of the Royal Society B: Biological Sciences*, 374, 20180055.

Caro, T., and Sherman, P.W. (2013). Eighteen reasons animal behaviourists avoid involvement in conservation. *Animal Behaviour*, 85, 305–312.

Carwardine, J., Martin, T.G., Firn, J., et al. (2019). Priority threat management for biodiversity conservation: a handbook. *Journal of Applied Ecology*, 56, 481–490.

Chan, A.A.Y-H., and Blumstein, D.T. (2011). Attention, noise, and implications for wildlife conservation and management. *Applied Animal Behavior Science*, 131, 1–7.

Chetkiewicz, C.L.B., St. Claire, C.C., and Boyce, M.S. (2006). Corridors for conservation: integrating pattern and process. *Annual Review of Ecology, Evolution, and Systematics*, 37, 317–342.

Chock, R.Y., Clucas, B., Peterson, E.K., et al. (2021). Evaluating potential effects of solar power facilities on wildlife from an animal behavior perspective. *Conservation Science and Practice*, 3, e319.

Clayton, S., and Myers, G. (2015). *Conservation Psychology: Understanding and Promoting Human Care for Nature*. John Wiley & Sons, West Sussex.

Conservation Measures Partnership. (2020). Open standards for the practice of conservation (Version 4.0). [online]. Available from: https://conservationstandards.org/download-cs/ [accessed on 6 July 2023].

Crooks, K.R., Burdett, C.L., Theobald, D.M., et al. (2017). Quantification of habitat fragmentation reveals extinction risk in terrestrial mammals. *Proceedings of the National Academy of Sciences*, 114, 7635–7640.

Crooks, K.R., and Sanjayan, M. (2006). *Connectivity Conservation*. Cambridge University Press, Cambridge.

Dominoni, D.M., Halfwerk, W., Baird, E., et al. (2020). Why conservation biology can benefit from sensory ecology. *Nature Ecology & Evolution*, 4, 502–511.

Doubek, J. (2018). There's no stopping Toronto's 'uber-raccoon'. NPR, retrieved 5 July 2022.

Dubois, N.S., Gomez, A., Carlson, S., and Russell, D. (2020). Bridging the research-implementation gap requires engagement from practitioners. *Conservation Science and Practice*, 2, e134.

Ehrlich, P.R., and Blumstein, D.T. (2018). The great mismatch. *BioScience*, 68, 844–846.

Evans, K.L., Newton, J., Gaston, K.J., et al. (2012). Colonisation of urban environments is associated with reduced migratory behaviour, facilitating divergence from ancestral populations. *Oikos*, 121, 634–640.

Fagan, W.F., Lewis, M.A., Auger-Methe, M., et al. (2013). Spatial memory and animal movement. *Ecology Letters*, 16, 1316–1329.

Forman, R.T.T., and Alexander, L.E. (1998). Roads and their major ecological effects. *Annual Review in Ecology, Evolution, and Systematics*, 29, 207–231.

Francis, C.D., and Barber, J.R. (2013). A framework for understanding noise impacts on wildlife: an urgent conservation priority. *Frontiers in Ecology and the Environment*, 11, 305–313.

Geffroy, B., Sadoul, B., Putman, B.J., et al. (2020). Evolutionary dynamics in the Anthropocene: life history and intensity of human contact shape antipredator responses. *PLOS Biology*, 18, e3000818.

Glista, D.J., DeVault, T.L., and DeWoody, J.A. (2009). A review of mitigation measures for reducing wildlife mortality on roadways. *Landscape and Urban Planning*, 91, 1–7.

Götz, T., and Janik, V.M. (2013). Acoustic deterrent devices to prevent pinniped depredation: efficiency, conservation concerns and possible solutions. *Marine Ecology Progress Series*, 492, 285–302.

Greggor, A.L., Berger-Tal, O., and Blumstein, D.T. (2020). The rules of attraction: the necessary role of animal cognition in explaining conservation failures and successes. *Annual Review in Ecology, Evolution, and Systematics*, 51, 483–503.

Gregory, R., Failing, L., Harstone, M., et al. (2012). *Structured Decision Making: A Practical Guide to Environmental Management Choices*. John Wiley & Sons, Chichester, West Sussex.

Griffin, A.S., Blumstein, D.T., and Evans, C.S. (2000). Training captive-bred or translocated animals to avoid predators. *Conservation Biology*, 14, 1317–1326.

Griffin, A.S., Evans, C.S., and Blumstein, D.T. (2001). Learning specificity in acquired predator recognition. *Animal Behaviour*, 62, 577–589.

Griffin, A.S., Evans, C.S., and Blumstein, D.T. (2002). Selective learning in a marsupial. *Ethology*, 108, 1103–1114.

Haddad, N.M., Brudvig, L.A., Clobert, J., et al. (2015). Habitat fragmentation and its lasting impact on Earth's ecosystems. *Science Advances*, 1, e1500052.

Hale, R., and Swearer, S.E. (2016). Ecological traps: current evidence and future directions. *Proceedings of the Royal Society B: Biological Sciences*, 283, 20152647.

Hale, R., and Swearer, S.E. (2017). When good animals love bad restored habitats: how maladaptive habitat selection can constrain restoration. *Journal of Applied Ecology*, 54, 1478–1486.

Halfwerk, W., Bot, S., Buikx, J., et al. (2011). Low-frequency songs lose their potency in noisy urban conditions. *Proceedings of the National Academy of Sciences*, 108, 14,549–14,554.

Hawlena, D., Saltz, D., Abramsky, Z., and Bouskila, A. (2010). Ecological trap for desert lizards caused by anthropogenic changes in habitat structure that favor predator activity. *Conservation Biology*, 24, 803–809.

Hebblewhite, M., and Merrill, E. (2008). Modelling wildlife-human relationships for social species with mixed-effects resource selection models. *Journal of Applied Ecology*, 45, 834–844.

Hemming, V., Camaclang, A.E., Afams, M.S., et al. (2022). An introduction to decision science for conservation. *Conservation Biology*, 36, e13868.

Hennessy, S.M., Wisinski, C.L., Ronan, C.L., et al. (2022). Release strategies and ecological factors influence mitigation translocation outcomes for burrowing owls: a comparative evaluation. *Animal Conservation*, 25, 614–626.

Hill, C.M. (2018). Crop foraging, crop losses, and crop raiding. *Annual Review of Anthropology*, 47, 377–394.

Hulme-Beaman, A., Dobney, K., Cucchi, T., and Searle, J.B. (2016). An ecological and evolutionary framework for commensalism in anthropogenic environments. *Trends in Ecology & Evolution*, 31, 633–645.

Jones, M.E., Bain, G.C., Hamer, R.P., et al. (2021). Research supporting restoration aiming to make a fragmented landscape 'functional' for native wildlife. *Ecological Management & Restoration*, 22, 65–74.

Jones, M.J., and Hale, R. (2020). Using knowledge of behaviour and optic physiology to improve fish passage through culverts. *Fish and Fisheries*, 21, 557–569.

Jones, P. (2018). Contexts of co-creation: designing with system stakeholders. In: P. Jones and K. Kijima (eds), *Systematic Design: Theory, Methods, and Practice*, pp. 3–52. Springer Nature Tokyo.

Kleindorfer, P.R., Kunreuther, H., and Schoemaker, P.J. (1993). *Decision Sciences: An Integrative Perspective*. Cambridge University Press, Cambridge.

Kotler, B.P., Morris, D.W., and Brown, J.S. (2016). Direct behavioral indicators as a conservation and management tool. In: O. Berger-Tal and D. Saltz (eds), *Conservation Behavior: Applying Behavioral Ecology to Wildlife Conservation and Management*, pp. 307–351. Cambridge University Press, Cambridge.

Kress, S.W. (1977). Establishing Atlantic puffins at a former breeding site. In: S. A. Temple (ed.), *Endangered Birds: Management Techniques for Preserving Endangered Species*, pp. 373–377. University of Wisconsin Press, Madison, WI.

Lim, M.L.M., Sodhi, N.S., and Endler, J.A. (2008). Conservation with sense. *Science*, 319, 281.

Linklater, W.L., Flamand, J., Rochat, Q., et al. (2006). Preliminary analyses of the free-release and scent-broadcasting strategies for black rhinoceros reintroduction. *Ecological Journal, Conservation Corporation Africa*, 7, 26–34.

Longcore, T., and Rich, C. (2004). Ecological light pollution. *Frontiers in Ecology and the Environment*, 2, 191–198.

Loss, R.L., Will, T., Loss, S.S., and Marra, P.P. (2014). Bird-building collisions in the United States: estimates of annual mortality and species vulnerability. *The Condor*, 116, 8–23.

Lumsden, L.F., Griffiths, S.R., Sillins, J.E., and Bennett, A.F. (2021). Roosting behaviour and the tree-hollow requirements of bats: insights from the lesser long-eared bat (*Nyctophilus geoffroyi*) and Gould's wattled bat (*Chalinolobus gouldii*) in south-eastern Australia. *Australian Journal of Zoology*, 68, 296–306.

Lürling, M. (2006). Effects of a surfactant (FFD-6) on Scenedesmus morphology and growth under different nutrient conditions. *Chemosphere*, 62, 1351–1358.

Marler, P., and Peters, S. (1987). A sensitive period for song acquisition in the song sparrow, *Melospiza melodia*: a case of age-limited learning. *Ethology*, 76, 89–100.

McClure, C.J.W., Ware, H.E., Carlisle, J., et al. (2013). An experimental investigation into the effects of traffic noise on distributions of birds: avoiding the phantom road. *Proceedings of the Royal Society B: Biological Sciences*, 280, 20132290.

Moseby, K.E., Blumstein, D.T., and Letnic, M. (2016). Harnessing natural selection to tackle the problem of prey naivete. *Evolutionary Applications*, 9, 334–343.

Muller, K.L., Stamps, J.A., Krishnan, V.V., and Willits, N.H. (1997). The effects of conspecific attraction and habitat quality on habitat selection in territorial birds (*Troglodytes aedon*). *The American Naturalist*, 150, 650–661.

Nathan, R., Getz, W.M., Revilla, E., et al. (2008). A movement ecology paradigm for unifying organismal movement research. *Proceedings of the National Academy of Sciences*, 105, 19,052–19,059.

Oro, D., Genovart, M., Tavecchia, G., et al. (2013). Ecological and evolutionary implications of food subsidies from humans. *Ecology Letters*, 16, 1501–1514.

Patten, M.A., and Kelly, J.F. (2010). Habitat selection and the perceptual trap. *Ecological Applications*, 20, 2148–2156.

Putman, B.J., and Blumstein, D.T. (2019). What is the effectiveness of using conspecific or heterospecific acoustic playbacks for the attraction of animals for wildlife management? A systematic review protocol. *Environmental Evidence*, 8, 6.

Radford, C., McNutt, J.W., Rogers, T., et al. (2020). Artificial eyespots on cattle reduce predation by large carnivores. *Communications Biology*, 3, 430.

Razgour, O., Taggart, J.B., Manel, S., et al. (2018). An integrated framework to identify wildlife populations under threat from climate change. *Molecular Ecology Resources*, 18, 18–31.

Riffell, J.A., Shlizerman, E., Sanders, E., et al. (2014). Flower discrimination by pollinators in a dynamic chemical environment. *Science*, 344, 1515–1518.

Robertson, B.A., and Blumstein, D.T. (2019). How to disarm an evolutionary trap. *Conservation Science and Practice*, 2019, e116.

Robertson, B.A., and Hutto, R.L. (2006). A framework for understanding ecological traps and an evaluation of existing evidence. *Ecology*, 87, 1075–1085.

Robertson, B.A., Ostfeld, R.S., and Keesing, F. (2017). Trojan females and Judas goats: evolutionary traps as tools in wildlife management. *BioScience*, 67, 983–994.

Rodriguez, A., Burgan, G., Dann, P., et al. (2014). Fatal attraction of Short-Tailed Shearwaters to artificial lights. *PLOS ONE*, 9, e110114.

Salmon, M. (2005). Protecting sea turtles from artificial night lighting at Florida's oceanic beaches. In: C. Rich and T. Longcore (eds), *Ecological Consequences of Artificial Night Lighting*, pp. 141–168. Island Press, Washington, DC.

Saltz, D., Berger-Tal, O., Motro, U., et al. (2018). Conservation implications of habituation in nubian ibex in response to ecotourism. *Animal Conservation*, 22, 220–227.

Schakner, Z.A., and Blumstein, D.T. (2021). The California sea lion: thriving in a human-dominated world. In: C. Campagna and R. Harcourt (eds), *Ethology and Behavioral Ecology of Otariids and the Odobenid*, pp. 347–365. Springer Nature Switzerland, Cham.

Schakner, Z.A., Buhnerkempe, M.G., Tennis, M.J., et al. (2016). Epidemiological models to control the spread of information in marine mammals. *Proceedings of the Royal Society B: Biological Sciences*, 283, 20162037.

Schakner, Z.A., Gotz, T., Janik, V.M., and Blumstein, D.T. (2017). Can fear conditioning repel California sea lions from fishing activities? *Animal Conservation*, 20, 425–432.

Schlaepfer, M.A., Runge, M.C., and Sherman, P.W. (2002). Ecological and evolutionary traps. *Trends in Ecology & Evolution*, 17, 474–480.

Schlaepfer, M.A., Sherman, P.W., Blossey, B., and Runge, M.C. (2005). Introduced species as evolutionary traps. *Ecology Letters*, 8, 241–246.

Senko, J.F., Peckham, S.H., Aguilar-Ramirez, D., and Wang, J.H. (2022). Net illumination reduces fisheries bycatch, maintains catch value, and increases operational efficiency. *Current Biology*, 32, 911–918.

Shamoon, H., Maor, R., Saltz, D., and Dayan, T. (2018). Increased mammal nocturnality in agricultural landscapes results in fragmentation due to cascading effects. *Biological Conservation*, 226, 32–41.

Shettleworth, S. (2010). *Cognition, Evolution, and Behaviour*. Oxford University Press, New York, NY.

Sih, A. (2013). Understanding variation in behavioral responses to human-induced rapid environmental change: a conceptual overview. *Animal Behaviour*, 85, 1077–1088.

Sih, A., Trimmer, P.C., and Ehlman, S.M. (2016). A conceptual framework for understanding behavioral responses to HIREC. *Current Opinion in Behavioral Sciences*, 12, 109–114.

Stamps, J.A. (1988). Conspecific attraction and aggregation in territorial species. *The American Naturalist*, 131, 329–347.

Stobo-Wilson, A.M., Murphy, B.P., Legge, S.M., et al. (2022). Counting the bodies: estimating the numbers and spatial variation of Australian reptiles, birds and mammals killed by two invasive mesopredators. *Diversity and Distributions*, 28, 976–991.

Sutherland, W.J., Pullin, A.S., Dolman, P.M., and Knight, T.M. (2004). The need for evidence-based conservation. *Trends in Ecology & Evolution*, 19, 305–308.

Swaddle, J.P. (2016). Evolution and conservation behavior. In: O. Berger-Tal and D. Saltz (eds), *Conservation Behavior: Applying Behavioral Ecology to Wildlife Conservation and Management*, pp. 36–65. Cambridge University Press, Cambridge.

Swaddle, J.P., Francis, C.D., Barber, J.R., et al. (2015). A framework to assess evolutionary responses to anthropogenic light and sound. *Trends in Ecology & Evolution*, 30, 550–560.

Tidewell, K.S., Carrothers, B.A., Blumstein, D.T., and Schakner, Z.A. (2021). Steller sea lion (*Eumetopias jubatus*) response to non-lethal hazing at Bonnevilee dam. *Frontiers in Conservation Science*, 2, 760866.

Tinbergen, N. (1963). On aims and methods in ethology. *Zeitschrift fur tierpsychologie*, 20, 410–433.

Todgham, A.E., and Stillman, J.H. (2013). Physiological responses to shifts in multiple environmental stressors: relevance in a changing world. *Integrative and Comparative Biology*, 53, 539–544.

Tuxbury, S.M., and Salmon, M. (2005). Competitive interactions between artificial lighting and natural cues during seafinding by hatchling marine turtles. *Biological Conservation*, 121, 311–316.

Uchida, K., Blakey, R.V., Burger, J.R., et al. (2021). Urban biodiversity and the importance of scale. *Trends in Ecology & Evolution*, 36, 123–131.

van Dyck, H. (2012). Changing organisms in rapidly changing anthropogenic landscapes: the significance of the 'Umwelt'-concept and functional habitat for animal conservation. *Evolutionary Applications*, 5, 144–153.

Vardi, R., and Berger-Tal, O. (2022). Environmental variability as a predictor of behavioral flexibility in urban environments. *Behavioral Ecology*, 33, 573–581.

Von Uexküll, J. (1909). *Umwelt und Innenwelt der Tiere*. Springer, Berlin.

Walsh, J.C., Dicks, L.V., and Sutherland, W.J. (2015). The effect of scientific evidence on conservation practitioners' management decisions. *Conservation Biology*, 29, 88–98.

White, T.B., Petrovan, S.O., Christie, A.P., et al. (2022). What is the price of conservation? A review of the status quo and recommendations for improving cost reporting. *BioScience*, 72, 461–471.

Williams, B.K. (2011). Adaptive management of natural resources—framework and issues. *Journal of Environmental Management*, 92, 1346–1353.

Winger, B.M., Weeks, B.C., Farnsworth, A., et al. (2019). Nocturnal flight-calling behaviour predicts vulnerability to artificial light in migratory birds. *Proceedings of the Royal Society B: Biological Sciences*, 286, 20190364.

Witherington, B.E. (1997). The problem of photopollution for sea turtles and other nocturnal animals. In: J.R. Clemmons and R. Buchholz (eds), *Behavioral Approaches to Conservation in the Wild*, pp. 303–328. Cambridge University Press, Cambridge.

Woodroffe, R., Donnelly, C.A., Cox, D.R., et al. (2006). Effects of culling on badger *Meles meles* spatial organization: implications for the control of bovine tuberculosis. *Journal of Applied Ecology*, 43, 1–10.

CHAPTER 12

Conservation breeding and translocation

Alison L. Greggor, Melissa J. Merrick, Colleen L. Wisinski, Rachel Y. Chock, and Debra M. Shier

Overview

An increasing number of species are reliant on conservation breeding and translocations for their recovery. These interventions occur in several stages, each of which can directly benefit from applying conservation behaviour theory. From encouraging breeding, to preparing animals for release, to staging the reintroduction process and managing the release site, there are ways to leverage animal behaviour as a tool to improve outcomes. Here we focus on common questions and pitfalls managers face when planning conservation breeding and releases, and the resulting strategies that have proven to increase success of such interventions. Taking this approach allows one to consider behavioural topics as important conservation tools, not merely fodder for theory.

12.1 Introduction

Biodiversity conservation is a multidisciplinary endeavour requiring research that spans social and biological sciences, with interventions that can involve regulation, policy, public interest, and political will. Behavioural ecology is one of many tools from the biological sciences that can help address conservation problems. Using an understanding of behaviour to address real-world conservation problems involves going beyond theory; application can be messier, require specific knowledge on understudied and rare species or habitats, and can require addressing socioeconomic drivers underlying the

problems (Chapter 11). Despite these constraints, there are certain conservation problems where conservation behaviour can have a major impact. For example, using behaviour in animal conservation breeding and translocation programmes can improve species' survival and recovery (Shier 2016). However, behavioural knowledge is still underused in these contexts (Berger-Tal et al. 2020).

Conservation breeding, the human-managed care and propagation of at-risk species to aid recovery, is a last resort conservation strategy, typically employed when efforts to promote species in their natural habitats fail. Bringing species into human care is expensive and can be risky, since many programmes do not meet their breeding or recovery objectives (Fischer and Lindenmayer 2000; Mathews et al. 2005; Bowkett 2009; Conde et al. 2013). Yet anthropogenic activities, such as human disturbance and unchecked development, are accelerating habitat loss and fragmentation (Chapter 8), degrading ecosystem function, facilitating the spread of invasive species and diseases (Chapters 6 and 17), and altering climate (Chapter 1), such that ~18% of the total evaluated vertebrate species are threatened with extinction (International Union for Conservation of Nature 2022), with more species added every year. As a result, an increasing number of species may become reliant on ex-situ conservation breeding. In fact, roughly 64% of recovery plans under the USA's Endangered Species Act list ex-situ population creation as a means for promoting recovery (Tear et al. 1993).

Alison L. Greggor et al., *Conservation breeding and translocation*. In: *Behavioural Responses to a Changing World*. Edited by: Bob B. M. Wong and Ulrika Candolin, Oxford University Press. © Oxford University Press (2024). DOI: 10.1093/oso/9780192858979.003.0012

Conservation breeding only supports species recovery if it produces animals that survive and breed in the wild following release. Mimicking environmental conditions sufficiently to promote breeding, while also encouraging the development and maintenance of wild-type behaviours, can be challenging, and sometimes requires opposing approaches to address stress and welfare (Clark et al. 2023; see also Chapter 15). For instance, unmitigated chronic stress (i.e. distress) often reduces reproductive potential, yet acute stressful experiences, such as interactions with predators, can facilitate the development of survival skills. Finding this balance is critical, especially for programmes with multigenerational breeding that must contend with artificial selection pressures and long-term consequences of ex-situ conditions and husbandry. Limiting the effects of captivity involves understanding wild-type behaviour and the extent to which behaviours are governed by experience and environmental or social cues. Only then can key ecologically relevant stimuli be identified and provided. Along the way are choices—e.g. should species be given experience with seasonal food restrictions or have ad libitum access to food?—which often expose key unknowns about how environment shapes behaviour. Additionally, for many at-risk species, we lack knowledge of the consequences of altered behavioural expression. For example, what number of food types does an omnivorous species need to consume in human care to forage competently? When are changes in the expression of socially learned behaviour problematic, and will those changes impact individuals' ability to integrate into wild populations (Greggor and Goldenberg 2023)? Ultimately the consequences of behavioural change in captivity are best understood in relation to their impact on fitness following translocation back to the wild.

The International Union for Conservation of Nature (IUCN) defines translocation as deliberate movement of animals between locations for conservation purposes, a process that covers many methods and approaches (Batson et al. 2015). Here we focus on strategies for improving success during the transition from human care to the wild, but similar issues can be applied for wild-to-wild translocations (e.g. animals may show detrimental homing or maladaptive habitat preferences). Animals that are head-started from the wild (brought into human care after wild conception or breeding) have fewer handicaps than those that come from multigenerational conservation breeding programmes, yet even short periods of rearing in artificial environments can affect brain development and behaviour (Clark et al. 2023). Additionally, upon release, animals are confronted with many types of novelty (e.g. weather, habitat, food, heterospecifics), with which they may be ill-prepared to interact. Accordingly, many translocation efforts fail, and a lack of behavioural preparedness is often a leading cause (Kleiman 1989; Berger-Tal et al. 2020). 'Soft-release' strategies aim to reduce stress and ease the transition between ex-situ and in-situ environments by providing various types of support. Soft-release strategies—such as temporarily holding animals at the release site, feeding them post-release, or providing wild-type experiences before release—are increasingly used. While gentler transitions may boost early survival, these strategies can have negative consequences too, if they limit the integration of released individuals into wild populations or become crutches that are challenging to remove. For instance, if animals are reintroduced into areas with predator-proof fencing, it is unclear how and when you remove those protections and if animals will avoid predators in an adaptive way (Blumstein et al. 2019). These unknowns can illustrate the ways in which transitioning animals back into the wild can be even more challenging than the transition into human care.

Addressing challenges across the conservation breeding/translocation pipeline requires connecting behaviour to concrete outcomes, such as how behaviour influences fitness metrics (e.g. survival and reproduction) following release. These challenges invoke theories on topics ranging from learning and perception to foraging and space use, often in surprising ways. Meaningfully applying conservation behaviour in these contexts requires adaptive management and a problem-solving approach, all while working within the complex landscape of on-the-ground conservation action (Figure 12.1; Chapter 11). Controlled experiments can be challenging with at-risk species due to sample sizes and strict regulations, but can provide critical

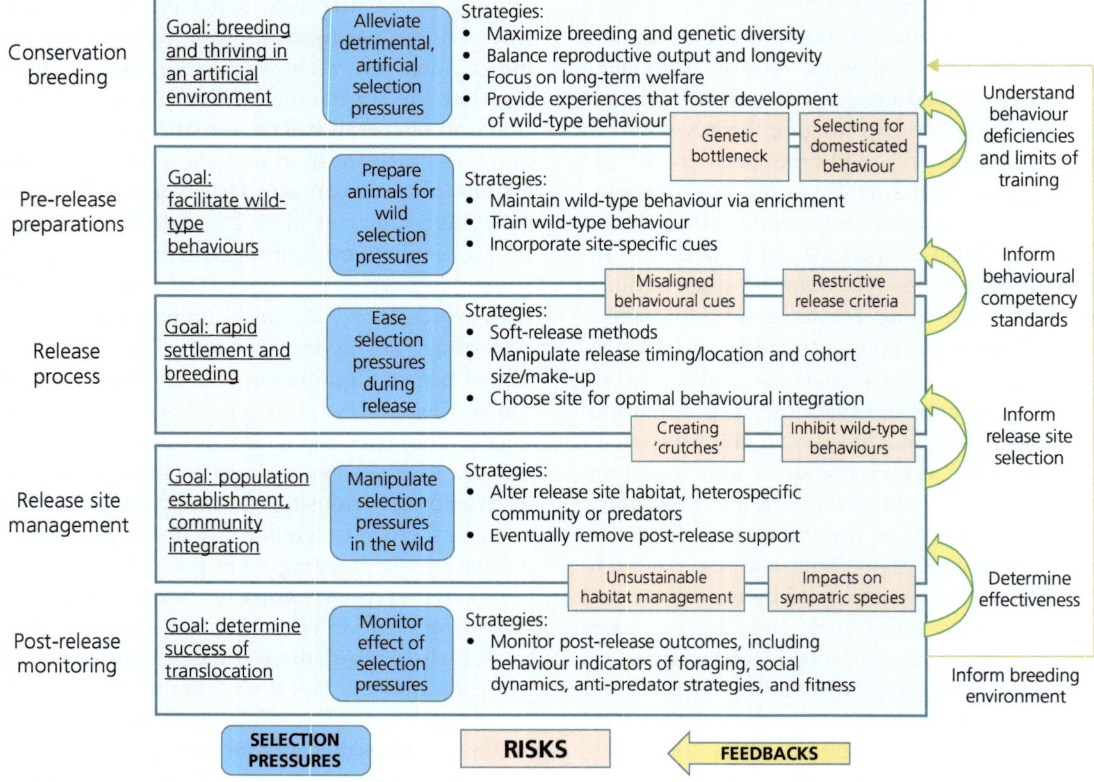

Figure 12.1 Targeted applications of conservation behaviour from conservation breeding to translocation. All stages of the process involve different goals, which require tinkering with selection pressures (circles) via various strategies. Risks arise that may hamper the subsequent stage and the overall success of the conservation programme (pale orange boxes). Each stage also offers opportunities for feedback and refining strategies at earlier stages (yellow arrows).

information about the factors that influence conservation outcomes. Where strict experimentation is not possible, incorporating 'lessons learned' from less robust designs to refine species' management can still be powerful.

In this chapter we walk through several stages of the conservation breeding and translocation process where behavioural applications have effectively improved outcomes. Specifically, we cover common questions managers face when planning breeding and releases and outline clusters of strategies. Taking this approach allows one to think about behavioural topics as tools, not just theory. For each question we address common pitfalls, where known, and place them in the wider context of non-behavioural interventions.

12.2 How do you preserve and breed species in human care?

Although one of the primary objectives when establishing ex-situ populations is preserving genetic diversity, a species is more than just its genes. Maintaining species' ecological function and their ability to survive and breed post-release with little to no human intervention requires preserving wild-type behaviour. Nearly all fitness-related benchmarks for successful translocation, including habitat and mate selection, rearing offspring, integrating with social groups, finding food, and avoiding predators, rely on underlying behavioural mechanisms that must be maintained, developed, or refined in managed care. Thus behaviour-centred breeding programmes that focus on maximizing genetic and behavioural diversity (McDougall et al. 2006) are necessary for successful conservation translocation outcomes (Tripovich et al. 2021). A major challenge for this phase of the conservation breeding/translocation process involves mitigating artificial selection pressures associated with ex-situ environments to maintain a behaviourally diverse population.

12.2.1 Encouraging animals to breed

Successful reproduction ex-situ requires understanding how reproductive biology, physiology, and social and mating systems intersect to produce species-specific behaviour and cues for natural mating, pregnancy, offspring rearing, and development (Martin-Wintle et al. 2019). This is no small feat and can take decades of committed science-based inquiry to identify the timing of physiological, social, and environmental cues that improve reproductive success (Martin-Wintle et al. 2019; Flanagan et al. 2020). Species' social and mating systems can inform housing arrangements and management of social interactions in human care. Enclosures designed to mimic native habitat features, support ecologically relevant behaviour and life-history events, and house groups at appropriate densities can increase both individual welfare and reproductive success. Narrow-headed garter snakes *Thamnophis rufipunctatus*, for instance, successfully reproduced when moved to outdoor group-living enclosures that allowed natural brumation as a social group (Blais et al. 2022). By contrast, species that are territorial or solitary may find it stressful to be housed near conspecifics, and including visual barriers or acoustically manipulating ambient noise may reduce stress and improve reproduction. For territorial 'alalā (Hawaiian crow) *Corvus hawaiiensis*, reproductive success increased dramatically with distance between aviaries and visual obstruction of other adult pairs (Flanagan et al. 2020). But, species-specific knowledge is key, as solid opaque barriers used to separate solitary kangaroo rats disrupted oestrous cycling (Yoerg 1999), presumably because scent and/or visual cues are important for triggering breeding. Solitary species may rely on olfactory or auditory cues to advertise receptivity

and require carefully timed introduction of these cues prior to pairings. For example, a successful giant panda *Ailuropoda melanoleuca* conservation breeding programme purposefully built facilities to behaviourally manage reproduction (Martin-Wintle et al. 2019). A circular arrangement of individual pens, connected via walkways with shift doors, facilitated moving individuals at critical oestrous periods, establishing scent cues, and assessing mate compatibility in species-appropriate ways. Importantly, giving females the opportunity to choose among mates during oestrus and rear offspring has increased mating and reproductive success (Martin-Wintle et al. 2019), and can be accomplished even within the limitations of studbook recommendations to maintain genetic diversity (e.g. Hartnett et al. 2018).

12.2.2 Keeping them wild

Artificial ex-situ environments present novel stimuli and stressors that differ from the wild, and can disproportionately select against individuals with certain behaviour types, reducing the behavioural diversity of release candidates (McDougall et al. 2006). Conversely, novel stressors in human care may produce individuals with increased behavioural and developmental plasticity that may fare better in highly modified landscapes (Mason et al. 2013). Rapid evolution in human care can trigger morphological changes that have downstream effects on behaviour, and decay ecologically relevant behaviours and skills, with implications for the survival of released individuals and their progeny (McDougall et al. 2006). Animals can lose motor skills required for hunting, foraging, and predator avoidance, as well as the ability to recognize essential food, social signals, and habitat cues. Further, gut microbiome may be substantially altered and simplified in captivity, which can negatively impact nutrient uptake upon release (Yao et al. 2019; Williams et al. 2023). There are several approaches, such as providing enrichment experiences, training, and opportunities for social learning, that have been successfully used to facilitate the development and maintenance of ecologically relevant behaviours in managed care. While they use different techniques, these approaches all attempt to minimize artificial selection pressures that captivity otherwise imposes.

12.2.3 Offer natal habitat-informed experiences

Early experience with habitat features and environmental cues can have lasting impacts on morphological and neuromuscular development, learned habitat and food associations, and subsequent performance in those environments (Aubret and Shine 2008; Hammond et al. 2022; Stamps and Swaisgood 2007). For many species, cues in natal environments (e.g. locations, habitat type, shelter, or nest material) can influence habitat preferences later in life and shape an individual's definition of 'home', a process called natal habitat preference induction (NHPI; Stamps and Swaisgood 2007). Ensuring that ex-situ environments provide critical environmental cues helps animals seek appropriate habitats post-release. For example, salmonid species imprint upon the olfactory signature of their natal stream. Unless populations slated for restocking experience odours from spawning locations, they will fail to return and reproduce there (Dittman et al. 2015). Providing NHPI-informed experience, that is the habitat features and cues similar to those at release sites (Hammond et al. 2022), can improve translocation outcomes and promote habitat-appropriate development and morphology. Australian tiger snakes *Notechis scutatus* reared in enclosures that mimicked different habitats (terrestrial, arboreal, and aquatic) consistently selected and exhibited more effective locomotion in the habitat component they had experienced (Aubret and Shine 2008). Exposing mountain yellow-legged frogs *Rana muscosa* to variable water flow regimes that mimicked stream conditions at the release site improved post-release site fidelity in subadults and apparent survival in juveniles (Hammond et al. 2022). Further, stream-experienced juveniles developed longer limb lengths and swam faster than control juveniles, but treatment effects were not observed in subadults, suggesting a sensitive developmental window (Hammond et al. 2022). The challenge for conservation practitioners is identifying which release site cues are most reliably tied to individual fitness, habitat selection and use, as well as the optimal timing for exposure. Like many behavioural

deficiencies, strategies for addressing issues associated with settlement span multiple stages of the release process (Figure 12.2).

12.2.4 Promote social learning and social fluency

Animals' social environments shape how they interact with conspecifics, which determines how they respond to competition, attract mates, gain social information, communicate, and integrate into wild social groups. Maintaining social housing in human care can help ensure animals acquire species-specific social skills (Greggor and Goldenberg 2023). However, doing so can be challenging since socially housed animals need sufficient space to resolve conflicts or engage in fission-fusion interactions, which can increase the cost of breeding programmes. Socially appropriate housing can also foster social learning of important culturally-transmitted information, including recognition of predator cues, habitat quality, and foraging skills (Shier and Owings 2007; Greggor et al. 2017; Morris et al. 2021). Parenting manipulations, such as cross fostering, can influence the development of food preferences, food handling, or selection of a foraging niche (e.g. when fostered across species, great tits *Parus major* and blue tits *Cyanistes caeruleus* adopt the preferred food size of foster parents; Slagsvold and Wiebe 2011). Later social exposure can also help maintain survival-relevant behaviour. For instance, having social learning opportunities with an experienced adult demonstrator increased vigilance, alarm calling, and post-release survival in captive-reared black-tailed prairie dogs *Cynomys ludovicianus* (Shier and Owings 2007). Therefore, ensuring captive-bred individuals have opportunities to be reared with or integrated into social groups at an early age promotes the maintenance of behaviours essential for reproduction and survival.

12.3 How do you prepare animals for life in the wild?

In order to survive and breed in the wild, animals raised in conservation breeding programmes need to select and settle in suitable habitat, move and navigate appropriately within it, acquire or build

Problem: Released animals fail to settle in appropriate habitat.

Actions:

Conservation breeding

- Give animals access to habitat cues during development
- Provide opportunities for species-specific social interactions

Pre-release prep

- Train or give experience with relevant habitat features and their associated risks/rewards
- Use habitat seeking behaviour as a criterion for release readiness

Release site management

- Determine if ecological traps are misdirecting animals' cue seeking
- Manipulate competition by heterospecifics if target species is being competitively excluded

Release strategies

- Manipulate conspecific cues at the release site
- Determine if release site offers necessary habitat for a reintroduced population

Research to inform action:

- What cues or proxies identify suitable habitat?
- What is the role of experience in honing these cues?
- Are there sensitive developmental windows for learning these cues?
- At what habitat scale (e.g. general habitat, microsite, preferences within habitat) and developmental stage did they fail to show appropriate behaviour?
- Do they show species-appropriate locomotion and social cue use?

Figure 12.2 Conservation actions and relevant research questions to address the problem of failed settlement at release sites with behavioural tools. Behavioural issues that occur during the release process can stem from deficiencies at different stages of breeding, release prep, release strategy, and wider site management. Research can help identify the stage where the problem arises, and what cues and experience can be used to address it. These unifying problems highlight the need to understand fundamental drivers of behaviour, their plasticity, and how they translate to on-the-ground conservation action.

Photos from the Pacific pocket mouse *Perognathus longimembris pacificus* conservation breeding and release programme ©SDZWA: Conservation breeding—group of littermates housed in species-typical social arrangement; pre-release prep—pocket mouse interacting with branches containing native seed; release strategies—acclimation chambers at the release site located within 5 m of each other, to facilitate rapid establishment of territories; release site management—predator exclusion fence used to control predation pressure and heterospecific densities

shelter, find and process or store food, and interact with conspecifics, heterospecifics, and predators effectively (Shier 2016). Learning theory can provide insight into the underlying mechanisms that influence the development of these behaviours. In practice, implementing these ideas involves providing animals with critical cues and experiences within developmental time frames that are species-appropriate, while considering that the current 'wild' conditions may be different from those experienced by the species historically. For example, animals' diet may need to shift if native prey species are no longer present or palatable novel prey species

have been introduced into their range. Additionally, animals may need new skills to cope with novel invasive predators or to avoid human conflict. In these cases, conservation practitioners may need to understand or predict the target species' flexibility to respond to novel stimuli.

12.3.1 Offer exposure, enrichment, or training

Behavioural deficiencies can be addressed with targeted exposures to relevant cues (such as foods), environmental enrichment (e.g. habitats designed to mirror aspects of wild environments, such as dynamic perches or stream flow), or explicit training (i.e. encouraging associations between stimuli or between behaviour and a given outcome; Shier 2016). While the set-up may appear similar between some of these approaches, the cognitive and learning mechanisms underlying their subsequent behavioural modification can differ. Regardless of the intervention, empirical support for their effectiveness is crucial (e.g. comparing outcomes with control groups). Yet scientific rigour can be challenging when withholding training that has the potential to improve survival may be viewed as unethical. Here we focus on training and explore several areas where effective training has been documented, noting that it is but one approach to altering behaviour, and is often blended with other approaches, even within single species (Figure 12.3).

Few species innately recognize and respond appropriately to their native predators without prior experience. Translocation failure is often attributed to predation, partially because the ability to respond appropriately to predators can degrade quickly or be lost in managed care (Berger-Tal et al. 2020; Edwards et al. 2021; Morris et al. 2021). For example, captive-bred Vancouver Island marmots *Marmota vancouverensis* lost the ability to recognize mountain lions in only five generations, contributing to high offspring mortality (Dixon-MacCallum et al. 2021). Improving anti-predator responses can be accomplished via direct exposure to predation pressure (e.g. releasing predatory cats into predator-free enclosures; Blumstein et al. 2019) or through structured training (Rowell et al. 2020; see Box 12.1).

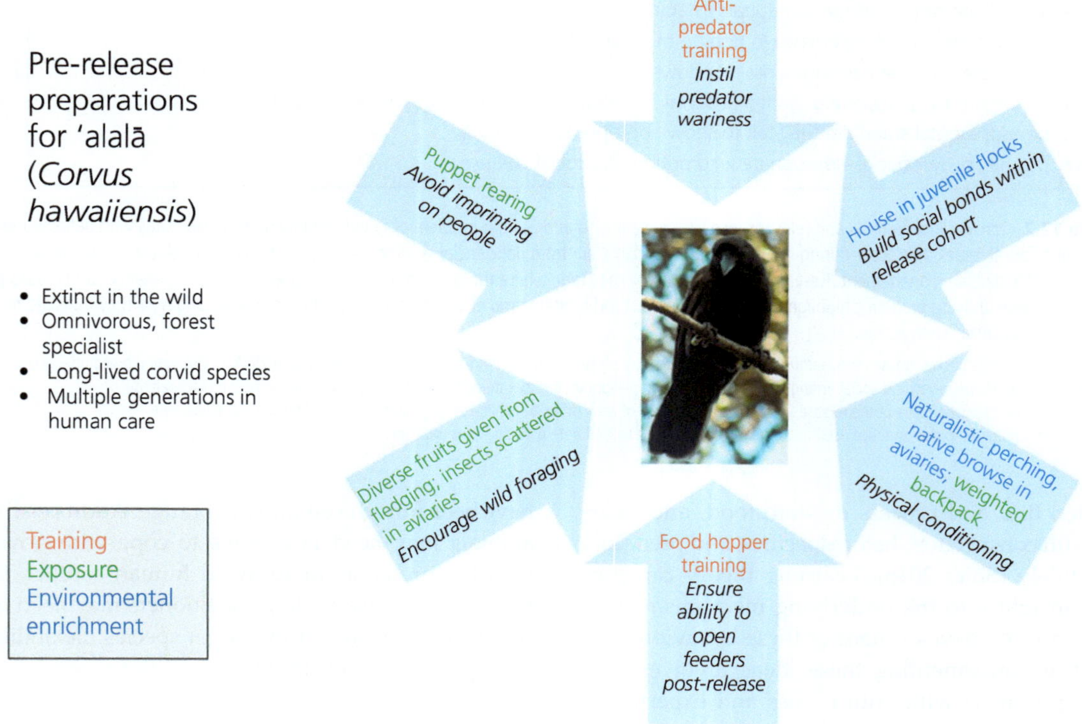

Figure 12.3 An example of behaviour-focused pre-release preparations. For 'alalā, different types of preparation are administered throughout development using training, exposure, and environmental enrichment methods. Each arrow represents an area where behavioural deficiencies are addressed to achieve a post-release outcome (in black italics). Species-specific perceptual and learning biases will influence how these approaches can be applied to other species.

Photo of 'alalā: ©SDZWA

Box 12.1 Improving effectiveness of anti-predator training

Anti-predator training embodies the practical application of learning theory (Greggor et al. 2014). Training can induce rapid associative learning of predator recognition or responses in otherwise predator-naïve species across taxonomic groups ranging from fish to birds and mammals (Rowell et al. 2020; Edwards et al. 2021). However, training is not always successful at provoking learning or improving survival outcomes after translocation. Understanding where and why it works is key to deploying it successfully, and for justifying its use, especially when it can be logistically challenging to execute and can involve welfare considerations, such as prompting short but intense stress. Fundamentally, anti-predator training creates a lasting association between an aversive experience (unconditioned stimulus; US) and a predator (conditioned stimulus; CS). The US can be created by a single cue or combinations of cues, including social alarm or distress cues (e.g. alarm calls or olfactory cues), and noxious or startling stimuli (e.g. loud sounds, squirts of water, or simulated capture). The US should provoke a highly aversive response without any prior training. Ideally, the predatory cues (CS) match the predator and its mode of hunting, from the perspective of the prey species (Edwards et al. 2021). For instance, broadcasting predatory scents may be more appropriate for prey species that rely on olfaction, while visual stimuli may be more evocative to visually biased prey (e.g. Figure 12.4). Combining CS cues into a multimodal presentation can also increase their predatory resemblance and effectiveness (Shier 2016). Unless data already exist on target species' responses toward candidate US and CS options, testing responses in comparison to control sounds, scents, or visual stimuli can increase the likelihood that training will attract attention and prompt learning.

Once stimuli have been chosen, there are key design considerations to be made. How many trainings to conduct, when to conduct training in relation to development and the release timeline, what social context will make training most effective, and what behaviours determine competency all need to be decided (Shier 2016). There is no simple solution here, but species-specific knowledge can help guide these decisions. Considering the species' memory retention would influence how close to release a training may need to occur. Details of their social system would inform if having access to social tutors (i.e. observing experienced individuals respond to predators) may help boost learning (Shier and Owings 2007). Finally, examining the species' natural anti-predator behaviour may narrow down appropriate responses to different predators.

Unless a training regime has already been tested in the target species, or a close relative that experiences similar predators, then the effectiveness of training should be measured. Effectiveness can be considered via leading indicators pre-release, such as documenting changes in behaviour as a result of training, or lagging indicators, such as post-release survival. In both cases, BACI (before-after-control-intervention) designs are preferable.

Figure 12.4 Example of live anti-predator training. (a) A two-striped garter snake *Thamnophis hammondii* occupies the rear portion of a tank housing mountain yellow-legged frog tadpoles. (b) Groups of tadpoles are separated from the snake by Plexiglas with holes that expose the tadpoles to visual and chemical predator cues (Hammond et al. 2023).

Photos: ©SDZWA

Species-typical foraging can involve multiple cognitive mechanisms, physical abilities, and learned associations. Animals need to identify suitable food, capture or collect it, process and consume it, and for some species, they need to know when, where, and how to store it. There can be sensitive periods that shape preferences for certain foods, and a reliance on social stimuli for guiding choices. Food availability and quality can change seasonally, and dynamic foraging strategies may involve different space use. While providing animals born and reared ex-situ with native foods to increase recognition and handling skills may seem simple, it can be challenging to provide experiences that foster more complex foraging strategies (e.g. resource acquisition across space and season). Additionally, little research exists on the consequences of potential cognitive deficits that may result from living in captivity (Clark et al. 2023). Yet, the implications of cognitive deficits for species that rely on their spatial memory, for example, could be dire, especially for species that cache food to survive. Different pre-release exposures and trainings may be needed depending upon which aspects of foraging behaviour are not expressed. For instance, managers may need to actively shape foraging behaviour by sequentially offering increasingly difficult-to-access food (e.g. opening seed pods). For active hunting species, individuals slated for release may need practice with group hunting or live prey. For generalist species, having exposure to multiple food types may help prevent food neophobia. Adaptive foraging post-release can also involve avoiding certain food types. Training animals in what not to eat, to help them avoid poisonous prey items (e.g. cane toads *Rhinella marinus*; Jolly et al. 2018; Chapter 6), for instance, by using conditioned taste aversion, can help but is not yet common in pre-release preparations. Overall, species-specific knowledge should drive training design, but implementation may be limited by the time, effort, and resources available to re-create wild-type foraging experiences.

12.3.2 Train skills, not just knowledge

Released animals increasingly need behavioural resiliency to survive in human-altered wild environments. Cognitive training, learning how to learn, is one approach that may be achieved through exposure to wild-type challenges (Clark et al. 2023), such as encouraging flexible foraging with changes in food cues. Not all resiliency is behavioural; for instance, vaccinating animals ahead of the arrival of disease. However, even disease avoidance can involve behaviour, such as habitat preferences that might reduce disease risk. For example, Panamanian golden frogs *Atelopus zeteki* that selected warmer microhabitats, increasing their body temperature, reduced their odds of infection by chytrid fungus (Richards-Zawacki 2010). Overall, pre-release preparations have focused on the known novelty animals face, but promoting wider flexibility may offer potential for survival long term amid changing environmental conditions.

12.4 How do you improve chances of survival and reproduction during the release process?

Release strategies are increasingly being developed for individual species and sites through adaptive management (Kemp et al. 2015). Research on release strategies has focused on release site selection, release timing, group size, age, and composition (McCallum et al. 1995), 'soft' vs 'hard' release (Letty et al. 2000), and source population environment (Bright and Morris 1994). Here we highlight how behaviour can be used when choosing release strategies.

12.4.1 Select an appropriate release site

The first step in a successful translocation is selecting an appropriate release site. While social, economic, and ecological factors need to be considered (IUCN/SSC 2013), understanding the target species' behaviour is important for predicting how it will interact with abiotic and biotic environments. Habitat quality is a key site metric (IUCN/SSC 2013; Stadtmann and Seddon 2020), but habitats vary across space and time. While we typically use the characteristics of extant sites as guidelines, with limited populations of endangered species, extant sites may no longer contain the conditions the species historically experienced (White et al. 2015). Meanwhile, effectively evaluating habitat suitability of candidate sites also involves knowing the target species' behaviours that drive habitat selection. Behavioural traits, such as communication mode, perceptual ability, and search strategy can influence an animal's ability to detect available resources, colonize or persist within habitat patches, or to tolerate edge effects (Bowler and Benton 2005). For example, translocated white-shouldered fire-eye *Pyriglena leucoptera* were unable to detect distant (80–90 m) rainforest patches following release, thereby changing their search strategy and travelling longer routes during dispersal, which increased energetic costs and mortality (Awade et al. 2017).

Even if release sites contain sufficient high-quality habitat for the target species, they may not be ideal for the chosen cohort or for released animals in general. Species become locally adapted to environmental conditions (Osborne and Seddon 2012) and what is suitable for one population may not be suitable for another, due to processes such as NHPI (Stamps and Swaisgood 2007; Section 12.2.3). Therefore, selecting release sites with habitat characteristics matching those of the ex-situ natal environment may improve release outcomes. Meanwhile, resident populations may need different habitat targets than reintroduced populations. Reintroductions can be considered 'forced dispersal' events (Stamps and Swaisgood 2007) whereby individuals are relocated from natal to novel habitat where they must find required resources, evaluate suitability, and settle if they are to survive and reproduce. Thus, selecting release sites with increased habitat structure (i.e. cover) or shelter may reduce predation pressure during settlement, and help the translocated population through the initial survival and establishment phase (Miller et al. 2014). This strategy may be especially important for reintroduced prey species that seek shelter to avoid predation, as opposed to those that rely on crypsis, aposematism, confusion, or dilution.

How the target species interacts with sympatric conspecifics and heterospecifics should also be considered during site selection. Whether it is preferable to select sites with conspecifics present depends on the social behaviour of the target species, habitat quality of the site, and density of resident animals.

For territorial species, the presence of resident conspecifics may restrict available habitat or resources and may impose high energetic and injury costs associated with settlement. For these species it may be beneficial to select sites where high-quality habitat remains unoccupied. By contrast, for social species like black-faced spider monkeys *Ateles chamek*, having residents at the release site improves establishment success (Pottie et al. 2021), and may be needed for transferring social information. Thus, if reintroducing a social species, understanding how individuals and social groups interact and use space will determine whether release sites with conspecific residents are suitable.

12.4.2 Optimize release type, timing, and support

Release strategies can be tailored to species-specific behaviour in ways that minimize dispersal and maximize survival and reproduction (Batson et al. 2015). Release type (e.g. hard/immediate, soft/delayed, stepping-stone), timing, conspecific attraction, release cohort composition (e.g. age, sex, familiarity), and translocation distance can all influence individual settlement decisions, with post-release survival and reproductive consequences. Dispersal away from the release site often compromises translocations, thus dampening post-release dispersal can improve success (Berger-Tal et al. 2020). How releases are structured can affect post-release settlement, survival, and reproduction across many taxa (e.g. Mitchell et al. 2011; Knox et al. 2017), with soft release generally outperforming hard release (Resende et al. 2021). However, the relationship between release type and success is not clear-cut and behavioural traits of the target species (along with other non-behavioural factors) may favour one strategy over another (Moseby et al. 2014; Morris et al. 2021). For example, species that habituate to release enclosures quickly may learn maladaptive anti-predator behaviours during holding, which would lower post-release survival, or species that become easily stressed while confined may suffer from elevated pre- or post-release mortality; in these cases, a hard/immediate release may be more appropriate. In some species, a mix of release types may maximize success across different demographic groups (e.g. Letty et al. 2000). Release timing can also have an effect on success. For translocated black-tailed prairie dogs, particularly juveniles, later season translocations were most successful, presumably because older juveniles had more time to learn anti-predator skills before being moved to a novel site (Shier and Owings 2006).

Conspecific attraction can reduce dispersal or foster settlement by signalling habitat quality. The presence of conspecifics or artificially planted cues (e.g. decoys, call playback, scents) can greatly increase settlement of translocated individuals. Hennessy et al. (2022) found that burrowing owls *Athene cunicularia* translocated to sites with conspecifics or artificial cues (vocalization playback and artificial droppings) were 20 times more likely to settle than those translocated without. Homing, where released individuals return to their source location, is another problematic type of movement. Translocating animals further away is an effective method for curtailing homing behaviour in some birds, mammals, and reptiles (e.g. Villaseñor et al. 2013; Hinderle et al. 2015; Hennessy et al. 2022).

12.4.3 Select release cohort

Research on optimizing release cohort selection is growing. While early work focused on founder group genetic diversity, size, and source (Griffith et al. 1989; Fischer and Lindenmayer 2000), more recently, research has shown that age and social behaviour of target species are important considerations when selecting a founder group. For example, reintroducing individuals of similar age and condition as natal dispersers may improve outcomes (Stamps and Swaisgood 2007). Marbled teals *Marmaronetta angustirostris* released just after fledging had higher survival than teals released immediately before the breeding season (Green et al. 2005). By contrast, modelling suggests that for long-lived birds with delayed reproductive maturity, releasing adults is more efficient than releasing juveniles for population creation because breeding happens faster on the landscape (e.g. Sarrazin and Legendre 2000).

Social relationships influence communication, territoriality, breeding, and survival. Accordingly, strategies to maintain social relationships and release intact social groups may influence post-release fitness (Goldenberg et al. 2019). For example, black-tailed prairie dogs (Shier and Owings 2006) and brown treecreepers *Climacteris picumnus* (Bennett et al. 2012) relocated within social groups were more likely to survive translocation. Similar effects occur even in solitary species, such as Stephens' kangaroo rats *Dipodomys stephensi*, where individuals relocated with familiar neighbours had higher survival and reproduction than those released without familiar neighbours (Shier and Swaisgood 2012). Nevertheless, for some species, maintaining social groups during relocation did not improve success (e.g. New Zealand saddlebacks *Philesturnus carunculatus*; Armstrong 1995), perhaps because release protocols can influence social cohesion of the founder group (Moseby et al. 2020). For instance, released bettongs *Bettongia lesueur* may have been unable to find social group members in large release areas (Moseby et al. 2018). Taken together, the results suggest that the effects of transferring animals in social groups are species-specific and possibly relate to the degree to which the species relies on social relationships for predator defence, food, or reproduction.

12.5 When and how should you manipulate animals or habitats at release sites?

Preparing the release cohort and release site are critical for reintroduction success. Despite best efforts, there are often still factors that may stunt or prevent colonization of a release site. The needs, as well as ecological role, of reintroduced populations may differ from extant populations due to small population size or behavioural preparedness. Coexistence and community assembly theory can explain how a species can be integrated—or why it may be excluded—from a community (e.g. HilleRisLambers et al. 2012). Manipulating the resident community to make it more favourable to the target species can be critical, especially when reintroducing small or subordinate species that will face competition.

12.5.1 Map and respond to community dynamics

Understanding community dynamics and behaviour can help direct habitat modifications. If species within a community exhibit different habitat preferences or spatial niche partitioning, habitat at release sites can be selected or managed to optimize the target species' preference or reduce suitability for heterospecifics to minimize competitive interactions (Chock et al. 2022). Meanwhile, knowledge of predator-prey interactions can help determine if manipulating the habitat would reduce predation risk (e.g. by adding cover piles, burrows, or shelters). The habitat can also be manipulated to make it less suitable for predators (e.g. by removing perches used by raptors or cover needed by ambush predators). Likewise, parasitic species may negatively affect outcomes and may require long-term management and/or changes to release strategy or location.

Interactions with predators, competitors, and parasites are complicated, but a first step is to document the native and invasive species present in potential release sites. Once known, systematically evaluating how the target and heterospecific species interact is required. We would expect the target species to have co-evolved with native heterospecific species and thus be able to coexist once established. However, re-establishing the target species into an intact heterospecific community may be challenging, especially if the target is a prey species, easily outcompeted, and/or the density of heterospecific species at the release site is high. Release candidates may need experience with heterospecifics before release to exhibit appropriate interactions. Other release strategies, such as reducing the number of predators (Moseby et al. 2016) or competitors prior to release (Clarke et al. 2002; Chock et al. 2022), or releasing the target species when densities of heterospecifics are low, can help (e.g. reducing the prevalence of noisy miners *Manorina melanocephala* increases habitat for endangered regent honeyeaters *Anthochaera phrygia*; Crates et al. 2023). If potential release sites have non-native heterospecifics, considering how the target species will interact with the non-native predators, competitors, parasites, or prey will help develop strategies to either remove or reduce the non-native species or facilitate community integration. For example, highly toxic cane toads invaded Australia and ingesting this species often proves fatal. Because they are widespread and unlikely to be controlled, selecting release sites for native predators without cane toads is unlikely, but this constraint justifies training the target species to avoid consuming cane toads following release, which has been successful on several occasions (Jolly et al. 2018; see Section 12.3.1).

12.5.2 Conduct predator manipulations

Predation is a leading cause of translocation failure (Blumstein et al. 2019). Excluding or removing predators through fencing, capture and translocation, or lethal removal can be effective in the short term, but is not always feasible or appropriate (DeCesare et al. 2009). Leveraging knowledge of predator behaviour can also provide opportunities to minimize predation. For example, aversion training can deter egg predation on seabird colonies by corvids using eggs treated with carbamate compounds and electrified eggs (Forys et al. 2020). In desert tortoise *Gopherus agassizii* habitat where ravens provisioning young prey on tortoise hatchlings, destroying raven *Corvus corax* nests reduces tortoise predation (Shields et al. 2019). However, removing eggs or destroying nests may prompt ravens to lay a second clutch. Oiling the eggs—a form of addling, which leaves the eggs intact while preventing them from hatching—can decrease predation pressure. Egg-oiling prolongs ravens' incubation, which reduces the potential for re-nesting and ultimately prevents successful fledging (Shields et al. 2019).

Finally, community dynamics may have shifted since the focal species was last present, which can prevent a translocated population from establishing even in previously occupied areas. Species introductions, range shifts, or local extirpations of predators, competitors, or prey species can all have unexpected consequences and cascading effects. For example, the establishment of feral pigs *Sus scrofa* on the California Channel Islands allowed golden eagles *Aquila chrysaetos* to colonize. Although the pigs did not directly compete with endemic island foxes *Urocyon littoralis*, the eagle populations, subsidized by pigs, were responsible for crashing the fox populations. Successful conservation breeding of island foxes prevented extinction, but reintroduction required removing and relocating golden eagles, eradicating the feral pigs that supported them, and restoring a bald eagle *Haliaeetus leucocephalus* population that helped prevent golden eagles returning. Complex and indirect interactions may be important to identify for small populations, whether they are in decline or in the early stages of translocation (DeCesare et al. 2009).

12.6 How do you measure the success of a translocation?

Measuring translocation outcomes is critical to assessing success and identifying opportunities for adaptive management. Post-release monitoring aims to assess whether individuals have remained at release sites, are surviving, and are reproducing. Understanding the life history and key behaviours of the target species helps define success and design effective post-release monitoring. Success criteria (e.g. population growth, survival, reproductive success) should be measured over biologically and ecologically relevant time frames and geographic scales, which may be determined using reproductive phenology and lifespan (IUCN/SSC 2013). For example, long-lived species with delayed reproductive maturity or longer inter-birth intervals require longer post-release monitoring than shorter-lived species that reproduce early and/or often. Effective and accurate monitoring depends on whether individuals can be detected after release. However, target species' behaviour, such as crypsis, can profoundly limit detectability (Nafus et al. 2015; Hammond et al. 2021). Appropriate wild-type behaviours should lead to higher survival and reproductive success. As such, monitoring plans should consider how to incorporate species-specific behaviours into monitoring targets and

aim to minimize biases stemming from behaviour (Hammond et al. 2021).

12.7 Conclusion and future directions: what behavioural innovations are on the horizon?

Many breakthroughs in conservation breeding and translocations have stemmed from innovation in conservation behaviour. However, successful applications have often proven to be species- or context-specific. There will always be unknowns about how theory applies to different systems, species, or situations, but identifying where we can reduce the uncertainty surrounding the likelihood of success is key. Documenting the variation, mechanisms, and consequences of behaviour across species can help because we still lack a fundamental understanding of behaviour in many endangered species, which are not otherwise studied until their populations crash. Having a proactive checklist of critical behavioural knowledge for currently non-threatened species could help conservation breeding or translocations if they were to need such tools in future. Otherwise, how do we know what wild-type behaviours to approximate if we do not know what wild animals should do?

In addition to what we covered in this chapter, there are emerging topics that have potential to increase the success of conservation breeding and translocations. For example, cross fostering from human care to wild clutches or litters could mitigate many effects of ex-situ environments on development, but best practice has yet to be developed. Also, better understanding the nuances of habitat cues for matching species' needs with appropriate sites would be helpful, since we currently have coarse metrics of habitat suitability. We also lack knowledge on disruptions in social learning within and between species. The larger knock-on effects of dealing with small, isolated populations that lack normal social structure or community dynamics are critically understudied. Meanwhile, most animal species are managed singly, with recovery plans and actions tailored to their needs that can compromise the success of other, perhaps less threatened species. However, community dynamics are complex and an approach that focuses on co-management of listed species may have more success in managing species interactions and restoring ecosystem function.

The field of conservation behaviour has produced many useful tools, but there is still incredible insight left to be translated into the concrete actions needed for species recovery. By thinking about where within the breeding and release process these insights can be used, we can best tailor them to species in need.

Acknowledgements

We thank Leah Jacobs for help with Figure 12.4.

References

Armstrong, D.P. (1995). Effects of familiarity on the outcome of translocations, II. A test using New Zealand robins. *Biological Conservation*, 71, 281–288.

Aubret, F., and Shine, R. (2008). Early experience influences both habitat choice and locomotor performance in tiger snakes. *The American Naturalist*, 171, 524–531.

Awade, M., Candia-Gallardo, C., Cornelius, C., and Metzger, J.P. (2017). High emigration propensity and low mortality on transfer drives female-biased dispersal of *Pyriglena leucoptera* in fragmented landscapes. *PLOS ONE*, 12, e0170493.

Batson, W.G., Gordon, I.J., Fletcher, D.B., and Manning, A.D. (2015). Translocation tactics: a framework to support the IUCN Guidelines for wildlife translocations and improve the quality of applied methods. *Journal of Applied Ecology*, 52, 1598–1607.

Bennett, V.A., Doerr, V.A.J., Doerr, E.D., et al. (2012). Habitat selection and post-release movement of reintroduced brown treecreeper individuals in restored temperate woodland. *PLOS ONE*, 7, e50612.

Berger-Tal, O., Blumstein, D.T., and Swaisgood, R.R. (2020). Conservation translocations: a review of common difficulties and promising directions. *Animal Conservation*, 23, 121–131.

Blais, B.R., Wells, S.A., Poynter, B.M., et al. (2022). Adaptive management in a conservation breeding: mimicking habitat complexities facilitates reproductive success in narrow-headed gartersnakes (*Thamnophis rufipunctatus*). *Zoo Biology*, 41, zoo.21682.

Blumstein, D., Letnic, M., and Moseby, K. (2019). In situ predator conditioning of naive prey prior to

reintroduction. *Philosophical Transactions of the Royal Society B: Biological Sciences*, 374, 20180058.

Bowkett, A.E. (2009). Recent captive-breeding proposals and the return of the ark concept to global species conservation. *Conservation Biology*, 23, 773–776.

Bowler, D.E., and Benton, T.G. (2005). Causes and consequences of animal dispersal strategies: relating individual behaviour to spatial dynamics. *Biological Reviews*, 80, 205–225.

Bright, P.W., and Morris, P.A. (1994). Animal translocation for conservation: performance of dormice in relation to release methods, origin and season. *Journal of Applied Ecology*, 31, 699.

Chock, R.Y., Shier, D.M., and Grether, G.F. (2022). Niche partitioning in an assemblage of granivorous rodents, and the challenge of community-level conservation. *Oecologia*, 198, 553–565.

Clark, F.E., Greggor, A.L., Montgomery, S.H., and Plotnik, J.M. (2023). The endangered brain: actively preserving ex-situ animal behaviour and cognition will benefit in-situ conservation. *Royal Society Open Science*, 10230707.

Clarke, R.H., Boulton, R.L., and Clarke, M.F. (2002). Translocation of the socially complex black-eared miner *Manorina melanotis*: a trial using hard and soft release techniques. *Pacific Conservation Biology*, 8, 223.

Conde, D.A., Colchero, F., Gusset, M., et al. (2013). Zoos through the lens of the IUCN Red List: a global metapopulation approach to support conservation breeding programs. *PLOS ONE*, 8, e80311.

Crates, R., McDonald, P.G., Melton, C.B., et al. (2023). Towards effective management of an overabundant native bird: the noisy miner. *Conservation Science and Practice*, 5, e12875.

DeCesare, N.J., Hebblewhite, M., Robinson, H.S., and Musiani, M. (2009). Endangered, apparently: the role of apparent competition in endangered species conservation: apparent competition and endangered species. *Animal Conservation*, 13, 353–362.

Dittman, A.H., Pearsons, T.N., May, D., et al. (2015). Imprinting of hatchery-reared salmon to targeted spawning locations: a new embryonic imprinting paradigm for hatchery programs. *Fisheries*, 40, 114–123.

Dixon-MacCallum, G.P., Rich, J.L., Lloyd, N., et al. (2021). Loss of predator discrimination by critically endangered Vancouver Island marmots within five generations of breeding for release. *Frontiers in Conservation Science*, 2, 1–11.

Edwards, M.C., Ford, C., Hoy, J.M., et al. (2021). How to train your wildlife: a review of predator avoidance training. *Applied Animal Behaviour Science*, 234, 105170.

Fischer, J., and Lindenmayer, D.B. (2000). An assessment of the published results of animal relocations. *Biological Conservation*, 96, 1–11.

Flanagan, A.M., Rutz, C., Farabaugh, S., et al. (2020). Inter-aviary distance and visual access influence conservation breeding outcomes in a territorial, endangered bird. *Biological Conservation*, 242, 108429.

Forys, E.A., Campo, J.A., Silva, E., and Douglass, N.J. (2020). A comparison of 2 methods to deter fish crows from depredating seabird eggs. *Wildlife Society Bulletin*, 44, 670–676.

Goldenberg, S.Z., Owen, M.A., Brown, J.L., et al. (2019). Increasing conservation translocation success by building social functionality in released populations. *Global Ecology and Conservation*, 18, e00604.

Green, A.J., Fuentes, C., Figuerola, J., et al. (2005). Survival of marbled teal (*Marmaronetta angustirostris*) released back into the wild. *Biological Conservation*, 121, 595–601.

Greggor, A.L., Clayton, N.S., Phalan, B., and Thornton, A. (2014). Comparative cognition for conservationists. *Trends in Ecology & Evolution*, 29, 489–495.

Greggor, A.L., and Goldenberg, S.Z. (2023). Manipulating animal social interactions to enhance translocation impact. *Trends in Ecology & Evolution*, 38, 316–319.

Greggor, A.L., Thornton, A., and Clayton, N.S. (2017). Harnessing learning biases is essential for applying social learning in conservation. *Behavioral Ecology and Sociobiology*, 71, 1–12.

Griffith, B., Scott, J.M., Carpenter, J.W. and Reed, C. (1989). Translocation as a Species Conservation Tool: Status and Strategy. *Science*, 245, 477–480.

Hammond, T.T., Curtis, M.J., Jacobs, L.E., et al. (2021). Behavior and detection method influence detection probability of a translocated, endangered amphibian. *Animal Conservation*, 24, 401–411.

Hammond, T.T., Jacobs, L.E., Curtis, M.J., et al. (2023). Early life experience with predators impacts development, behavior, and post-translocation outcomes in an endangered amphibian. *Animal Conservation*, 28 May 2023, acv.12880.

Hammond, T.T., Swaisgood, R.R., Jacobs, L.E., et al. (2022). Age-dependent effects of developmental experience on morphology, performance, dispersal and survival in a translocated, endangered species. *Journal of Applied Ecology*, 59, 1745–1755.

Hartnett, C.M., Parrott, M.L., Mulder, R.A., et al. (2018). Opportunity for female mate choice improves reproductive outcomes in the conservation breeding program of the eastern barred bandicoot (*Perameles gunnii*). *Applied Animal Behaviour Science*, 199, 67–74.

Hennessy, S.M., Wisinski, C.L., Ronan, N.A., et al. (2022). Release strategies and ecological factors influence mitigation translocation outcomes for burrowing owls: a comparative evaluation. *Animal Conservation*, 25, 614–626.

HilleRisLambers, J., Adler, P.B., Harpole, W.S., et al. (2012). Rethinking community assembly through the lens of coexistence theory. *Annual Review of Ecology, Evolution, and Systematics*, 43, 227–248.

Hinderle, D., Lewison, R.L., Walde, A.D., et al. (2015). The effects of homing and movement behaviors on translocation: desert tortoises in the western Mojave Desert: tortoise homing and movement after translocation. *The Journal of Wildlife Management*, 79, 137–147.

International Union for Conservation of Nature. (2022). IUCN Red List version 2022–2: Table 1a. [online]. Available from: https://www.iucnredlist.org/resources/summary-statistics [accessed 29 May 2022].

International Union for Conservation of Nature and Natural Resources and Species Survival Commission. (2013). Guidelines for reintroductions and other conservation translocations. [online]. Available from: http://data.iucn.org/dbtw-wpd/edocs/2013-009.pdf [accessed 12 August 2022].

Jolly, C.J., Kelly, E., Gillespie, G.R., et al. (2018). Out of the frying pan: reintroduction of toad-smart northern quolls to southern Kakadu National Park. *Austral Ecology*, 43, 139–149.

Kemp, L., Norbury, G., Groenewegen, R., and Comer, S. (2015). The roles of trials and experiments in fauna reintroduction programmes. In: D. Moro et al. (eds), *Advances in Reintroduction Biology of Australian and New Zealand Fauna*, p.18. CSIRO Publishing, Clayton South, Australia.

Kleiman, D.G. (1989). Reintroduction of captive mammals for conservation. *BioScience*, 39, 152–161.

Knox, C.D., Jarvie, S., Easton, L.J., and Monks, J.M. (2017). Soft-release, but not cool winter temperatures, reduces post-translocation dispersal of jewelled geckos. *Journal of Herpetology*, 51, 490–496.

Letty, J., Marchandeau, S., Clobert, J., and Aubineau, J. (2000). Improving translocation success: an experimental study of anti-stress treatment and release method for wild rabbits. *Animal Conservation*, 3, 211–219.

Martin-Wintle, M.S., Kersey, D.C., Wintle, N.J.P., et al. (2019). Comprehensive breeding techniques for the giant panda. In: P. Comizzoli, J.L. Brown, and W.V. Holt (eds), *Reproductive Sciences in Animal Conservation. Advances in Experimental Medicine and Biology*, pp. 275–308. Springer International Publishing, Cham.

Mason, G., Burn, C.C., Dallaire, J.A., et al. (2013). Plastic animals in cages: behavioural flexibility and responses to captivity. *Animal Behaviour*, 85, 1113–1126.

Mathews, F., Orros, M., McLaren, G., et al. (2005). Keeping fit on the ark: assessing the suitability of captive-bred animals for release. *Biological Conservation*, 121, 569–577.

McCallum, H., Timmers, P., and Hoyle, S. (1995). Modelling the impact of predation on reintroductions of bridled nailtail wallabies. *Wildlife Research*, 22, 163–171.

McDougall, P.T., Réale, D., Sol, D., and Reader, S.M. (2006). Wildlife conservation and animal temperament: causes and consequences of evolutionary change for captive, reintroduced, and wild populations. *Animal Conservation*, 9, 39–48.

Miller, K.A., Bell, T.P., and Germano, J.M. (2014). Understanding publication bias in reintroduction biology by assessing translocations of New Zealand's herpetofauna: publication bias in reintroduction biology. *Conservation Biology*, 28, 1045–1056.

Mitchell, A.M., Wellicome, T.I., Brodie, D., and Cheng, K.M. (2011). Captive-reared burrowing owls show higher site-affinity, survival, and reproductive performance when reintroduced using a soft-release. *Biological Conservation*, 144, 1382–1391.

Morris, V., Pitcher, B.J., and Chariton, A. (2021). A cause for alarm: increasing translocation success of captive individuals through alarm communication. *Frontiers in Conservation Science*, 2, 1–9.

Moseby, K., Lollback, G., and Lynch, C. (2018). Too much of a good-thing, successful reintroduction leads to overpopulation in a threatened mammal. *Biological Conservation*, 219, 78–88.

Moseby, K.E., Blumstein, D.T., and Letnic, M. (2016). Harnessing natural selection to tackle the problem of prey naïveté. *Evolutionary Applications*, 9, 334–343.

Moseby, K.E., Blumstein, D.T., Letnic, M., and West, R. (2020). Choice or opportunity: are post-release social groupings influenced by familiarity or reintroduction protocols? *Oryx*, 54, 215–221.

Moseby, K.E., Hill, B.M., and Lavery, T.H. (2014). Tailoring release protocols to individual species and sites: one size does not fit all. *PLOS ONE*, 9, e99753.

Nafus, M.G., Germano, J.M., Perry, J.A., et al. (2015). Hiding in plain sight: a study on camouflage and habitat selection in a slow-moving desert herbivore. *Behavioral Ecology*, 26, 1389–1394.

Osborne, P.E., and Seddon, P.J. (2012). Selecting suitable habitats for reintroductions: variation, change and the role of species distribution modelling. In: J.G. Ewen et al. (eds), *Reintroduction Biology*, pp. 73–104. 1st ed. Wiley-Blackwell, Chichester.

Pottie, S., Bello, R., and Donati, G. (2021). Factors influencing establishment success in reintroduced black-faced spider monkeys *Ateles chamek*. *Primates*, 62, 1031–1036.

Resende, P.S., Viana-Junior, A.B., Young, R.J., and Azevedo, C.S. (2021). What is better for animal conservation translocation programmes: soft- or hard-release?

A phylogenetic meta-analytical approach. *Journal of Applied Ecology*, 58, 1122–1132.

Richards-Zawacki, C.L. (2010). Thermoregulatory behaviour affects prevalence of chytrid fungal infection in a wild population of Panamanian golden frogs. *Proceedings of the Royal Society B: Biological Sciences*, 277, 519–528.

Rowell, T.A.A.D., Magrath, M.J.L., and Magrath, R.D. (2020). Predator-awareness training in terrestrial vertebrates: progress, problems and possibilities. *Biological Conservation*, 252, 108740.

Sarrazin, F., and Legendre, S. (2000). Demographic approach to releasing adults versus young in reintroductions. *Conservation Biology*, 14, 488–500.

Shields, T., Currylow, A., Hanley, B., et al. (2019). Novel management tools for subsidized avian predators and a case study in the conservation of a threatened species. *Ecosphere*, 10, e02895.

Shier, D.M. (2016). Manipulating animal behavior to ensure reintroduction success. In: O. Berger-Tal and D. Saltz (eds), *Applying Behavioral Ecology to Wildlife Conservation and Management*, pp. 275–304. Cambridge University Press, Cambridge.

Shier, D., and Owings, D.H. (2006). Effects of predator training on behavior and post-release survival of captive prairie dogs (*Cynomys ludovicianus*). *Biological Conservation*, 132, 126–135.

Shier, D.M., and Owings, D.H. (2007). Effects of social learning on predator training and postrelease survival in juvenile black-tailed prairie dogs, *Cynomys ludovicianus*. *Animal Behaviour*, 73, 567–577.

Shier, D.M., and Swaisgood, R.R. (2012). Fitness costs of neighborhood disruption in translocations of a solitary mammal: social effects in translocation. *Conservation Biology*, 26, 116–123.

Slagsvold, T., and Wiebe, K.L. (2011). Social learning in birds and its role in shaping a foraging niche. *Philosophical Transactions of the Royal Society B: Biological Sciences*, 366, 969–977.

Stadtmann, S., and Seddon, P.J. (2020). Release site selection: reintroductions and the habitat concept. *Oryx*, 54, 687–695.

Stamps, J.A., and Swaisgood, R.R. (2007). Someplace like home: experience, habitat selection and conservation biology. *Applied Animal Behaviour Science*, 102, 392–409.

Tear, T.H., Scott, J.M., Hayward, P.H., et al. (1993). Status and prospects for success of the Endangered Species Act: a look at recovery plans. *Science*, 262, 976–977.

Tripovich, J.S., Popovic, G., Elphinstone, A., et al. (2021). Born to be wild: evaluating the zoo-based regent honeyeater breed for release programme to optimise individual success and conservation outcomes in the wild. *Frontiers in Conservation Science*, 2, 1–16.

Villaseñor, N.R., Escobar, M.A.H., and Estades, C.F. (2013). There is no place like home: high homing rate and increased mortality after translocation of a small mammal. *European Journal of Wildlife Research*, 59, 749–760.

White, T.H., de Melo Barros, Y., Develey, P.F., et al. (2015). Improving reintroduction planning and implementation through quantitative SWOT analysis. *Journal for Nature Conservation*, 28, 149–159.

Williams, C.L., Williams, C.E., King S.N.D., and Shier, D.M. (2023). Environmental change drives multi-generational shifts in the gut microbiome that mirror changing animal fitness. *BioRxiv*.

Yao, R., Xu, L., Hu, T., et al. (2019). The 'wildness' of the giant panda gut microbiome and its relevance to effective translocation. *Global Ecology and Conservation*, 18, e00644.

Yoerg, S.I. (1999). Solitary is not asocial: effects of social contact in kangaroo rats (Heteromyidae: *Dipodomys heermanni*). *Ethology*, 105, 317–333.

CHAPTER 13

Mitigating human-wildlife impacts

Jeremy J. Cusack and Rocío A. Pozo

Overview

Rapid anthropogenic changes to natural landscapes are increasing the frequency and severity of human-wildlife impacts (HWIs), which occur when the behaviours of wildlife and humans result in recurring costs to both. Here, we decompose the HWI sequence into three key components (spatiotemporal overlap, direct costs, and learning), which we subsequently use to derive a typology of associated human and animal behaviours. We then map existing mitigation strategies aimed at reducing HWIs onto each component, illustrating their application using a range of examples and case studies. We conclude that fostering long-term human-wildlife coexistence in a rapidly changing world will be dependent upon our ability to identify, understand, and manage the behavioural processes underlying instances of HWI.

13.1 Human-wildlife impacts in a changing world

Whether as a result of the rapid expansion of anthropogenic activities into previously undisturbed environments, or the recovery and recolonization of once threatened species, human contact with wildlife is increasing (Williams et al. 2020; Cimatti et al. 2021). Although many of the resulting human-wildlife interactions are considered positive, such as insect species pollinating the crops we grow (Basset and Lamarre 2019), birds and mammals visiting our garden feeders (Mumaw and Mata 2022), and scavengers consuming the waste we generate (O'Bryan et al. 2018), many others are regarded

as negative and detrimental to both humans and wildlife.

So-called 'human-wildlife conflicts' arise from the real or perceived impacts that wild animals cause to humans or their activities, and which, in turn, prompt pre-emptive or retaliatory actions against wildlife by humans (Nyhus 2016). Repetition of these impacts over time can result in considerable costs to both wildlife and humans, often leading to severe clashes between the interests of conservation and those of affected livelihoods (Redpath et al. 2013). Although the term 'human-wildlife conflict' continues to be widely used in the literature, we take the view expressed by many authors that its inherent implication that wildlife is a conscious human antagonist is problematic (Peterson et al. 2010). For this reason, we hereafter use the term 'human-wildlife impacts' (HWIs), a label that does not necessarily imply that costs are inflicted with purpose.

Both the frequency and severity of HWIs are intricately linked to the ever-expanding human footprint on both natural ecosystems and the climate (Abrahms et al. 2023). In this rapidly changing world, situations of HWI have become all too common, involving a diverse range of species and human activities (Nyhus 2016). While most are centred around agricultural activities, and in particular the loss of crops and livestock to wild herbivores (Gordon 2009) and carnivores (Van Eeden et al. 2018; Wilkinson et al. 2020), respectively, they are also increasingly being associated with urban and peri-urban areas (Schell et al. 2021), recreational activities (Perona et al. 2019), transport networks (Blackwell et al. 2016), and energy production

Jeremy J. Cusack and Rocío A. Pozo, *Mitigating human-wildlife impacts*. In: *Behavioural Responses to a Changing World*. Edited by: Bob B. M. Wong and Ulrika Candolin, Oxford University Press. © Oxford University Press (2024). DOI: 10.1093/oso/9780192858979.003.0013

(Thaker et al. 2018). In effect, reducing and managing HWIs for the benefit of both wildlife and humans has become one of conservation's greatest challenges.

13.2 A typology of HWI-related behaviours

HWI situations are varied and complex (Dickman 2010), yet all share a number of key components, of which three constitute the focus of this chapter. First, the occurrence of HWIs is dependent on a degree of overlap in space and/or time between wildlife and human activities (Lischka et al. 2018), which influences the probability of contact or interaction between the two. Second, HWIs involve interactions that result in direct costs to both wildlife *and* humans. This requirement excludes from consideration situations in which only wildlife is impacted by humans (e.g. through hunting or habitat destruction; see Chapters 7 and 8, respectively), as well as those in which impacted humans do not seek retaliatory action against wildlife. On the contrary, the latter scenario may well be indicative of high human tolerance and coexistence (Pooley et al. 2021). Third and last, costly interactions must be recurrent; that is, they must be repeated over time or replicated across space, thus perpetuating a state of apparent 'conflict'.

Driving this sequence of HWI components are complex and dynamic behavioural processes occurring at different spatiotemporal scales (Figure 13.1). Making sense of these behavioural processes is not only fundamental to our understanding of why and how HWIs occur, but also to our ability to devise effective ways of mitigating them (Greggor et al. 2016; see also Chapters 11 and 12). Importantly, the present chapter considers the behaviours of both wildlife and humans, thus acknowledging the important role each party plays in shaping the HWI sequence.

13.2.1 Predisposing behaviours

Within the context of HWIs, predisposing behaviours can be said to increase the spatiotemporal overlap between human and wildlife activities, thus increasing the potential for interactions—including both positive and negative ones—to occur (Lischka et al. 2018). As such, they increase the probability for wildlife and humans to come into contact, but crucially, do not directly cause impacts.

In the case of wildlife, predisposing behaviours include behaviours relating to foraging and movement across human-modified landscapes, which are by far the most well-documented drivers of increased contact with humans (Wilson et al. 2020;

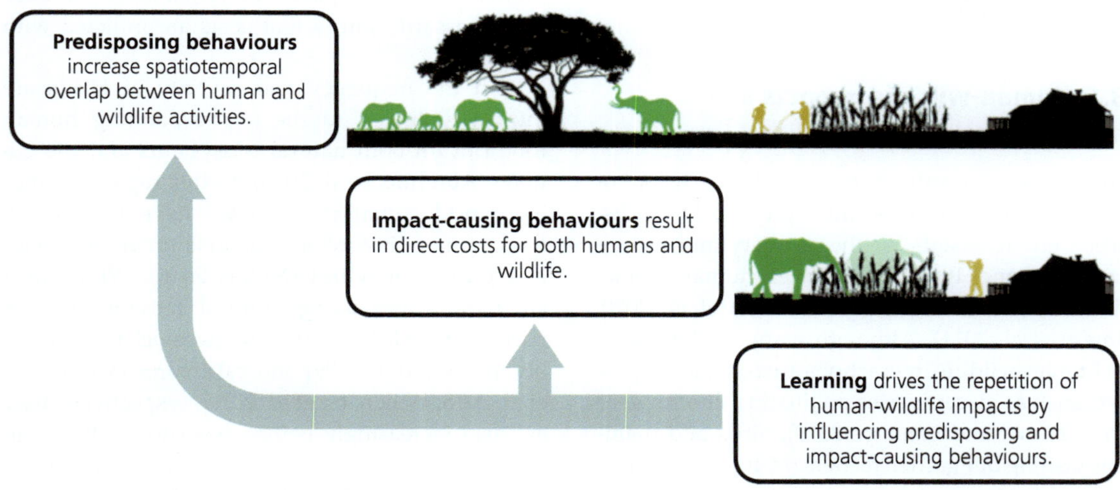

Figure 13.1 Behavioural processes driving the human-wildlife impact sequence.

Fehlmann et al. 2021). For instance, the large-scale movement behaviour of many grazing bird species, such as geese, cranes, and swans, interacts with the distribution and management of agricultural activities, increasing the risk of HWIs at spatially predictable locations (Nilsson et al. 2019).

Other relevant, predisposing behaviours associated with wildlife include territoriality, reproductive activities such as nesting, and social behaviour involving large gatherings of individuals. For example, Herr et al. (2009) found that stone marten *Martes foina* damage to parked car engines in Luxembourg was associated with scent-marking and patrolling tendencies, leading the authors to suggest that marten-car contact is driven by territorial behaviour. For many bird species, the search for suitable nesting sites is increasingly influenced by the availability of human-made structures, such as pylons and utility poles (Mainwaring 2015). The resulting potential for interference with power lines, which can lead to fires, power outages, and increased bird mortality from collision and electrocution, has become an important cause of HWIs worldwide (e.g. white storks *Ciconia ciconia*, Moreira et al. 2018).

On the human side, it is also possible to identify a range of predisposing behaviours that facilitate contact with wildlife without directly causing impact. The most commonly reported behaviours relate to the management of food resources that are, or can also be, used by wildlife (Nyhus 2016). These include decisions and actions pertaining to where, when, and how to plant crops, keep livestock, use water resources, and dispose of garbage, amongst others. For example, in the urban area of Aspen, Colorado, USA, Lewis et al. (2015) found that 57% of the bear-resistant garbage containers were not properly secured, allowing black bears *Ursus americanus* to access this anthropogenic food source, thus increasing the risk of more serious human-bear impacts.

Human behaviours associated with recreational activities also play an important role in increasing the likelihood of negative encounters with wildlife. Penteriani et al. (2016) highlighted a strong positive correlation between the number of recorded 'attacks' on humans by large carnivores and the number of visitors to protected areas in the USA.

Importantly, they highlight a number of common human behaviours influencing the risk of attack, including leaving children unattended, walking with an unleashed dog, and engaging in outdoor activities during nighttime hours. In the marine environment, risky human behaviours related to coastal recreational activities (e.g. bathing and surfing) are often implicated in the occurrence of shark bite incidents (Chapman and McPhee 2016).

13.2.2 Impact-causing behaviours

The second behavioural component relates to specific actions carried out by wildlife or humans that impose a direct cost on the other party. These impact-causing behaviours—sometimes referred to as 'problem' behaviours—have traditionally constituted the focus of HWI-related research. They can generally be reduced to specific actions, such as jumping into an enclosure to kill livestock, crossing a road in front of a moving vehicle, entering and feeding on a field of crops, or poisoning a water source, the final example exemplifying a retaliatory action against wildlife by humans.

In contrast to predisposing behaviours, impact-causing behaviours typically occur at finer spatial and temporal scales, making it possible to pinpoint specific contextual factors influencing their incidence. For example, Siljander et al. (2020) highlighted early mornings and late afternoons as being the most problematic times of the day regarding primate crop-foraging around Taita Hills, southeast Kenya. Similarly, coyote *Canis latrans* and black bear attacks on humans in urban areas of the USA and Canada are associated with specific environmental factors, such as building density and artificial lighting (Bombieri et al. 2018).

The sequence and complexity of impact-causing behaviours is likely to depend on a range of demographic (e.g. age and sex of individuals involved), biophysical (e.g. the structure of the landscape and characteristics of the human infrastructure), and anthropogenic factors (e.g. cultural norms), highlighting the importance of considering both social and ecological contexts (Wilkinson et al. 2020). Recently, Schultz et al. (2021) showed that female brown skuas *Catharacta antarctica lonnbergi* in the Chatham Island Archipelago, New Zealand, had a higher propensity to scavenge sheep from farmlands than males and, as a consequence, were more likely to be culled to reduce HWIs.

Most definitions of HWI emphasize the coupling of wildlife-induced threats or damage with human retaliation (Nyhus 2016), reflecting a sequence of impact-causing behaviours that can be seen as indicative of 'conflict'. For example, in a recent study, Ontiri et al. (2019) found that livestock depredation was the most important driver behind the killing of lions *Panthera leo* by Maasai pastoralists, further showing that such retaliatory behaviour was not indiscriminate (i.e. it is targeted at lions specifically, and not other wild carnivores). In some instances, retaliatory behaviour can give way to pre-emptive actions by humans against wildlife, in an attempt to avoid costly impacts. Rather than considering this as a form of mitigation, we argue that both pre-emptive and retaliatory actions aimed at harming wildlife should be considered as impact-causing behaviours.

13.2.3 Learning

HWI situations are characterized by the repetition of impact-causing behaviours, thereby leading to recurring costs for humans and wildlife. The tendency for such behaviours to be repeated over time, or replicated across space, has been linked to various forms of learning, broadly defined here as a change in knowledge or behaviour caused by experience.

Chief amongst these is associative learning through reinforcement, whereby a wild animal repeats a behaviour that results in a reward. In many cases of HWI, anthropogenic food sources such as crops, livestock, or garbage offer nutritious and profitable rewards to wildlife, often resulting in significant fitness benefits to those individuals adopting problem behaviours. For example, crop- and urban-foraging chacma baboons *Papio ursinus* invest less time in foraging, have improved body condition, and exhibit higher reproductive success than their non-raiding counterparts (Fehlmann et al. 2017). Learning may also lead to intricate behavioural adjustments aimed at optimizing access to anthropogenic resources

at specific locations and times of the day or year (i.e. time-place learning). For instance, Valeix et al. (2012) found that lions in Botswana's Magkadikgadi ecosystem switched to hunting more abundant and readily available livestock in periods of wild herbivore prey scarcity.

Another form of learning that has been shown to influence the recurrence of impact-causing behaviours is habituation, whereby an animal's response to a given stimulus weakens following prolonged or repeated exposure to it (Blumstein 2016). Of particular relevance to HWI situations is the loss of fear that wild animals may experience towards humans, which may promote or encourage impact-causing behaviours. Recently, Cui et al. (2021) reported that habituation to humans brought about by increased tourism had led to an escalation of rhesus macaque *Macaca mulatta* raids on crop farms in China's Nanwan peninsula. Indeed, re-establishing the fear of humans as a mitigation strategy to avoid repetition of impact-causing behaviour offers a promising, albeit challenging, route to improved coexistence (see the section on *Aversive conditioning* below; Miller and Schmitz 2019).

Repetition of impact-causing behaviours in time and space may also come about through social learning, that is, learning from the actions of others or as a by-product of others' behaviour (Donaldson et al. 2012). Indeed, a growing number of studies have implicated social learning in aiding the spread of problem behaviours (Greggor et al. 2016). Examples include the depredation of fishing catches by sperm whales *Physeter macrocephalus* (Schakner et al. 2014), crop foraging behaviours by African elephants *Loxodonta africana* (Chiyo et al. 2012), and the spread of reliance on anthropogenic foods in dolphins (Donaldson et al. 2012). Morehouse et al. (2016) showed that offspring of problem grizzly bear *Ursus arctos horribilis* mothers were significantly more likely to be involved in conflict behaviours than offspring from non-problem mothers. Understanding the role of horizontal (i.e. learning from peers) and vertical (i.e. learning from parents) forms of social learning in the recurrence of impact-causing behaviours is not only vital to predicting the spread of such behaviours in a given population, but also crucial to ensuring the long-term effectiveness of mitigation strategies (Barrett et al. 2019).

13.3 Understanding behaviour to effectively mitigate HWIs

This chapter has so far presented a typology of HWI-related behaviours based on whether they influence the spatiotemporal proximity between humans and wildlife, lead to direct impacts, or involve the repetition of impacts over time and across space. Aside from its conceptual value, distinguishing between these different types of behaviour also serves a practical purpose, in that it can help identify relevant behaviours, as well as guide the implementation of effective mitigation strategies that promote coexistence in a constantly changing world. For example, the following three questions could be used to identify the human and animal behaviours involved in a given HWI situation:

- What are the behaviours that increase the likelihood of interaction between wildlife and humans (or their activities)?
- What are the behaviours that cause impacts on wildlife and humans?
- Is there evidence for behaviour repetition through learning?

Whilst it may not always be possible to identify— let alone understand—all of the behaviours involved in a given HWI scenario, the answers to the above questions may nevertheless reveal key behaviours in the HWI sequence at which appropriate mitigation strategies can be targeted. Mitigation strategies may seek to reduce contact between wildlife and human activities by targeting predisposing behaviours (i.e. preventative strategies), avert impact-causing behaviours (i.e. protective strategies), or generate a long-lasting change in behaviour (i.e. learning-based strategies) (Figure 13.2). In the following sections, we illustrate the application of these different approaches to a range of HWI settings, focusing first on methods that specifically target wildlife behaviour.

13.3.1 Mitigation strategies targeted at wildlife behaviour

Preventative strategies

An important subset of mitigation strategies seeks to minimize the likelihood of wildlife coming into contact with humans (or their activities), thereby preventing costly impacts from occurring. These methods generally target predisposing behaviours, aiming to modify the actions of wildlife at broad spatial scales. As such, they tend to require landscape-level planning and broad stakeholder engagement, rendering their implementation more challenging.

One of the most common forms of preventative strategy is spatial zoning, whereby different land management regimes are given priority in different areas, thus collectively minimizing overlap between wildlife and human activities (Linnell et al. 2005). Depending on the context, management regimes may prioritize the conservation of species and their habitats (e.g. protected areas or wildlife refuges), seek to control the abundance or space use of wildlife populations (e.g. hunting blocks or buffer zones), or focus on promoting human activities (e.g. agricultural or residential areas), amongst other objectives. The Carnivore Management Zones (CMZs) implemented in Norway provide a good example of the use of large-scale spatial zoning to mitigate HWIs (Strand et al. 2019). Within CMZs, carnivore abundance (e.g. brown bear *Ursus arctos*, lynx *Lynx lynx*, wolverine *Gulo gulo*, and grey wolf *Canis lupus*) is managed through both protection and hunting to ensure populations remain close to regional and national targets (Cusack et al. 2022). Areas located outside of CMZs are prioritized for livestock, which includes the implementation of mitigation strategies at finer spatial scales (see the section on *Protective strategies* below).

A different form of spatial zoning involves creating or safeguarding habitat corridors that enable wildlife to move across human-modified landscapes, thus limiting the risk of negative contact with human activities. A related approach involves evaluating the spatial overlap between human activities and wildlife distribution or use of natural

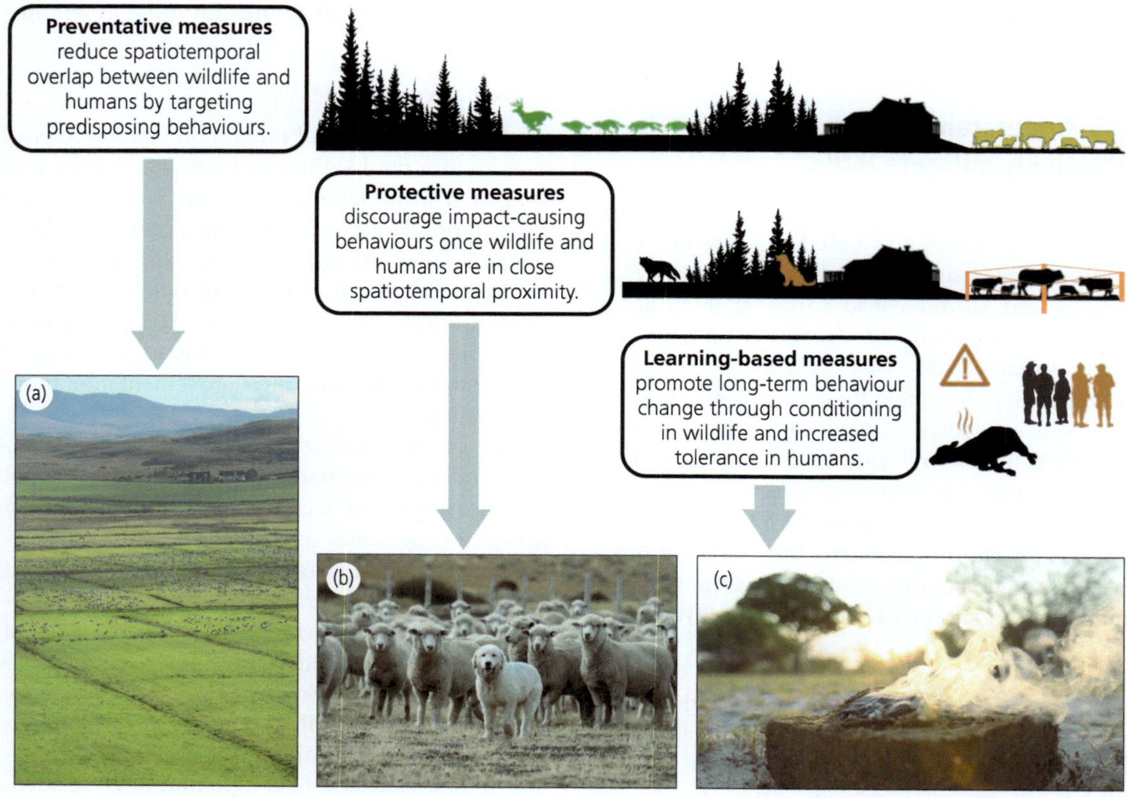

Figure 13.2 Strategies used to mitigate human-wildlife impacts, classified according to the type of behaviour targeted. Images represent examples of each type of strategy: (a) sacrificial pasture fields for Greenland barnacle geese *Branta leucopsis* on the island of Islay, Scotland; (b) livestock-guarding dogs used to protect sheep from pumas *Puma concolor* in Estancia Cerro Guido, Chilean Patagonia; and (c) a burning chilli-briquette used to deter African elephants from crop fields in Botswana's eastern Okavango Delta Panhandle.

Photos: (a) Jeremy Cusack; (b) Nicolás Lagos; and (c) Jeremy Cusack

resources. This is usually done by determining the spatial factors influencing wildlife occurrence or foraging, and subsequently mapping overlap with human activities, thereby highlighting potential areas of risk or competition (Kuiper et al. 2022). For instance, Redfern et al. (2013) combined whale-habitat models and ship-traffic patterns to predict the risk of costly strikes for humpback *Megaptera novaeangliae*, blue *Balaenoptera musculus*, and fin *Balaenoptera physalus* whales off the coast of Southern California.

The success of spatial zoning is dependent on a detailed understanding of wildlife behaviour and its drivers, including migratory habits, home range size, resource distribution, social dynamics, and fine-scale responses to different management regimes (e.g. fear). In the case of Norway, carnivores migrating outside of CMZs create a spillover effect resulting in substantial loss of livestock in neighbouring areas (Strand et al. 2019), a pattern that reflects what commonly occurs on the outskirts of protected areas that are too small to contain the movements of large, highly mobile species. Addressing these shortcomings requires the collection and analysis of data relating to wildlife movement behaviour, which can then be used to inform zoning strategies that incorporate both human and wildlife needs (see Box 13.1).

Box 13.1 A behavioural focus on HWIs: crop foraging elephants in northern Botswana

Botswana's eastern Okavango Delta Panhandle (ODP) spans an area of approximately 8000 km² and is home to an estimated 16,000 people living in and around 14 villages (Redmore et al. 2020). Subsistence agriculture in the form of crop production and livestock herding is the main livelihood in the area. Crop fields are spread out along the fertile banks of the Okavango River, reaching up to 14 km inland. The planting of crops is dependent on rainfall and mainly occurs between November and January, with harvest taking place between the months of April and June. Throughout these periods, crop damage and foraging are the most prevalent causes of negative interactions between humans and the estimated 18,000 African elephants (*Loxodonta africana*) that also inhabit the ODP (Figure 13.3; Songhurst et al. 2016; Pozo et al. 2017).

Figure 13.3 (a) Spatiotemporal interactions between elephants and (b) humans in Botswana's eastern Okavango Delta Panhandle, including an elephant herd crossing a road in an agricultural area close to the Okavango River, (c) a map of the main elephant corridors and predicted probability of elephant crop-raiding damage across the study area, and (d) monthly village trends in the number of elephant damage reports per 100 people, showing increased impacts between January and April.

Photos: (a) and (b) Jeremy Cusack; (c) and (d) based on Pozo et al. (2018) and Pozo et al. (2021), respectively

---

Box 13.1 *Continued*

Research and conservation efforts carried out over the past two decades in the eastern ODP have revealed important behavioural processes linked to human-elephant impacts. In the first instance, they have identified key behaviours that increase the likelihood of contact between elephants and agricultural activities. For humans, it is the poorly controlled intensification and expansion of unprotected agricultural activities into previously undisturbed elephant habitat, a process driven by the opportunistic allocation of agricultural land by the country's Land Board Office (Pozo et al. 2017). This tendency interacts with the movement behaviour of elephant herds along well-defined pathways linking the Okavango River and foraging areas in dry woodland habitat further inland, a to-and-fro substantiated by both ground surveys of elephant tracks and Global Positioning System telemetry of males and females (Songhurst et al. 2016; Buckholtz et al. 2021). Thus, behavioural processes occurring at a landscape level play a key role in bringing humans and wildlife closer together.

However, the emergence of actual impacts typically depends on behaviours that take place at a finer spatial scale. In the case of elephants, these involve entering and feeding on crop fields or grain stores, which is often accompanied by extensive trampling and damage to infrastructure (Pozo et al. 2021). Impact events are associated with opportunistic behaviour occurring in areas of low elephant density (Pozo et al. 2018), but are also sex- and season-specific (Vogel et al. 2020). In retaliation for the losses caused, humans will chase, injure, or kill elephants, including individuals other than that or those causing the problem (Velempini 2021). Both of these behaviours—crop foraging and retaliation—are repeated over time, as is reflected in the high number of crop damage incidents by elephants reported in the area every year (Pozo et al. 2021), and the documented deaths of both humans and elephants as a result (Redmore et al. 2020).

To date, a wide variety of mitigation strategies have been implemented to reduce human-elephant impacts in the eastern ODP, many of them spearheaded by EcoExist, a non-governmental organization dedicated to fostering coexistence between elephants and people (http://www.ecoexistproject.org/). At the landscape scale, information on elephant movement patterns has enabled the formal designation of 13 corridors linking the Okavango River to inland foraging areas (Songhurst et al. 2016; Pozo et al. 2018). Not only has this designation enabled agricultural activities to be distributed outside of areas of high elephant traffic, but it has also contributed towards reducing harmful encounters between elephants and local villagers. For example, a local bus service has been set up to ferry children and healthcare workers across elephant corridors. In addition, work with local communities has focused on improving the resilience of agricultural activities, for example by increasing the yield of smaller fields, which are easier to patrol and protect, and encouraging the early harvest of crops, thereby avoiding crop loss during periods of high elephant abundance.

At a finer spatial scale, protective mitigation strategies involving the use of chilli peppers have been tested and implemented to protect crop fields from elephants. For example, the burning of 'chilli-briquettes' (i.e. bricks made up of dry chilli, elephant dung, and water) was found to have a repellent effect on elephants, but only during the time that briquettes were giving off smoke (Pozo et al. 2017). Using a randomized block design experiment implemented in one of the eastern ODP corridors, Matsika et al. (2020) showed that the buffering of more palatable crops with chilli plants significantly reduced crop losses to elephants. Together with an improved understanding of farmer uptake of different mitigation strategies (Vogel et al. 2022), these studies have highlighted the potential for chilli-based measures to facilitate human-elephant coexistence in the area.

While spatial zoning need not require impermeable barriers between different management areas, it can, at its most extreme, involve confining wildlife to certain areas, for example by fencing off entire protected areas. As well as being very costly to apply at broad spatial scales—one study estimated the cost of erecting fences around protected areas in Africa to lie between US$1816 to $33,090 per kilometre (Pekor et al. 2019)—fences indiscriminately modify movement behaviours, but do not necessarily prevent impact-causing behaviours from occurring elsewhere. For example, Osipova et al. (2018) found that large-scale fencing led to shifts in the location of elephant crop-foraging and overgrazing to previously unaffected areas. Similar problems have been found to plague large-scale veterinary fences aimed at preventing disease transmission between wildlife and domestic animals (Mysterud and Rolandsen 2019).

A more active set of preventative strategies seek to divert wildlife away from valuable human activities or resources by providing artificial or sacrificial alternatives. Diversionary feeding techniques, whereby an alternative food source is provided as a distraction from valuable crops, livestock, or game, has shown varying levels of success in reducing the likelihood of HWIs. In their review of the approach, Kubasiewicz et al. (2016) highlight the fundamental role that individual variation in behaviour plays in the effectiveness of diversionary feeding, pointing to examples in which a subset of individuals were targeted to influence population-level behaviour (e.g. female hen harriers *Circus cyaneus* predating red grouse *Lagopus lagopus* in the UK; Redpath et al. 2001).

Whereas the provision of diversionary or supplementary resources to reduce impacts is often considered a short-term strategy, the restoration of wild food sources, on the other hand, has emerged as a potential win-win approach, albeit one that has received surprisingly little attention (Kubasiewicz et al. 2016). Bagchi et al. (2020) compared livestock consumption by snow leopards *Panthera uncia* between areas with and without efforts to recover wild ungulate prey populations in northern India. They found that an increase in the abundance of wild prey was associated with a decrease in livestock consumption, despite snow leopard numbers having also increased during the same 10-year period. In an 18-month behavioural study of chimpanzee *Pan troglodytes verus* feeding dynamics around the village of Bossou in southeastern Republic of Guinea, Hockings et al. (2009) found that the consumption of many cultivated fruits increased in periods of wild fruit scarcity. They caution, however, that key crops, such as maize, were consumed according to their availability despite a high abundance of wild foods, highlighting the importance of studying and understanding wildlife foraging decisions.

Protective strategies

In contrast to preventative measures, protective strategies seek to avert impact-causing behaviours once wildlife and humans are in close spatiotemporal proximity. As such, they tend to address a more specific set of behaviours and are often applied more locally to suit a given socioecological context. Whilst an exhaustive review of the myriad strategies on offer is beyond the scope of this chapter, it is important to highlight that their success in preventing impacts from occurring is conditional on an understanding of animal behaviour and cognition (Barrett et al. 2018).

The most common form of protection is the implementation of passive exclusionary devices that impede wildlife access to humans or their resources. Examples include enclosures built to protect livestock from wild carnivores (Van Eeden et al. 2018), nets placed around salmon farms to protect them from pinniped species (Heredia-Azuaje et al. 2022), wildlife-proof containers used to store garbage (Lewis et al. 2015), and small-scale fences surrounding crop fields to protect them from wild herbivores (Gordon 2009). Though conceptually simple, the design of such devices needs to account for the range of impact-causing behaviours that a species, or group of species, might exhibit. For example, a study carried out by Kolowski and Holekamp (2006) in villages adjacent to the Masai Mara National Reserve, Kenya, found that livestock enclosures built of sturdy pole timber were more effective at preventing entry by spotted hyenas *Crocuta crocuta* than by leopards *Panthera pardus* owing to the latter's ability to scale the structure. In contrast, enclosures made of bush material, whilst providing fewer footholds for leopards, were less likely to resist forceful entry by spotted hyenas.

Protective strategies may also actively deter wildlife from approaching or using human resources. In its most traditional form, such a strategy involves people watching over their resources. Despite the safety risks involved, particularly related to encounters with large herbivores or carnivores, active guarding of crops and livestock by people remains widespread (Gross et al. 2019). Its effectiveness, however, is dependent on an understanding of key aspects of wildlife behaviour, such as the timing of foraging events (is nighttime guarding required?), group size (how feasible is it for a person to chase off a large herd of elephants?), and foraging dynamics (are there dominant individuals who can be targeted to influence the group?).

Increasingly, trained livestock-guarding dogs are being used to deter carnivores from attacking livestock, with potential benefits to human-wildlife coexistence (Spencer et al. 2020). Kinka et al. (2021) recently reported a significant reduction in the probability of detecting both large carnivores and deer during and after the passage of sheep bands (sheep accompanied by humans and livestock-guarding dogs), highlighting the potential displacement effect on both predators and competitors. In contrast, coyotes were significantly more likely to be detected when a sheep band was present, emphasizing the importance of considering species-specific responses to mitigation measures.

In many cases, guarding of human resources involves chasing or scaring off unwanted wildlife. Research into the effectiveness of wildlife scaring (also known as hazing) methods has greatly benefitted from high-resolution tracking of animal movement behaviour. Indeed, it is now possible to study in ever increasing detail behavioural responses to hazing events in a wide variety of HWI settings. One common finding is that hazing works best when combined with the provision of wildlife refuges and/or diversionary food sources (Pekarsky et al. 2021), thereby minimizing knock-on impacts in neighbouring areas. For instance, based on an analysis of high-resolution satellite telemetry, Corriveau et al. (2022) suggested providing magpie geese *Anseranas semipalmata* in northern Australia with regional sanctuaries to increase the effectiveness of farm-level hazing efforts.

Advances in technology continue to pave the way for ever more creative ways of averting impact-causing behaviours, from proximity loggers used to study the risk of disease transmission between wild herbivores and cattle (Lavelle et al. 2016) to artificial raptors used in airports to avoid collisions between birds and aeroplanes (Storms et al. 2022). Such advances contrast with the less costly, but equally as effective, methods that are often more readily accessible to affected communities. In a recent study, Radford et al. (2020) showed that artificial eyespots painted on cattle rumps were associated with reduced attacks by ambush predators, such as lions and leopards. Not only do such

approaches provide cost-effective means to avert impact-causing behaviours, but they illustrate the importance of adapting strategies to the behaviour and ecology of the target species.

Aversive conditioning

Whilst both preventative and protective strategies have had notable successes in alleviating HWIs in a range of settings, implementation of such methods in the long term is often costly and impractical. This has motivated a renewed interest in animal conditioning—changing the behaviour of an animal in the long term through associative learning—as a potential strategy for promoting durable (and more affordable) human-wildlife coexistence (Snijders et al. 2019). In the case of HWIs, the general idea involves creating an association between an impact-causing behaviour and a specific stimulus, thereby promoting or discouraging repetition of that behaviour in the future. In other words, the approach seeks to hijack the problem animal's capacity to learn, in turn influencing predisposing and/or impact-causing behaviours in the long term (Figure 13.1).

One of the most common forms of conditioning applied to HWIs is aversive conditioning, whereby an animal associates an impact-causing behaviour with a negative stimulus or outcome (Snijders et al. 2021; see also Chapter 11). Such a stimulus may occur before, during, or after an unwanted behaviour, in accordance with different types of conditioning theories (e.g. classical or operant). For example, playbacks of tiger growls were used to deter Asian elephants *Elephas maximus* from entering crop fields around the Nilgiri Biosphere Reserve, India, exploiting the existing negative association between predator and potential prey (Thuppil and Coss 2016). Other methods use pain or discomfort to 'punish' the animal as it undertakes an impact-causing behaviour, including electric and bee-hive fences (King et al. 2017), or predator-adapted shock collars (Bruns et al. 2020). Traumatic experiences such as capturing and releasing, or translocating, a problem individual in response to a recent impact-causing behaviour have also been trialled as a form of aversive conditioning (Kidd-Weaver et al. 2022).

The challenges of applying aversive conditioning to reduce or extinguish impact-causing behaviours in wildlife are clearly illustrated in the case of conditioned taste aversion (CTA). This technique relies on the avoidance of a given food item after experiencing nausea or discomfort following its ingestion. Since early attempts to cause nausea in grey wolves and coyotes following ingestion of meat baits or livestock carcasses (Gustavson et al. 1974), CTA has shown mixed results across a range of settings. In a recent review, Snijders et al. (2021) highlight key considerations from the perspective of learning theory that must be taken into account when applying CTA to HWI scenarios. These include the evolutionary relevance and salience of the chosen stimulus, the intensity and frequency with which it is applied, and the probability that the aversion will be extinguished over time through other types of learning (e.g. habituation or social learning).

Whether through repetitive hazing, audio playbacks, electric shocks, or nauseating food, aversive conditioning relies on fear as the fundamental mechanism driving wildlife avoidance of humans (Wilson et al. 2020). Just as the reintroduction of wolves to Yellowstone National Park in the 1990s is said to have re-established a 'landscape of fear' for ungulate prey (Laundré et al. 2001), many impact mitigation techniques aim to impose a similar pressure on the behaviour of potentially problematic species (Miller and Schmitz 2019). For instance, the use of falconry as a mitigation method was found to increase the vigilance and decrease the abundance of nuisance Egyptian geese *Alopochen aegyptiaca* on South African golf courses (Atkins et al. 2017). However, the effect of introduced falcons was short-lived, with goose vigilance and abundance returning to pre-treatment levels within two months. The latter observation represents what is perhaps the principal challenge of current attempts at applying aversive conditioning to mitigate HWIs: how can the associated change in behaviour be maintained and reinforced in the long term? This remains an open question and one of considerable interest to researchers, managers, and communities working to mitigate HWIs in a changing world.

13.3.2 Mitigation strategies targeted at human behaviour

As we have seen, behavioural conceptualizations of HWIs have moved beyond a largely wildlife-centric view, to one that acknowledges the fundamental role human actions play in perpetuating them. Given this, it would be unfair not to highlight the diverse range of HWI mitigation strategies aimed at modifying human behaviour, the vast majority of which are grounded in the social and psychological sciences (neither of which the authors of this chapter declare themselves experts in).

Perhaps one of the most popular ways of influencing predisposing behaviours in humans is education. This can take many forms, including awareness-raising campaigns highlighting the plight of conflict-prone species, citizen science programmes aimed at gathering socioecological information pertaining to HWIs, and the implementation of novel approaches to managing valuable resources. Examples of this latter approach include the implementation of safer livestock-herding practices that minimize the risk of depredation by large carnivores (Kuiper et al. 2022), and the early ploughing and harvesting of crops to avoid periods of high herbivore abundance (see Box 13.1).

Yet, despite its widespread appeal as an HWI mitigation strategy, there have been few robust tests of the effectiveness of education campaigns in influencing human behaviour towards problem wildlife. In one of the few reported experimental tests, Baruch-Mordo et al. (2011) found that education had little impact in changing human behaviour related to the proofing of garbage containers against black bears. Rather, they found that proactive enforcement, such as dispensing warning notices, was more effective, thus highlighting the potential of complementary 'carrot and stick' approaches to influencing human behaviour.

Whilst the enforcement of laws that seek to protect wildlife from human persecution—for example, by making it illegal to kill or harass a protected species—can effectively reduce the likelihood of retaliatory action, it is by itself unlikely to foster coexistence in the long term. In many cases, monetary compensation schemes are put in place to alleviate the costs of impact-causing behaviours by wildlife, though how the resulting payments are determined and administered, and whether they are sufficient to effectively dissuade humans from taking retaliatory action in the long term, remains uncertain (Ravenelle and Nyhus 2017). Using eight years of retaliatory lion killing data collected in southern Kenya, Hazzah et al. (2014) showed that a community-based programme employing local Maasai as 'Lion Guardians' was more cost-effective at reducing lion killings than a concurrent predator compensation scheme. As well as supporting the implementation of mitigation strategies and contributing towards behaviour change from within the community, Lion Guardians prevented lion hunts from occurring in response to livestock depredation events, thereby fulfilling the role of wildlife protectors.

As highlighted by Pooley et al. (2021), tolerance or acceptance of wildlife plays a fundamental role in determining the behavioural responses of humans to situations of HWI. It could therefore be argued that many of the strategies targeting human behaviour do so with the objective of increasing tolerance of wildlife or of the impacts that wildlife may cause. Decades of research have shown that tolerance of wildlife is dependent on myriad factors, including social, cultural, economic, and historical components. Understanding the link between tolerance and behaviour in the context of HWI therefore requires the application of interdisciplinary approaches that can be adapted to suit specific contexts (Redpath et al. 2013). In this sense, focusing on the different behavioural processes put forward in this chapter may help to evaluate how changes in tolerance or acceptance can in practice lead to changes in the spatiotemporal proximity between humans and wildlife, as well as the recurrence of impact-causing behaviours by humans.

13.4 Fostering human-wildlife coexistence in a changing world

The importance of addressing human-wildlife impacts in a rapidly changing world was highlighted at the 15th Conference of the Parties to the Convention on Biological Diversity held in Montreal in December 2022, in which participating countries agreed to 'effectively manage human-wildlife interactions to minimize human-wildlife conflict for coexistence' by the year 2030 (Target 4). As we have seen, achieving this ambitious target will be dependent on a thorough understanding of—and ability to influence—the behavioural processes underlying situations of HWI. The framework put forward in this chapter has the potential to contribute towards this goal in three important ways. First, by helping to identify both the animal *and* human behaviours involved in situations of HWI. Second, by evaluating the role these behaviours might play in the development of HWIs, and third, by linking them to appropriate and effective mitigation strategies that target both wildlife and human actions in the short- and long term.

One of the biggest challenges in the application of behavioural research to HWIs is accounting for the different levels at which variation in behaviour might occur. Whilst many of the examples in the present chapter emphasize variation across species due to socioecological and evolutionary factors, differences in the behavioural responses of both wildlife and humans between populations and individuals are also widespread (Swan et al. 2017; Zimmermann et al. 2021). To complicate matters further, behavioural plasticity, which includes different forms of learning, can also contribute to variation in behaviour over the course of an individual's life (see Chapters 14 and 19). Importantly, variation at any of the aforementioned ecological levels can influence the manifestation of predisposing, impact-causing, and learning behaviours underlying HWIs. The resulting complexity likely imposes limits on the extent to which behavioural patterns and responses across HWI scenarios can be generalized (Blackwell et al. 2016; Zimmermann et al. 2021), thus highlighting the importance of not only carrying out, but also synthesizing, empirical tests of the behavioural responses underlying HWIs and their mitigation.

Acknowledgements

J.J.C. acknowledges funding from the Chilean Agencia Nacional de Investigación y Desarrollo (ANID) for project 11220713 (Fondecyt de

Iniciación). R.A.P. is grateful to the Proyecto Anillo de Intensificación Ecológica, ANID—ACT192027.

References

Abrahms, B., Carter, N.H., Clark-Wolf, T.J., et al. (2023). Climate change as a global amplifier of human–wildlife conflict. *Nature Climate Change*, 13, 224–234.

Atkins, A., Redpath, S.M., Little, R.M., and Amar, A. (2017). Experimentally manipulating the landscape of fear to manage problem animals. *The Journal of Wildlife Management*, 81, 610–616.

Bagchi, S., Sharma, R.K., and Bhatnagar, Y.V. (2020). Change in snow leopard predation on livestock after revival of wild prey in the Trans-Himalaya. *Wildlife Biology*, 2020, 1–11.

Barrett, L.P., Stanton, L.A., and Benson-Amram, S. (2019). The cognition of 'nuisance' species. *Animal Behaviour*, 147, 167–177.

Baruch-Mordo, S., Breck, S.W., Wilson, K.R., and Broderick, J. (2011). The carrot or the stick? Evaluation of education and enforcement as management tools for human-wildlife conflicts. *PLOS ONE*, 6, e15681.

Basset, Y., and Lamarre, G.P. (2019). Toward a world that values insects. *Science*, 364, 1230–1231.

Blackwell, B.F., DeVault, T.L., Fernández-Juricic, E., et al. (2016). No single solution: application of behavioural principles in mitigating human–wildlife conflict. *Animal Behaviour*, 120, 245–254.

Blumstein, D.T. (2016). Habituation and sensitization: new thoughts about old ideas. *Animal Behaviour*, 120, 255–262.

Bombieri, G., Delgado, M.D.M., Russo, L.F., et al. (2018). Patterns of wild carnivore attacks on humans in urban areas. *Scientific Reports*, 8, 1–9.

Bruns, A., Waltert, M., and Khorozyan, I. (2020). The effectiveness of livestock protection measures against wolves (*Canis lupus*) and implications for their co-existence with humans. *Global Ecology and Conservation*, 21, e00868.

Buchholtz, E.K., Spragg, S., Songhurst, A., et al. (2021). Anthropogenic impact on wildlife resource use: spatial and temporal shifts in elephants' access to water. *African Journal of Ecology*, 59, 614–623.

Chapman, B.K., and McPhee, D. (2016). Global shark attack hotspots: identifying underlying factors behind increased unprovoked shark bite incidence. *Ocean and Coastal Management*, 133, 72–84.

Chiyo, P.I., Moss, C.J., and Alberts, S.C. (2012). The influence of life history milestones and association networks on crop-raiding behavior in male African elephants. *PLOS ONE*, 7, e31382.

Cimatti, M., Ranc, N., Benítez-López, A., et al. (2021). Large carnivore expansion in Europe is associated with human population density and land cover changes. *Diversity and Distributions*, 27, 602–617.

Corriveau, A., Klaassen, M., Garnett, S.T., et al. (2022). Seasonal space use and habitat selection in magpie geese: implications for reducing human-wildlife conflicts. *The Journal of Wildlife Management*, 86, e22289.

Cui, Q., Ren, Y., and Xu, H. (2021). The escalating effects of wildlife tourism on human–wildlife conflict. *Animals*, 11, 1378.

Cusack, J.J., Nilsen, E.B, Israelson, M.F., et al. (2022). Quantifying the checks and balances of collaborative governance systems for adaptive carnivore management. *Journal of Applied Ecology*, 59(4), 1038–1049.

Dickman, A.J. (2010). Complexities of conflict: the importance of considering social factors for effectively resolving human–wildlife conflict. *Animal Conservation*, 13, 458–466.

Donaldson, R., Finn, H., Lusseau, D., and Calver, M. (2012). The social side of human–wildlife interaction: wildlife can learn harmful behaviours from each other. *Animal Conservation*, 15, 427–435.

Fehlmann, G., O'Riain, J.M., Fürtbauer, I., and King, A.J. (2021). Behavioral causes, ecological consequences, and management challenges associated with wildlife foraging in human-modified landscapes. *BioScience*, 71, 40–54.

Fehlmann, G., O'Riain, M.J., Kerr-Smith, C., et al. (2017). Extreme behavioural shifts by baboons exploiting risky, resource-rich, human-modified environments. *Scientific Reports*, 7, 1–8.

Gordon, I.J. (2009). What is the future for wild, large herbivores in human-modified agricultural landscapes? *Wildlife Biology*, 15, 1–9.

Greggor, A.L., Berger-Tal, O., Blumstein, D.T., et al. (2016). Research priorities from animal behaviour for maximising conservation progress. *Trends in Ecology & Evolution*, 31, 953–964.

Gross, E.M., Lahkar, B.P., Subedi, N., et al. (2019). Does traditional and advanced guarding reduce crop losses due to wildlife? A comparative analysis from Africa and Asia. *Journal for Nature Conservation*, 50, 125712.

Gustavson, C.R., Garcia, J., Hankins, W.G., and Rusiniak, K.W. (1974). Coyote predation control by aversive conditioning. *Science*, 184, 581–583.

Hazzah, L., Dolrenry, S., Naughton, L., et al. (2014). Efficacy of two lion conservation programs in Maasailand, Kenya. *Conservation Biology*, 28, 851–860.

Heredia-Azuaje, H., Niklitschek, E.J., and Sepúlveda, M. (2022). Pinnipeds and salmon farming: threats, conflicts and challenges to co-existence after 50 years of industrial growth and expansion. *Reviews in Aquaculture*, 14, 528–546.

Herr, J., Schley, L., and Roper, T.J. (2009). Stone martens (*Martes foina*) and cars: investigation of a common human–wildlife conflict. *European Journal of Wildlife Research*, 55, 471–477.

Hockings, K.J., Anderson, J.R., and Matsuzawa, T. (2009). Use of wild and cultivated foods by chimpanzees at Bossou, Republic of Guinea: feeding dynamics in a human-influenced environment. *American Journal of Primatology*, 71, 636–646.

Kidd-Weaver, A.D., Rainwater T.R., Murphy, T.M., and Bodinof Jachowski, M. (2022). Evaluating the efficacy of capture as aversive conditioning for American alligators in human-dominated landscapes. *The Journal of Wildlife Management*, 86, e22259.

King, L.E., Lala, F., Nzumu, H., et al. (2017). Beehive fences as a multidimensional conflict-mitigation tool for farmers coexisting with elephants. *Conservation Biology*, 31, 743–752.

Kinka, D., Schultz, J.T., and Young, J.K. (2021). Wildlife responses to livestock guard dogs and domestic sheep on open range. *Global Ecology and Conservation*, 31, e01823.

Kolowski, J.M., and Holekamp, K.E. (2006). Spatial, temporal, and physical characteristics of livestock depredations by large carnivores along a Kenyan reserve border. *Biological Conservation*, 128, 529–541.

Kubasiewicz, L.M., Bunnefeld, N., Tulloch, A.I., et al. (2016). Diversionary feeding: an effective management strategy for conservation conflict? *Biodiversity and Conservation*, 25, 1–22.

Kuiper, T., Loveridge, A.J., and Macdonald, D.W. (2022). Robust mapping of human–wildlife conflict: controlling for livestock distribution in carnivore depredation models. *Animal Conservation*, 25, 195–207.

Laundré, J.W., Hernández, L., and Altendorf, K.B. (2001). Wolves, elk, and bison: reestablishing the 'landscape of fear' in Yellowstone National Park, USA. *Canadian Journal of Zoology*, 79, 1401–1409.

Lavelle, M.J., Kay, S.L., Pepin, K.M., et al. (2016). Evaluating wildlife-cattle contact rates to improve the understanding of dynamics of bovine tuberculosis transmission in Michigan, USA. *Preventive Veterinary Medicine*, 135, 28–36.

Lewis, D.L., Baruch-Mordo, S., Wilson, K.R., et al. (2015). Foraging ecology of black bears in urban environments: guidance for human-bear conflict mitigation. *Ecosphere*, 6, 1–18.

Linnell, J.D., Nilsen, E.B., Lande, U.S., et al. (2005). Zoning as a means of mitigating conflicts with large carnivores: principles and reality. In: A. Rabinowitz, R. Woodroffe, and S. Thirgood (eds), *People and Wildlife, Conflict or Co-existence?*, pp. 162–175. Cambridge University Press, Cambridge.

Lischka, S.A., Teel, T.L., Johnson, H.E., et al. (2018). A conceptual model for the integration of social and ecological information to understand human-wildlife interactions. *Biological Conservation*, 225, 80–87.

Mainwaring, M.C. (2015). The use of man-made structures as nesting sites by birds: a review of the costs and benefits. *Journal for Nature Conservation*, 25, 17–22.

Matsika, T.A., Adjetay, J.A., Motshwari, O., et al. (2020). Alternative crops as a mitigation measure for elephant crop raiding in the eastern Okavango Panhandle. *Pachyderm*, 61, 140–152.

Miller, J.R., and Schmitz, O.J. (2019). Landscape of fear and human-predator coexistence: applying spatial predator-prey interaction theory to understand and reduce carnivore-livestock conflict. *Biological Conservation*, 236, 464–473.

Morehouse, A.T., Graves, T.A., Mikle, N., and Boyce, M.S. (2016). Nature vs. nurture: evidence for social learning of conflict behaviour in grizzly bears. *PLOS ONE*, 11, e0165425.

Moreira, F., Martins, R.C., Catry, I., and D'Amico, M. (2018). Drivers of power line use by white storks: a case study of birds nesting on anthropogenic structures. *Journal of Applied Ecology*, 55, 2263–2273.

Mumaw, L., and Mata, L. (2022). Wildlife gardening: an urban nexus of social and ecological relationships. *Frontiers in Ecology and the Environment*, 20, 379–385.

Mysterud, A., and Rolandsen, C.M. (2019). Fencing for wildlife disease control. *Journal of Applied Ecology*, 56, 519–525.

Nilsson, L., Bunnefeld, N., Persson, J., et al. (2019). Conservation success or increased crop damage risk? The Natura 2000 network for a thriving migratory and protected bird. *Biological Conservation*, 236, 1–7.

Nyhus, P.J. (2016). Human–wildlife conflict and coexistence. *Annual Review of Environment and Resources*, 41, 143–171.

O'Bryan, C.J., Braczkowski, A.R., Beyer, H.L., et al. (2018). The contribution of predators and scavengers to human well-being. *Nature Ecology & Evolution*, 2, 229–236.

Ontiri, E.M, Odino, M., Kasanga, A., et al. (2019). Maasai pastoralists kill lions in retaliation for depredation of livestock by lions. *People and Nature*, 1, 59–69.

Osipova, L., Okello, M.M., Njumbi, S.J., et al. (2018). Fencing solves human-wildlife conflict locally but shifts problems elsewhere: a case study using functional connectivity modelling of the African elephant. *Journal of Applied Ecology*, 55, 2673–2684.

Pekarsky, S., Schiffner, I., Markin, Y., and Nathan, R. (2021). Using movement ecology to evaluate the effectiveness of multiple human-wildlife conflict management practices. *Biological Conservation,* 262, 109306.

Pekor, A., Miller, J.R.B., Flyman, M.V., et al. (2019). Fencing Africa's protected areas: costs, benefits, and management issues. *Biological Conservation,* 229, 67–75.

Penteriani, V., Delgado, M.D.M., Pinchera, F., et al. (2016). Human behaviour can trigger large carnivore attacks in developed countries. *Scientific Reports,* 6, 1–8.

Perona, A.M., Urios, V., and López-López, P. (2019). Holidays? Not for all. Eagles have larger home ranges on holidays as a consequence of human disturbance. *Biological Conservation,* 231, 59–66.

Peterson, M.N., Bickhead, J.L., Leong, K., et al. (2010). Rearticulating the myth of human–wildlife conflict. *Conservation Letters,* 3, 74–82.

Pooley, S., Bhatia, S., and Vasava, A. (2021). Rethinking the study of human–wildlife coexistence. *Conservation Biology,* 35, 784–793.

Pozo, R.A., Coulson, T., McCulloch, G., et al. (2017). Determining baselines for human-elephant conflict: a matter of time. *PLOS ONE,* 12, e0178840.

Pozo, R.A., Cusack, J.J., McCulloch, G., et al. (2018). Elephant space-use is not a good predictor of crop-damage. *Biological Conservation,* 228, 241–251.

Pozo, R.A., LeFlore, E.G., Duthie, A.B., et al. (2021). A multispecies assessment of wildlife impacts on local community livelihoods. *Conservation Biology,* 35, 297–306.

Radford, C., McNutt, J.W., Rogers, T., et al. (2020). Artificial eyespots on cattle reduce predation by large carnivores. *Communications Biology,* 3, 1–8.

Ravanelle, J., and Nyhus, P.J. (2017). Global patterns and trends in human–wildlife conflict compensation. *Conservation Biology,* 31, 1247–1256.

Redfern, J.V., McKenna, M.F., Moore, T.J., et al. (2013). Assessing the risk of ships striking large whales in marine spatial planning. *Conservation Biology,* 27, 292–302.

Redmore, L., Stronza, A., Songhurst, A., and McCulloch, G. (2020). Where elephants roam: perceived risk, vulnerability, and adaptation in the Okavango Delta. *Ecology and Society,* 25.

Redpath, S.M., Clarke, R., Madders, M., and Thirgood, S.J. (2001). Assessing raptor diet: comparing pellets, prey remains, and observational data at hen harrier nests. *The Condor,* 103, 184–188.

Redpath, S.M., Young, J., Evely, A., et al. (2013). Understanding and managing conservation conflicts. *Trends in Ecology & Evolution,* 28, 100–109.

Schakner, Z.A., Lunsford, C., Straley, J., et al. (2014). Using models of social transmission to examine the spread of longline depredation behavior among sperm whales in the Gulf of Alaska. *PLOS ONE,* 9, e109079.

Schell, C.J., Stanton, L.A., Young, J.K., et al. (2021). The evolutionary consequences of human–wildlife conflict in cities. *Evolutionary Applications,* 14, 178–197.

Schultz, H., Chang, K., Bury, S.J., et al. (2021). Sex-specific foraging of an apex predator puts females at risk of human–wildlife conflict. *Journal of Animal Ecology,* 90, 1776–1786.

Siljander, M., Kuronen, T., Johansson, T., et al. (2020). Primates on the farm—spatial patterns of human–wildlife conflict in forest-agricultural landscape mosaic in Taita Hills, Kenya. *Applied Geography,* 117, 102185.

Snijders, L., Greggor, A.L., Hilderink, F., and Doran, C. (2019). Effectiveness of animal conditioning interventions in reducing human–wildlife conflict: a systematic map protocol. *Environmental Evidence,* 8, 1–10.

Snijders, L., Thierij, N.M., Appleby, R., et al. (2021). Conditioned taste aversion as a tool for mitigating human-wildlife conflicts. *Frontiers in Conservation Science,* 2, 744704.

Songhurst, A., McCullock, G., and Coulson, T. (2016). Finding pathways to human–elephant coexistence: a risky business. *Oryx,* 50, 713–720.

Spencer, K., Sambrook, M., Bremner-Harrison, S., et al. (2020). Livestock guarding dogs enable human-carnivore coexistence: first evidence of equivalent carnivore occupancy on guarded and unguarded farms. *Biological Conservation,* 241, 108256.

Storms, R.F., Carere, C., Musters, R., et al. (2022). Deterrence of birds with an artificial predator, the RobotFalcon. *Journal of the Royal Society Interface,* 19, 20220497.

Strand, G.H., Hansen, I., de Boon, A., and Sandström, C. (2019). Carnivore management zones and their impact on sheep farming in Norway. *Environmental Management,* 64, 537–552.

Swan, G.J., Redpath, S.M., Bearhop, S., and McDonald, R.A. (2017). Ecology of problem individuals and the efficacy of selective wildlife management. *Trends in Ecology & Evolution,* 32, 518–530.

Thaker, M., Zambre, A., and Bhosale, H. (2018). Wind farms have cascading impacts on ecosystems across trophic levels. *Nature Ecology & Evolution,* 2, 1854–1858.

Thuppil, V., and Coss, R.G. (2016). Playback of felid growls mitigates crop-raiding by elephants *Elephas maximus* in southern India. *Oryx,* 50, 329–335.

Valeix, M., Hemson, G., Loveridge, A.J., et al. (2012). Behavioural adjustments of a large carnivore to access secondary prey in a human-dominated landscape. *Journal of Applied Ecology,* 49, 73–81.

Van Eeden, L.M., Crowther, M.S., Dickman, C.R., et al. (2018). Managing conflict between large carnivores and livestock. *Conservation Biology*, 32, 26–34.

Velempini, K. (2021). About the human–elephant conflict in Botswana, what did people in the Okavango Delta panhandle have to say from their experience? *Socio-Ecological Practice Research*, 3, 411–425.

Vogel, S.M., Lambert, B., Songhurst, A.C., et al. (2020). Exploring movement decisions: can Bayesian movement-state models explain crop consumption behaviour in elephants (*Loxodonta africana*)? *Journal of Animal Ecology*, 89, 1055–1068.

Vogel, S.M., Songhurst, A.C., McCulloch, G., and Stronza, A. (2022). Understanding farmers' reasons behind mitigation decisions is key in supporting their coexistence with wildlife. *People and Nature*, 4, 1305–1318.

Wilkinson, C.E., McInturff, A., Miller, J.R.B, et al. (2020). An ecological framework for contextualizing carnivore–livestock conflict. *Conservation Biology*, 34, 854–867.

Williams, B.A., Venter, O., Allan, J.R., et al. (2020). Change in terrestrial human footprint drives continued loss of intact ecosystems. *One Earth*, 3, 371–382.

Wilson, M.W., Ridlon, A.D., Gaynor, K.M., et al. (2020). Ecological impacts of human-induced animal behaviour change. *Ecology Letters*, 23, 1522–1536.

Zimmermann, A., Johnson, P., de Barros, A.E., et al. (2021). Every case is different: cautionary insights about generalisations in human-wildlife conflict from a range-wide study of people and jaguars. *Biological Conservation*, 260, 109185.

CHAPTER 14

Plasticity and adaptation

Emilie Snell-Rood and Amy-Charlotte Devitz

Overview

The behavioural responses of animals to anthropogenic environments are often led by plasticity and followed by genetic change. This chapter first reviews examples of how developmental and evolutionary responses allow organisms to adjust to variable environments, with a focus on behavioural traits. Then, we consider how the structure and design of the built environment could promote population resilience through such plastic and evolutionary change. This approach combines an understanding of a species' ancestral environments with the novel dimensions of the anthropogenic environment—a combination of abiotic and biotic factors, from resources and predators to temperature and noise. The landscape and habitat structure of the new environment will affect both the development of cognition and evolutionary processes. Patterns of colonization, gene flow, and adaptation will likely vary across types of built environments, from patchy cities to linear roadsides. These perspectives offer opportunities for collaboration between biologists, designers, and stakeholders in conservation.

14.1 Introduction

Humans have created a diversity of novel conditions drastically different from the environments in which most organisms evolved, including light pollution (Chapter 4), acoustic noise (Chapter 2), new chemicals (Chapter 3), elevated nutrient levels, and entirely new habitats, such as cities and vast agricultural monocultures (Chapters 8 and 9; Ellis and Ramankutty 2008; Sih et al. 2011; Hobbs et al.

2013). Across different research fields—including biology, conservation, agriculture, and even urban planning—people are interested in understanding which organisms will adjust to human environments, and which will not. For instance, why do some species thrive in cities, and will this suite of species look the same 1000 years from now (Tuomainen and Candolin 2011; Dunn et al. 2022; Szulkin et al. 2020)? Flexible behavioural responses and developmental plasticity are key components of organismal resilience in the face of anthropogenic change (Sih 2013; Buchholz et al. 2019).

An important piece in predicting responses to environmental change involves understanding to what extent responses to novel environments stem from immediate trait adjustments (plasticity) or underlying evolutionary change. As we discuss here, most instances of organisms coping with novel human environments stem from a combination of plasticity and genetic change over time (Volis et al. 2015; Lambert et al. 2021; Martin et al. 2021). Much recent emphasis has been placed on the importance of plasticity because developmental flexibility, and behaviour itself, often initially leads the way in novel conditions (Section 14.3; Gross et al. 2010; Corl et al. 2018; Radersma et al. 2020). The evolution of phenotypic plasticity tends to be favoured in species that experience environmental variation, whether because they live in more inherently variable areas (Figure 14.1a), or because they choose a greater range of resources or microhabitats (e.g. generalists, reviewed in Snell-Rood and Steck 2019; Snell-Rood and Ehlman 2021). Plasticity may come as the development of different traits depending on early life conditions, with such developmental

Emilie Snell-Rood and Amy-Charlotte Devitz, *Plasticity and adaptation*. In: *Behavioural Responses to a Changing World*. Edited by: Bob B. M. Wong and Ulrika Candolin, Oxford University Press. © Oxford University Press (2024). DOI: 10.1093/oso/9780192858979.003.0014

plasticity often favoured when the environment varies between generations, but less so within generations ('coarse-grained variation'; Figure 14.1b). Plasticity may also refer to behaviour itself, for instance when different behavioural responses are expressed in different environments; such 'activational plasticity' or context-dependent behaviour is favoured when the environment changes more rapidly (Figure 14.1c).

In this chapter, we first review what we mean by a 'novel' environment. When thinking of adaptation to human environments, we must first consider just how different the human environment is from the ancestral environment of the species in question, what we term here as 'environment matching' (Section 14.2). Then, we discuss examples of the importance of both plastic and evolutionary responses to novel environments (Section 14.3), before shifting the question to ask how we can structure human environments to promote resilience of a wide range of species in the Anthropocene (Section 14.4; Smith et al. 2014). We are at a point where theory from evolutionary biology and behavioural ecology can offer suggestions of how human environments might be constructed to promote adaptive plasticity and the evolution of species that can deal with humans, but such ideas need to be tested, offering exciting and applicable future directions (Section 14.5; Johnson and Munshi-South 2017).

14.2 Environment matching: the emergence of stress in novel environments

Consider a city centre, possibly the most drastically divergent habitat animals face today relative to their evolutionary history: pavement, skyscrapers, human disturbance, cars, pollution, noise, light,

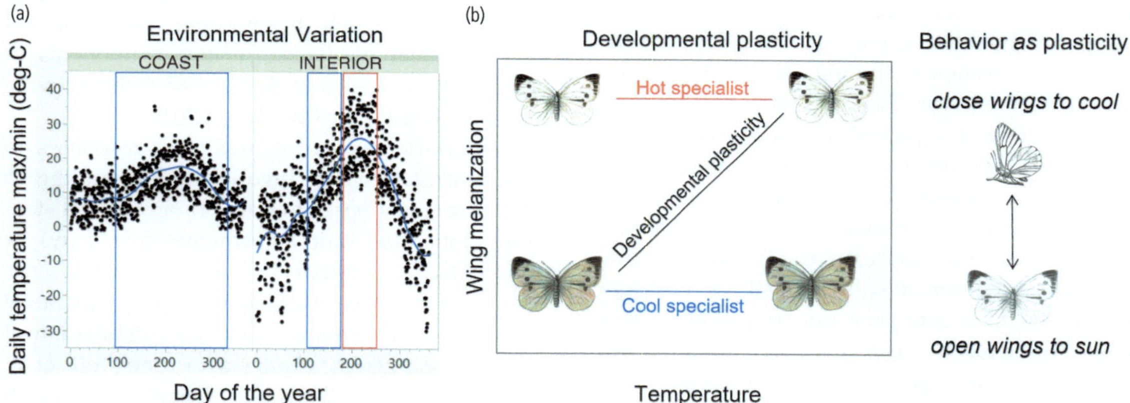

Figure 14.1 An overview of plasticity. (a) Environmental variation tends to favour the evolution of plasticity. In this example, we consider variation in temperature. Shown here are two plots from weather stations from coastal (left) and interior (right) locations in the USA (data from 2022, RAWS weather stations). Each plot goes from January to December and includes daily temperature minima and maxima; both seasonal and daily temperature swings can be seen. A species specialized on hot conditions may time their life cycle during the red boxed period, while a species specialized on cool conditions may time their life cycle during the blue boxed periods. A generalist would make use of both the hot and cool periods, experiencing greater environmental variation relative to the specialists. (b) Seasonal variation in temperature is a form of predictable environmental variation that is correlated with reliable cues, such as photoperiod. Such patterns of environmental variation tend to select for developmental plasticity when environmental conditions vary between generations, in contrast to specialists that experience more constant conditions within a generation. For instance, a butterfly whose two-week flight period corresponds to spring conditions will likely evolve darker wing coloration relative to a species that is only out during the summer. A generalist species with multiple flight periods will likely evolve developmental plasticity, where wing coloration depends on rearing photoperiod. However, in an environment with less seasonal variation in temperature (the coastal environment), the costs of this plasticity (see Snell-Rood and Ehlman 2021) might result in shifts to a fixed phenotype. (c) Fine-grained environmental variation (variation within an individual's lifespan), such as temperature swings within a day, select for forms of plasticity that are reversible. Behaviour and physiology represent forms of such 'activational plasticity', where underlying behaviours are expressed as the conditions change. For instance, when it is cool, insects often open their wings to capture warmth from the sun, or vibrate to warm up. When it is warm, they seek shade or close their wings and change their orientation to be parallel to the sunlight.

Images are Creative Commons: closed wings—from *The Book of the Garden* (1853); open wings—from *Blitz-Lexikon* (1932)

and trash (Grimm et al. 2008). And yet, there are species that thrive in cities, sometimes benefitting from the novel resources (Evans et al. 2011). One explanation offered for the ecological success of some city-dwellers is that their ancestral environment happens to match aspects of this human environment, sometimes termed 'habitat analogues' (Lundholm and Richardson 2010). For example, peregrine falcons *Falco peregrinus* and pigeons *Columba livia* historically nested on cliffs, a possible pre-adaptation, or exaptation, for exploitation of city skyscrapers (Francis 2020; Winchell et al. 2023). This initial matching of new conditions allows populations to survive in the new environment, paving the way for subsequent adaptation (Johnson and Munshi-South 2017; Lambert et al. 2021).

In thinking about how species vary in their response to novel ecosystems, we first need to consider the overlap between the novel environment and a species' ancestral environments (Heger et al. 2019). The ancestral environment of a species might include everything from their tolerance to high temperatures to the resources they use. This degree of overlap—the evolutionary match or mismatch (Pollack et al. 2022)—should determine how stressful human environments are to the organisms there (Figure 14.2a), and thus whether a species can gain

a foothold in a novel environment in the first place (Wikelski and Cooke 2006). Evolutionary responses to novel conditions will not happen if the population goes extinct before adaptation occurs (Carlson et al. 2014). Thus, any consideration of responses to new conditions, such as evolutionary rescue, must first consider this degree of environment matching and resulting stress (Figure 14.2b). Similarly, many models that seek to predict range shifts consider a species' ancestral climactic niche relative to future projections, but we must expand this idea to all aspects of an organism's environments (Kearney and Porter 2009).

14.3 Responses to novel environments result from both plasticity and evolution

In general, responses of animals to human environments include both plastic and evolutionary components (e.g. Boutin and Lane 2014), with plastic responses often dominating early in the process of moving into a new environment. More broadly, across behavioural and other traits, meta-analyses tend to find that both plasticity and evolution are important in responses to new and varying environments. Indeed, plastic and evolutionary responses are often in parallel directions (Radersma et al. 2020) with plasticity explaining approximately double

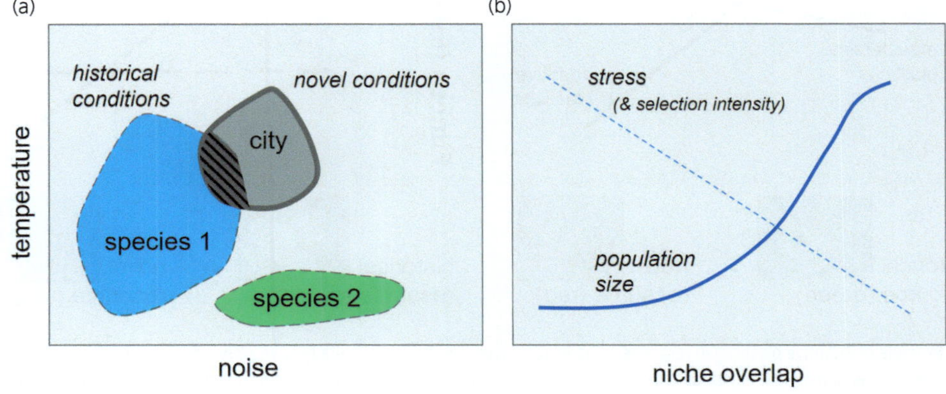

Figure 14.2 (a) This hypothetical example shows two axes of a species' ancestral environments or 'niche'—temperature and noise levels (e.g. for an acoustically communicating bird). The shapes with dashed outlines show historical conditions for two species, while the dark grey shape outlines the conditions of a novel environment such as a city. We would expect that only species 1 (in blue) would move into the city initially. (b) With increases in the degree of niche overlap between ancestral and novel conditions comes a decline in stress and, in theory, an increase in the potential for further adaptation to the novel environment as a larger and more fecund population can gain a foothold in the new environment.

the amount of phenotypic variation in population differences (Stamp and Hadfield 2020). In this section, we discuss a series of examples with respect to behavioural plasticity and evolution (see also Box 14.1 for an additional example on responses to artificial light).

14.3.1 Responses to novel resources and predators

In a changing environment, encounters with novel resources and predators are inevitable, and both plastic and genetic modifications in behaviour are important in adaptive responses. The exploitation of novel resources is seen in the use of human resources in urban environments, the acceptance of invasive plants as resources, or shifts to feed on crops. In urbanized regions, animals from a number of taxonomic groups take advantage of food sources such as gardens, garbage, or artificial feeders (Sol et al. 2013; Chapter 9). Cognition plays a large role in animals learning to exploit

new resources; for instance, large-brained birds show innovative feeding behaviours and thrive in cities (Maspons et al. 2019; Sayol et al. 2020). When plastic responses facilitate the use of novel resources, subsequent evolutionary change may follow. For example, urban and rural finches have diverged in beak shape in association with changes in the hardness of resources in cities (Badyaev et al. 2008). In butterfly species, such as *Pieris rapae*, individuals readily learn to search for rare or novel colours or shapes associated with host or nectar resources (Figure 14.3), allowing them to use novel resources (Graves and Shapiro 2003; Snell-Rood and Papaj 2009). At the same time, species show differences in innate colour preferences, suggesting potential for evolutionary change in search biases towards new resources (Bernard and Remington 1991).

We see a similar mixture of plastic and evolutionary responses in looking at novel interactions between animals and novel predators in anthropogenic environments. Many species show fear

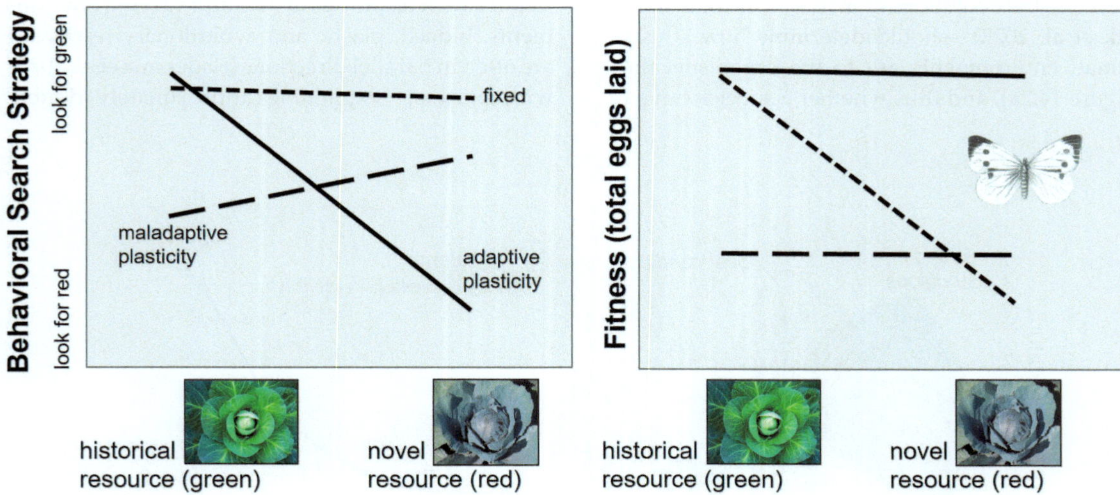

Figure 14.3 Example of adaptive plasticity in response to novel resources. The cabbage white butterfly *Pieris rapae* lays eggs on plants in the mustard family, such as cabbage. Most mustards are green in colour, and these butterflies have a bias to search for green colours during host search. However, they can learn to search for other colours. When presented with a novel, red cabbage host plant, some genotypes can more effectively shift to search for red colours, maintaining high performance across different host plants (solid line). Fixed genotypes (fine dotted line) are specialized to search for green and do well at finding only green host plants while genotypes with maladaptive plasticity (dashed line) express the wrong phenotype in the new environment and do poorly across environments.

Diagram is a stylized version of data presented in Snell-Rood and Papaj (2009); images are Creative Commons: butterfly—from *Blitz-Lexikon* (1932); red cabbage—Jamain (2012); green cabbage—Via Tsuji (2012)

responses to people, but animals also learn to adjust these responses when certain people or situations are less threatening (e.g. Greggor et al. 2016). For instance, comparative studies of birds that contrast the flight initiation distance of urban and rural populations suggest adaptive loss of these fear responses in urban areas (Moller 2008). Responses to cars provide another example of interactions between plastic and evolutionary responses (see also Chapter 9). Cars in many ways are novel 'predators' in human environments, for which most species have no innate avoidance mechanisms (apart from general avoidance of loud and looming stimuli; Lima et al. 2015). Plastic behavioural responses play a role in avoidance of collisions. For instance, larger-brained bird species are more likely to make the 'correct' choice when a car is approaching and fly away from the road (Husby and Husby 2014). Many vertebrates show immediate behavioural responses to roads, such as higher vigilance and avoidance (McCorquodale 2013). In longer-lived species, many of these strategies may be learned, such as avoiding high-traffic roads, or crossing at times when collision risk is less (Zeller et al. 2020). At the same time, mortality from cars is causing evolutionary changes in roadside populations. Bridge-nesting swallows *Petrochelidon pyrrhonota* show shifts in wing morphology over a 30-year period to have more manoeuvrable wing shapes, and roadkill rates in the area have declined (Brown and Brown 2013). In addition, the relative brain size of these populations also seems to be increasing, which has been suggested as cognitive increases to avoid collisions (Wagnon 2019).

14.3.2 Responses to climate change

Responses to climate change offer another well-studied area where organisms show a wide range of plastic and evolutionary behavioural changes (Parmesan 2006; see also Chapter 1). At the level of temperature, shifts in microhabitat choice allow animals to buffer their response to temperature extremes, for instance a lizard choosing to hide in shade or a caterpillar moving up a plant to avoid heat stress (Huey et al. 2003; Nielsen and Papaj 2015). Such ability to behaviourally thermoregulate may even be responsible for population stability of burrowing desert mammals in the face of

climate change (Riddell et al. 2021). Changes in microhabitat use associated with thermoregulation in turn affect morphological evolution (Munoz and Losos 2018). In some cases, initial plastic responses to environmental change are maladaptive (perhaps up to 20% of the time; Stamp and Hadfield 2020), but subsequent selection may dampen such plasticity. For instance, in anoles *Anolis cristatellus*, initial patterns of gene expression in heat stress are maladaptive, but urban populations show a signature of selection to reduce such plasticity (Campbell-Staton et al. 2021).

Animals also show plastic and evolutionary responses to climate change through shifts in phenology, sometimes adaptively tracking shifts in the timing of annual cycles. For instance, a meta-analysis showed that birds are advancing their migratory schedules at a rate to keep track with climate change, but it is unclear whether this stems from evolutionary or plastic responses (Charmantier and Gienapp 2014), although it is likely a combination of both (Akesson and Helm 2020). In mammals, the handful of studies that can tease apart the causes of phenological shifts suggest much of the changes are at least initially due to plasticity (Boutin and Lane 2014). Species are also shifting migration routes in response to climate change: some songbirds (e.g. *Anthus richardi*) are moving their southward routes westward (Dufour et al. 2021).

In cities, the presence of pavement, metals, and other materials elevates temperatures in a parallel manner to climate change (Youngsteadt et al. 2015). In response to this 'urban heat island' effect, we see both plastic and genetic changes playing a role in responses to elevated temperature. With respect to behaviour, microhabitat choice of irrigated and shady parts of the city can mitigate temperature stress in lizards and other species (Ackley et al. 2015). With respect to physiology, rearing temperature results in pronounced plastic changes in heat tolerance in ants, but urban and rural ants also show evolutionary differences in heat tolerance (Martin et al. 2019). Across many taxonomic groups, habitat types, and fitness contexts, we see both plastic and evolutionary shifts in behaviour that allow animals to adjust to new conditions. Next, we consider building human environments in a way that promotes such adaptive change.

14.4 How can the human environment promote adaptive responses of populations?

Evolutionary biology and behavioural ecology have a rich theoretical history, and recent research has begun to layer this onto rapid environmental change to make predictions on how we might promote adaptive responses, for instance through evolutionary rescue (Carlson et al. 2014). We discuss and expand this framework by considering how the built environment affects evolutionary dynamics and cognitive development.

14.4.1 Landscape considerations: what conditions promote adaptation to novel environments

Classic evolutionary models of adaptation often consider allopatric populations, perhaps separated by a mountain range, adapting to contrasting environments (e.g. Lande 1980). Rates of gene flow between populations, patterns of genetic variation, and environmental differences between the populations will all influence rates of adaptation to the contrasting environments (Orr and Smith 1998). In some cases, novel human environments have a similar structure to allopatric or parapatric models. One example is human-caused elevation of nitrogen, an important macronutrient that can affect the development of cognition and behaviour (Snell-Rood et al. 2015). Human activity causes elevated nitrogen deposition from the atmosphere over vast spatial areas (Figure 14.4a), with geographic barriers skewing patterns of deposition (e.g. Harmens et al. 2011). However, in many cases, human environments follow more complex landscape patterns; for instance, nitrogen deposition can also be narrow and linear, through deposition from cars along roads (Fenn et al. 2018). In other words, many human environments are less analogous to the allopatric model of population divergence and may be more likely to have gene flow between novel and ancestral conditions. While evolutionary divergence can still proceed in the face of gene flow (Nosil 2008), the dynamics are more complex than in the case of allopatric divergence. Here, we consider cities as a case study to investigate this idea further (Johnson and Munshi-South 2017).

Urban environments have been likened to islands, at both the within-city scale of patches of green space within the city, and at the between-city

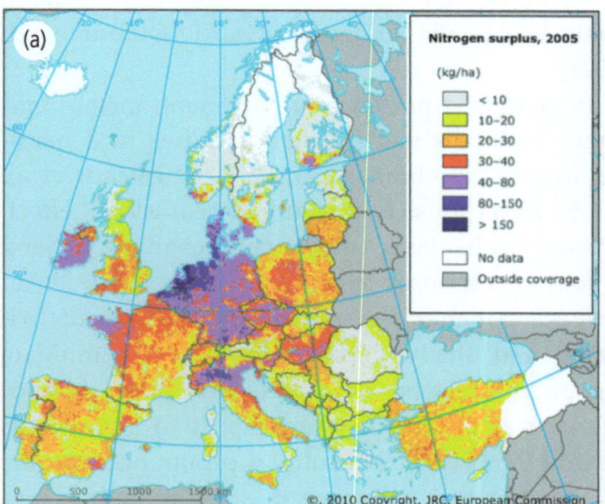

Figure 14.4 Examples of the spatial landscape of anthropogenic change. (a) In the case of nitrogen deposition, there are relatively large contiguous spatial areas of elevated nitrogen, determined in part by geologic barriers that may also affect gene flow. Such a pattern may be more analogous to models of parapatric or allopatric evolutionary processes. (b) In the case of urban areas (here highlighted by nighttime light pollution), spatial variation is much patchier, with islands of urban space separated by ancestral habitat, agriculture, and other non-urban habitat. This pattern may be more analogous to island models of ecology and evolution.

Images: (a) CC-4, European Environment Agency; (b) CC-2 NASA Goddard Space Flight Center

scale, in terms of cities separated by non-city habitats (Davis and Glick 1978; Fattorini et al. 2018). Patterns of connectivity between the novel environments and habitats more typical of ancestral conditions will affect patterns of dispersal, gene flow, and adaptation (Figure 14.5; Johnson and Munshi-South 2017; Parris et al. 2018; see also Chapter 8). Novel environments that have high connectivity with ancestral environments (Figure 14.5b) should result in high environmental variability experienced by this population, which should favour the evolution of developmental plasticity if the alternate environments are experienced in a coarse-grained manner, or behavioural plasticity if the pattern of variation is more fine-grained (Snell-Rood and Ehlman 2021). In contrast, when connectivity to the ancestral environment is lower (Figure 14.5a), environmental variation and gene flow are reduced, lowering selection on plasticity, and increasing the likelihood of adaptation to the novel environment and, over time, population divergence in the novel conditions (Johnson and Munshi-South 2017). Which of these scenarios is 'better' from a conservation or management perspective is a separate question, but one might surmise that higher connectivity not only promotes colonization of the new conditions, but also promotes more robust populations through plasticity (Smith et al. 2014). Regardless, it is important to note that the relevant scale of environmental heterogeneity will depend on the animal in consideration: animals that disperse by walking (versus flying) may show relatively stronger population differentiation in a city landscape (e.g. foxes *Vulpes vulpes*; Robinson and Marks 2001).

Cities also occur at a much larger landscape scale that will affect gene flow and evolution. At a regional scale, cities may occur as 'islands' separated by seas of agriculture or less developed land (Figure 14.4b; Davis and Glick 1978), although in some cases, growing cities may merge creating a more interconnected novel environment, more analogous to an 'archipelago'. The theory of island biogeography can be used to predict what would happen if a mutant that improved city performance arose in one city; this gene may be more likely to spread across city populations with little distance between them. Larger cities, or those with more green space, should be more likely to host larger populations, fuelling larger effective population sizes, and increasing the chances of an origin of a beneficial mutation. These ideas apply very differently across species, depending on their size, life history, and mechanism of dispersal (Table 14.1). For instance, smaller species are likely to be more limited in cross-city dispersal than larger species, or those that may disperse in association with humans or roads (Chytry et al. 2012). Species that can fly, like pigeons, may be able to readily disperse between cities, especially in interconnected 'mega-cities', resulting in a more homogeneous city population (Carlen and Munshi-South 2021). And if species can hitch a ride with humans, adaptations to cities may spread rapidly across city populations.

14.4.2 Changing the human environment to increase environment matching

If we alter human environments to be more similar to ancestral environments, animals are more likely to immediately and plastically adjust their behaviour and settle in these new environments, which may then act as stepping-stones to more extreme human environments. In very basic models of adaptation to two environments, we often think of ancestral and novel habitats as two categories of habitat (Figure 14.5). However, habitats fall on a continuum, such that some habitats are more closely matched to ancestral habitats in several dimensions (Figure 14.2). This results in more individuals surviving in the new conditions, fuelling subsequent evolution through 'evolutionary rescue' (Carlson et al. 2014). Increasingly, designers of human environments are partnering with ecologists and biologists to create habitat refuges within novel environments and to creatively reduce the novelty or harshness of the built environment (Parris et al. 2018; see also Chapter 16). This parallels the idea of 'reconciliation ecology', where we encourage biodiversity in human environments through the rewilding of human spaces such as roadsides, golf courses, yards, or roofs (Rosenzweig 2003).

The concepts of 'green infrastructure' and 'nature-based solutions' capture a wide range of efforts in urban design to build with nature (Hobbie

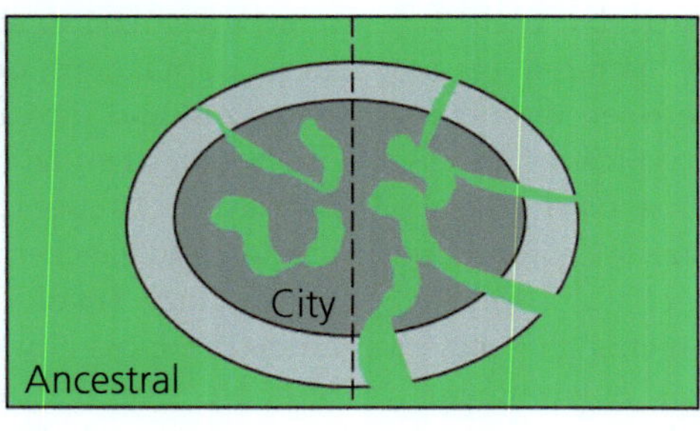

Figure 14.5 How scenarios of habitat connectivity may affect adaptation and plasticity. Consider a city nested within ancestral habitat; the more urban city centre (in dark grey) is surrounded by developed, but less dense, suburban areas (light grey). Within the city, there are pockets of green space that resemble the ancestral environment (enough for individuals to gain a foothold, but different enough that they are novel; see Figure 14.2). Habitat corridors connect these microhabitats to ancestral conditions. As habitat connectivity increases, gene flow and environmental variation increase, reducing the likelihood of population divergence and increasing selection on plasticity.

Figure 14.6 Examples of green infrastructure. (a) Stormwater retention, SUNY campus; (b) Green Roof at Walter Reed; (c) Lurie Garden in Chicago's Millennium Park; (d) Rain garden in residential neighbourhood in St Paul, MN; (e) Roadside restoration for pollinators along county highway, St Paul, MN.

Photos: (a) Creative Commons, D.A. Sonnenfeld; (b) Creative Commons, Arlington County; (c) Creative Commons, J.R.P.; (d) E. Snell-Rood; (e) E. Snell-Rood

and Grimm 2020). Green infrastructure includes building with biology, such as increasing canopy cover to reduce heat islands, incorporating more space for urban agriculture, or relying on vegetated ponds and rain gardens in stormwater management (Figure 14.6; Lundholm 2015). Broadly defined, green infrastructure also includes more life-friendly materials, such as permeable pavement, and biophilic design, such as reliance on natural light and living walls (Gaffin et al. 2012). Increasing the amount of green space in cities, sometimes even to just 20%, and decreasing housing density can support more abundant and diverse mammal populations (Fidino et al. 2021). Increasing the size of green spaces, and reducing the distance between them, also has positive effects on urban biodiversity, as predicted by island biogeography theory (Fattorini et al. 2018). Analogies exist in the aquatic space as well. For instance, eco-engineering of seawalls, through increasing crevice complexity and seeding with habitat-forming species, increases the use of this anthropogenic habitat by fish (Ushiama et al. 2019).

Shifting the human-built environment towards biological systems increases the match between novel and ancestral environments for many species. However, the range of species supported by such efforts will also depend on the range of biomes incorporated into the design. As an example, a city such as Minneapolis, USA, that sits at the intersection of three major biomes (prairie, deciduous forest, and boreal forest) should consider habitat elements of each of these biomes if a goal of urban planning is to foster resilient populations of local biodiversity, similar to elements of 'bio-regional design' that look to landscape context (Fisk 2021). Eco-engineering of urban infrastructure to create habitat refuges must similarly consider the diversity of body sizes across target species, as benefits of such interventions are size-dependent (Parris et al. 2018). Connectivity of terrestrial spaces should be complemented by connectivity of aquatic spaces: for instance, irrigation ditches and hydroelectric canals can act as dispersal corridors for fish (Guivier et al. 2019; see also Chapter 8). Habitat interventions that increase the match between ancestral and novel environments will allow more species to gain a foothold in novel environments, often with larger population sizes that can fuel subsequent adaptation; a range of within-city habitat types could also act as 'stepping-stones' of adaptation to more extreme human environments.

14.4.3 The importance of habitat choice in adaptation to human environments

Many novel habitats created by humans are patchy, linear, or other configurations that do not conform to a simple 'allopatric' model of evolutionary and ecological processes. In such cases, habitat choice may significantly increase rates of adaptation to novel environments (Ravigne et al. 2009; Snell-Rood and Steck 2019). In evolutionary models, specialization to a habitat evolves much faster when an organism is able to choose that habitat. Such

choice is particularly relevant to animals that can move around and make active choices, but even plants exert aspects of habitat choice that can influence adaptation, for instance through the timing of growth or the actions of their dispersers (Bazzaz 1991).

To illustrate these ideas in more detail, let us consider a novel habitat—roadsides—and a natural analogy—riparian habitat. In the USA alone, there are over 4 million miles of public roads, 76% of which are two-lane rural highways with a 10 to 15-m adjacent right of way (Federal Highway Administration 2000). For comparison, there is *less* linear distance of rivers and streams in the USA (3.5 million miles; Environmental Protection Agency 2023). Roads are interconnected in a vast network, as are streams and rivers (Figure 14.7), although the structure of such networks is different in terms of the flow of large and small arteries. Roadside habitat is novel. While its open vegetation resembles ancestral prairies or woodland edges, the nutrition of plants in this habitat is complex, with elevated levels of certain nutrients (nitrogen, salts), but also sometimes toxic levels of salts, metals, and pesticides (Phillips et al. 2020). At the same time, there is a risk of wildlife collisions with cars (Keilsohn et al. 2018; see also Chapter 8) and roadsides often contain high numbers of disturbance-adapted species (e.g. invasive plants; Forman and Alexander 1998). In considering responses to novel roadside habitat, we might first focus on plastic, behavioural responses, such as avoidance of cars through immediate reactions (Husby and Husby 2014; Moller and Erritzoe 2017) or learning (e.g. crossing at a particular time; Zeller et al. 2020). But how might a species *evolve* a response to roadsides given their highly linear and diffuse nature over the landscape? Possibly by coupling such adaptation to habitat choice. The analogous linear riparian habitat has many specialist species after millions of years of adaptation. These species are likely cuing into reliable indicators of such habitat, such as tree height, certain plant species, or the presence of water (Rockwell and Stephens 2018). One might hypothesize that over time, species adapted to human habitats—human 'commensals' such as fruit flies *Drosophila melanogaster*, and house sparrows *Passer domesticus* (Hulme-Beaman et al. 2016)—may evolve preferences for cues reliably associated

(a) Road networks

(b) Stream networks

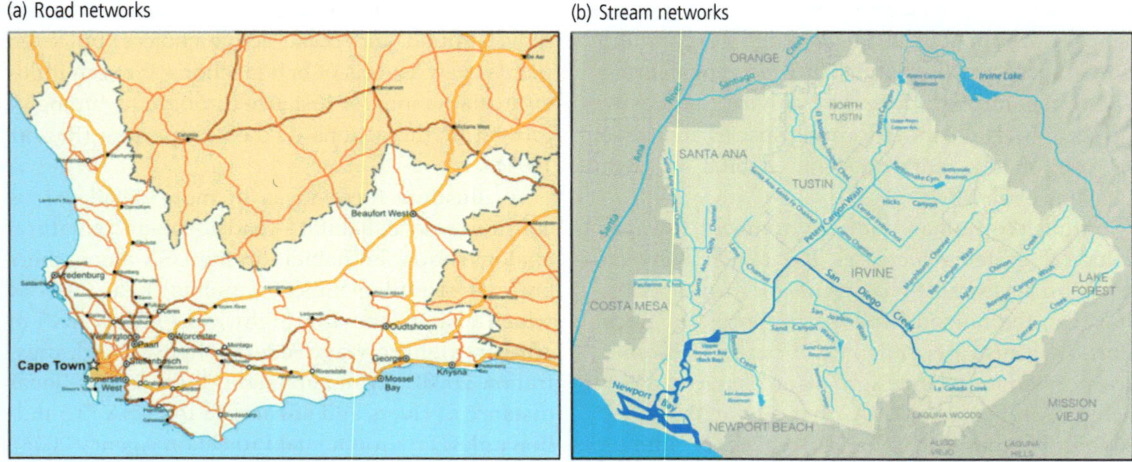

Figure 14.7 Parallels between networks of roads and streams. (a) A map of roads in the Cape Town, South Africa region shows a linear network of large arteries and smaller connecting roads. (b) A map of stream networks in the Newbort Bay watershed in Orange County, California, USA showing smaller stream tributaries joining to larger river drainages.

Images: (a) Creative Commons license, created by Adrian Frith; (b) Creative Commons license, created by Shannon1

Table 14.1 Summary of factors that promote adaptive plastic and evolutionary responses to novel environments.

Plastic responses	Evolutionary responses
Environmental variation within generations	Large effective population sizes
Environmental variation between generations	Rapid generation times
Reliable and predictive cues	Barriers between contrasting habitats
Developmental movement, stimulation	High connectivity between similar habitats
Developmental and social support	Standing genetic variation
Resource availability	Assortative mating through habitat choice
Minimal toxin exposure	Assortative mating through mate choice
Species with long generation times	Plastic responses aligned with variation

with humans. For example, some mosquitoes (e.g. *Anopheles gambiae*) have evolved attraction to human odours (Dormont et al. 2021). Time will tell if the early colonists of roadside habitats diverge to 'roadside specialists' that take advantage of this novel habitat through habitat choice.

14.4.4 Promoting behavioural plasticity in novel environments

Behavioural plasticity is a primary mechanism by which animals adaptively adjust to novel, human environments. We can draw from the literature on the evolution of behavioural plasticity to begin

to hypothesize how the structure of built environments might promote plasticity (Table 14.1). Behavioural plasticity includes both immediate behavioural responses to new conditions ('see a car, fly away') and learned behavioural responses ('observe conspecific dead on road, learn to avoid road'; Snell-Rood 2013). Adaptive plasticity stems from a few key factors. First, making the right decision in new environments depends on reliable cues of the environment, and the associated right decision. Mismatch of evolved cue-response systems in novel environments can create ecological or evolutionary traps (Robertson et al. 2013; see also Chapter 8): how can animals adjust? If cues are unpredictable or changing, it is hard to learn

Box 14.1 Promoting adaptation to artificial light

To illustrate the ideas covered in this chapter, we focus here on the example of artificial light at night. This topic is covered in detail in Chapter 4; here, we focus on plastic and evolutionary responses to artificial light, and how this knowledge can increase robust and adaptive responses to light pollution. For many animals, artificial light at night represents an ecological trap—a once useful cue is no longer correlated with environments where the animals thrive, such as when streetlights attract sea turtles away from the sea, or trap moths or migrating birds in their glare. In other words, attraction to artificial light is a maladaptive behavioural response.

However, some animals have adjusted their behaviour in a seemingly adaptive way to artificial light. Some wading birds that use visual cues in hunting make use of artificial night light to feed for longer stretches of time, increasing their food intake rates by 83% (Santos et al. 2010). In some songbirds, artificial light allows earlier dawn singing and advances in reproductive timing (Kempenaers et al. 2010). Predators such as bats and spiders often feast on trapped insects at streetlights (Gomes 2020). At least some of these adaptive responses represent plasticity, whether an immediate response to prey availability or a learned response to seek out prey at artificial lights. Some of these adaptive responses are a result of evolutionary change. Naïve spiders (*Larinioides sclopetarius*) show heritable variation in attraction to light for web-building (Heiling 1999), and urban spiders (*Steatoda triangulosa*) show reduced light avoidance relative to rural spiders when reared in a common garden (Czaczkes et al. 2018).

How can we use this knowledge to promote conservation of diverse species in the face of artificial light? First, we should consider a range of interventions to reduce the negative impacts of artificial light on species where it is an ecological trap. Populations have a greater chance of 'evolutionary rescue' if their decline is less steep, as larger populations have greater rates of evolutionary change. Indeed, for insects, the spread of artificial light has been a suggested factor in the 'insect apocalypse' (Boyes et al. 2021). Fortunately, there are changes to artificial lighting that are less detrimental to insects, such as the use of high-pressure sodium bulbs (Plummer et al. 2016). In another example, dimming of artificial lights allows more light-averse species of bats to use the area (Rowse et al. 2018), and halving building window-lighting could reduce collisions of migrating birds by 6–11 times (Van Doren et al. 2021). Second, we might consider interventions to increase the chance that individuals may learn to avoid lights. For instance, intermittent flashing lights are much less detrimental in attracting migrating birds to towers (Longcore et al. 2008), and could potentially give birds an opportunity to learn, as they can 'escape' from the light. Any intervention that reduces mortality from lights also increases the chance of learning: some studies suggest that over a third of insects that are attracted to streetlights perish (Eisenbeis 2006), and individuals cannot learn if they are dead. Over time, evolutionary changes may allow further adaptation to such artificial light and population rebounds. One could test such ideas by contrasting parts of the country with 'dark sky laws', as these areas tend to have less total light and rely disproportionally on sodium bulbs. We might expect that larger populations could colonize the less stressful 'dark sky' regions, which could then provide a stepping-stone of adaptation to more extreme artificial light environments.

the optimal response (McLinn and Stephens 2006). Ensuring the availability of detectable and reliable cues can increase the chance animals can adjust their behaviour to human environments. Paved surfaces, metals, and machines are a few reliable indicators of human environments, which animals could use as cues. However, technological advancements mean these cues are sometimes a moving target. For example, the presence of a car coming down the road generally has both a visual cue (shiny, fast-moving object) and a noise, from both the tyres and the car engine (Lima et al. 2015). Improvements to car tyres and pavements, along with conversion to electric vehicles, has resulted in quieter cars (Sandberg 2012), which may make roadsides less stressful in terms of noise. However, quieter cars may present a problem for animals learning to avoid cars, as there are fewer detectable and reliable cues associated with a car approaching.

Second, adaptive behavioural plasticity is often associated with larger relative brain size across species (Sol et al. 2005; Benson-Amram et al. 2016) and conditions that promote neural development within species (Sale et al. 2009). For instance, bird species with larger brains show more innovative foraging behaviours, are more likely to fly away from oncoming cars, and do better in cities than species with smaller brains (Maklakov et al. 2011; Husby and Husby 2014). It is possible that we could promote cognitive development of animals living in human environments through changes in diet, activity, and social support that stimulate cognitive development (Johnson et al. 2016; Snell-Rood and Snell-Rood 2020). Many human environments, while novel, are also simplified, such as a monoculture lawn. Simplification tends to have negative effects on cognitive development (van Praag et al. 2000; Sale et al. 2009), suggesting that more diverse urban landscapes could also support more behaviourally plastic wildlife, although this is a hypothesis that needs to be tested. Many human environments also contain elevated levels of heavy metals that negatively affect neural development (lead, cadmium, mercury; Liu and Lewis 2014; Karri et al. 2016; see also Chapter 3). Efforts to reduce pollutant levels will likely have corresponding benefits to wildlife cognition (Burger and Gochfeld 1993). Finally, perception of risk, and associated social or parental support, plays a major role in cognitive development (Snell-Rood and Snell-Rood 2020; Tooley et al. 2021). This suggests that habitat features that reduce predation risk (e.g. availability of cover) or improve contact with social groups (e.g. reduce noise to increase parent-offspring communication) may be beneficial for cognitive development. These ideas represent hypotheses in need of future research.

14.5 Conclusions and next steps

In this chapter, we have joined theory from behavioural ecology and evolutionary biology to consider the importance of plasticity in human environments, and how the design of human environments may promote plasticity and adaptation. In trying to promote adaptive plasticity and evolution in novel, human habitats, we have hypothesized that a range of factors may be important (Table 14.1), and many of these factors have implications for conservation biology, urban planning, and landscape architecture. For instance, the hypothesis that cognitive development may be promoted in urban environments through more complex and biodiverse landscapes could be tested through mesocosm plots of lawns versus more biodiverse prairies where parental insects are introduced into large cages to control for genetic differences. The impact of linking this research to conservation biology will be greatly enhanced by direct collaboration with relevant stakeholders, such as urban planners or natural resource agencies. Many urban designers are eager to develop and test new methods, offering exciting partnerships with biologists (Felson and Ellison 2021). For example, one might hypothesize that green walls will support larger populations of an insect of interest in urban parks, fostering evolutionary change over time. Biologists could work with urban designers to sample populations before and following installation of such structures.

In the present discussion, we have focused primarily on animals and animal behaviour. However, many of the most powerful studies and meta-analyses that tease apart the effects of plastic and genetic responses to new environments come, not from animal behaviour, but from studies of plants or microbes (Radersma et al. 2020). It is important to consider to what extent we can generalize across all systems. Do we expect plants and animals to be different? It is possible that plants may indeed be different from a plasticity perspective—they are generally more sessile, which could increase selection on plasticity through coarse-grained environmental variation, and they tend to be modular, which may decrease the costs of plasticity (Snell-Rood and Ehlman 2021). Animals may be more likely to match their environment through movement and habitat choice than through plasticity (Holtmann et al. 2017). It is unclear what this means for comparative responses to human environments, but underscores that studying a diversity of systems is important.

In conclusion, many animals are showing both adaptive plastic responses to novel human environments, and subsequent evolutionary adaptation. However, not all species are responding in the same way. The degree of initial niche overlap with novel conditions will affect which species can move into new human environments. Capacity for developmental plasticity (such as learning) and behavioural flexibility (such as context-dependent behavioural expression) will further influence who does well in novel conditions. Populations that gain a foothold in human environments often undergo subsequent adaptation. The landscape and built structure of the new environment will affect patterns of cognitive development and also factors that influence evolutionary change. Humans can consider how altering the built environment—from incorporating green infrastructure, to planting a diversity of resources in habitat corridors—can increase the likelihood of species adjusting to novel conditions.

Acknowledgements

We are grateful to the intellectual community offered by the MSP Urban LTER, which provided much inspiration for these ideas (NSF-DEB 2045382). Interactions with collaborators in the College of Design (Mary Guzowski, William Weber, and Jessica Rossi-Mastracci) also provided important food-for-thought while developing this chapter (funded by the Templeton Foundation 62220).

References

Ackley, J.W., Angilletta, M.J., DeNardo, D., et al. (2015). Urban heat island mitigation strategies and lizard thermal ecology: landscaping can quadruple potential activity time in an arid city. *Urban Ecosystems*, 18, 1447–1459.

Akesson, S., and Helm, B. (2020). Endogenous programs and flexibility in bird migration. *Frontiers in Ecology and Evolution*, 8, 78.

Badyaev, A.V., Young, R.L., Oh, K.P., and Addison, C. (2008). Evolution on a local scale: developmental, functional, and genetic bases of divergence in bill form and associated changes in song structure between adjacent habitats. *Evolution*, 62, 1951–1964.

Bazzaz, F.A. (1991). Habitat selection in plants. *The American Naturalist*, 137, S116–S130.

Benson-Amram, S., Dantzer, B., Stricker, G., et al. (2016). Brain size predicts problem-solving ability in mammalian carnivores. *Proceedings of the National Academy of Sciences*, 113, 2532–2537.

Bernard, G.D., and Remington, C.L. (1991). Color vision in Lycaena butterflies—spectral tuning of receptor arrays in relation to behavioral ecology. *Proceedings of the National Academy of Sciences*, 88, 2783–2787.

Boutin, S., and Lane, J.E. (2014). Climate change and mammals: evolutionary versus plastic responses. *Evolutionary Applications*, 7, 29–41.

Boyes, D.H., Evans, D.M., Fox, R., et al. (2021). Is light pollution driving moth population declines? A review of causal mechanisms across the life cycle. *Insect Conservation and Diversity*, 14, 167–187.

Brown, C.R., and Brown, M.B. (2013). Where has all the road kill gone? *Current Biology*, 23, R233–R234.

Buchholz, R., Banusiewicz, J.D., Burgess, S., et al. (2019). Behavioural research priorities for the study of animal response to climate change. *Animal Behaviour*, 150, 127–137.

Burger, J., and Gochfeld, M. (1993). Lead and behavioral development in young herring gulls—effects of timing and exposure on individual recognition. *Fundamental and Applied Toxicology*, 21, 187–195.

Campbell-Staton, S.C., Velotta, J.P., and Winchell, K.M. (2021). Selection on adaptive and maladaptive gene expression plasticity during thermal adaptation to urban heat islands. *Nature Communications*, 12, 6195.

Carlen, E., and Munshi-South, J. (2021). Widespread genetic connectivity of feral pigeons across the Northeastern megacity. *Evolutionary Applications*, 14, 150–162.

Carlson, S.M., Cunningham, C.J., and Westley, P.A.H. (2014). Evolutionary rescue in a changing world. *Trends in Ecology & Evolution*, 29, 521–530.

Charmantier, A., and Gienapp, P. (2014). Climate change and timing of avian breeding and migration: evolutionary versus plastic changes. *Evolutionary Applications*, 7, 15–28.

Chytry, M., Lososova, Z., Horsak, M., et al. (2012). Dispersal limitation is stronger in communities of microorganisms than macroorganisms across Central European cities. *Journal of Biogeography*, 39, 1101–1111.

Corl, A., Bi, K., Luke, C., et al. (2018). The genetic basis of adaptation following plastic changes in coloration in a novel environment. *Current Biology*, 28, 2970–2977.

Czaczkes, T.J., Bastidas-Urrutia, A.M., Ghislandi, P., and Tuni, C. (2018). Reduced light avoidance in spiders from populations in light-polluted urban environments. *Science of Nature*, 105, 11–12.

Davis, A.M., and Glick, T.F. (1978). Urban ecosystems and island biogeography. *Environmental Conservation*, 5, 299–304.

Dormont, L., Mulatier, M., Carrasco, D., and Cohuet, A. (2021). Mosquito attractants. *Journal of Chemical Ecology*, 47, 351–393.

Dufour, P., de Franceschi, C., Doniol-Valcroze, P., et al. (2021). A new westward migration route in an Asian passerine bird. *Current Biology*, 31, 5590–5596.

Dunn, R.R., Burger, J.R., Carlen, E.J., et al. (2022). A theory of city biogeography and the origin of urban species. *Frontiers in Conservation Science*, 3, 761449.

Eisenbeis, G. (2006). Artificial night lighting and insects: attraction of insects to streetlamps in a rural setting in Germany. *Ecological Consequences of Artificial Night Lighting*, 2, 191–198.

Ellis, E.C., and Ramankutty, N. (2008). Putting people in the map: anthropogenic biomes of the world. *Frontiers in Ecology and the Environment*, 6, 439–447.

Environmental Protection Agency. (2023). National Rivers and Streams Assessment. [online]. Available from: https://www.epa.gov/national-aquatic-resource-surveys/nrsa [accessed 12 August 2023].

Evans, K.L., Chamberlain, D.E., Hatchwell, B.J., et al. (2011). What makes an urban bird? *Global Change Biology*, 17, 32–44.

Fattorini, S., Mantoni, C., De Simoni, L., and Galassi, D.M.P. (2018). Island biogeography of insect conservation in urban green spaces. *Environmental Conservation*, 45, 1–10.

Federal Highway Administration. (2000). Our nation's highways. [online]. Available from: https://www.fhwa.dot.gov/ohim/onh00/our_ntns_hwys.pdf [accessed 12 August 2023].

Felson, A.J., and Ellison, A.M. (2021). Designing (for) urban food webs. *Frontiers in Ecology and Evolution*, 9, 582041.

Fenn, M.E., Bytnerowicz, A., Schilling, S.L., et al. (2018). On-road emissions of ammonia: an underappreciated source of atmospheric nitrogen deposition. *Science of the Total Environment*, 625, 909–919.

Fidino, M., Gallo, T., Lehrer, E.W., et al. (2021). Landscape-scale differences among cities alter common species' responses to urbanization. *Ecological Applications*, 31, e02253.

Fisk, P. (2021). Bioregional design: the design science of the future. In: M. Neuman and W. Zonneveld (eds), *The Routledge Handbook of Regional Design*, pp. 356–373. Routledge, London.

Forman, R.T.T., and Alexander, L.E. (1998). Roads and their major ecological effects. *Annual Review of Ecology and Systematics*, 29, 207–231.

Francis, R.A. (2020). Urban cliffs. In: I. Douglas, P.M.L. Anderson, D. Goode, et al. (eds), *The Routledge Handbook of Urban Ecology*, pp. 278–288. Routledge, London.

Gaffin, S.R., Rosenzweig, C., and Kong, A.Y.Y. (2012). Adapting to climate change through urban green infrastructure. *Nature Climate Change*, 2, 704.

Gomes, D.G.E. (2020). Orb-weaving spiders are fewer but larger and catch more prey in lit bridge panels from a natural artificial light experiment. *PeerJ*, 8, e8808.

Graves, S.D., and Shapiro, A.M. (2003). Exotics as host plants of the California butterfly fauna. *Biological Conservation*, 110, 413–433.

Greggor, A.L., Clayton, N.S., Fulford, A.J.C., and Thornton, A. (2016). Street smart: faster approach towards litter in urban areas by highly neophobic corvids and less fearful birds. *Animal Behaviour*, 117, 123–133.

Grimm, N.B., Faeth, S.H., Golubiewski, N.E., et al. (2008). Global change and the ecology of cities. *Science*, 319, 756–760.

Gross, K., Pasinelli, G., and Kunc, H.P. (2010). Behavioral plasticity allows short-term adjustment to a novel environment. *The American Naturalist*, 176, 456–464.

Guivier, E., Gilles, A., Pech, N., et al. (2019). Canals as ecological corridors and hybridization zones for two cyprinid species. *Hydrobiologia*, 830, 1–16.

Harmens, H., Norris, D.A., Cooper, D.M., et al. (2011). Nitrogen concentrations in mosses indicate the spatial distribution of atmospheric nitrogen deposition in Europe. *Environmental Pollution*, 159, 2852–2860.

Heger, T., Bernard-Verdier, M., Gessler, A., et al. (2019). Towards an integrative, eco-evolutionary understanding of ecological novelty: studying and communicating interlinked effects of global change. *Bioscience*, 69, 888–899.

Heiling, A.M. (1999). Why do nocturnal orb-web spiders (Araneidae) search for light? *Behavioral Ecology and Sociobiology*, 46, 43–49.

Hobbie, S.E., and Grimm, N.B. (2020). Nature-based approaches to managing climate change impacts in cities. *Philosophical Transactions of the Royal Society B: Biological Sciences*, 375, 20190124.

Hobbs, R.J., Higgs, E.S., and Hall, C. (2013). *Novel Ecosystems: Intervening in the New Ecological World Order*. John Wiley & Sons, Hoboken, NJ.

Holtmann, B., Santos, E.S.A., Lara, C.E., and Nakagawa, S. (2017). Personality-matching habitat choice, rather than behavioural plasticity, is a likely driver of a phenotype-environment covariance. *Proceedings of the Royal Society B: Biological Sciences*, 284.

Huey, R.B., Hertz, P.E., and Sinervo, B. (2003). Behavioral drive versus behavioral inertia in evolution: a null model approach. *The American Naturalist*, 161, 357–366.

Hulme-Beaman, A., Dobney, K., Cucchi, T., and Searle, J.B. (2016). An ecological and evolutionary framework for commensalism in anthropogenic environments. *Trends in Ecology & Evolution*, 31, 633–645.

Husby, A., and Husby, M. (2014). Interspecific analysis of vehicle avoidance behavior in birds. *Behavioral Ecology*, 25(3), 504–508.

Johnson, M.T.J., and Munshi-South, J. (2017). Evolution of life in urban environments. *Science*, 358, eaam8327.

Johnson, S.B., Riis, J.L., and Noble, K.G. (2016). State of the art review: poverty and the developing brain. *Pediatrics*, 137, e20153075.

Karri, V., Schuhmacher, M., and Kumar, V. (2016). Heavy metals (Pb, Cd, As and MeHg) as risk factors for cognitive dysfunction: a general review of metal mixture mechanism in brain. *Environmental Toxicology and Pharmacology*, 48, 203–213.

Kearney, M., and Porter, W. (2009). Mechanistic niche modelling: combining physiological and spatial data to predict species' ranges. *Ecology Letters*, 12, 334–350.

Keilsohn, W., Narango, D.L., and Tallamy, D.W. (2018). Roadside habitat impacts insect traffic mortality. *Journal of Insect Conservation*, 22, 183–188.

Kempenaers, B., Borgstrom, P., Loes, P., et al. (2010). Artificial night lighting affects dawn song, extra-pair siring success, and lay date in songbirds. *Current Biology*, 20, 1735–1739.

Lambert, M.R., Brans, K.I., Des Roches, S., et al. (2021). Adaptive evolution in cities: progress and misconceptions. *Trends in Ecology & Evolution*, 36, 239–257.

Lande, R. (1980). Genetic variation and phenotypic evolution during allopatric speciation. *The American Naturalist*, 116, 463–479.

Lima, S.L., Blackwell, B.F., DeVault, T.L., and Fernandez-Juricic, E. (2015). Animal reactions to oncoming vehicles: a conceptual review. *Biological Reviews*, 90, 60–76.

Liu, J.H., and Lewis, G. (2014). Environmental toxicity and poor cognitive outcomes in children and adults. *Journal of Environmental Health*, 76, 130–138.

Longcore, T., Rich, C., and Gauthreaux, S.A. (2008). Height, guy wires, and steady-burning lights increase hazard of communication towers to nocturnal migrants: a review and meta-analysis. *Auk*, 125, 485–492.

Lundholm, J.T. (2015). The ecology and evolution of constructed ecosystems as green infrastructure. *Frontiers in Ecology and Evolution*, 3, 106.

Lundholm, J.T., and Richardson, P.J. (2010). Habitat analogues for reconciliation ecology in urban and industrial environments. *Journal of Applied Ecology*, 47, 966–975.

Maklakov, A.A., Immler, S., Gonzalez-Voyer, A., et al. (2011). Brains and the city: big-brained passerine birds succeed in urban environments. *Biology Letters*, 7, 730–732.

Martin, P.R., Burke, K.W., and Bonier, F. (2021). Plasticity versus evolutionary divergence: what causes habitat partitioning in urban-adapted birds? *The American Naturalist*, 197, 60–74.

Martin, R.A., Chick, L.D., Yilmaz, A.R., and Diamond, S.E. (2019). Evolution, not transgenerational plasticity, explains the adaptive divergence of acorn ant thermal tolerance across an urban-rural temperature cline. *Evolutionary Applications*, 12, 1678–1687.

Maspons, J., Molowny-Horas, R., and Sol, D. (2019). Behaviour, life history and persistence in novel environments. *Philosophical Transactions of the Royal Society B: Biological Sciences*, 374, 20180056.

McCorquodale, S.M. (2013). *A Brief Review of the Scientific Literature on Elk, Roads, and Traffic*. Washington Department of Fish and Wildlife, Olympia, WA.

McLinn, C.M., and Stephens, D.W. (2006). What makes information valuable: signal reliability and environmental uncertainty. *Animal Behaviour*, 71, 1119–1129.

Moller, A.P. (2008). Flight distance of urban birds, predation, and selection for urban life. *Behavioral Ecology and Sociobiology*, 63, 63–75.

Moller, A.P., and Erritzoe, J. (2017). Brain size in birds is related to traffic accidents. *Royal Society Open Science*, 4, 161040.

Munoz, M.M., and Losos, J.B. (2018). Thermoregulatory behavior simultaneously promotes and forestalls evolution in a tropical lizard. *The American Naturalist*, 191, E15–E26.

Nielsen, M.E., and Papaj, D.R. (2015). Effects of developmental change in body size on ectotherm body temperature and behavioral thermoregulation: caterpillars in a heat-stressed environment. *Oecologia*, 177, 171–179.

Nosil, P. (2008). Ernst Mayr and the integration of geographic and ecological factors in speciation. *Biological Journal of the Linnean Society*, 95, 26–46.

Orr, M.R., and Smith, T.B. (1998). Ecology and speciation. *Trends in Ecology & Evolution*, 13, 502–506.

Parmesan, C. (2006). Ecological and evolutionary responses to recent climate change. *Annual Review of Ecology, Evolution, and Systematics*, 37, 637–669.

Parris, K.M., Amati, M., Bekessy, S.A., et al. (2018). The seven lamps of planning for biodiversity in the city. *Cities*, 83, 44–53.

Phillips, B.B., Wallace, C., Roberts, B.R., et al. (2020). Enhancing road verges to aid pollinator conservation: a review. *Biological Conservation*, 250, 108687.

Plummer, K.E., Hale, J.D., O'Callaghan, M.J., et al. (2016). Investigating the impact of street lighting changes on garden moth communities. *Journal of Urban Ecology*, 2, 1–10.

Pollack, L., Munson, A., Savoca, M.S., et al. (2022). Enhancing the ecological realism of evolutionary mismatch theory. *Trends in Ecology & Evolution*, 37, 233–245.

Radersma, R., Noble, D.W.A., and Uller, T. (2020). Plasticity leaves a phenotypic signature during local adaptation. *Evolution Letters*, 4, 360–370.

Ravigne, V., Dieckmann, U., and Olivieri, I. (2009). Live where you thrive: joint evolution of habitat choice and local adaptation facilitates specialization and promotes diversity. *The American Naturalist*, 174, E141–E169.

Riddell, E.A., Iknayan, K.J., Hargrove, L., et al. (2021). Exposure to climate change drives stability or collapse of desert mammal and bird communities. *Science*, 371, 633–636.

Robertson, B.A., Rehage, J.S., and Sih, A. (2013). Ecological novelty and the emergence of evolutionary traps. *Trends in Ecology & Evolution*, 28, 552–560.

Robinson, N.A., and Marks, C.A. (2001). Genetic structure and dispersal of red foxes (*Vulpes vulpes*) in urban Melbourne. *Australian Journal of Zoology*, 49, 589–601.

Rockwell, S.M., and Stephens, J.L. (2018). Habitat selection of riparian birds at restoration sites along the Trinity River, California. *Restoration Ecology*, 26, 767–777.

Rosenzweig, M.L. (2003). Reconciliation ecology and the future of species diversity. *Oryx*, 37, 194–205.

Rowse, E.G., Harris, S., and Jones, G. (2018). Effects of dimming light-emitting diode street lights on light-opportunistic and light-averse bats in suburban habitats. *Royal Society Open Science*, 5, 180205.

Sale, A., Berardi, N., and Maffei, L. (2009). Enrich the environment to empower the brain. *Trends in Neurosciences*, 32, 233–239.

Sandberg, U. (2012). Recent development in Europe aiming at the reduction of exterior vehicle noise by producing quieter vehicles, tires and pavements. *SAE Technical Paper Series*, SAE International, 9, 2012–36–0645.

Santos, C.D., Miranda, A.C., Granadeiro, J.P., et al. (2010). Effects of artificial illumination on the nocturnal foraging of waders. *Acta Oecologica*, 36, 166–172.

Sayol, F., Sol, D., and Pigot, A.L. (2020). Brain size and life history interact to predict urban tolerance in birds. *Frontiers in Ecology and Evolution*, 8, 58.

Sih, A. (2013). Understanding variation in behavioural responses to human-induced rapid environmental change: a conceptual overview. *Animal Behaviour*, 85, 1077–1088.

Sih, A., Ferrari, M.C.O., and Harris, D.J. (2011). Evolution and behavioural responses to human-induced rapid environmental change. *Evolutionary Applications*, 4, 367–387.

Smith, T.B., Kinnison, M.T., Strauss, S.Y., et al. (2014). Prescriptive evolution to conserve and manage biodiversity. *Annual Review of Ecology, Evolution, and Systematics*, 45, 1–22.

Snell-Rood, E.C. (2013). An overview of the evolutionary causes and consequences of behavioural plasticity. *Animal Behaviour*, 85, 1004–1011.

Snell-Rood, E., Cothran, R., Espeset, A., et al. (2015). Life history evolution in the Anthropocene: effects of increasing nutrients on traits and tradeoffs. *Evolutionary Applications*, 8, 635–649.

Snell-Rood, E.C., and Ehlman, S.M. (2021). Ecology and evolution of plasticity. In: D.W. Pfennig (ed.), *Phenotypic Plasticity and Evolution*, pp. 139–160. CRC Press, Boca Raton, FL.

Snell-Rood, E.C., and Papaj, D.R. (2009). Patterns of phenotypic plasticity in common and rare environments: a study of host use and color learning in the cabbage white butterfly *Pieris rapae*. *The American Naturalist*, 173, 615–631.

Snell-Rood, E., and Snell-Rood, C. (2020). The developmental support hypothesis: adaptive plasticity in neural development in response to cues of social support. *Philosophical Transactions of the Royal Society B: Biological Sciences*, 375, 20190491.

Snell-Rood, E.C., and Steck, M.K. (2019). Behaviour shapes environmental variation and selection on learning and plasticity: review of mechanisms and implications. *Animal Behaviour*, 147, 147–156.

Sol, D., Duncan, R.P., Blackburn, T.M., et al. (2005). Big brains, enhanced cognition, and response of birds to novel environments. *Proceedings of the National Academy of Sciences*, 102, 5460–5465.

Sol, D., Lapiedra, O., and Gonzalez-Lagos, C. (2013). Behavioural adjustments for a life in the city. *Animal Behaviour*, 85, 1101–1112.

Stamp, M.A., and Hadfield, J.D. (2020). The relative importance of plasticity versus genetic differentiation in explaining between population differences; a meta-analysis. *Ecology Letters*, 23, 1432–1441.

Szulkin, M., Munshi-South, J., and Charmantier, A. (2020). *Urban Evolutionary Biology*. Oxford University Press, New York, NY.

Tooley, U.A., Bassett, D.S., and Mackey, A.P. (2021). Environmental influences on the pace of brain development. *Nature Reviews Neuroscience*, 22, 372–384.

Tuomainen, U., and Candolin, U. (2011). Behavioural responses to human-induced environmental change. *Biological Reviews*, 86, 640–657.

Ushiama, S., Mayer-Pinto, M., Bugnot, A.B., et al. (2019). Eco-engineering increases habitat availability and utilisation of seawalls by fish. *Ecological Engineering*, 138, 403–411.

Van Doren, B.M., Willard, D.E., Hennen, M., et al. (2021). Drivers of fatal bird collisions in an urban center. *Proceedings of the National Academy of Sciences*, 118, e2101666118.

van Praag, H., Kempermann, G., and Gage, F.H. (2000). Neural consequences of environmental enrichment. *Nature Reviews Neuroscience*, 1, 191–198.

Volis, S., Ormanbekova, D., and Yermekbayev, K. (2015). Role of phenotypic plasticity and population differentiation in adaptation to novel environmental conditions. *Ecology and Evolution*, 5, 3818–3829.

Wagnon, G.S. (2019). *Temporal Changes in Morphology of the Cliff Swallow (Petrochelidon pyrrhonota*. The University of Tulsa, Tulsa, OK.

Wikelski, M., and Cooke, S.J. (2006). Conservation physiology. *Trends in Ecology & Evolution*, 21, 38–46.

Winchell, K.M., Losos, J.B., and Verrilli, B.C. (2023). Urban evolutionary ecology brings exaptation back into focus. *Trends in Ecology & Evolution*, 38, 719–726.

Youngsteadt, E., Dale, A.G., Terando, A.J., et al. (2015). Do cities simulate climate change? A comparison of herbivore response to urban and global warming. *Global Change Biology*, 21, 97–105.

Zeller, K.A., Wattles, D.W., and Destefano, S. (2020). Evaluating methods for identifying large mammal road crossing locations: black bears as a case study. *Landscape Ecology*, 35, 1799–1808.

CHAPTER 15

Animal welfare

Culum Brown and Sarah Wahltinez

Overview

Our rapidly changing world has considerable repercussions for animal welfare. Here we provide a brief introduction to the history of animal welfare science and consider the impacts of a changing world on animal welfare, including food production systems, climate change, pollution, and invasive species, as well as identifying solutions. We then outline how we can harness our growing knowledge of animal behaviour to generate effective welfare interventions. Finally, we identify areas where future research needs to be concentrated, such as responding to animal welfare during human conflict, increasing our capacity to provide positive welfare experience for animals in our care, and recognizing sentience and welfare in invertebrates.

15.1 Introduction

Animal welfare considers the emotional and physical health and well-being of animals. Although the origins of animal welfare science are in the treatment of animals under human care (Broom 2011), animal welfare can, and should, be applied to all animals. It is important to recognize the grass-roots origins of animal welfare, since it arose, not through government or industry movements, but out of the general public's concern for the well-being of animals. This public concern ultimately forced governments to act through the creation of animal welfare and protection laws. Initially, the concept arose out of concern for working animals, such as horses *Equus caballus* and dogs *Canis lupus familiaris*, but it quickly spread to cover domesticated farm

animals following the industrialization of food production systems. The modern movement arguably started in the 1960s in the UK after the publication of Ruth Harrison's *Animal Machines* (1964) but quickly spread throughout Europe (Broom 2011). The EU now has some of the most stringent animal welfare regulations found anywhere in the world, with animal welfare being a core principle of all business conducted within the EU (Article 13 of the Lisbon Treaty). Other nations have followed suit with varying degrees of animal welfare regulations (for a review see Schmid and Kilchsperger 2010).

At its inception, the concept of animal welfare was based on the premise that animals are capable of suffering and that we have a moral obligation to prevent this suffering. As our understanding of animal welfare has evolved over time, a growing emphasis has been placed on positive experiences and emotional states for animals. In 1789, Bentham (1789, p. 283n) asked 'the question is not, Can they reason? nor, Can they talk? but, Can they suffer'. In a utilitarian viewpoint of the world championed by Singer and others, we have an obligation to reduce suffering wherever it may be encountered (e.g. Singer 2005). In order to suffer, animals must have the capacity to 'feel'; that is, they have the capacity for positive and negative effective states; the animal must be sentient. Acknowledging that positive or poor welfare is largely about quantifying an animal's feelings presents us with a seemingly insurmountable problem: how can one know what others are thinking or feeling? For this reason, there is great debate over exactly which animals are considered sentient because sentience is ultimately a private experience. While we cannot quantify how an animal feels directly, we can garner some ideas by looking for signs and symptoms, which is most often achieved by recording changes in behaviour and physiology (Dawkins 1993). There is also growing recognition that affective states can influence an animal's decision-making processes (Paul et al. 2005). Despite these issues, it is widely acknowledged that all vertebrates are likely sentient, as are at least some invertebrates, such as cephalopods and decapod crustaceans. In the context of sentient invertebrates, Hymenopterans are arguably at the forefront of the debate at present, with mounting evidence that bees, wasps, and ants show traits that suggest they may also be sentient (Barron and Klein 2016; Mikhalevich and Powell 2020; see Box 15.1). This view suggests that sentience may have arisen more than once in the Animal Kingdom (Brown 2020). While the concept of sentience underlies all notions of animal welfare, not all welfare legislation specifically states that animals are sentient.

The Five Freedoms model provided guidance on how to prevent poor welfare of animals in our care. The model initially sprang from the UK after a parliamentary enquiry recognized that farm animals must be able to stand, turn around, lie down, stretch, and groom themselves (Brambell 1965). In the late 1970s this was extended to include the psychological well-being of animals, such as freedom from fear, by the Farm Animal Welfare Council. The modern concept of the Five Freedoms was further refined in the early 1990s (Farm Animal Welfare Council 1993):

(1) Freedom from hunger and thirst: by ready access to fresh water and a diet to maintain full health and vigour.
(2) Freedom from discomfort: by providing an appropriate environment including shelter and a comfortable resting area.
(3) Freedom from pain, injury, or disease: by prevention through rapid diagnosis and treatment.
(4) Freedom to express normal behaviour: by providing sufficient space, proper facilities, and company of the animal's own kind.
(5) Freedom from fear and distress: by ensuring conditions and treatment that avoid mental suffering.

Although the Five Freedoms concept provided the foundations of animal welfare, it arguably did not provide sufficient attention to the mental health of animals, nor did it place much emphasis on the need for positive welfare: a life worth living (Mellor 2016). This line of thinking ultimately gave rise to the Five Domains model initially outlined by Mellor and Reid (1994) and updated by Mellor et al. (2020). The components of the Five Domains model are similar to the Five Freedoms but, in the Five Domains model, nutrition, the physical environment, health, and behavioural interactions all, ultimately, influence the mental status of the animal (Figure 15.1).

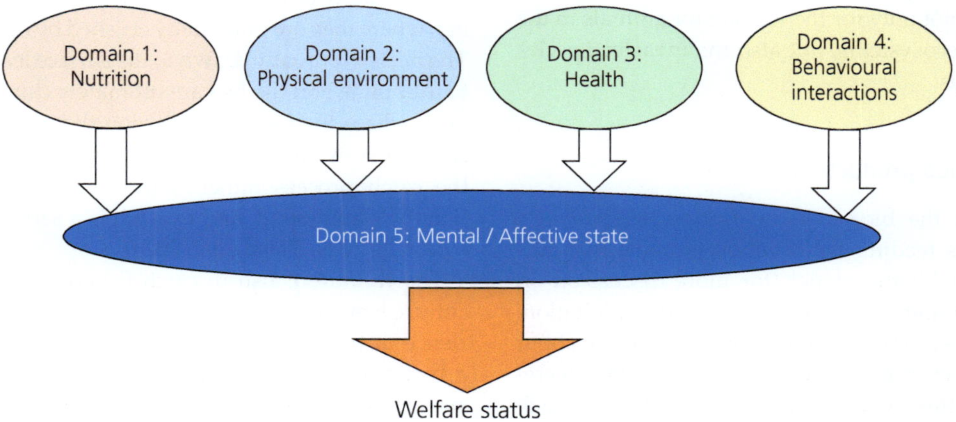

Figure 15.1 The Five Domains model.
After Mellor et al. 2020

15.2 The impact of a changing world on animal welfare

The world is changing at unprecedented rates thanks largely to anthropogenic impacts. The fundamental underlying issue is that the human population continues to grow and is projected to reach 10.4 billion people by 2100 (UN 2022). Changes associated with a growing human population have profound impacts on the natural and artificial environment in which animals live, and will have widespread implications for animal welfare. The increasing human population will mean that we have more interactions with animals as food sources, working animals, pets, and companions, and increased wildlife conflict (see also Chapter 13). In the context of environmental change and human impacts, the number of topics that could be addressed from an animal welfare perspective are almost limitless. Here, we have chosen to focus on just four. First, we consider the welfare implications of food production, which is intimately linked to human population growth, and is also responsible for a considerable proportion of land clearing. Second, we discuss the animal welfare consequences of climate change, arguably the most insidious threat of our time. Third, we address pollution, which is a major source of environmental degradation, but is seldom viewed through an animal welfare lens. Lastly, we address the welfare impacts of managing invasive species, which is an emerging environmental issue that has ramifications, not only for the welfare for animals in the recipient ecosystems, but also the invasive species themselves.

15.2.1 Food production

Arguably, the biggest issue from a welfare perspective is feeding the 8 billion humans that currently inhabit the planet (for more discussion on behavioural effects of harvesting and exploitation, see Chapter 7). Developing countries derive about 60% of their protein from vegetables (primarily cereals), but this is declining and expected to reach 50% by 2030. By contrast, industrialized countries receive only 30% of their protein from cereals. In the meantime, world meat consumption has

doubled since the 1960s, but has tripled in developing countries over the same period (WHO 2002). Most animal protein for human consumption is generated from industrial meat production systems, which have been the centre of attention for animal welfare improvement and innovation for decades. Gathering food from wild animals is almost entirely restricted to commercial fishing practices, which are proving to be unsustainable and for which there is little to no welfare protection (Metcalfe 2009). Most terrestrial animal production systems, by contrast, are covered by existing animal welfare legislation. For this reason, here we focus on welfare concerns surrounding fishing and aquaculture practices.

Fishing practices vary considerably, and so too do the animal welfare implications (Veldhuizen et al. 2018; Breen et al. 2020; Figure 15.2). Some tuna fishing industries in the Indian and Pacific Oceans use line and pole catch methods, which are arguably the best in terms of fish welfare. This is because the fish are captured on a barbless hook, rapidly brought on board the vessel, and killed quickly before being put on ice. Tuna is a very high-value product and care is taken to ensure each fish gets to market in the best possible condition. Pole and line fishing methods tend to have low levels of bycatch, but many also rely on bait species (e.g. sardines and herring) that are often captured using seine or trawl nets, which have poor welfare outcomes for the bait. At the other end of the extreme, long-haul demersal trawl nets have multiple welfare concerns. Fish are effectively chased to exhaustion and fall back into the net, where they are potentially crushed by other fish. The net is eventually drawn to the surface leading to barotrauma, and the fish are ultimately dumped on deck where they suffocate during the sorting process. Most fish die or are seriously injured through the catch process rather than from being intentionally slaughtered, and bycatch is a serious issue (Metcalfe 2009; Breen et al. 2020).

The welfare of fish in catch fisheries and aquaculture has more implications than solely the experience of the fish. Ensuring good welfare leads to a better, higher-quality product and can enhance sustainability. Research shows that reduced stress during capture can also prolong the shelf life and enhance flesh texture (Daskalova 2019; Anders et al. 2022). While there is ongoing research trying to

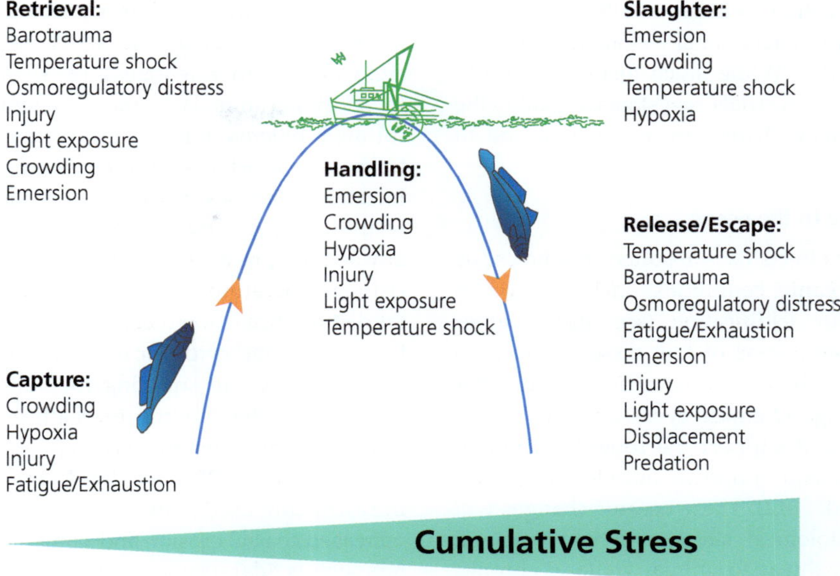

Figure 15.2 Some of the welfare impacts of the wild fisheries capture process.
From Breen et al. 2020

reduce the welfare impacts of commercial fishing practices (Metcalfe 2009; Breen et al. 2020), the reality is that shifts away from current practices and the decline of fish stocks (Food and Agriculture Organization of the United Nations 2022) will likely cause the industry to become economically inviable. Aquaculture has grown rapidly and has recently overtaken wild capture in terms of supplying protein for human consumption (Food and Agriculture Organization of the United Nations 2022). There is ongoing effort to ensure animal welfare during fish production (Daskalova 2019), and multiple welfare certification schemes are emerging (e.g. RSPCA, Friends of the Sea, Aquaculture Stewardship Council, etc.).

One possible solution to the welfare issues faced by animal production is to remove live animals from the production system entirely. There is increasing interest in producing animal protein through lab-based methods, and there are multiple companies investing in the technology (Waltz 2021). Cultured meats are still in their early stages, but were approved for sale in Singapore in 2020 and in the USA in 2022 (Chodkowska et al. 2022). While technically feasible, the production of 'cell-based'

seafood is currently expensive and has to compete with wild-caught and aquaculture-derived products. Given that cell-based seafood is still some way off, in the meantime, there is still an urgent need to address welfare issues in wild caught and aquaculture industries.

15.2.2 Climate change

The environment, human well-being, and animal welfare are interconnected, as recognized by the emerging discipline of One Welfare, which extends the concept of One Health and seeks to improve both human and animal health (García Pinillos et al. 2016). The changing environment (see also Chapter 1) will invariably impact animal health and welfare through many avenues, including physiological distress from the environmental change itself, impacts on animal behaviour, and emerging diseases. Even climate change mitigation strategies can affect animal welfare. As campaigns focus on decreasing greenhouse gas emissions through changes to the management of livestock, little attention has been given to the animal welfare outcomes of these strategies, such as altering feed and genetic selection to improve efficiency, which may negatively impact animal welfare (Shields and Orme-Evans 2015). While much focus to date has been directed at terrestrial ecosystems, arguably the impacts of climate change are more profound in aquatic systems.

Climate change in the ocean

Anthropogenic effects have resulted in wholescale changes to oceanic conditions, including ocean warming, ocean acidification (see also Chapter 5), and increasing areas of low dissolved oxygen concentrations. These oceanic changes will alter the complex biogeochemical cycles and ecosystems (Gruber 2011) that support the roughly 1 million marine species and countless individual animals (Appeltans et al. 2012). These oceanic changes will result in physiological stress, and while not all stresses pose a threat to animal welfare, chronic, unavoidable physiological stress can result in distress, which negatively impacts animal welfare. Due to the inverse relationship between water temperature and the solubility of oxygen, increasing ocean temperatures will result in decreasing dissolved oxygen. For aquatic organisms, hypoxic water is of particular concern for the potential to negatively impact animal welfare through the feeling of breathlessness (Garcia et al. 2005; Beausoleil and Mellor 2015). Human activities further exacerbate this issue, as nutrient inputs result in eutrophication and increasing hypoxic dead zones (Howarth et al. 2011). Exposure to decreased pH has been shown to impact the nervous system in many species of fish and marine invertebrates (Moya et al. 2016; Thomas et al. 2020; Chapter 5). In fish, behavioural changes appear to be caused by alterations to GABA (γ-aminobutyric acid; Nilsson et al. 2012; Hamilton et al. 2014). Some of these behavioural states, such as anxiety, represent poor welfare, and these alterations to GABA may further impact welfare as it plays an important role in the modulation of pain (Enna and McCarson 2006).

Climate change and aquaculture

Effects of climate change can impact the health and welfare of aquatic animals worldwide, including those in fisheries and aquaculture. As above, ocean warming, ocean acidification, changes in rainfall patterns, and hypoxia can affect the productivity and welfare of invertebrates and fish in aquaculture. In addition, animals in these systems risk harmful algal blooms, the impacts of rising sea levels, freshwater shortages, and increased pressure from infectious diseases. Harmful algal blooms (HABs) result in losses of over US \$8 billion per year globally and are predicted to increase in frequency with rising temperatures, nutrient enrichment, and habitat disturbance (Brown et al. 2020). The neurotoxic, haemolytic, and cytotoxic toxins produced by algae in HABs result in suffering for humans and animals alike, often leading to death in fish and the birds and marine mammals that predate upon them (Brown et al. 2020). Sea level rise will damage infrastructure used in near-shore production, cause increased coastal erosion and vulnerability to disasters such as tidal surges, and will result in decreased water quality with saline intrusion into brackish and freshwater systems (Rosa et al. 2012; Dabbadie et al. 2018)—with profound impacts on offshore fish farms. In freshwater systems, decreasing availability of freshwater from changes in weather patterns and increased pond evaporation will result in struggles to maintain water quality and production (Rosa et al. 2012; Dabbadie et al. 2018; Figure 15.3). Poor water quality results in poor animal welfare through physiological stress, increased susceptibility to disease, and decreased nutritional status (Ellis et al. 2002). Climate change will have effects on the host range, pathogen range, immune competence of the host, virulence of the pathogen, and pathogen transmission rates (Marcogliese 2014). The transmission rates and virulence of pathogens is expected to increase with increasing temperatures (Marcogliese 2014). Combined with environmental stressors from climate change, it is likely there will be increased morbidity and mortality due to pathogens and parasites in aquacultured fish and invertebrates (Marcogliese 2014).

Heat

In both aquatic and terrestrial ecosystems, the increasing—and increasingly variable—temperatures associated with climate change can result in stress, with consequences for animal welfare. Ambient temperature plays an important

Figure 15.3 Salmonid aquaculture both onshore and offshore will be impacted by climate change having various impacts on fish welfare.
Photo: Shuttershock.com

role in animal physiology, behaviour, survival, reproduction, and distribution (Elmore et al. 2017). While not all changes associated with increasing temperatures will result in poor welfare, living outside an animal's thermal preference zone alone can challenge homeostasis and act as a stressor (Moberg 2000). Animals will need to use their cognitive processes to respond and adjust their behaviour as the climate around them changes. In experimental studies, heat stress negatively impacts cognition; however, more research is needed to evaluate the link between heat stress and cognition in free-ranging wildlife (Soravia et al. 2021). Each species has a thermal tolerance that results from its physiology and behavioural capacity (Elmore et al. 2017). As the ambient temperature changes, the distribution of species will change, as wildlife move to remain in their thermal tolerance envelope. This movement of animals will have ramifications for food webs, ecological communities, and ecosystems (Walther 2010; Zhang et al. 2017). Some animals may thrive in their new environment, thus providing opportunities for positive welfare, while others may struggle, and experience poor welfare.

Additionally, due to habitat fragmentation, animals may not be able to shift their distributions enough to remain in an appropriate habitat. Freshwater fishes are a good example. In such cases, species translocations, or assisted migrations, can be a useful conservation tool to safeguard vulnerable species from extinction (Thomas 2011; Butt et al. 2020). Conservation efforts will certainly face many challenges as climate, ecosystems, and biological communities change. Conservation strategies will need to adapt to be effective, and cross-discipline collaborations will be necessary to adequately plan conservation actions in the face of climate change (Hannah et al. 2002; see also Chapter 11).

Heat stress is already becoming an increasing welfare and production concern for livestock in both extensive and intensive systems as ambient temperatures rise. Domestic ruminants suffering from heat stress have reduced metabolic rates, oxidative stress, immune suppression, and decreased feed intake, with resulting changes in productivity (Slimen et al. 2016; Lacetera 2019). Heat-induced death has also been reported during

Figure 15.4 Dairy cows kept cool by water sprays and fans.
Photo: Shuttershock.com

the hottest months and during extreme weather events (Lacetera 2019). When determining the welfare impacts of temperatures, measurement of the ambient temperature is insufficient. There are many factors that govern heat exchange and the heat load experienced by the animal; animal behaviour can also modulate these environmental impacts, and there are differences in the susceptibility to heat stress depending on species, breed, and even individuals (Silanikove 2000). To minimize welfare impacts of increased temperatures, livestock should be provided with shade and water, limitations on transportation should be instituted, and appropriate breeds and species should be selected (Silanikove 2000). For example, dairy goats *Capra hircus* are more resilient to heat stress and can be used as a replacement for cows *Bos* spp. for milk production in areas with temperatures outside the thermal preference zone for cows (Silanikove and Koluman 2015; Figure 15.4).

Climate change and disease

Animal health (see also Chapter 17) is an important aspect of welfare. When an animal is diseased, it will have worse welfare than when healthy; conversely, an animal that has poor welfare may have a compromised immune system and be predisposed to disease (Broom 1991). As the concepts of One Welfare and One Health indicate, animal health and welfare, human health and welfare, and environmental health are inextricably linked together (García Pinillos et al. 2016; Shaheen 2022). When humans change the environment with loss of biodiversity, urbanization, and the expansion and intensification of agriculture, zoonotic viruses are able to spill over from animals to humans—and can lead to pandemics (Goldstein et al. 2022). Of the roughly 1400 infectious organisms that are known to cause disease in humans, 61% can be transmitted between people and animals through zoonosis, and 75% of emerging pathogens are zoonotic (Taylor et al. 2001). Pathogens are transmitted through hunting, preparation, and consumption of wildlife (e.g. bushmeat), arthropod vectors (e.g. ticks and mosquitoes), and contact with wildlife (e.g. ecotourism, as pets, or preventing crop raiding) (Muehlenbein 2013). Forest fragmentation can create an 'edge effect', which can increase the chance of zoonotic pathogen transmission (Muehlenbein 2013).

15.2.3 Pollution

The impacts of pollution on animal behaviour are reasonably well established (Bertram et al. 2022) and there are multiple ways in which it might have welfare consequences. Plastics, heavy metals, and chemicals, including pharmaceuticals, all have impacts on the livelihoods, health, and well-being of animals (see also Chapter 3). Classic examples of the impacts of plastics include entanglement by marine animals, and ingestion by livestock, seabirds, and turtles (Butterworth 2016; Anderssen et al. 2022; Figure 15.5). Like humans, animals are also exposed to chemical pollution, whether that be in the air they breathe or through water they drink or live in. There is growing realization, for example, that air pollution can have massive impacts on the welfare of food production animals through declining health (Ni et al. 2021). Of course, there have been many examples of anthropogenic chemical catastrophes throughout human history (e.g. Agent Orange, nuclear fallout, and, more recently, per- and polyfluoroalkyl substances) for which the animal welfare impacts are simply too great to list.

Arguably, one of the more insidious sources of chemical pollution stems from pharmaceuticals. Humans increasingly rely on pharmaceuticals to address all kinds of physical and mental health problems. After ingestion, much of the active ingredients or by-products are excreted and enter wastewater systems that are not designed to treat these chemicals. Those chemicals then enter aquatic ecosystems, having a wide range of impacts on wildlife behaviour (Brodin et al. 2014) and the entire ecosystem (Jones et al. 2001). Some of the key traits that can be altered include activity (movement and

migration), predator avoidance, foraging, sexual function, mate choice, parental care, and even personality (Saaristo et al. 2018). The intensity of the impact is dose dependent, and can range from acute toxicity to subtle shifts in behaviour. While much work has been done on examining the impacts of particular drugs on the behaviour of certain model taxa (e.g. zebrafish *Danio rerio*), little is known about how cocktails of drugs in the environment might interact with one another or the broader impacts on ecological function (e.g. Udebuani et al. 2021). More research is needed to understand the extent and impacts of chemical pollution before effective solutions can be fully implemented. Some strategies that may be useful include environmental education for prevention (Karataş and Karataş 2016), nanomaterials for remediation (Trujillo-Reyes et al. 2014), and ecological engineering, such as vegetated filter strips and constructed wetlands (Wu et al. 2013).

Some of the less obvious impacts of pollution on animals are those stemming from anthropogenic noise (Chapter 2) and light pollution (Chapter 4) both in aquatic and terrestrial environments. Of the two, far more is known about the impacts of noise on animals and there are a multitude of impacts that are likely of concern to animal welfare. Ironically, some of this noise pollution is generated by our attempts to address climate change, for instance through the installation of offshore wind turbines (Haelters et al. 2013). Noise is often generated by machinery, such as cars and aeroplanes, and not only causes shifts in animal distributions and behaviour, but can also interfere with animal communication and, in extreme cases, cause deafness (Duquette et al. 2021). This is exacerbated in aquatic environments because sound travels further and faster underwater (Weilgart 2007). Stranding in cetaceans is often attributed to marine prospecting, for example, especially where seismic surveys are conducted to identify gas and oil deposits (Weilgart 2007). While whales are often the centre of attention, there are also impacts on other animals, including turtles (Nelms et al. 2016) and fishes (Radford et al. 2014). Solutions for noise pollution in the ocean include the creation of acoustic refuges, development of quieter propellers, reducing the speed of ships, and the development and implementation of

Figure 15.5 Plastic is pervasive in the environment causing considerable animal welfare problems ranging from (a) entanglement to (b) ingestion.
Photo: Shuttershock.com

quieter methods for pile driving and exploring the ocean floor (Weilgart 2023).

In the context of light pollution, there is increased urgency to address the likely impacts on wildlife (Hölker et al. 2010) given that about 23% of the Earth already experiences artificial light at night (Komine et al. 2020), with the problem growing at about 6% per annum (Silva et al. 2017). As with sound pollution, light pollution can have important welfare implications. At the fundamental level, most animals' activity patterns, behaviour, growth, and even metabolic rate are controlled by circadian rhythms or clocks. Changing light levels—particularly at night when it is supposed to be dark—has profound effects, not just on individual species, but on entire ecosystems. Some classic examples of how light can change behaviour include insects being drawn to light around which they hover until they die of exhaustion. Bats, reptiles, and amphibians (e.g. invasive cane toads

Figure 15.6 Animals are attracted to lights, which can significantly enhance predation risk.
Photo: Shuttershock.com

Rhinella marina) are drawn to lights knowing that their insect prey will be available (Komine et al. 2020; Figure 15.6). Light pollution also interferes with migration. In birds, for example, changes to the circadian clock cause disorientation during flight, which can have dramatic consequences during this critical stage of their life cycle (Cabrera-Cruz et al. 2018). Birds are also drawn towards artificial lights in buildings, causing death following collisions (Van Doren et al. 2021). Likewise, light pollution impacts the nesting behaviour of sea turtles that must pull out onto sandy beaches to lay their eggs. Adult females are startled and disorientated by artificial light, and light therefore reduces the number of suitable laying locations. The greatest impacts, however, are on the nestlings that emerge from the sand and must scurry to the safety of the sea to avoid predation. Hatchlings are normally attracted to light scattered off the sea surface and are disorientated by artificial lights, and the enhanced illumination greatly increases predation (Silva et al. 2017). To mitigate light pollution, the duration, trespass, intensity, and spectrum of lighting can be changed to reduce the ecological impact, while unlit areas should be preserved as dark refuges for animals (Gaston et al. 2012).

15.2.4 Biological invasions

Invasive species (Chapter 6) are a growing problem around the world. Biological invasion can have welfare implications for the native animals that are negatively affected by the invasion, as well as through attempts to control the invasive species. When a non-native animal arrives with human assistance and establishes a self-sustaining population, it is known as invasive (Simberloff 2013). Some definitions include a provision that the species must cause harm to the environment, economy, or human health (Simberloff 2013). Of course, invasive species can have significant impacts on the welfare of native species via competition for resources, aggression, predation, and even cross-infection of parasites. In many instances, native species have not evolved to cope with the invasive animal. A classic example of this is the introduction of cane toads in Australia, where native predators were naïve to the toxic invaders, which, in turn, caused widespread devastation (Shine 2010). A mainstay of invasive species management is lethal control, which can take the form of hunting, physical methods (e.g. traps or snares), chemicals (e.g. insecticides, 1080, rotenone, anticoagulant rodenticides), or biological methods (e.g. natural enemies or pathogens) (Cowan et al. 2011; Simberloff 2013). However, emphasis should be placed on rigorous exclusion and biosecurity practices to prevent invasions, the solution with the best animal welfare for all species involved. When that fails, invasive species management often involves the death of many animals, including both the target species and non-target by-kill. These methods raise animal welfare concerns over the pain and suffering caused by the method employed, as well as indirect flow-on effects. When possible, the control methods chosen should be as humane as possible, the minimum number of animals should be killed, and lethal control should only be utilized when there is a high likelihood of success in controlling the invasive species (Cowan et al. 2011). Naturally, this will need to be counterbalanced against broader welfare impacts of invasive species on native wildlife.

15.3 Harnessing behaviour for effective welfare interventions

Animal behaviour and animal welfare are intertwined. The behaviours of an animal can indicate their welfare state, the ability to perform certain behaviours may impact welfare, and behaviour can be used to evaluate the wants and needs of animals. Behaviour is often easily observable and can be evaluated quickly, and without the need for expensive equipment, which makes it an attractive animal welfare indicator. For relatively low cost, behavioural observations can be automated, which facilitates its use in intensive animal production systems (Rushen et al. 2012). For behaviour to be a meaningful animal welfare indicator, the repertoire of normal behaviours needs to be established first, typically in the form of an ethogram that contains information on which behaviours are being performed, as well as their context. Given that many domestic animals are maintained in artificial environments that do not provide an opportunity for all natural behaviours to be performed, care should be taken to provide a more extensive environment when establishing ethograms. However, when comparing animals in extensive and intensive management, or free-ranging and captive animals, it is important to remember that animals can display remarkable

plasticity in behaviour and a behavioural change alone does not necessarily imply anything about the animal's welfare. This is one of the criticisms of the use of behavioural integrity as a welfare indicator (Dawkins 2008). Nevertheless, despite such criticisms, behavioural integrity can still be a useful starting point, and can be applied to internally motivated behaviours and behaviours where the eliciting stimuli are present where the animals are kept (Würbel 2009). Since outside factors can impact behaviour, covariates including husbandry, weather, enrichment, and feeding practices should also be collected at the same time as behavioural observations (Watters et al. 2009).

It is important to realize that simply being a 'natural behaviour' does not indicate that animals must be able to perform that behaviour to have the opportunity for good welfare. For example, a prey animal avoiding a predator may actually have a negative impact on the predator's welfare. Rather, an animal-centric view of animal welfare should focus on what behaviours or resources are important to that animal (Dawkins 2023). Behaviours that an animal is highly motivated to perform will have the greatest impact on welfare. Examples of highly internally motivated behaviours include nest building in sows *Sus scrofa domesticus* (Arey et al. 1991), dust bathing in chickens *Gallus gallus domesticus* (Hogan and van Boxel 1993), and swimming in mink *Mustela vison* (Warburton and Mason 2003). If animals are not given the resources to allow them to perform these highly motivated behaviours, frustration (an aversive emotional state) may occur. Evidence to support this frustration include animals remaining motivated to perform these behaviours even when not given the resources to do so (Rushen and Passillé 1992) and being willing to pay a cost to access them (e.g. pushing a weighted door; Mason et al. 2001).

One way to determine what behaviours or resources are important—or aversive—to an animal is to allow the animal to inform us by using a preference or choice test. Preference testing evaluates the choice an animal makes and can provide information about what the animal likes, dislikes, and their motivations behind the choice (Fraser and Nicol 2018). This testing procedure allows the animal to indicate their preference when choosing between two or more options, and allows researchers to use this information to determine what behaviours or resources are important to the animal. This technique has been used to evaluate cage flooring in hens (Hughes and Black 1973), substrates for pigs (Beattie et al. 1998), bedding materials for mice *Mus musculus* (Kawakami et al. 2007), lighting for rats *Rattus norvegicus* (Blom et al. 1995), and water temperature for lobsters *Homarus americanus* (Nielsen and McGaw 2016) among many others. With sufficient replication and comparisons, one can build a hierarchy of preferences. When combined with assays that determine how much animals are willing to work to access certain needs (Duncan and Kite 1987), which provides a 'common currency', scientists can build a complex picture of the animal's needs. While a well-designed preference testing experiment can help refine husbandry and improve animal welfare, it is important to remember that it does not necessarily specify what is best for the animal; rather, the animal is choosing what is preferable, or least aversive, given the options provided. Moreover, preferences may change over time (e.g. season or ontogeny) and with context. Another limitation is that animals may not choose what is best for their long-term interests. For example, animals may indicate a preference for a certain food type, but this may not represent a healthy, balanced diet (Prebble and Meredith 2014).

Behavioural responses can be evaluated to determine an animal's emotional state. As recognized in the Five Domains, positive and negative subjective experiences impact an animal's emotional state, which together make up the animal's welfare status (Mellor et al. 2020). In judgement bias testing, the impact of an animal's emotional state on cognitive processes is evaluated by measuring the response to ambiguous cues. Reduced anticipation of a positive event to ambiguous stimuli is observed in depressed or anxious humans and is thought to similarly correspond to negative affective states in animals (Harding et al. 2004), and has been linked to poor welfare (Henry et al. 2017). Changes to an animal's affective state have been noted to alter the judgement bias in multiple species, and although the effect is small and responses can be variable, judgement bias tests are widely recognized as a valid method to measure affect in animals (Lagisz et al. 2020). Tests to evaluate fear and anxiety focus on the animal's response to something novel; a new situation in the novel arena test, a new inanimate object in the novel object test, a new person in the voluntary or involuntary approach test, or handling in the tonic immobility test or restraint test. While these tests were initially developed for use in rodents (Hall 1936), they have been applied to livestock (Forkman et al. 2007), rainbow trout *Oncorhynchus mykiss* (Sneddon et al. 2003), captive corn snakes *Pantherophis guttatus* (Hoehfurtner et al. 2021), hornbills *Bucerotidae* spp. (Garcia-Pelegrin et al. 2022), and even Port Jackson sharks *Heterodontus portusjacksoni* (Byrnes et al. 2016). These tests can provide informative results when well designed; however, it is important to account for the complexity of emotions and potential confounding factors.

Behaviour can be a useful indicator for both positive and negative animal welfare. In fact, behaviour is one of the few indicators that can currently be used to identify positive welfare in animals. Play behaviours, behavioural diversity, behaviour synchrony, and prosocial behaviours in social species have been proposed as indicators of positive welfare (Napolitano et al. 2009; Held and Špinka 2011; Rault 2019; Miller et al. 2020). It is important to note that the absence of behaviours indicative of negative welfare does not indicate positive welfare. Behaviours associated with negative welfare, such as pain, illness, and distress, have received much attention. The facial expressions of animals have been used to evaluate pain as 'grimace scores', which have been developed for a variety of species including mice, cats *Felis catus*, cows, and pigs. To create a baseline, animals should be scored for at least three days before the painful event and trends, rather than absolute scores, tend to be the most useful (Cohen and Beths 2020). Another well-established behavioural indicator for pain is gait scoring in dairy cattle *Bos taurus*. A five-point scale using back posture, head position, and presence of a limp was able to identify cows with sole ulcers, a painful condition (Flower and Weary 2006) and improved with analgesic medications (Flower et al. 2008), which indicates that gait scoring can be a reliable behavioural indicator of pain.

Stereotypies have attracted considerable attention as a behavioural indicator of welfare, especially in animals housed in zoos. Stereotypies are repetitive behaviours that have no clear function for the animal performing them, and while they have previously been defined as invariant, they may have slightly flexible motor patterns (Mason et al. 2007). Since they are often observed in situations thought to be aversive and increase in stressful situations, the presence of stereotypies is often thought to indicate poor welfare. If injury results from the stereotypy, there is a clear negative impact on welfare. However, there are many factors that influence the development and frequency of stereotypies and the relationship between stereotypic behaviour and welfare is complex. Stereotypies almost certainly indicate that an animal has experienced poor welfare; however, they do not always indicate the animal is currently suffering (Mason 1991). It is also important to differentiate between stereotypies and anticipatory behaviour, which can both present as a repetitive locomotor activity such as pacing. Anticipatory behaviour is goal-oriented behaviour that occurs in the appetitive phase prior to reward acquisition and may generate a positive affective state for the animal (Watters 2014). As such, it is important to evaluate the animal's motivation and carefully consider the context of behaviours to correctly assess their relationship to an animal's welfare.

Behaviour can also be a valuable tool in assessing the welfare of free-ranging wildlife. As with domestic animals, it is important to evaluate the context of the observed behaviour, note external factors that may influence behaviour, and remember that behaviour is plastic. When assessing the welfare of wildlife, it is important to use species-specific measurable or observable indicators that are validated, be able to identify individual animals, and use a grading system (Harvey et al. 2020). Camera traps can be used to non-invasively collect data on welfare indicators (Harvey et al. 2021) such as flight initiation distance, breeding success, and mate selection (Tarlow and Blumstein 2007).

While it may not be as obvious as with animals under human care, human activities can negatively impact the behaviour and welfare of wildlife by preventing animals from performing important behaviours or interactions. As stated by Heini Hediger, 'however paradoxical it may sound, the truth is actually this: the free animal does not live in freedom: neither in space nor as regards its behaviour

towards other animals' (Hediger 1950: 4). Human activities have changed the habitat and distribution of wildlife, as well as adding domestic and invasive species to the landscape. As with captive animals, the behavioural response of wildlife can also mitigate these human impacts on animal welfare (Yahner and Mahan 1997; Cale 2003). Human interactions are often stressful to wildlife. Behaviour has been used to evaluate animal welfare during population control of feral horses (Hampton et al. 2017) and wild boar *Sus scrofa* (Fahlman et al. 2020), the responses of whales to anthropogenic sound (Tyack 2009), and the stress responses of wild guanaco to being handled and sheared (Taraborelli et al. 2017). Wildlife tourism is growing in popularity and represents an important way to reconnect people with the natural environment (Curtin and Kragh 2014). However, some wildlife tourism activities can have negative impacts on wildlife, ranging from transient changes in behaviour due to human presence to increased mortality in the population (Green and Giese 2004). One activity that has particularly large impacts on animal behaviour and welfare is cub petting, where lion *Panthera leo* or tiger *Panthera tigris* cubs are kept for photographs with tourists (Cohen 2012; Chorney et al. 2022). These cubs exhibit altered activity budgets and stress behaviours including avoidance and aggression (Chorney et al. 2022; Wilson and Phillips 2023). Other types of wildlife tourism, such as gorilla trekking to see mountain gorillas *Gorilla gorilla beringei*, visiting sea turtle or crocodile farms, and wild dolphin interactions, have negative impacts on animal welfare and conservation (Moorhouse et al. 2015).

Observations of behaviours can be a useful tool to evaluate the welfare of both domestic and free-ranging animals, the ability to perform certain behaviours may impact welfare, and behaviour can also be used to evaluate what animals want and need. However, it is not without limitations. There is no single behavioural indicator of optimal or poor welfare; in fact, there can be considerable overlap. The use of multiple welfare indicators is far better than overreliance on a few, and these indicators should be ground-truthed to known physiological indicators (e.g. cortisol, heart rate, lactate, body condition, etc.). Animal behaviour should be evaluated holistically, with the context of the behaviour in mind and covariates about other factors collected alongside behavioural observations. It is also important to take into account observer effects, where animal behaviour may change due to the presence of the human observer (Metcalfe et al. 2022). When possible, blinding should be used in behavioural evaluation, as the observer's expectations may impact results (Tuyttens et al. 2014), and behaviours should be carefully screened for consistency and repeatability to help avoid gender bias in the observer (Marsh and Hanlon 2004).

15.4 Future directions

At the end of the Second World War, the new world order was supposed to bring prolonged peace and stability to the world. However, as we have seen repeatedly since, large-scale wars and conflicts continue apace, as demonstrated most recently during Russia's invasion of Ukraine. As climate change enforces new pressures on society, it is predicted that conflicts will likely increase in future. Animals are one of the seldom recognized victims of conflict, whether they be directly used in conflict, such as dogs, sea lions *Zalophus californianus*, bottlenose

dolphins *Tursiops truncatus*, and horses, or live in the areas of conflict, such as companion animals, farm animals, animals used in research, or those held in zoos or aquariums. In all cases, the welfare of these animals often takes second place over self-preservation, although there are often images of people desperately trying to look after their animals in conflict zones (Figure 15.7). Some refuse to evacuate unless their pets come with them; others are reunited with their livestock post-conflict—displaying all the emotions of being reunited with long-lost friends. There is a clear and growing need to pay more attention to animal welfare in war zones.

As the world around us changes, so should our concept of animal welfare. Since its inception, animal welfare science has tended to focus on reducing negative welfare experiences. However, as the switch from the Five Freedoms to the Five Domains model highlights, there is a growing need to focus on providing positive experiences for animals. Such a move will necessarily involve generating greater understanding of the needs and desires of animals in our care and recognition of positive affective states. Examples of this might include providing

Figure 15.7 A man holds his pet cat in Ukraine after Russian bombardment.

Photo: Shuttershock.com

individuals with the capacity to make choices to enhance the feeling that they can exercise some control over their lives, or providing the opportunity for play and meaningful companionship.

Invertebrate welfare is an emerging frontier in animal welfare science (Box 15.1) and has recently been gaining increasing attention. Invertebrates make up > 95% of the species on Earth (Ruppert et al. 2004), and are involved in many processes that are vital to the health of ecosystems—and of humans. Invertebrates pollinate plants (Chapter 16),

control vegetation, break down detritus, form coral reefs that protect shorelines, and serve as important food sources for humans and other animals (Wilson 1987). Due to their role in food production, aquatic invertebrates are at the forefront of many discussions regarding what, if any, welfare considerations should be extended to invertebrate taxa. According to the Food and Agriculture Organization of the United Nations, 5.6 million tonnes of crustaceans, 2.5 million tonnes of cephalopod molluscs, 3.4 million tonnes of non-cephalopod

---

Box 15.1 Leveraging behavioural research to improve invertebrate welfare

Beyond having appropriate physiology (e.g. integrative brain regions, nociceptors, and connections between these parts of the nervous system), behaviour can be used to evaluate the sentience of animals. Crump and colleagues (2022) have proposed the following behavioural criteria for sentience:

- Responses to noxious stimuli are affected by anaesthetic and analgesic agents
- Flexible decision-making with motivational trade-offs
- Self-protective behaviours are performed in response to noxious stimuli (e.g. rubbing or guarding the affected area)
- Associative learning beyond habituation and sensitization
- The animal indicates that it values anaesthetic and analgesic agents following noxious stimuli

Selected behavioural evidence for sentience in cephalopod molluscs

- Bock's pygmy octopus *Octopus bocki* demonstrated protective behaviours following acetic acid application, which were extinguished with lidocaine application, and demonstrated preference for chambers in the testing arena where they were given lidocaine following injury (Crook 2021).
- Common cuttlefish *Sepia officinalis* passed the 'marshmallow test' and were willing to forgo immediate gratification for a preferred, but delayed, reward (Schnell et al. 2021).
- Algae octopus *Abdopus aculeatus* and lesser octopus *Eledone cirrhosa* showed protective behaviours (Polglase et al. 1983; Alupay et al. 2014) and squid *Loligo pealeii* demonstrated long-term defensive behaviours following arm injury (Crook et al. 2011).
- Cephalopods are highly trainable and several species have demonstrated associative learning in the

'prawn-in-the-tube' experiment, where they learn to not attack an inaccessible prey item (Zepeda et al. 2017), chambered nautilus *Nautilus pompilius* have been trained on a classical conditioning task (Crook and Basil 2008), and octopus *Octopus vulgaris* can perform observational learning (Fiorito and Scotto 1992).

Selected behavioural evidence for sentience in decapod crustaceans

- Application of local anaesthetics changed the behavioural response of shrimp to eyestalk ablation (Taylor et al. 2004; Diarte-Plata et al. 2012).
- Hermit crabs *Pagurus bernhardus* require a higher voltage shock to leave a shell they find more desirable and show grooming behaviours after a shock (Appel and Elwood 2009).
- Shore crabs *Carcinus maenas* show directed behaviours to their mouth and eyes following acetic acid application (Elwood et al. 2017) and edible crabs *Cancer pagurus* show wound-directed behaviours following manual declawing (McCambridge et al. 2016).
- Mud crabs *Eurypanopeus depressus* are capable of shock-avoidance learning (Punzo 1983) and burrowing crabs *Neohelice granulatus* can be trained using both positive and negative reinforcers (Dimant and Maldonado 1992; Fernandez-Duque et al. 1992).

The behavioural evidence was evaluated in a report compiled by Birch and colleagues (2021) at the London School of Economics and Political Science and led to cephalopod molluscs and decapod crustaceans being recognized under the Animal Welfare (Sentience) Bill in November 2021. This paves the way for increased oversight of practices that may negatively impact the welfare of these invertebrates in aquaculture, fisheries, and research.

molluscs, and 0.5 million tonnes of other invertebrates including jellyfish and sea urchins were harvested from capture fisheries worldwide in 2020 (Food and Agriculture Organization of the United Nations 2022). An additional 525,000 tonnes of invertebrates were reared in aquaculture (Food and Agriculture Organization of the United Nations 2022). This represents a staggering number of individual invertebrate animals that are harvested for consumption by people each year. The complexity of the nervous system and behavioural repertoire of cephalopod molluscs (e.g. octopus, squid, cuttlefish) and decapod crustaceans (e.g. crabs, lobsters, shrimp) have led to recognition that they may be sentient (see Box 15.1 and Crump et al. 2022 for proposed criteria for sentience), and thus have resulted in increased animal welfare protections. However, proven sentience is not a prerequisite for consideration of an animal's welfare, especially with the overwhelming lack of research with many species. Given the challenges in interpreting the emotional state of fellow humans, much less animals with a very different life history, more research that involves creative approaches and a wide range of species is needed. Absence of evidence is not evidence of absence. In the meantime, the welfare of invertebrates can be improved through the use of effective anaesthesia, analgesia, and euthanasia techniques; implementation of less invasive diagnostic and research sampling methods; use of humane slaughter methods; and reducing the impacts of invasive procedures in aquaculture and fisheries, such as eyestalk ablation of shrimp and declawing of crabs (Wahltinez et al. 2022).

15.5 Conclusions

Animal welfare regards the emotional and physical health and well-being of animals, and has evolved from the prevention of suffering to the promotion of positive emotional states such as play and social behaviours. Human activities have had a profound effect on animal welfare, particularly through intensive agriculture, climate change, pollution, and invasive species. Animal welfare and behaviour are inextricably linked. Behavioural observation can be used to evaluate welfare, the ability to perform certain behaviours may impact welfare, and

behaviour can also be leveraged to determine what animals want and need to promote better treatment of animals. There are many areas in need of further research, including how to account for animal welfare during human conflict, increasing the capacity for animals to experience positive welfare, and improving the welfare of invertebrate animals. While many of the topics within this chapter are challenging and may seem insurmountable, with human ingenuity, hard work, and the resilience of animals, we can find our way forward. However, we must take into account the welfare of all animals we come into contact with as we plan solutions and advance onward.

References

Alupay, J.S., Hadjisolomou, S.P., and Crook, R.J. (2014). Arm injury produces long-term behavioral and neural hypersensitivity in octopus. *Neuroscience Letters*, 558, 137–142.

Anders, N., Breen, M., Skåra, T., et al. (2022). Effects of capture-related stress and pre-freezing holding in refrigerated sea water (RSW) on the muscle quality and storage stability of Atlantic mackerel (*Scomber scombrus*) during subsequent frozen storage. *Food Chemistry*, 405, 134819.

Anderssen, K.E., Gabrielsen, G.W., Kranz, M., and Collard, F. (2022). Magnetic resonance imaging for non-invasive measurement of plastic ingestion in marine wildlife. *Marine Pollution Bulletin*, 185, 114334.

Appel, M., and Elwood, R.W. (2009). Motivational trade-offs and potential pain experience in hermit crabs. *Applied Animal Behaviour Science*, 119, 120–124.

Appeltans, W., Ahyong, S.T., Anderson, G., et al. (2012). The magnitude of global marine species diversity. *Current Biology*, 22, 2189–2202.

Arey, D.S., Petchey, A.M., and Fowler, V.R. (1991). The preparturient behaviour of sows in enriched pens and the effect of pre-formed nests. *Applied Animal Behaviour Science*, 31, 61–68.

Barron, A.B., and Klein, C. (2016). What insects can tell us about the origins of consciousness. *Proceedings of the National Academy of Sciences*, 113, 4900–4908.

Beattie, V.E., Walker, N., and Sneddon, I.A. (1998). Preference testing of substrates by growing pigs. *Animal Welfare*, 7, 27–34.

Beausoleil, N.J., and Mellor, D.J. (2015). Introducing breathlessness as a significant animal welfare issue. *New Zealand Veterinary Journal*, 63, 44–51.

Bentham, J. (1789). *An Introduction to the Principles of Morals and Legislation*. Republished 1907 by Clarendon Press, Oxford.

Bertram, M.G., Martin, J.M., McCallum, E.S., et al. (2022). Frontiers in quantifying wildlife behavioural responses to chemical pollution. *Biological Reviews*, 97, 1346–1364.

Birch, J., Burn, C., Schnell, A., et al. (2021). *Review of the evidence of sentience in cephalopod molluscs and decapod crustaceans*. London School of Economics and Political Science, London.

Blom, H.J.M., Van Tintelen, G., Baumans, V., et al. (1995). Development and application of a preference test system to evaluate housing conditions for laboratory rats. *Applied Animal Behaviour Science*, 43, 279–290.

Brambell, F.W.R. (1965). *Report on the Technical Committee to Enquire into the Welfare of Livestock Kept Under Intensive Conditions*. Her Majesty's Stationery Office, London.

Breen, M., Anders, N., Humborstad, O.-B., et al. (2020). Catch welfare in commercial fisheries. In: T.S. Kristiansen, A. Ferno, M.A. Pavlidis, and H. van de Vis (eds), *The Welfare of Fish*, pp. 401–437. Springer Nature, Cham.

Brodin, T., Piovano, S., Fick, J., et al. (2014). Ecological effects of pharmaceuticals in aquatic systems—impacts through behavioural alterations. *Philosophical Transactions of the Royal Society B: Biological Sciences*, 369, 20130580.

Broom, D.M. (1991). Animal welfare: concepts and measurement. *Journal of Animal Science*, 69, 4167–4175.

Broom, D.M. (2011). A history of animal welfare science. *Acta Biotheoretica*, 59, 121–137.

Brown, A.R., Lilley, M., Shutler, J., et al. (2020). Assessing risks and mitigating impacts of harmful algal blooms on mariculture and marine fisheries. *Reviews in Aquaculture*, 12, 1663–1688.

Brown, C. (2020). Convergent evolution of sentience? *Animal Sentience*, 5(29), 25. https://www.wellbeingintlstudiesrepository.org/animsent/vol5/iss29/25/

Butt, N., Chauvenet, A.L.M., Adams, V.M., et al. (2020). Importance of species translocations under rapid climate change. *Conservation Biology*, 35, 775–783.

Butterworth, A. (2016). A review of the welfare impact on pinnipeds of plastic marine debris. *Frontiers in Marine Science*, 3, 1–10.

Byrnes, E.E., Vila-Pouca, C., and Brown, C. (2016). Laterality strength is linked to stress reactivity in Port Jackson sharks (*Heterodontus portusjacksoni*). *Behavioural Brain Research*, 305, 239–246.

Cabrera-Cruz, S.A., Smolinsky, J.A., and Buler, J.J. (2018). Light pollution is greatest within migration passage areas for nocturnally-migrating birds around the world. *Scientific Reports*, 8, 4–11.

Cale, P.G. (2003). The influence of social behaviour, dispersal and landscape fragmentation on population structure in a sedentary bird. *Biological Conservation*, 109, 237–248.

Chodkowska, K.A., Wódz, K., and Wojciechowski, J. (2022). Sustainable future protein foods: the challenges and the future of cultivated meat. *Foods*, 11, 4008.

Chorney, S., DeFalco, A., Jacquet, J., et al. (2022). Poor welfare indicators and husbandry practices at lion (*Panthera leo*) 'cub-petting' facilities: evidence from public YouTube videos. *Animals*, 12, 2767.

Cohen, E. (2012). Tiger tourism: from shooting to petting. *Tourism Recreation Research*, 37, 193–204.

Cohen, S., and Beths, T. (2020). Grimace scores: tools to support the identification of pain in mammals used in research. *Animals*, 10, 1–20.

Cowan, P., Warburton, B., and Fisher, P. (2011). Welfare and ethical issues in invasive species management. In: *8th European Vertebrate Pest Management Conference*, Julius-Kühn-Archiv, pp. 44–45. http://www.cabi.org/ISC/FullTextPDF/2012/20123018858.pdf

Crook, R., and Basil, J. (2008). A biphasic memory curve in the chambered nautilus, *Nautilus pompilius* L. (Cephalopoda: Nautiloidea). *Journal of Experimental Biology*, 211, 1992–1998.

Crook, R.J. (2021). Behavioral and neurophysiological evidence suggests affective pain experience in octopus. *iScience*, 24, 102229.

Crook, R.J., Lewis, T., Hanlon, R.T., and Walters, E.T. (2011). Peripheral injury induces long-term sensitization of defensive responses to visual and tactile stimuli in the squid *Loligo pealeii*, Lesueur 1821. *Journal of Experimental Biology*, 214, 3173–3185.

Crump, A., Browning, H., Schnell, A., et al. (2022). Sentience in decapod crustaceans: a general framework and review of the evidence. *Animal Sentience*, 7, 1.

Curtin, S., and Kragh, G. (2014). Wildlife tourism: reconnecting people with nature. *Human Dimensions of Wildlife*, 19, 545–554.

Dabbadie, L., Aguilar-manjarrez, J., Beveridge, M.C.M., et al. (2018). Effects of climate change on aquaculture: drivers, impacts and policies. In: M. Barange, T. Bahri, M.C.M. Beveridge, et al. (eds), *Impacts of Climate Change on Fisheries and Aquaculture: Synthesis of Current Knowledge, Adaptation, and Mitigation Options*, pp. 449–463. FAO, Rome.

Daskalova, A. (2019). Farmed fish welfare: stress, post-mortem muscle metabolism, and stress-related meat quality changes. *International Aquatic Research*, 11, 113–124.

Dawkins, M.S. (1993). *Through our Eyes Only? The Search for Animal Consciousness*. Oxford University Press, Oxford.

Dawkins, M.S. (2008). The science of animal suffering. *Ethology*, 114, 937–945.

Dawkins, M.S. (2023). Farm animal welfare: beyond 'natural' behavior. *Science*, 379, 326–328.

Diarte-Plata, G., Sainz-Hernández, J.C., Aguiñaga-Cruz, J.A., et al. (2012). Eyestalk ablation procedures to minimize pain in the freshwater prawn *Macrobrachium americanum*. *Applied Animal Behaviour Science*, 140, 172–178.

Dimant, B., and Maldonado, H. (1992). Habituation and associative learning during exploratory behavior of the crab Chasmagnathus. *Journal of Comparative Physiology A: Neuroethology, Sensory, Neural, and Behavioral Physiology*, 170, 749–759.

Duncan, I.J.H., and Kite, V.G. (1987). Some investigations into motivation in the domestic fowl. *Applied Animal Behaviour Science*, 18, 387–388.

Duquette, C.A., Loss, S.R., and Hovick, T.J. (2021). A meta-analysis of the influence of anthropogenic noise on terrestrial wildlife communication strategies. *Journal of Applied Ecology*, 58, 1112–1121.

Ellis, T., North, B., Scott, A.P., et al. (2002). The relationships between stocking density and welfare in farmed rainbow trout. *Journal of Fish Biology*, 61, 493–531.

Elmore, R.D., Carroll, J.M., Tanner, E.P., et al. (2017). Implications of the thermal environment for terrestrial wildlife management. *Wildlife Society Bulletin*, 41, 183–193.

Elwood, R.W., Dalton, N., and Riddell, G. (2017). Aversive responses by shore crabs to acetic acid but not to capsaicin. *Behavioural Processes*, 140, 1–5.

Enna, S.J., and McCarson, K.E. (2006). The role of GABA in the mediation and perception of pain. *Advances in Pharmacology*, 45, 1–27.

Fahlman, Å., Lindsjö, J., Norling, T.A., et al. (2020). Wild boar behaviour during live-trap capture in a corral-style trap: implications for animal welfare. *Acta Veterinaria Scandinavica*, 62, 1–11.

Farm Animal Welfare Council. (1993). Report on Priorities for Animal Welfare Research and Development. FAWC, Surbiton, UK.

Fernandez-Duque, E., Valeggia, C., and Maldonado, H. (1992). Multitrial inhibitory avoidance learning in the crab Chasmagnathus. *Behavioral and Neural Biology*, 57, 189–197.

Fiorito, G., and Scotto, P. (1992). Observational learning in *Octopus vulgaris*. *Science*, 256, 545–547.

Flower, F.C., Sedlbauer, M., Carter, E., et al. (2008). Analgesics improve the gait of lame dairy cattle. *Journal of Dairy Science*, 91, 3010–3014.

Flower, F.C., and Weary, D.M. (2006). Effect of hoof pathologies on subjective assessments of dairy cow gait. *Journal of Dairy Science*, 89, 139–146.

Food and Agriculture Organization of the United Nations. (2022). The State of World Fisheries and Aquaculture 2022. Towards Blue Transformation. FAO, Rome.

Forkman, B., Boissy, A., Meunier-Salaün, M.C., et al. (2007). A critical review of fear tests used on cattle, pigs, sheep, poultry and horses. *Physiology & Behavior*, 92, 340–374.

Fraser, D., and Nicol, C.J. (2018). Preference and motivation research. In: M.C. Appleby, I.A.S. Olsson, and F. Galindo (eds), *Animal Welfare*, 3rd edition, pp. 213–231. CABI, Wallingford.

Garcia, H.E., Boyer, T.P., Levitus, S., et al. (2005). On the variability of dissolved oxygen and apparent oxygen utilization content for the upper world ocean: 1955 to 1998. *Geophysical Research Letters*, 32, 1–4.

Garcia-Pelegrin, E., Clark, F., and Miller, R. (2022). Increasing animal cognition research in zoos. *Zoo Biology*, 41, 281–291.

García Pinillos, R., Appleby, M.C., Manteca, X., et al. (2016). One Welfare—A platform for improving human and animal welfare. *Veterinary Record*, 179, 412–413.

Gaston, K.J., Davies, T.W., Bennie, J., and Hopkins, J. (2012). Reducing the ecological consequences of nighttime light pollution: options and developments. *Journal of Applied Ecology*, 49, 1256–1266.

Goldstein, J.E., Budiman, I., Canny, A., and Dwipartidrisa, D. (2022). Pandemics and the human-wildlife interface in Asia: land use change as a driver of zoonotic viral outbreaks. *Environmental Research Letters*, 17, 063009.

Green, R., and Giese, M. (2004). Negative effects of wildlife tourism on wildlife. In: K. Higgenbottom (ed.), *Wildlife Tourism: Impacts, Management and Planning*, pp. 81–97. Common Ground Publishing Pty Ltd, Altona.

Gruber, N. (2011). Warming up, turning sour, losing breath: ocean biogeochemistry under global change. *Philosophical Transactions of the Royal Society A: Mathematical, Physical and Engineering Sciences*, 369, 1980–1996.

Haelters, J., Debusschere, E., Botteldooren, D., et al. (2013). The effects of pile driving on marine mammals and fish in Belgian waters. In: S. Degraer, R. Brabant, and B. Rumes (eds), *Environmental Impacts of Offshore Wind Farms in the Belgian Part of the North Sea: Learning From the Past to Optimize Future Monitoring Programs*, pp. 70–77. Royal Belgian Institute of Natural Sciences (RBINS) Operational Directorate Natural Environment, Marine Ecology and Management Section, Brussels. http://www.vliz.be/imisdocs/publications/252750.pdf.

Hall, C.S. (1936). Emotional behavior in the rat. III. The relationship between emotionality and ambulatory activity. *Journal of Comparative Psychology*, 22, 345–352.

Hamilton, T.J., Holcombe, A., and Tresguerres, M. (2014). CO_2-induced ocean acidification increases anxiety in Rockfish via alteration of GABAA receptor functioning. *Proceedings of the Royal Society B: Biological Sciences*, 281, 20132509.

Hampton, J.O., Edwards, G.P., Cowled, B.D., et al. (2017). Assessment of animal welfare for helicopter shooting of feral horses. *Wildlife Research*, 44, 97–105.

Hannah, L., Midgley, G.F., and Millar, D. (2002). Climate change-integrated conservation strategies. *Global Ecology and Biogeography*, 11, 485–495.

Harding, E.J., Paul, E.S., and Mendl, M. (2004). Cognitive bias and affective state. *Nature*, 427, 312.

Harrison, R. (1964). *Animal Machines*. Vincent Stuart Publishers LTD, London.

Harvey, A.M., Beausoleil, N.J., Ramp, D., and Mellor, D.J. (2020). A ten-stage protocol for assessing the welfare of individual non-captive wild animals: free-roaming horses (*Equus ferus caballus*) as an example. *Animals*, 10, 148.

Harvey, A.M., Morton, J.M., Ramp, D., et al. (2021). Use of remote camera traps to evaluate animal-based welfare indicators in individual free-roaming wild horses. *Animals*, 11, 2101.

Hediger, H. (1950). *Wild Animals in Captivity*. Butterworths Scientific Publications, London, UK.

Held, S.D.E., and Špinka, M. (2011). Animal play and animal welfare. *Animal Behaviour*, 81, 891–899.

Henry, S., Fureix, C., Rowberry, R., et al. (2017). Do horses with poor welfare show 'pessimistic' cognitive biases? *The Science of Nature*, 104, 8.

Hoehfurtner, T., Wilkinson, A., Nagabaskaran, G., and Burman, O.H.P. (2021). Does the provision of environmental enrichment affect the behaviour and welfare of captive snakes? *Applied Animal Behaviour Science*, 239, 105324.

Hogan, J.A., and van Boxel, F. (1993). Causal factors controlling dustbathing in Burmese red junglefowl: some results and a model. *Animal Behaviour*, 46, 627–635.

Hölker, F., Wolter, C., Perkin, E.K., and Tockner, K. (2010). Light pollution as a biodiversity threat. *Trends in Ecology & Evolution*, 25, 681–682.

Howarth, R., Chan, F., Conley, D.J., et al. (2011). Coupled biogeochemical cycles: eutrophication and hypoxia in temperate estuaries and coastal marine ecosystems. *Frontiers in Ecology and the Environment*, 9, 18–26.

Hughes, B.O., and Black, A.J. (1973). The preference of domestic hens for different types of battery cage floor. *British Poultry Science*, 14, 615–619.

Jones, O.A.H., Voulvoulis, N., and Lester, J.N. (2001). Human pharmaceuticals in the aquatic environment: a review. *Environmental Technology (United Kingdom)*, 22, 1383–1394.

Karataş, A., and Karataş, E. (2016). Environmental education as a solution tool for the prevention of water pollution. *Journal of Survey in Fisheries Sciences*, 3, 61–70.

Kawakami, K., Shimosaki, S., Tongu, M., et al. (2007). Evaluation of bedding and nesting materials for laboratory mice by preference tests. *Experimental Animals*, 56, 363–368.

Komine, H., Koike, S., and Schwarzkopf, L. (2020). Impacts of artificial light on food intake in invasive toads. *Scientific Reports*, 10, 3–5.

Lacetera, N. (2019). Impact of climate change on animal health and welfare. *Animal Frontiers*, 9, 26–31.

Lagisz, M., Zidar, J., Nakagawa, S., et al. (2020). Optimism, pessimism and judgement bias in animals: a systematic review and meta-analysis. *Neuroscience & Biobehavioral Reviews*, 118, 3–17.

Marcogliese, D.J. (2014). The impact of climate change on the parasites and infectious diseases of aquatic animals. *Revue Scientifique et Technique*, 27, 467–484.

Marsh, D.M., and Hanlon, T.J. (2004). Observer gender and observation bias in animal behaviour research: experimental tests with red-backed salamanders. *Animal Behaviour*, 68, 1425–1433.

Mason, G., Clubb, R., Latham, N., and Vickery, S. (2007). Why and how should we use environmental enrichment to tackle stereotypic behaviour? *Applied Animal Behaviour Science*, 102, 163–188.

Mason, G.J. (1991). Stereotypies and suffering. *Behavioural Processes*, 25, 103–115. https://www.sciencedirect.com/science/article/pii/037663579190014Q%0Ahttps://www.sciencedirect.com/science/article/pii/037663579190013P

Mason, G.J., Cooper, J., Clarebrough, C. (2001). Frustrations of fur-farmed mink. *Nature*, 410, 35–36.

McCambridge, C., Dick, J.T.A., and Elwood, R.W. (2016). Effects of autotomy compared to manual declawing on contests between males for females in the edible crab *Cancer pagurus*: implications for fishery practice and animal welfare. *Journal of Shellfish Research*, 35, 1037–1044.

Mellor, D.J. (2016). Updating animal welfare thinking: moving beyond the 'Five Freedoms' towards 'a Life Worth Living'. *Animals*, 6, 1–20.

Mellor, D.J., Beausoleil, N.J., Littlewood, K.E., et al. (2020). The 2020 Five Domains model: including human–animal interactions in assessments of animal welfare. *Animals*, 10, 1–24.

Mellor, D.J., and Reid, C.S.W. (1994). Concepts of animal well-being and predicting the impact of procedures on experimental animals. *Improving the Well-being of Animals in the Research Environment*, 3–18. http://org.uib.no/dyreavd/harm-benefit/Conceptsofanimalwellbeingandpredicting.pdf

Metcalfe, C.A., Yaicurima, A.Y., and Papworth, S. (2022). Observer effects in a remote population of large-headed capuchins, Sapajus macrocephalus. *International Journal of Primatology*, 43, 216–234.

Metcalfe, J.D. (2009). Welfare in wild-capture marine fisheries. *Journal of Fish Biology*, 75, 2855–2861.

Mikhalevich, I., and Powell, R. (2020). Minds without spines: evolutionarily inclusive animal ethics. *Animal Sentience*, 329, 1–26.

Miller, L.J., Vicino, G.A., Sheftel, J., and Lauderdale, L.K. (2020). Behavioral diversity as a potential positive indicator of animal welfare. *Animals*, 10, 1–17.

Moberg, G.P. (2000). Biological response to stress: implications for animal welfare. In: *The Biology of Animal Stress: Basic Principles and Implications for Animal Welfare*, pp. 1–22. CABI, Wallingford.

Moorhouse, T.P., Dahlsjö, C.A.L., Baker, S.E., et al. (2015). The customer isn't always right—conservation and animal welfare implications of the increasing demand for wildlife tourism. *PLOS ONE*, 10, 1–15.

Moya, A., Howes, E.L., Lacoue-Labarthe, T., et al. (2016). Near-future pH conditions severely impact calcification, metabolism and the nervous system in the pteropod Heliconoides inflatus. *Global Change Biology*, 22, 3888–3900.

Muehlenbein, M.P. (2013). Human-wildlife contact and emerging infectious diseases. In: E.S. Brondizio and E.F. Moran (eds), *Human-Environment Interactions: Current and Future Directions*, pp. 79–94. Springer, Dordrecht.

Napolitano, F., Knierim, U., Grasso, F., and de Rosa, G. (2009). Positive indicators of cattle welfare and their applicability to on-farm protocols. *Italian Journal of Animal Science*, 8, 355–365.

Nelms, S.E., Piniak, W.E.D., Weir, C.R., and Godley, B.J. (2016). Seismic surveys and marine turtles: an underestimated global threat? *Biological Conservation*, 193, 49–65.

Ni, J.-Q., Erasmus, M.A., Croney, C.C., et al. (2021). A critical review of advancement in scientific research on food animal welfare-related air pollution. *Journal of Hazardous Materials*, 408, 124468.

Nielsen, T.V., and McGaw, I.J. (2016). Behavioral thermoregulation and trade-offs in juvenile lobster Homarus americanus. *Biology Bulletin*, 230, 35–50.

Nilsson, G.E., Dixson, D.L., Domenici, P., et al. (2012). Near-future carbon dioxide levels alter fish behaviour by interfering with neurotransmitter function. *Nature Climate Change*, 2, 201–204.

Paul, E.S., Harding, E.J., and Mendl, M. (2005). Measuring emotional processes in animals: the utility of a cognitive approach. *Neuroscience & Biobehavioral Reviews*, 29, 469–491.

Polglase, J.L., Bullock, A.M., and Roberts, R.J. (1983). Wound healing and the haemocyte response in the skin of the lesser octopus *Eledone cirrhosa* (Mollusca: Cephalpoda). *Journal of Zoology*, 201, 185–204.

Prebble, J.L., and Meredith, A.L. (2014). Food and water intake and selective feeding in rabbits on four feeding regimes. *Journal of Animal Physiology and Animal Nutrition*, 98, 991–1000.

Punzo, F. (1983). Localization of brain function and neurochemical correlates of learning in the mud crab, *Eurypanopeus depressus* (Decapoda). *Comparative Biochemistry and Physiology Part A: Physiology*, 75, 299–305.

Radford, A.N., Kerridge, E., and Simpson, S.D. (2014). Acoustic communication in a noisy world: can fish compete with anthropogenic noise? *Behavioral Ecology*, 25, 1022–1030.

Rault, J.L. (2019). Be kind to others: prosocial behaviours and their implications for animal welfare. *Applied Animal Behaviour Science*, 210, 113–123.

Rosa, R., Marques, A., and Nunes, M.L. (2012). Impact of climate change in Mediterranean aquaculture. *Reviews in Aquaculture*, 4, 163–177.

Ruppert, E.E., Fox, R.S., and Barnes, R.D. (2004). *Invertebrate Zoology*. Brooks/Cole-Thomson Learning, Belmont, CA.

Rushen, J., Chapinal, N., and De Passillé, A.M. (2012). Automated monitoring of behavioural-based animal welfare indicators. *Animal Welfare*, 21, 339–350.

Rushen, J., and Passillé, A.M.B. de. (1992). The scientific assessment of the impact of housing on animal welfare: a critical review. *Canadian Journal of Animal Science*, 72, 721–743.

Saaristo, M., Brodin, T., Balshine, S., et al. (2018). Direct and indirect effects of chemical contaminants on the behaviour, ecology and evolution of wildlife. *Proceedings of the Royal Society B: Biological Sciences*, 285, 20181297.

Schmid, O., and Kilchsperger, R. (2010). Overview of animal welfare standards and initiatives in selected EU and third countries. In: *Deliverable No. 1.2 of EconWelfare Project*, pp. 1–240. Research Institute of Organic Agriculture (FiBL), Frick, Switzerland.

Schnell, A.K., Boeckle, M., Rivera, M., et al. (2021). Cuttlefish exert self-control in a delay of gratification task. *Proceedings of the Royal Society B: Biological Sciences*, 288, 20203161.

Shaheen, M.N.F. (2022). The concept of one health applied to the problem of zoonotic diseases. *Reviews in Medical Virology*, 32, e2326.

Shields, S., and Orme-Evans, G. (2015). The impacts of climate change mitigation strategies on animal welfare. *Animals*, 5, 361–394.

Shine, R. (2010). The ecological impact of invasive cane toads (*Bufo marinus*) in Australia. *The Quarterly Review of Biology*, 85, 253–291.

Silanikove, N. (2000). Effects of heat stress on the welfare of extensively managed domestic ruminants. *Livestock Production Science*, 67, 1–18.

Silanikove, N., and Koluman, N. (2015). Impact of climate change on the dairy industry in temperate zones: predications on the overall negative impact and on the positive role of dairy goats in adaptation to earth warming. *Small Ruminant Research*, 123, 27–34.

Silva, E., Marco, A., da Graça, J., et al. (2017). Light pollution affects nesting behavior of loggerhead turtles and predation risk of nests and hatchlings. *Journal of Photochemistry and Photobiology B: Biology*, 173, 240–249.

Simberloff, D. (2013). *Invasive Species: What Everyone Needs to Know*. Oxford University Press, Oxford.

Singer, P. (2005). *In Defense of Animals: The Second Wave*. Wiley-Blackwell, Hoboken, New Jersey.

Slimen, I.B., Najar, T, Ghram, A., and Abdrrabba, M. (2016). Heat stress effects on livestock: molecular, cellular and metabolic aspects, a review. *Journal of Animal Physiology and Animal Nutrition*, 100, 401–412.

Sneddon, L.U., Braithwaite, V.A., and Gentle, M.J. (2003). Novel object test: examining nociception and fear in the rainbow trout. *The Journal of Pain*, 4, 431–440.

Soravia, C., Ashton, B.J., and Ridley, A.R. (2021). The impacts of heat stress on animal cognition: implications for adaptation to a changing climate. *WIREs Climate Change*, 12, e713.

Taraborelli, P., Mosca Torres, M.E., Gregorio, P.F., et al. (2017). Different responses of free-ranging wild guanacos (*Lama guanicoe*) to shearing operations: implications for better management practices in wildlife exploitation. *Animal Welfare*, 26, 49.

Tarlow, E.M., and Blumstein, D.T. (2007). Evaluating methods to quantify anthropogenic stressors on wild animals. *Applied Animal Behaviour Science*, 102, 429–451.

Taylor, J., Vinatea, L., Ozorio, R., et al. (2004). Minimizing the effects of stress during eyestalk ablation of *Litopenaeus vannamei* females with topical anesthetic and a coagulating agent. *Aquaculture*, 233, 173–179.

Taylor, L.H., Latham, S.M., and Woolhouse, M.E.J. (2001). Risk factors for human disease emergence. *Philosophical Transactions of the Royal Society B: Biological Sciences*, 356, 983–989.

Thomas, C.D. (2011). Translocation of species, climate change, and the end of trying to recreate past ecological communities. *Trends in Ecology & Evolution*, 26, 216–221.

Thomas, J.T., Munday, P.L., and Watson, S.A. (2020). Toward a mechanistic understanding of marine invertebrate behavior at elevated CO_2. *Frontiers in Marine Science*, 7, 345.

Trujillo-Reyes, J., Peralta-Videa, J.R., and Gardea-Torresdey, J.L. (2014). Supported and unsupported nanomaterials for water and soil remediation: are they a useful solution for worldwide pollution? *Journal of Hazardous Materials*, 280, 487–503.

Tuyttens, F.A.M., de Graaf, S., Heerkens, J.L.T., et al. (2014). Observer bias in animal behaviour research: can we believe what we score, if we score what we believe? *Animal Behaviour*, 90, 273–280.

Tyack, P. (2009). Acoustic playback experiments to study behavioral responses of free-ranging marine animals to anthropogenic sound. *Marine Ecology Progress Series*, 395, 187–200.

Udebuani, A.C., Pereao, O., Akharame, M.O., et al. (2021). Acute toxicity of piggery effluent and veterinary pharmaceutical cocktail on freshwater organisms. *Environmental Monitoring and Assessment*, 193, 293.

United Nations Department of Economic and Social Affairs Population Division. (2022). World population prospects 2022: Summary of results. UN DESA/POP/2022/TR/NO.3. UN, New York, NY.

Van Doren, B.M., Willard, D.E., Hennen, M., et al. (2021). Drivers of fatal bird collisions in an urban center. *Proceedings of the National Academy of Sciences*, 118, e2101666118.

Veldhuizen, L.J.L., Berentsen, P.B.M., and de Boer, I.J.M. (2018). Fish welfare in capture fisheries: a review of injuries and mortality. *Fisheries Research*, 204, 41–48.

Wahltinez, S.J., Stacy, N.I., Hadfield, C.A., et al. (2022). Perspective: opportunities for advancing aquatic invertebrate welfare. *Frontiers in Veterinary Science*, 9, 973376.

Walther, G.-R. (2010). Community and ecosystem responses to recent climate change. *Philosophical Transactions of the Royal Society B: Biological Sciences*, 365, 2019–2024.

Waltz, E. (2021). No bones, no scales, no eyeballs: appetite grows for cell-based seafood. *Nature Biotechnology*, 39, 1483–1485.

Warburton, H., and Mason, G. (2003). Is out of sight out of mind? The effects of resource cues on motivation in mink, Mustela vison. *Animal Behaviour*, 65, 755–762.

Watters, J.V. (2014). Searching for behavioral indicators of welfare in zoos: uncovering anticipatory behavior. *Zoo Biology*, 33, 251–256.

Watters, J.V., Margulis, S.W., and Atsalis, S. (2009). Behavioral monitoring in zoos and aquariums: a tool for guiding husbandry and directing research. *Zoo Biology*, 28, 35–48.

Weilgart, L. (2023). Ocean noise pollution. In: F. Obaidullah (ed.), *The Ocean and Us*, pp. 153–160. Springer, Cham.

Weilgart, L.S. (2007). A brief review of known effects of noise on marine mammals. *International Journal of Comparative Psychology*, 20, 159–168.

WHO. (2002). Diet, nutrition and the prevention of chronic diseases. *World Health Organization Technical Report Series* 916. https://www.fao.org/3/ac911e/ac911e00.htm#Contents

Wilson, A., and Phillips, C.J.C. (2023). Behaviour and welfare of African lion (*Panthera leo*) cubs used in contact wildlife tourism. *Animal Welfare*, 32, 1–13.

Wilson, E.O. (1987). The little things that run the world (the importance and conservation of invertebrates). *Conservation Biology*, 1, 344–346.

Wu, M., Tang, X., Li, Q., et al. (2013). Review of ecological engineering solutions for rural non-point source water pollution control in Hubei Province, China. *Water, Air, & Soil Pollution*, 224, 1–18.

Würbel, H. (2009). Ethology applied to animal ethics. *Applied Animal Behaviour Science*, 118, 118–127.

Yahner, R.H., and Mahan, C.G. (1997). Behavioral considerations in fragmented landscapes. *Conservation Biology*, 11, 569–570.

Zepeda, E.A., Veline, R.J., and Crook, R.J. (2017). Rapid associative learning and stable long-term memory in the squid Euprymna scolopes. *Biology Bulletin*, 232, 212–218.

Zhang, L., Takahashi, D., Hartvig, M., and Andersen, K.H. (2017). Food-web dynamics under climate change. *Proceedings of the Royal Society B: Biological Sciences*, 284, 20171772.

CHAPTER 16

Pollinator management and food security

Scarlett R. Howard and Zong-Xin Ren

Overview

Humans rely heavily on animal-mediated pollination for crop production and native plant reproduction. Recent declines in mammal, bird, and particularly insect pollinators caused by environmental change at regional and global scales threaten food security and ecosystem services. There is a noticeable paucity of knowledge regarding how behavioural alterations in animal pollinators caused by environmental change impact pollination, food security, and ecosystem function. Here we discuss the importance of studying the impact of environmental change on pollinator behaviour, what critical knowledge gaps exist, and the current policies or interventions to support pollinators in a changing world. We conclude with a suggested strategic plan for addressing the paucity of research on the effects of environmental change on pollinator behaviour.

16.1 Introduction: the impact of environmental change on pollinators

In recent years, mammal, bird, and particularly insect pollinator declines at local, regional, and global scales have been reported (Hallmann et al. 2017; Lister and Garcia 2018; Regan et al. 2015; Şekercioğlu et al. 2004), and this poses a significant threat to crop pollination and food security. There is some debate over the extent and specific causes of global and regional insect decline (Montgomery et al. 2020). The implicated drivers of these reported declines include climate change, habitat loss and fragmentation, urbanization, landscape change, pesticide use, invasive species introductions, and a variety of other anthropogenic activities (Lister and Garcia 2018; Outhwaite et al. 2022; Wagner et al. 2021). For example, significant insect declines have been observed in long-term studies in Germany (76% decline over 27 years; Hallmann et al. 2017) and Puerto Rico (75% and 98% over 35 years; Lister and Garcia 2018). Pollinator function has also been reported to be in decline, with evidence of lower visitation rates and decreased fidelity to individual plant species observed (Burkle et al. 2013). One of the biggest issues is the concept of 'death by a thousand cuts', which describes the theory that pollinators face not just one or two threats in isolation but, rather, face multiple stressors and threats simultaneously, with impacts that can interact to cause more damaging consequences (Goulson et al. 2015; Wagner et al. 2021; Chapter 10). For example, a recent study demonstrated that the interaction between historical climate warming and intensive agriculture is associated with declines in insect abundance and species richness in tropical areas (Outhwaite et al. 2022). Insects can be impacted by interactions between fungicides, insecticides, disease, and diet (Goulson et al. 2015). For example, in wild and managed bees, exposure to certain fungicides can increase the toxicity of insecticides, and exposure to such toxicants can lower their resistance to disease. Furthermore,

Scarlett R. Howard and Zong-Xin Ren, *Pollinator management and food security*. In: *Behavioural Responses to a Changing World*. Edited by: Bob B. M. Wong and Ulrika Candolin, Oxford University Press. © Oxford University Press (2024). DOI: 10.1093/oso/9780192858979.003.0016 dietary stress impedes the capacity of bees to tolerate the impacts of toxins and pathogens (Goulson et al. 2015), thus highlighting that interacting stressors can have varied and complex impacts (Chapter 10).

The impact of environmental change and anthropogenic activities on animal phenology, abundance, species richness, distribution, and demography are generally well studied. By contrast, the impact of such threats on animal behaviour has received less attention, and constitutes a significant gap in knowledge. As changes to behaviour can be a vital indicator of stress, adaptation, and health in animals, a lack of data in this area inhibits our ability to understand the full impacts of environmental change on species and ecosystems (Wong and Candolin 2015). As pollinators are keystone species in many ecosystems, and insects are the most abundant group of animal pollinators, there is a general consensus that we need more information on insect populations (Montgomery et al. 2020). In this chapter we argue that understanding changes to animal behaviour as a result of changing environments is particularly significant for pollinators. Environmental change threatens to cause pollination deficits through alterations to behaviour impacting pollinator effectiveness, efficiency, and abundance during pollination of both crops and native plants. Such changes would pose significant threats to food security, crop production, ecosystem services, and biodiversity.

In this chapter, we will consider how environmental change impacts pollinator behaviour with a particular emphasis on the consequential effects on pollination and food security. After outlining the process of pollination and pollinator behaviour (Section 16.2), we will address how changing environments impact pollinator behaviour (Section 16.3), discuss the changing relationships between plants and pollinators (Section 16.4), explore examples of mitigation strategies currently in place for supporting pollination and food security (Section 16.5), and identify where the gaps in knowledge are for developing policy and conservation plans to support pollinator management and protect our food security (Sections 16.3 and 16.6).

16.2 Pollination, food security, and the role of pollinators in human life

16.2.1 What is pollination?

Flowering plants, or angiosperms, reproduce through the transfer of pollen from a flower to a receptive stigma. Pollen transfer between plants can be achieved through water, wind, or animal transport, with some plants reproducing by self-pollination. In this regard, animal transport is among the most common methods of plant reproduction (Ollerton 2021), which is generally achieved by cross-pollination. In fact, animals contribute to the reproduction of 87.5% of flowering plants (Ollerton et al. 2011). There are about 352,000 species of angiosperm (Paton et al. 2008) pollinated by about 350,000 animal species (Kearns et al. 1998; Ollerton 2017). Animal pollination is essential to food security, with about 75% of crops relying on animal pollination to some extent (Klein et al. 2007). The process of a successful animal pollination event begins with an animal visiting a flower and ends in the production of a seed.

Pollination is generally considered a form of mutualism as both the plant and animal benefit from the interaction. Through animal pollination, plants are able to reproduce and cross-pollinate, thereby improving genetic diversity, compared with self-pollination. The pollination process can be beneficial to animals as they often receive energy and nutrition, in the form of nectar or pollen in most cases, which they can eat or use to feed their offspring or larva of their colony. There are cases where the mutualism breaks down, such as with nectar robbers or deceptive flowers.

16.2.2 Who are the pollinators?

Animals that contribute to pollination include invertebrates such as bees, flies, beetles, moths, butterflies, wasps, cockroaches, ants, spiders, and vertebrates such as primates, lizards, birds, bats, and other mammals (Rader et al. 2016; Figure 16.1). However, not all flower visitors are effective

Figure 16.1 Pollinator diversity, including (a) bees, (b) hoverflies, (c) beetles, (d) butterflies, (e) hawkmoths, (f) birds, (g) bats, (h) lizards, and (i) mammals.

Photo credits: (a, e) Zong-Xin Ren; (b, c, f) Mani Shrestha; (d) Yuansheng Fu, (g) Paulo Oliveira and Christiano Coelho; (h, i) Petra Wester

pollinators. It is tempting to see an animal visiting a flower and assume that it is a pollinator, but in fact, animals have to fulfil a very specific process (as discussed above) to be considered pollinators, rather than just flower visitors or floral resource robbers (Ollerton 2021).

Animal pollinators can be specialists, generalists, or somewhere in between. Specialist pollinators, such as pollen specialist bees, forage for pollen on a single plant species, or a few closely related plants. Generalist pollinators are less picky and will forage on a wider range of plant species. Specialist species may be more at risk from the impacts of environmental change than generalists (Dixon 2009) as they are less behaviourally flexible in terms of what plants they can forage on and their offspring may only develop on the pollen of specific plant species.

16.2.3 The role of pollinators in human life and food security

Pollinators are closely tied to human survival, culture, economics, society, and food security. Pollinators and the services they provide are particularly important to food production, agriculture, ecosystem health, and biodiversity.

One of the most significant and important aspects of pollinators for human society is their value to crop pollination, food production, and the agricultural industry (Figure 16.2). Over one-third of food production requires animal pollination and up to 75% of major crops need animal pollination to some extent, with those crops making up 35% of worldwide crop yields (Klein et al. 2007). Animal pollination also improves crop quality and aesthetics. For example, bee pollination is known to improve the crop yield, quality, shelf life, and commercial value of fruits, such as strawberries (Klatt et al. 2014). Subsequently, animal pollination alone is estimated to contribute US$235–577 billion per year to the economy, and accounts for 5–8% of crop production globally (Potts et al. 2016). However, this figure is surely an underestimate of the true value of pollinators, as pollinators also contribute to the production of home-grown food (Ollerton 2021). In fact, it would cost US$2.16 billion for farmers to employ staff to hand-pollinate crops just in the UK (Ollerton 2021).

Often the nutritional value of the crop is overlooked for the financial value, but it is important to consider the nutritional value that animal-pollinated crops offer (Ellis et al. 2015). Animal-pollinated fruits and seeds provide high proportions of essential minerals and vitamins, such as vitamins A, C, and E. Many oils are from seeds of crops that require insect pollination, including rapeseed, oil palm, and many others. If we suffered a complete loss of all pollinators, 262 million people could become vitamin A and B deficient (Smith et al. 2015). Such an event would also cause a reduction of 25% to fruit supplies, 15% in vegetables, and > 20% for nuts and seeds—resulting in an increase of global deaths from malnutrition diseases by 1.4 million people a year (Smith et al. 2015).

Different pollinators are responsible for the successful pollination and, thus, crop harvest of specific agricultural plants (Figure 16.2). Bats are important to the tequila industry due to their role as the main pollinators of agaves (Trejo-Salazar et al. 2016) and may also be the main pollinators of

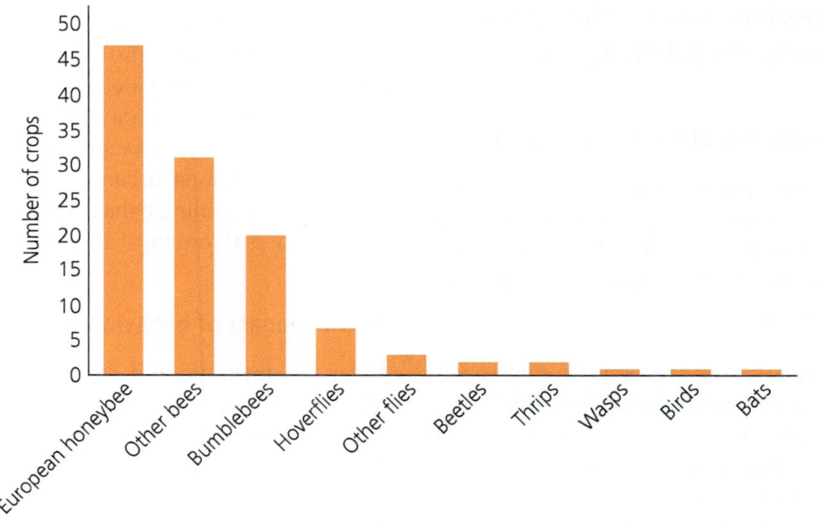

Figure 16.2 Contribution of different pollinators to crop plants around the world. The figure shows the global frequency with which different pollinator groups are effective pollinators of different crops.

Adapted from Ollerton (2021)

socioeconomically important crops such as durians, mangos, and pitayas (Requier et al. 2022). Bats and birds are pollinators of fruits such as bananas (Itino et al. 1991). Stingless bees (Apidae: Meliponini) are known to visit approximately 90 crop species (Heard 1999), with native stingless bees in Australia *Tetragonula carbonaria* pollinating macadamia and capsicum crops (Heard 1999). Honeybees are known to visit a range of crops and improve crop quality and yield (Klatt et al. 2014). Bumblebees are involved in the pollination of field beans, berries, tomatoes, and oilseed rape (Stanley et al. 2015). Thus, pollinators are strongly tied to food security and facilitate the benefits of having food diversity in our diets, such as micronutrient supply (Smith et al. 2015).

The value of pollinators to humans is substantial. Animal pollinators have huge importance for food security, human health, and the economy. They also hold great cultural and artistic value across time and regions (Prendergast et al. 2021). However, pollinators are living in a changing world and these changes appear to be having catastrophic negative impacts on pollination through changes to pollinator diversity, abundance, behaviour, and distribution, leading to decreases in food security.

16.3 The importance and challenges of monitoring behavioural change in pollinators

16.3.1 The importance of behavioural data

It is critical to understand how pollinator behaviours that underpin pollination services and enable food security may be altered or diminished by various threats. These behaviours include, but are not limited to:

- flower discrimination and recognition during foraging (learning and memory requirements)
- flower trait learning and recall (learning and memory of specific flower colours, scents, patterns, shapes, and/or locations)
- navigation (memorization of landmarks and pathways from nests/hives to flowers for resource collection)
- decision-making (choosing a flower to visit)

- resource collection (e.g. pollen, nectar, oil, resin)
- foraging strategies and movements (how many flowers visited per foraging trip, deciding when to leave a flower, deciding when to leave a patch of flowers, movement between landings)
- flower constancy (memory of specific flower traits to continually visit flowers with certain traits associated with rewards)
- other behaviours that can result in pollination (e.g. sheltering in flowers, oviposition and mating with sexually deceptive orchids).

Without knowledge of the baseline behaviours of pollinators before environmental change occurs (e.g. as temperatures increase, or before and after landscape changes), we cannot predict changes in foraging behaviours, flower preferences, and pollination capability, which could threaten food production, native plant reproduction, and overall ecosystem function. Unfortunately, few efforts to collect this important baseline data have occurred, partially as environmental change is often difficult to predict. Therefore, to appreciate the full impact of environmental and anthropogenic change on pollinators, we need to specifically design data collection efforts to understand how pollinator behaviour may be impacted by changing environments. Efforts to collect these data would include measuring flower preferences, learning ability, and foraging strategies in relatively undisturbed conditions and then again following simulated or real environmental change. For example, data could be collected before and after landscape change by sampling pollinator behaviour along urban gradients (Tüzün et al. 2017), or determining behavioural traits at different temperatures (Norgate et al. 2010).

16.3.2 Impact of behavioural change

Learning, memory, and cognition capabilities are critical for pollinator survival. To locate, learn, and remember rewarding flowers, pollinators must have undisrupted cognitive function, as learning and memory are critical to flower trait learning, flower recognition, and navigation. Furthermore, learning and recognizing rewarding flower traits enables pollinators to reduce energy and time costs associated with visiting deceptive or unrewarding flowers. A reduction in pollinator cognitive function from environmental stressors can directly impact pollination effectiveness and efficiency, leading to threats to food security. Neonicotinoid pesticides are the most widely used insecticides to kill pests, often to protect crops, but increasing evidence has demonstrated a range of lethal and sublethal impacts on pollinators. Bumblebees *Bombus terrestris* are essential pollinators of many important crops (Figure 16.2), including being major pollinators of apples, a globally important crop (Garratt et al. 2014). In 2012, 75 million tonnes of apples were harvested from 95 countries at an estimated export value of US$71 billion (Stanley et al. 2015). After exposure to environmentally realistic levels of a pesticide, thiamethoxam, bumblebee pollination behaviour was impaired, resulting in reduced crop pollination services (Stanley et al. 2015). Exposure of bumblebee colonies to the pesticide resulted in lower crop visitation rates, fewer foraging trips resulting in pollen collection, trees with increased abortion of fruit, and fewer seeds in apples—an indicator of reduced fruit quality. The exposure also modified the floral preferences and flower constancy (loyalty to specific flower trait/s, improving plant pollination efficiency) of bumblebees. Thus, exposure to a neonicotinoid impaired the apple crop pollination abilities of bumblebees. As bumblebees are essential pollinators of many other important crops, such as field beans, berries, tomatoes, and oilseed rape (Stanley et al. 2015), it is likely that pesticides would also diminish pollination services to other important crops. Changes to pollinator behaviour threaten crop pollination and, therefore, food security.

16.3.3 What are the major issues in behavioural research?

Currently there are some major limitations surrounding how we understand pollinator behaviour in changing environments. Here we wish to highlight three major challenges that need to be resolved: (1) a lack of data on how pollinator behaviour is impacted by environmental change, (2) support for long-term data collection efforts, and (3) recognition and linking experiments to address the mismatch between lab and field research on the one hand, and real-world scenarios on the other.

Knowledge gaps

There are clear knowledge gaps when it comes to understanding how pollinators behave under changing environments. As the process of pollination and plant-pollinator interactions are already so complex, it can be challenging to determine which are the most important gaps to address. We propose the following categories of behaviours that should be an initial focus when trying to gauge the impact of environmental change on pollinator behaviour due to their importance in foraging and pollination: (1) shifts in flower preference, (2) changes to cognition, learning, and memory (including navigation and flower constancy), and (3) alterations to pollinator effectiveness (e.g. amount of pollen collected, number of visits to flowers). These three categories overlap; for example, learning and memory deficits will inevitably impact flower preference and pollination effectiveness.

Long-term data collection

A major gap in behavioural research—but also in pollinator diversity, abundance, and distribution monitoring—is access to long-term data or support for such projects. Many studies are short-term due to limited time, funding, and expertise. There have been recent calls to improve pollinator monitoring over longer time frames (Harvey et al. 2020). We suggest that these calls include long-term behavioural monitoring across generations, and particularly for long-lived pollinator species, such as birds and mammals. Long-term behavioural monitoring would show us changes in behaviour across longer time frames to determine how pollinators respond, adapt, or are impaired over short, mid, and long periods of time.

Of particular interest for long-term behavioural data collection is the monitoring of plant-pollinator interactions and networks. We should aim to design long-term observations of interacting species and their populations at the community level by monitoring their functional traits and the changing pollination networks through time and environmental change events. Data on pre- and post-environmental change events would be particularly valuable, but are often challenging to accumulate and predict.

A mismatch in experimental design and real-world problems

Behavioural studies on pollinators enable us to understand their cognitive abilities (learning and memory), vision, olfactory use, flower preferences, navigational abilities, and more. A major barrier to these experiments is determining to what extent we can extrapolate lab, greenhouse, and field-based behavioural studies directly to complex natural settings in real-word scenarios (Howard and Symonds 2023). In behavioural experiments, it is key to critically evaluate what is tested and how it was tested to determine if behavioural data can be applied to natural contexts. This is a vital consideration when looking to apply experimental and observational data to real-world scenarios and problems.

Overall, more controlled lab-based experiments have many advantages, such as less noise, more control over variables and confounds, consistent data collection, accuracy of treatments, replicable results, and potentially larger sample sizes. However, they also lead to results that *may* be less generalizable and ecologically relevant. Conversely, field-based experiments can yield more ecologically applicable conclusions, which may more closely resemble what happens in real-world situations. However, more naturalistic settings can also make it more difficult to control different variables, collect data, replicate the experiment under the exact same conditions, and collect large datasets. Compromises, such as conducting pollination and flower choice experiments in greenhouses, can aid in this process but still suffer from issues on both sides, with some pollinators less willing to cooperate and forage inside a greenhouse environment (Howard et al. 2021). While we may never fully solve these issues, or have the ability to generalize our results, we can be aware of those limitations when designing and conducting experiments, analysing results, drawing conclusions, and applying the knowledge to real-world scenarios or suggesting solutions to environmental or agricultural problems. We then need to ensure that we collaborate widely with those who can bring value to this conversation from a range of sectors, including researchers (behaviourists, psychophysicists, neurobiologists, botanists, ecologists), industry, community, non-governmental

organizations (NGOs), and policy-makers, to aid them in understanding how ecologically applicable different methods and results are, and how to critically evaluate generalizations from lab and field experiments to address real-world problems, such as food security.

16.4 The interaction of plants and pollinator behaviour

Environmental change, such as climate change (Chapter 1), can have interacting effects on plants and their pollinators. Pollinator behaviour and pollination do not exist in a vacuum, and indeed, the varied impacts of environmental change on plant traits, such as their distribution, blooming time and duration, morphology (Figure 16.3), physiology, and nectar and pollen quality, quantity, and composition, have major implications for pollinator behaviour, the process of pollination, and the attraction of pollinators to crops. Consequently, while changes to plant traits can impact pollinator behaviour, changed pollinator behaviour can cause changes to plant traits. As a result, both of these sides need to be considered when discussing pollinator behaviour in changing environments. In this section, we discuss the impact of climate change on plants and the subsequent effect on pollinators as an example of why we need to consider both plants and pollinators when addressing the issues caused by changing environments.

Spatial and temporal mismatches between interacting plants and pollinators are proposed by species distribution models and other model simulations (Hegland et al. 2009), but rarely detected in natural systems due to a lack of long-term monitoring, and potentially because generalist pollinators can be flexible. Migration and shifting of phenology are two main components for such mismatches. Uneven changing of plant and pollinator distributions following climate warming results in changes to population dynamics, and even leads to the extinction of some populations or species. Phenological changes result in shifts in flowering periods and/or peaks in pollinator activity, which may cause pollination deficits. For example, Kudo and Ida (2013) found that the flowering of a

Figure 16.3 Some nectar is only accessible by long-tongued pollinators and thus nutrition access, visitation, and pollen transport are restricted to specific species. (a) A tabanid fly *Philoliche longirostris* with long proboscis pollinates for an alpine ginger *Roscoea purpurea*. (b) The fly foraging for nectar.

Photo: Babu Ram Paudel

spring ephemeral *Corydalis ambigua* tended to temporally occur ahead of the first bumblebee pollinator detection when spring came early, resulting in lower seed production owing to low pollination services, suggesting that phenological mismatches between peak flower blooms and pollinator activity are a major limiting factor for plant reproduction (Kudo and Ida 2013).

In most cases, pollination is a by-product of foraging. Floral foraging determines the benefits for both partners involved—the plant and pollinator. A pollinator forages on flowers to collect floral resources and to find resources of optimal quality. In contrast, the highest benefit for plants is to invest fewer resources to attract more pollinators to visit and move among flowers for pollination and reproduction. Thus, plants and pollinators have an asymmetric mutualism that impacts their investments into the pollination process and the benefits they each receive, which can be influenced by climate change. Greenhouse and field studies found that drought reduced both pollen and nectar production of flowers in the plant *Ipomopsis aggregate*. Low nectar availability caused hummingbirds to visit flowers at a higher rate, and low pollen availability caused them to deposit less pollen per visit (Waser and Price 2016). In this case, drought caused a high visitation rate of hummingbirds, but the increased foraging

rate did not transfer into pollination success when measured as amount of pollen deposited.

Pollinator flower choices and foraging decisions may shift under increased temperatures due to changes in floral temperatures. Flower surface temperature has been shown to be a driver of flower choice in bees. Australian stingless bees *T. carbonaria* (previously *Trigona carbonaria*) are known to seek out flowers with warmer nectar when the temperature is under ~32°C but will seek out cooler nectar under higher temperatures (Norgate et al. 2010). We currently do not know how flower choice will be impacted by increased temperatures and how widespread flower temperature preferences are across pollinators. This is a key gap to address because, as temperatures increase in some regions, many species may seek out cooler flower surfaces and/or nectar temperatures at high ambient temperatures and ignore plants without cooling mechanisms (Shrestha et al. 2018). These choices would directly impact crop and native plant pollination, with future shifts in flower preferences and decision-making causing unknown ecosystem changes. Thus, it is imperative to understand how decision-making and flower preferences may change in response to altered climatic conditions and changes in flower traits.

16.5 Pollinator decline, mitigation, and policy

Over the past decades, we have witnessed an increase in pollinator awareness and concern among the public leading to increased policy and conservation strategies in different regions (Box 16.1). Many of these plans were inspired by declines in pollinator abundance, species richness, and distribution, which threaten human health and food security. Pollinator behaviour is sometimes incorporated into strategic plans and policy, but it appears to rarely be the driver of such policies or mitigation plans. In this section, we discuss three major threats to pollinators and food security: pesticide use, urbanization and habitat loss/fragmentation, and invasive species introductions, and how policy and mitigation plans have been developed in different regions.

16.5.1 Pesticides

As discussed above, neonicotinoids are some of the most widely used pesticides and are known to have detrimental impacts on important insect pollinators, particularly bees (see also Chapter 3). Pesticides are often targeted to crops to protect them from pest

---

Box 16.1 Architecture, bee bricks, and biodiversity in Brighton, UK

Insects are the largest group of animal pollinators, with bees contributing a substantial amount to the pollination of crops and native plants. While the most well-known bees are eusocial (highly social), such as honeybees and bumblebees, most bee species are solitary. Solitary bees tend to nest in the ground, or can also nest in cavities including inside stems, tree trunks, bricks, wood, and more.

Environmental changes such as urbanization (Chapter 9) and habitat fragmentation and loss (Chapter 8) can be detrimental to biodiversity. Recently, it was noted by Brighton council in the UK that the construction of new buildings was potentially reducing the nesting habitat of solitary cavity-nesting bees.

These bees were often found nesting in the cavities of old buildings, such as crumbling mortar work and old brickwork. A company, Green&Blue, has developed the 'bee brick'—a brick the size of a regular building brick that can be incorporated into buildings to give solitary bees a place to nest in urban environments. It consists of differently sized holes to support different species of bees and wasps (Figure 16.4).

The city of Brighton and Hove, England, has created a planning law for all new buildings above 5 metres in height to include bee bricks and bird nesting boxes for swifts to support nesting behaviour. The policy was introduced for all planning permissions after 1 April 2020.

This measure has not been without its critics and supporters. Some have said that the policy is a step in the right direction and could have beneficial biodiversity impacts.

Others have criticized the move, saying that the bricks risk increased disease transmission, may host mites, are poorly designed for nesting, and could be used for greenwashing by developers. They also add that the bricks may not, in fact, have any impact on biodiversity. There is a suggestion that the new policy will allow an assessment of this implementation over a mid- to long-term period. While it may be unsuccessful in increasing or protecting pollinator biodiversity, the project enables an opportunity to study the risks and benefits on a relatively large scale, and to assess the effectiveness of such a strategy for implementation in other regions.

While the policy was created to support the nesting behaviour of solitary bees, research was also conducted to determine behavioural preferences for nests, such as preferred nesting heights and colours. This further incorporation of bee behaviour and preferences into the policy is a rare, but welcome, addition to enhancing pollinator support strategies. Bees appeared to prefer to nest in red- and yellow-coloured bricks and also prefer a nesting height of 0.6 metres (Shaw et al. 2021).

The initiative will be an interesting experiment into what mitigation strategies may or may not work to encourage bee nesting behaviour within urban environments. Whether a positive outcome is observed—or whether there is no change in pollinator diversity and nesting occupancy—the policy will enable the assessment of an implanted change and potentially garner more interest in future projects supporting pollinators in cities.

Figure 16.4 Bee bricks created to provide pollinators with nesting sites in urban areas. Bee bricks and bird nesting boxes are now required by policy to be included in all new buildings above 5 metres in height in some cities in the UK. (a) The variety of colours available; (b) an installed bee brick; (c) a leafcutter bee (genus: Megachile) returning to the bee brick; (d) a red mason bee *Osmia bicornis* entering the cavity provided by the bee brick.

Photos: Green&Blue Nature Ltd, used with permission

species, but have serious consequences for beneficial insects, such as pollinators. The lethal impacts on pollinating insects have been widely publicized, but the subsequent research into the sublethal impacts of neonicotinoids has led to changes in policy and conservation strategies. While neonicotinoids are known to cause mortality in insects, field-realistic levels of these pesticides can cause behavioural impairments as well. Studies show that neonicotinoids can cause sublethal effects in bees such as cognitive (Decourtye et al. 2004), navigational (e.g. homing ability; Henry et al. 2012), and foraging impairments (Stanley et al. 2016), including a reduction in crop pollination services

(Stanley et al. 2015). Neonicotinoids cause a reduction in wild bee density, solitary bee nesting, reproduction (Rundlöf et al. 2015), sociality (Crall et al. 2018), nursing (Crall et al. 2018), and colony growth (Rundlöf et al. 2015; Whitehorn et al. 2012). Recently, the impacts of neonicotinoids on bees and other insect pollinators have gained public attention leading to legislative change in many countries (Goulson et al. 2015).

One of the most prevalent examples of significant and tangible action to reduce the effect of pesticides on pollinator decline and behaviour is the EU ban on the use of three major neonicotinoids. Specifically, in 2013, the EU put a moratorium on the use of clothianidin, imidacloprid, and thiamethoxam, which forbid their use in flower crops visited by insect pollinators. In 2018, these three neonicotinoids were banned in the EU on all outdoor crops to protect insect pollinators, thus improving food security. The ban is binding in all member states, though exceptions are granted (Butler 2018).

In Finland, the Finnish Ministry of the Environment is funding a two-year study into the effect of pesticide exposure on pollinator behaviour. The project has been coordinated by LUKE (Natural Resources Institute Finland) and will examine how the pesticide exposure in oilseed rape fields impacts honeybees and wild pollinators in field conditions. In particular, the funding will support a project to specifically measure how pesticide residue levels affect bumblebee cognition and foraging behaviour. The information will be available to key parties, including farmers and beekeepers, to aid in the preservation and support of food security.

Pesticide banning can be the most effective way to prevent unexpected adverse impacts on pollinators; however, it has been suggested that pesticide bans are not a solution to sustain agricultural production and food security. Recently, researchers proposed the Integrated Pest and Pollinator Management (IPPM) framework by jointly considering the impact of pesticides on pests as well as pollinator health (Egan et al. 2020; Lundin et al. 2021). IPPM encourages non-chemical management—such as agroecosystem diversification, cultural practices, preventative measures, and biocontrol—to be considered ahead of pesticides, the use of which should only be taken as a last-resort, curative action (Egan et al. 2020; Lundin et al. 2021). That said, there is an urgent need for more empirical work to be done on IMMP, especially in diverse agricultural ecosystems with different agricultural practices, including the effects of the different management practices on pollinator behaviours.

16.5.2 Urbanization, habitat loss, and habitat fragmentation

Stressors, such as urbanization (Chapter 9), habitat loss and fragmentation (Chapter 8), and reductions in floral resource abundance and diversity are major

issues that also require urgent action. The impacts on pollinator behaviour from these stressors can be challenging to identify as multiple threats can act on pollinators individually or in interacting ways. For example, urbanization is a major threat to pollinators because it results in increases in impervious surfaces, habitat loss and fragmentation, temperature increases, pollution, more non-native flora and fauna, and an increase in pesticide and herbicide use (Faeth et al. 2012). The putative consequences of urbanization on bees, in particular, include lower flower visitation rates, lower species richness deriving from the loss of rare species, and homogenization of species pools (Baldock 2020).

To combat the effects of urbanization and habitat loss on pollinators, the Netherlands launched a combination of two national projects funded by the Dutch postal code lottery (3.2 million Euro) called 'The Netherlands Buzz' in 2017. The projects included the initiatives known as 'The Wild Beeline', which was launched in partnership with six NGOs and scientific organizations, and 'Wild Bee in the Lead Role', which is coordinated by Natuur & Milieu with the Dutch independent Institute for Nature and Sustainability Education (IVN). Importantly, The Wild Beeline aims to connect new and existing habitat for bees as well as increase awareness of the value of wild bees. Interestingly, the project will scale up the Dutch 'Bee Radar' programme, which uses citizen science to map wild bee behaviour. These programmes thus tackle the issue of habitat fragmentation and loss for pollinators.

16.5.3 Invasion of a parasite: *Varroa destructor*

As anthropogenic influences shape environments around the world, we have seen the invasion of many pest species into new regions. Varroa mites *Varroa destructor* are parasites causing global declines in honeybee populations in Africa, Europe, North America, and South America (Johnson 2007), with the mite recently reaching Australian shores (Chapman et al. 2023). Varroa mites feed on larvae, pupae, and the haemolymph of adult honeybees, while transmitting many honeybee viruses, such as deformed wing virus. The mites reproduce within honeybee brood cells (Iwasaki et al. 2015). The invasion and establishment of Varroa mites cause significant ecological, agricultural, and economic challenges, with significant threats to food security due to honeybee population collapses. For example, when Varroa mites reached New Zealand in 2000, the economic impact of the invasion and commercial costs were estimated to be US$227–380 million over 35 years (Iwasaki et al. 2015). The invasion has substantially impacted agriculture and beekeeping operations in New Zealand, with little known about the long-term impacts on feral colonies.

Varroa mites evolved with the Asian honeybee *A. cerana* and spread into the European honeybee *A. mellifera*, which lacked resistance to the mite, resulting in high mortality of both commercial hives and wild colonies (Iwasaki et al. 2015). Interestingly, African populations of *A. mellifera* honeybees in North and South America demonstrate behaviours and traits that counteract lethal levels of Varroa mites (Martin and Medina 2004), including hygienic behaviour, grooming, and uncapping brood cells containing mites.

A range of mitigation and elimination strategies have been implemented, including miticide treatments, selective breeding, and allowing natural resistance to occur through no intervention. Natural resistance results in smaller colonies and more frequent swarming behaviour and so is not ideal for beekeepers. Thus, selective breeding programmes have been established in many affected regions (Büchler et al. 2010; Rinderer et al. 2010). Many of these breeding programmes focus on selecting for mite-resistant behaviours such as grooming and hygienic behaviours, as these are believed to aid colonies impacted by Varroa mites. The hope is to breed bees with Varroa-resistant behaviours to ensure the mite has fewer negative impacts on honeybee populations.

Pollinator behaviour—such as flower preferences, nesting behaviours, cognition, navigation, foraging, and the act of pollination itself—has been classically overlooked in pollinator management and support programmes, as well as mitigation strategies to address pollinator decline. Furthermore, there are also a lack of programmes taking a holistic ecosystem approach to their plans, with a greater need for plant-pollinator interactions to be considered in such strategies. That said, there are certainly programmes and policies in existence that *have* chosen to base decisions and strategies solely on behaviour—or at least with an incorporation of behavioural data and knowledge. With further information on pollinator behaviour and better communication between scientists, industry, government, the community, and NGOs, we can hopefully improve policies to support pollinators in the future. Accordingly, in the final section of our chapter, we suggest potential pathways to incorporating behavioural knowledge into pollinator policies and management strategies.

16.6 Looking to the future: how can we incorporate behaviour into conservation?

16.6.1 Roadmap to behavioural monitoring and mitigation plans

Following reports of unprecedented declines in insects, a group of scientists published a roadmap for insect conservation and recovery in 2020 (Harvey et al. 2020). The strategic plan included immediate and short-term action, which would have benefits regardless of further planning or data collection, and were termed 'no-regret' solutions (Harvey et al. 2020). Mid-term action consisted of designing new research to fill important gaps in our knowledge, and leveraging existing data in private, museum, or academic collections, which were being underutilized. Long-term action consisted of developing public-private partnerships with the aim of restoring habitats and promoting a standardized monitoring programme.

We recommend a plan similar to that proposed by Harvey et al. (2020) in response to reductions in insect and other invertebrate abundance, diversity, and biomass, but adapted for addressing gaps in understanding and addressing the impacts of environmental change on behaviour. Similarly, our plan proposes action at short-, mid-, and long-term timescales to ensure effectiveness, efficiency, and to prevent delays to strategies and action that can be enacted now (Figure 16.5). As noted above, the three major issues with our current knowledge of pollinator behaviour are significant knowledge gaps, a lack of long-term data, and a paucity of linking experiments for lab and field research.

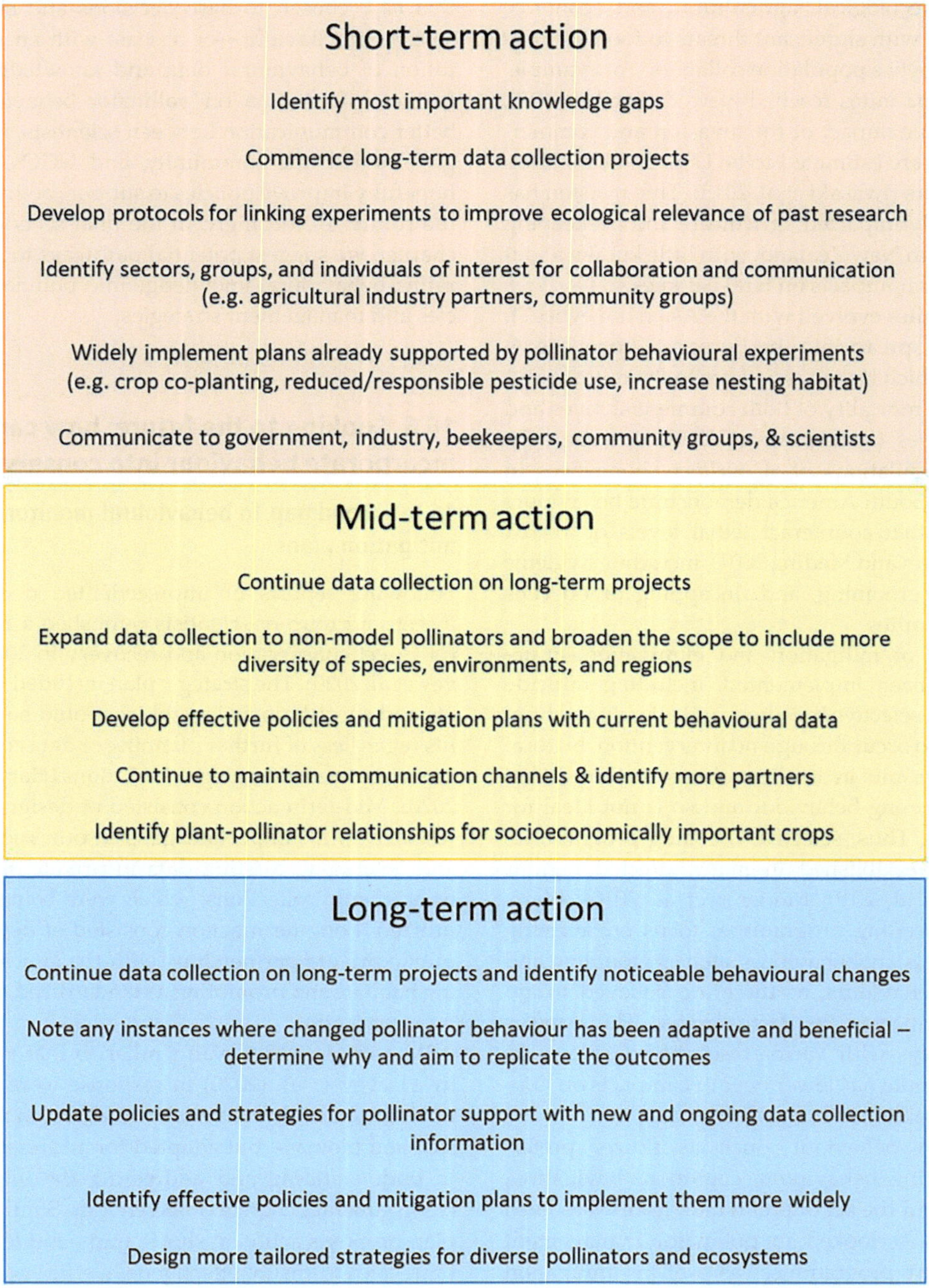

Figure 16.5 A roadmap for ongoing pollinator behavioural monitoring, improved data collection, and creation of policies that incorporate behavioural data. The roadmap suggests actions at short-, mid-, and long-term timescales.

There are immediate actions we can put in place to start to better understand and incorporate pollinator behaviour into policies and plans, while aiming to commence long-term data collection, fill knowledge gaps, plan linking experiments, and develop effective policies. This will ensure that we don't lose time waiting for more data before making the vital decisions needed to support and conserve pollinators. As time goes on, we can update policies and plans with more recent data and broaden our research scope as new questions are identified.

In the immediate and short term, we should focus on identifying the most significant knowledge gaps, commencing long-term data collection, designing linking experiments to improve ecological relevance of previous work, implementing plans and developing policies based on current behavioural data, and widely communicating such plans to important partners and groups of interest, such as governments, industry, community and hobby groups (e.g. beekeepers), and collaborators.

Mid-term action involves a continuation of long-term data collection, plans to expand data collection to more diverse species, ecosystems, and regions, updating policies and plans with new data collected, and importantly, maintaining communication and partnerships with interested parties. We should also be identifying threatened plant-pollinator interactions where socioeconomically important crops are involved for further research and support efforts.

Long-term action will involve continuing and maintaining the long-term data collection projects, identifying successes and failures in policy and experiments, updating policies and plans with new results, and designing and implementing more tailored strategies for different pollinators and ecosystems as projects started at the mid-term timescale begin to yield useful insights.

16.7 Conclusion

Environmental and anthropogenic changes are causing significant declines in pollinator abundance, species richness, and diversity, with admirable efforts from many sectors to try to mitigate these declines. Despite concern for pollinator decline, we still lack a deep understanding of the impacts of environmental change on pollinator behaviour. It is essential to determine how changing environments may alter behaviour, which can have important implications for pollination and food security through potential impairment to cognitive function, flower choices, pollination

effectiveness, and plant-pollinator interactions. In this chapter, we have established the threats facing pollinators and their importance to humans. We have outlined the importance of behavioural research and identified the major issues we currently believe need to be addressed, specifically, major gaps in knowledge, a lack of long-term studies, and a need to design more ecologically relevant experiments. We then provided examples of where pollinator behaviour has been incorporated into mitigation plans and policies to support pollinators. Finally, we have provided our roadmap of the next step in studying pollinator behaviour and incorporating it into pollinator management and conservation. Our hope is that, by beginning with immediate and short-term action, we can start to fill the gaps in pollinator behavioural knowledge and develop research and policies that will help us to conserve pollinators and the essential pollination services they provide. As changes to pollinator behaviour can negatively impact pollination, food production, and food security, with consequences for human health (Smith et al. 2015), it is critical to ensure that behavioural research is used as an opportunity to better understand pollinator needs and inform pollinator management and support strategies.

References

Baldock, K.C. (2020). Opportunities and threats for pollinator conservation in global towns and cities. *Current Opinion in Insect Science*, 38, 63–71.

Büchler, R., Berg, S., and Le Conte, Y. (2010). Breeding for resistance to *Varroa destructor* in Europe. *Apidologie*, 41, 393–408.

Burkle, L.A., Marlin, J.C., and Knight, T.M. (2013). Plant-pollinator interactions over 120 years: loss of species, co-occurrence, and function. *Science*, 339, 1611–1615.

Butler, D. (2018). Scientists hail European ban on bee-harming pesticides. *Nature*, 27.

Chapman, N.C., Colin, T., Cook, J., et al. (2023). The final frontier: ecological and evolutionary dynamics of a global parasite invasion. *Biology Letters*, 19, 20220589.

Crall, J.D., Switzer, C.M., Oppenheimer, R.L., et al. (2018). Neonicotinoid exposure disrupts bumblebee nest behavior, social networks, and thermoregulation. *Science*, 362, 683–686.

Decourtye, A., Armengaud, C., Renou, M., et al. (2004). Imidacloprid impairs memory and brain metabolism in the honeybee (*Apis mellifera* L.). *Pesticide Biochemistry and Physiology*, 78, 83–92.

Dixon, K.W. (2009). Pollination and restoration. *Science*, 325, 571–573.

Egan, P.A., Dicks, L.V., Hokkanen, H.M.T., and Stenberg, J.A. (2020). Delivering Integrated Pest and Pollinator Management (IPPM). *Trends in Plant Science*, 25, 577–589.

Ellis, A.M., Myers, S.S., and Ricketts, T.H. (2015). Do pollinators contribute to nutritional health? *PLOS One*, 10, e114805.

Faeth, S.H., Saari, S., and Bang, C. (2012). Urban biodiversity: patterns, processes and implications for conservation. In: *eLS*. John Wiley & Sons, Ltd: Chichester.

Garratt, M.P.D., Truslove, L., Coston, D., et al. (2014). Pollination deficits in UK apple orchards. *Journal of Pollination Ecology*, 12, 9–14.

Goulson, D., Nicholls, E., Botías, C., and Rotheray, E.L. (2015). Bee declines driven by combined stress from parasites, pesticides, and lack of flowers. *Science*, 347, 1255957.

Hallmann, C.A., Sorg, M., Jongejans, E., et al. (2017). More than 75 percent decline over 27 years in total flying insect biomass in protected areas. *PLOS ONE*, 12, e0185809.

Harvey, J.A., Heinen, R., Armbrecht, I., et al. (2020). International scientists formulate a roadmap for insect conservation and recovery. *Nature Ecology & Evolution*, 4, 174–176.

Heard, T.A. (1999). The role of stingless bees in crop pollination. *Annual Review of Entomology*, 44, 183–206.

Hegland, S.J., Nielsen, A., Lázaro, A., et al. (2009). How does climate warming affect plant-pollinator interactions? *Ecology Letters*, 12, 184–195.

Henry, M., Béguin, M., Requier, F., et al. (2012). A common pesticide decreases foraging success and survival in honey bees. *Science*, 336, 348–350.

Howard, S.R., Dyer, A.G., Garcia, J.E., et al. (2021). Naïve and experienced honeybee foragers learn normally configured flowers more easily than non-configured or highly contrasted flowers. *Frontiers in Ecology and Evolution*, 9, 662336.

Howard, S.R., and Symonds, M.R.E. (2023). Complex preference relationships between native and non-native angiosperms and foraging insect visitors in a suburban greenspace under field and laboratory conditions. *The Science of Nature*, 110, 16.

Itino, T., Kato, M., and Hotta, M. (1991). Pollination ecology of the two wild bananas, *Musa acuminata* subsp. *halabanensis* and *M. salaccensis*: chiropterophily and ornithophily. *Biotropica*, 23, 151–158.

Iwasaki, J.M., Barratt, B.I.P., Lord, J.M., et al. (2015). The New Zealand experience of varroa invasion highlights research opportunities for Australia. *Ambio*, 44, 694–704.

Johnson, R. (2007). *Recent honey bee colony declines. CRS Report for Congress*. Defense Technical Information Center, Fort Belvoir, VA.

Kearns, C.A., Inouye, D.W., and Waser, N.M. (1998). Endangered mutualisms: the conservation of plant-pollinator interactions. *Annual Review of Ecology and Systematics*, 29, 83–112.

Klatt, B.K., Holzschuh, A., Westphal, C., et al. (2014). Bee pollination improves crop quality, shelf life and commercial value. *Proceedings of the Royal Society B: Biological Sciences*, 281, 20132440.

Klein, A.-M., Vaissière, B.E., Cane, J.H., et al. (2007). Importance of pollinators in changing landscapes for world crops. *Proceedings of the Royal Society B: Biological Sciences*, 274, 303–313.

Kudo, G., and Ida, T.Y. (2013). Early onset of spring increases the phenological mismatch between plants and pollinators. *Ecology*, 94, 2311–2320.

Lister, B.C., and Garcia, A. (2018). Climate-driven declines in arthropod abundance restructure a rainforest food web. *Proceedings of the National Academy of Sciences*, 115, E10397–E10406.

Lundin, O., Rundlöf, M., Jonsson, M., et al. (2021). Integrated pest and pollinator management—expanding the concept. *Frontiers in Ecology and the Environment*, 19, 283–291.

Martin, S.J., and Medina, L.M. (2004). Africanized honeybees have unique tolerance to Varroa mites. *Trends in Parasitology*, 20, 112–114.

Montgomery, G.A., Dunn, R.R., Fox, R., et al. (2020). Is the insect apocalypse upon us? How to find out. *Biological Conservation*, 241, 108327.

Norgate, M., Boyd-Gerny, S., Simonov, V., et al. (2010). Ambient temperature influences Australian native stingless bee (*Trigona carbonaria*) preference for warm nectar. *PLOS ONE*, 5, e12000.

Ollerton, J. (2017). Pollinator diversity: distribution, ecological function, and conservation. *Annual Review of Ecology, Evolution and Systematics*, 48, 353–376.

Ollerton, J. (2021). *Pollinators and Pollination: Nature and Society*. Pelagic Publishing Ltd, Exeter, UK.

Ollerton, J., Winfree, R., and Tarrant, S. (2011). How many flowering plants are pollinated by animals? *Oikos*, 120, 321–326.

Outhwaite, C.L., McCann, P., and Newbold, T. (2022). Agriculture and climate change are reshaping insect biodiversity worldwide. *Nature*, 605, 97–102.

Paton, A.J., Brummitt, N., Govaerts, R., et al. (2008). Towards Target 1 of the Global Strategy for Plant Conservation: a working list of all known plant species—progress and prospects. *Taxon*, 57, 602–611.

Potts, S.G., Imperatriz-Fonseca, V., Ngo, H., et al. (2016). *Summary for policymakers of the assessment report of the Intergovernmental Science-Policy Platform on Biodiversity and Ecosystem Services (IPBES) on pollinators, pollination and food production.* Secretariat of the Intergovernmental Science-Policy Platform on Biodiversity and Ecosystem Services, Bonn, Germany.

Prendergast, K.S., Garcia, J.E., Howard, S.R., et al. (2021). Bee representations in human art and culture through the ages. *Art & Perception*, 10, 1–62.

Rader, R., Bartomeus, I., Garibaldi, L.A., et al. (2016). Non-bee insects are important contributors to global crop pollination. *Proceedings of the National Academy of Sciences*, 113, 146–151.

Regan, E.C., Santini, L., Ingwall-King, L., et al. (2015). Global trends in the status of bird and mammal pollinators. *Conservation Letters*, 8, 397–403.

Requier, F., Pérez-Méndez, N., Andersson, G.K., et al. (2022). Bee and non-bee pollinator importance for local food security. *Trends in Ecology & Evolution*, 38, 196–205.

Rinderer, T.E., Harris, J.W., Hunt, G.J., and De Guzman, L.I. (2010). Breeding for resistance to *Varroa destructor* in North America. *Apidologie*, 41, 409–424.

Rundlöf, M., Andersson, G.K., Bommarco, R., et al. (2015). Seed coating with a neonicotinoid insecticide negatively affects wild bees. *Nature*, 521, 77–80.

Şekercioğlu, Ç.H., Daily, G.C., and Ehrlich, P.R. (2004). Ecosystem consequences of bird declines. *Proceedings of the National Academy of Sciences*, 101, 18042–18047.

Shaw, R.F., Christman, K., Crookes, R., et al. (2021). Effect of height and colour of bee bricks on nesting occupancy of bees and wasps in SW England. *Conservation Evidence Journal*, 18, 10–17.

Shrestha, M., Garcia, J.E., Bukovac, Z., et al. (2018). Pollination in a new climate: assessing the potential influence of flower temperature variation on insect pollinator behaviour. *PLOS ONE*, 13, e0200549.

Smith, M.R., Singh, G.M., Mozaffarian, D., and Myers, S.S. (2015). Effects of decreases of animal pollinators on human nutrition and global health: a modelling analysis. *The Lancet*, 386, 1964–1972.

Stanley, D.A., Garratt, M.P., Wickens, J.B., et al. (2015). Neonicotinoid pesticide exposure impairs crop pollination services provided by bumblebees. *Nature*, 528, 548–550.

Stanley, D.A., Russell, A.L., Morrison, S.J., et al. (2016). Investigating the impacts of field-realistic exposure to a neonicotinoid pesticide on bumblebee foraging, homing ability and colony growth. *Journal of Applied Ecology*, 53, 1440–1449.

Trejo-Salazar, R.E., Eguiarte, L.E., Suro-Piñera, D., and Medellin, R.A. (2016). Save our bats, save our tequila: industry and science join forces to help bats and agaves. *Natural Areas Journal*, 36, 523–530.

Tüzün, N., Op de Beeck, L., Brans, K.I., et al. (2017). Microgeographic differentiation in thermal performance curves between rural and urban populations of an aquatic insect. *Evolutionary Applications*, 10, 1067–1075.

Wagner, D.L., Grames, E.M., Forister, M.L., et al. (2021). Insect decline in the Anthropocene: death by a thousand cuts. *Proceedings of the National Academy of Sciences*, 118, e2023989118.

Waser, N.M., and Price, M.V. (2016). Drought, pollen and nectar availability, and pollination success. *Ecology*, 97, 1400–1409.

Whitehorn, P.R., O'connor, S., Wackers, F.L., and Goulson, D. (2012). Neonicotinoid pesticide reduces bumble bee colony growth and queen production. *Science*, 336, 351–352.

Wong, B.B., and Candolin, U. (2015). Behavioral responses to changing environments. *Behavioral Ecology*, 26, 665–673.

CHAPTER 17

Parasites, disease, and behaviour

Iain Barber

Overview

Human activity is altering the environments in which host organisms and agents of infections interact at an unprecedented rate, with significant implications for the establishment of novel interactions and patterns of disease. Many of these anthropogenic impacts are known to alter the behavioural interactions between hosts and the organisms that serve as agents of disease, with profound implications for the establishment and spread of both novel and existing infections. This chapter provides a summary of contemporary behavioural research that is providing insights into the mechanisms involved. This knowledge is now being used to develop mitigation and control strategies for some diseases of human, veterinary, or wildlife importance. However, the diversity of life cycles of the organisms that can cause disease, and the myriad ways in which they respond to a bewildering array of anthropogenic pressures—which can also interact—mean that it is currently challenging to develop a comprehensive, general, or predictive understanding of the role of behaviour in determining patterns of disease in a rapidly changing world. There is therefore an urgent need to improve this understanding through further research, to protect global health and welfare of human, domestic animal, and wildlife populations.

17.1 Introduction

The COVID-19 pandemic served as a stark reminder that infectious diseases present a considerable threat to human survival, health, well-being, and livelihoods. The pandemic also emphasized the critical role of human and animal behaviour in shaping disease outbreaks. While the precise origins of SARS-CoV-2 remain enigmatic, it seems likely that the virus invaded the human population following zoonotic transfer from a wildlife host, a process most probably facilitated by increased frequency and intensity of novel human-animal interactions at a food market (Worobey et al. 2022). The rapid subsequent global spread of the virus was facilitated and promoted by mass intercontinental travel of an increasingly globalized human population (Barouki et al. 2021), and the pandemic was stabilized—at least in part—through reduced human social contact and restrictions on mobility (Tian et al. 2020). Finally, through an extraordinary global collaborative effort of human scientific endeavour and industry to develop and deploy multiple effective vaccines (Alhashemi et al. 2023), the pandemic phase was eventually brought under control.

Despite the inevitable focus on the global impact of the COVID-19 pandemic and its ongoing socio-economic challenges, non-pandemic infectious diseases—such as malaria, tuberculosis, and a host

Iain Barber, *Parasites, disease, and behaviour*. In: *Behavioural Responses to a Changing World*. Edited by: Bob B. M. Wong and Ulrika Candolin, Oxford University Press. © Oxford University Press (2024). DOI: 10.1093/oso/9780192858979.003.0017

Table 17.1 Definitions of key terms in relation to infectious diseases and their causative agents used in the chapter.

Term	Definition
Infectious diseases	Arise following the transmission of infectious, replicating, disease-causing agents from one individual host to another, allowing the disease to spread among and between host populations. Infectious agents include viruses, bacteria, fungi, and both single-celled (protozoan) and multicellular (metazoan) animals. Transmission and infection are ecological phenomena, with the host environment(s) providing a critical resource shaping the nature of the interaction.
Zoonotic diseases	Infectious diseases of animals that can cause disease when transmitted to humans, often following 'spillover' from reservoir wildlife or domestic animal hosts.
Spillover	The transmission of novel pathogens from non-human (animal) hosts to human hosts, when factors are favourable to the pathogen and allow the crossing of barriers between species. The predominant cause of emergence of new human infectious diseases.
Host	Any organism that is infected by a parasite, pathogen, parasitoid, or pest species, normally suffering detriment as a consequence of the interaction.
Parasite	An organism living on or in a host, from which it derives its nutrients, with hosts being negatively impacted by the association. May be parasitic across all life cycle stages or only in certain stages. Some are obligate parasites, while others might adopt a parasitic mode of life facultatively.
Pathogen	A microorganism (virus, bacterium, or fungus) that causes, or can cause, disease in a host organism.
Pest	A term typically used to refer to an animal—often an insect—that damages crops, livestock, and forestry, or otherwise causes a nuisance to people.
Parasitoid	An insect (typically a species of wasp or fly) whose larval stages infect a host, which is generally also an insect. Developing larval parasitoids feed on the host organism, eventually killing it.

of neglected tropical diseases—have long exerted equivalent levels of annual mortality on the global human population. Infectious diseases also pose a significant threat to livestock production (Lefèvre et al. 2010) and play a key role in regulating wildlife populations (Smith et al. 2006), with far-reaching implications for communities and ecosystems (Tompkins et al. 2011). As anthropogenic global changes increasingly impact the environments in which human and non-human hosts interact with a wide range of parasites, pathogens, and pests,[1] the threats posed by infectious diseases are expected to increase, along with the potential for novel zoonotic infections to arise as infectious agents cross over from animal to human populations (Baker et al. 2022). Understanding the impacts of global environmental change on patterns of infection has therefore become a high priority for human and animal health agencies and for wildlife ecologists,

[1] For simplicity, throughout the remainder of the chapter, I use 'parasites' as a catch-all term to include any causative agent of infectious disease, including those more frequently referred to as pathogens, and insect pests. See Table 17.1 for further definitions of key terms.

and there is an urgent need to develop a better understanding of how interactions between host organisms and the causative agents of infectious disease are affected, including the role played by behaviour (Wells and Flynn 2022).

17.1.1 Host-parasite interactions in a changing world: behavioural and other mechanisms

A diverse range of anthropogenic activities that impact natural environments can interfere with the interactions of hosts and parasites and thereby shape the prevalence and impact of infectious diseases (see Table 17.2 for a summary). While this chapter primarily focuses on the behavioural mechanisms that can underpin these effects, it is important to recognize that other, non-behavioural factors also play a key role in determining the nature, extent, and outcome of host-parasite interactions under conditions subject to anthropogenic change. These include ecological changes—such as the large-scale shifts in the distribution of organisms resulting from global climate change (Hulme 2017)—and the anthropogenic redistribution of organisms by deliberate or accidental species introductions, both of which can decouple existing host-parasite associations, or create new ones. Environmental changes can also reduce the effective functioning of host immune systems through their impacts on host stress responses—impairing the capacity of hosts to withstand infections following disease challenge—or induce other physiological or metabolic changes that affect the outcome of infections and disease phenotypes. As these factors often interact with behavioural changes to shape observed patterns of infection in impacted populations, they are included in the chapter where they are considered relevant.

17.1.2 The role of behaviour in host-parasite interactions

Parasitism and behaviour have long been studied side-by-side, and a rich literature exists on behavioural aspects of parasite transmission (Canning and Wright 1972), on infection-associated behavioural change in the context of 'adaptive manipulation', whereby parasites have been proposed to actively change host behaviour to maximize transmission success (Hughes et al. 2012), on the role played by parasites in the maintenance of consistent individual differences in behaviour ('animal personalities'; Barber and Dingemanse 2010), and on the neurochemical basis of behaviour change (Adamo and Webster 2013). In recent years, however, there has been a greater emphasis on understanding the role of behaviour in the context of parasite transmission under conditions of anthropogenic change.

Behaviour can be considered as the primary way in which organisms interact with one another and with both abiotic and biotic components of their ecosystems. Consequently, the way in which individual organisms behave can impact their chances of finding food, mates, or shelter, but it can also determine the risk of encountering parasites, vectors, or infectious conspecifics. The latent threat posed by the existence of parasites in the environment has therefore shaped adaptive patterns of animal behaviour, through both evolutionary and non-evolutionary processes, such that normally observed patterns of behaviour often serve—at least in part—to minimize the risk of being exposed

to, or acquiring, harmful infections (Hart 1990). Changes to normal behaviour patterns, therefore, by extension, have the potential to alter the risk from infectious agents.

It is now clear that anthropogenic global change is impacting significantly on behaviour in ways that can increase the challenges posed by the agents of disease. Chapters 1 to 10 of this book, and the earlier companion volume (Candolin and Wong 2012), together provide compelling evidence of the range and extent of animal behaviours that can be impacted in changing environments. Disruptions to normal patterns of behaviour arising as a response to global change can therefore have implications for the frequency and nature of host-parasite interactions, for the prevalence and severity of infections, and/or for the impact of disease.

The behaviour of hosts in anthropogenically-altered environments is not the only aspect that needs to be considered. Parasites themselves, and the vectors of pathogens that act as the causative agents of disease, also exhibit behaviours that have evolved to maximize the likelihood that they will encounter susceptible hosts (Lewis et al. 2002; Haas 2003). Consequently, if their behaviour is also impacted by anthropogenic change, there are likely to be implications for their interactions with hosts, and for the patterns of disease that emerge.

This chapter focuses on how interactions between hosts and parasites are being impacted by anthropogenic environmental changes, and the consequences these changed interactions have for the progression of disease in individuals and the spread of disease among and between populations, with a particular focus on the behavioural mechanisms involved. After briefly reviewing how different forms of anthropogenic change can impact disease (Table 17.2), I consider how these changes can affect interactions between hosts and the agents of disease at each of five stages required for the establishment of infections, focusing on the role of behaviour (Section 17.2). I then examine how knowledge gained from studies investigating the behavioural basis of host-parasite interactions is being used to inform management practices to limit the spread of diseases in anthropogenically altered environments (Section 17.3). Concluding thoughts, including some of the challenges to developing a comprehensive and predictive understanding, are then offered in Section 17.4.

17.2 Behavioural mechanisms underpinning altered patterns of infection and disease

The likelihood that a parasite becomes established within a host population can be viewed as the outcome of a five-stage process (Figure 17.1). Environmental changes arising from anthropogenic activity can impact the frequency or nature of interactions between host organisms and disease-causing agents at each of these stages, through a range of mechanisms, and may either increase or decrease the likelihood of occurrence. In this section, the mechanisms by which the interactions between hosts and the agents of infectious diseases can be impacted in a changing world are outlined and illustrated using examples drawn from laboratory and field studies. Throughout this section, the focus will be primarily on the role of behaviour, though non-behavioural mechanisms are included where they play a particularly key role and potentially interact with behaviour.

17.2.1 Stage I: Impacts of a changing world on the spatiotemporal overlap of hosts and parasites

A fundamental prerequisite for infectious diseases to become established—and hence their likelihood of impacting at an ecological level—is that hosts and parasites coexist at a broad scale, with overlapping biogeographic and ecological distributions in both space and time (Combes 2001). Any anthropogenic perturbation that impacts the biogeographic or ecological distribution of host or parasite taxa therefore creates opportunities for novel host-parasite associations to arise, as well as to impact, decouple, or destroy established interactions (Torchin et al. 2003).

Impact of species introductions

The introduction of species into environments where they have not previously existed generates the potential for novel host-parasite associations to arise. Such introductions may be pre-planned

and intentional—as in the case of biological control programmes, whereby parasites are released with the goal of establishing an association that is detrimental to a problematic target species—or may arise through accidental, reckless, or negligent behaviour. For example, the unintentional transfer of ectoparasitic monogeneans *Gyrodactylus salaris* along with Atlantic salmon *Salmo salar* from hatchery fish of Baltic Sea origin to the west coast of Norway in the 1970–1980s expanded the range of the parasites, bringing them into contact with a novel and highly susceptible host genotype, and decimating salmon stocks in many Norwegian rivers; further subsequent movements of infected rainbow trout *Oncorhynchus mykiss* appear to have been responsible for the continued spread of the parasite throughout mainland Europe (Paladini et al. 2021). In the UK, the crossover of squirrelpox virus (SQPV) from tolerant grey squirrels *Sciurus carolinensis*, introduced from North America from the 1890s, to highly susceptible native red squirrels *Sciurus vulgaris*, has acted in concert with competitive displacement to strongly facilitate the significant population decline of the native species (Chantrey et al. 2014).

Shifts in species distributions resulting from global climate change

Earth's changing climate—including increasing mean temperatures and more frequent extreme weather events—is exerting a profound effect on the behaviour of wildlife and the ecological communities they inhabit, with far-reaching implications for the interactions of hosts and parasites (Altizer et al. 2013). Range shifts arising as a direct response to climate change mean that novel host-parasite associations are being formed as new zones of overlap are created, while established associations are being lost, emancipating hosts from infections (Lafferty 2012). A wide range of vector-transmitted human diseases, including malaria, Chagas disease, and leishmaniasis, are predicted to change as the distributions of anolpheline mosquitoes, triatomine bugs, and sandflies respond to changing climates, as are parasitic diseases linked to the distribution of lymnaeid snail intermediate hosts, such as fascioliasis (Short et al. 2017).

Table 17.2 Overview of the potential for different forms of anthropogenic environmental and ecological change, reviewed in the first nine chapters of this volume, to impact patterns of infectious disease by influencing host-parasite interactions. Details of selected key review papers are provided in the final column.

Environmental or ecological change	Potential to impact host-parasite/pathogen/pest interactions and patterns of disease	Key review(s)
Global climate change (Chapter 1)	Changing climatic conditions have unparalleled potential to directly impact the distribution and behaviour of species and hence the frequency and nature of interactions between hosts and parasites; fundamental biological relationships between temperature, metabolic rate, immune function, growth, and development mean that thermal change can play a particularly important role. Indirect environmental and ecosystem impacts of changing climates can also affect the behavioural interactions of hosts and parasites, and hence the impact and spread of disease.	Altizer et al. 2013; Harvell et al. 2002
Noise pollution (Chapter 2)	Anthropogenic noise impacts the physiology and behaviour of animals in terrestrial and aquatic ecosystems. Evidence is now accumulating that anthropogenic noise can alter observed patterns of parasite infection, by altering the relative spatial or temporal distributions of hosts and parasites, changing the nature of their behavioural interactions, impacting the likelihood of infection following exposure, or by modulating the effects of infections once established.	Berkhout et al. 2023
Chemical pollution and environmental degradation (Chapter 3)	The input of anthropogenic chemicals into natural ecosystems, and the subsequent degradation of environments, presents organisms with considerable physiological challenges that can impact behaviour by affecting sensory systems. Organisms that live in aquatic environments are often particularly sensitive to such chemical input. Such chemicals have the potential to impair the behavioural interactions of hosts and parasites through lethal or sublethal toxic effects.	Sures et al. 2017
Light pollution (Chapter 4)	Artificial light at night (ALAN) creates a major anthropogenic challenge to terrestrial, coastal, and aquatic ecosystems because it alters two key aspects of natural environments—enhancing the 'brightness' of nocturnal environments while masking natural periodic light cycles, which entrain many circadian and circannual patterns of organismal activity and behaviour. Because light intensity and photoperiod play a key role in habitat selection and in the timing of behaviour of hosts and parasites, ALAN has considerable potential to impact their interactions, with implications for disease.	Coetzee et al. 2022
Ocean acidification (Chapter 5)	The uptake of anthropogenically derived CO_2 from the atmosphere means that marine organisms are being exposed to lower pH than at any other time in the Earth's history. Such changes potentially disrupt the physiology of marine host and parasite taxa, with implications for infections. Most of the relatively few studies undertaken to date have been restricted to studying experimentally amenable trematode and molluscan host systems, though more recent studies have begun to investigate other commercially important diseases of fish and shellfish.	Byers 2021
Introduced species and biological invasions (Chapter 6)	The translocation of animals and plants—either intentionally through planned introductions, or accidentally through negligent actions—constitutes a major cause of environmental change and biodiversity loss, with significant implications for animal behaviour and for the interactions between naïve host populations and novel parasites, pathogens, and pests.	Young et al. 2017
Harvesting and exploitation (Chapter 7)	The selective removal of individuals from populations through harvesting, hunting, or culling changes the composition and density of populations and can impose artificial selection on traits including behaviour, growth rates, and sexual development, which can drive rapid evolutionary change in exploited populations. In some cases, harvesting may mean that individuals with the highest or lowest parasite loads are selectively removed. Such changes have considerable potential to impact host-parasite interactions and patterns of disease.	Choisy and Rohani 2006
Land use changes and habitat loss and fragmentation (Chapter 8)	Changes in land use, for example the conversion of natural ecosystems to agricultural land, can lead to increased contact between domestic animals, humans, and wildlife, while habitat loss and fragmentation can affect local species' densities, the behaviour of individuals, and the nature and frequency of inter- and intraspecific interactions. When natural habitats are changed, the behaviour of hosts can be altered, with consequences for the transmission of the infectious agents they carry and the risk of disease spreading to humans and livestock.	White and Razgour 2020
Urbanization (Chapter 9)	Urban environments are simultaneously impacted by many of the factors detailed separately above, including noise, light, and chemical pollution and habitat degradation. They are characterized by skewed communities that are typically depauperate but contain high densities of a small number of species that are well adapted to living in close proximity to people, with exotic and introduced species being frequent (Chapter 9). While urbanization is often associated with reduced parasite biodiversity, species with life cycles that can be supported may find increased opportunities for transmission.	Bradley and Altizer 2007

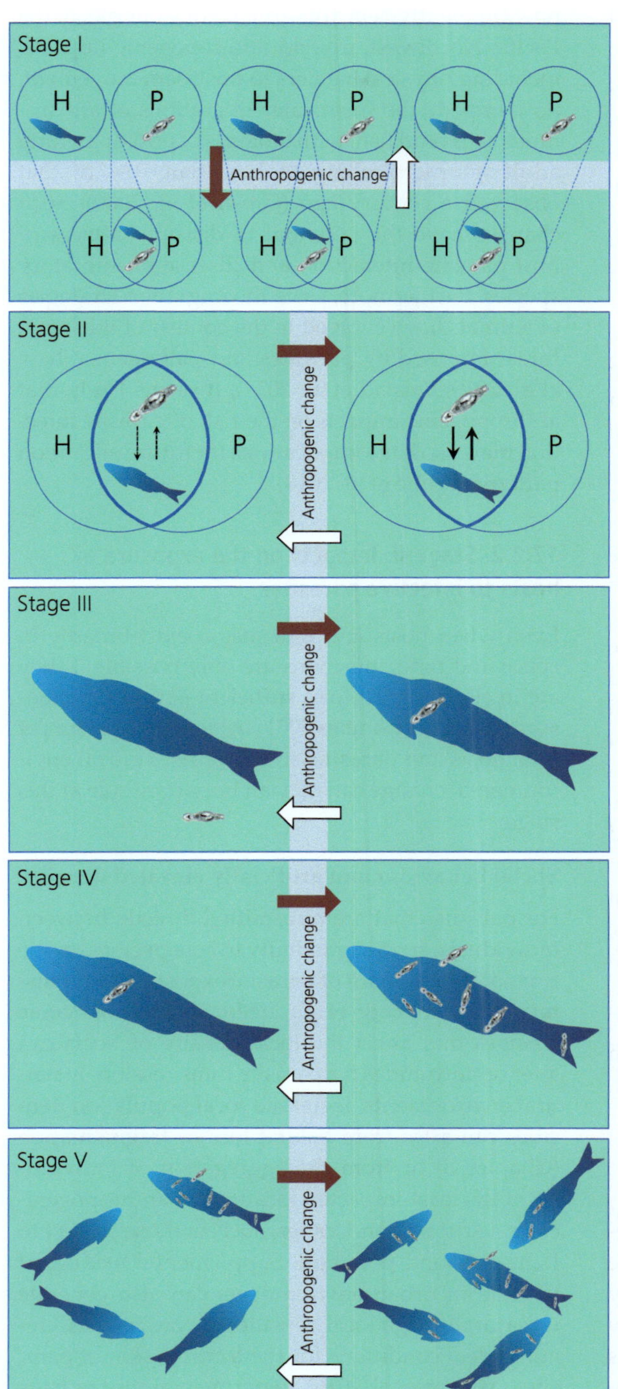

Figure 17.1 An illustration of the five stages at which the interactions between hosts (H) and parasites, pathogens, or pests (P) can be influenced by anthropogenic factors. Anthropogenic changes can act at Stage I, influencing the broad-scale spatiotemporal distributions of hosts and/or parasites, which can either create the potential for novel interactions to occur, or break existing interactions. At Stage II, behavioural and ecological effects of anthropogenic change can alter the likelihood that coexisting hosts and parasites interact closely for long enough to allow exposures to occur. Anthropogenic effects can then impact Stage III, which determines the likelihood that exposures lead to the establishment of infections. Once infected, anthropogenic changes can affect the progression of disease, which may relate to the proliferation, growth, or development of the infectious agent(s) in or on the host (Stage IV). Finally, anthropogenic factors might influence the probability that infections spread to other members of the population (Stage V) and therefore the potential for population, community, and ecosystem effects of infectious disease. At each stage the action of anthropogenic factors can operate in either direction, either promoting (brown arrows) or reducing (white arrows) the probability of widespread infections under changed conditions.

Novel species interactions arising from land use changes, population encroachment, and increased human-animal contact

The majority of infections infecting human populations have non-human animal origins, with an estimate of 75% of new or emerging infectious diseases being zoonotic in origin (Jones et al. 2008). 'Spillover' events from non-human to human populations can only occur when pathogens circulating within animal populations undergo sufficient genetic change through mutation to permit successful invasion of human hosts. Such events are normally expected to be very rare, but in a changing world, a greater likelihood of spillover is expected as the frequency and intensity of human-animal contact increases as a result of large-scale environmental and socio-economic changes, land use change, deforestation, increasing trade in wildlife, the expansion of human settlements into wildlife habitat, and the intensification of agriculture (Jones et al. 2013).

Influenza A viruses (IAVs) are the causative agents of significant disease in domestic poultry and mammals, including pigs, dogs, and horses, with waterbirds serving as the major wildlife reservoir of infection. The high density of gallinaceous species (including chickens, turkeys, quail) in intensive commercial rearing facilities facilitates the rapid spread of IAVs following the transfer of infection from local wild birds, and provides a platform for avian IAV spillovers into susceptible mammals, including humans. Such spillovers cause a substantial global health burden (Yoon et al. 2014), including the human influenza pandemics of 1918, 1957, 1968, and 2009, which all involved viruses with genetic segments of avian origin (Harrington et al. 2021).

Migrations and movement ecology

The large-scale, mass movements exhibited by animals through their seasonal migrations provide considerable opportunities for the parasites they carry to be transferred between geographically disparate regions. However, changes to patterns of animal migration as a consequence of anthropogenic environmental impacts—in terms of the timing, route, or pace of movement, or even its

existence at all (Wilcove and Wikelski 2008)—are now frequently being observed. For example, in the Pacific Northwest, myriad anthropogenic impacts, including physical barriers to upstream movement, now impede the freshwater migrations of anadromous salmonid fish, with biomass of post-spawning adult returners falling to less than 10% of that observed a century ago (Gresh et al. 2000). Such changes have the potential to disrupt patterns of host-parasite interaction and disease in migratory species. Furthermore, since the migration strategies of animals have evolved in the context of the latent threat imposed by parasites and pathogens (Shaw et al. 2019; Balstad et al. 2021), it seems likely that anthropogenic impacts on the risk of parasite infection may also have implications for future migration patterns (White et al. 2018).

17.2.2 Stage II: Impacts on the exposure of hosts to infective parasites

Even when hosts and parasites coexist broadly in space and time, infections are only possible if they are in close contact for a sufficient period for exposures to occur (Combes 2001). Altered behaviours of both hosts and parasites in response to global environmental change can play an important role at this stage.

Social behaviour and artificially elevated density

For parasites that are transmitted directly between individuals, spatial proximity to conspecifics can be a major determinant of parasite exposure and transmission (Arneberg et al. 1998), so anthropogenic changes that affect the local density of hosts can alter opportunities for parasite transmission. In natural environments, increased local population densities can arise from habitat loss or fragmentation (Chapter 8) or from the aggregation of individuals at the least-impacted locations in anthropogenically disrupted landscapes, such as those subject to light or sound pollution. Supplemental feeding of wildlife in urban environments can also generate unnaturally high local densities of conspecifics, creating ideal conditions for the transmission of parasites (Murray et al. 2016). Experimental studies have demonstrated that clumping of supplementary food in urban environments generates increased aggregation behaviour and elevated social contact among raccoons *Procyon lotor* and is associated with increases in both endoparasite prevalence and intra-community richness (Wright and Gompper 2005; Figure 17.2). High population densities in fragmented environments can also generate increased levels of intraspecific aggression (MacDonald et al. 2004), facilitating the transmission of infectious diseases that are spread through biting or other intimate contact, including infectious facial tumour disease in Tasmanian devils *Sarcophilus harisii* (Hamede et al. 2013).

Altered sensory ecology of hosts and parasites

Because obligate parasites have an absolute requirement to infect hosts, they are expected to have evolved—and in many cases have been shown to exhibit—sophisticated sensory and behavioural mechanisms for detecting, approaching, and making contact with susceptible hosts (Haas 2003). Similarly, many indirectly transmitted parasites induce behavioural changes in infected intermediate hosts that are expected to bring them into closer contact with susceptible subsequent hosts in their life cycles (Poulin 2010). On the other hand, hosts have evolved a wide range of behaviours to detect and avoid contact with infectious agents (Hart 1990), for example, by employing habitat selection choices, foraging preferences, social behaviours, and mate choice decisions. Anthropogenic changes—including pollution by chemicals, light, or sound, or exposure to altered thermal regimes—can have far-reaching consequences for the sensory performance of organisms in impacted environments (Candolin and Wong 2019), disrupting the capacity of hosts, parasites, or vector organisms to detect and/or avoid one another.

If altered environmental conditions impair the performance of sensory systems or degrade the quality of the environments in which sensory systems operate (or in which behavioural decisions are made) then the likelihood that exposures occur can be altered. Such perturbations are particularly prevalent in urban environments, where chemical, light, and acoustic pollution frequently disrupt normal patterns of behaviour. For example, in environments subject to anthropogenic noise, the capacity of the parasitoid fly *Ormia ochracea* to use acoustic

cues to successfully locate crickets—which act as oviposition sites for females and hosts for developing fly larvae—is significantly reduced (Phillips et al. 2019; Figure 17.2).

Although the impacts of light pollution on host-parasite interactions are only beginning to be explored, evidence is accruing that such mechanisms are potentially important (Kernbach et al. 2018) and may have consequences for human disease (Kernbach et al. 2021). For example, artificial light at night (ALAN) can attract certain insect vectors, leading to an increased susceptibility to being bitten, with consequences for the diffusion of serious vector-borne diseases including leishmaniasis and malaria (Barghini and de Medeiros 2010). ALAN can also disrupt natural patterns of circadian activity, with ecological implications for host-parasite interactions. The nocturnal beetle *Dastarcus helophoroides* is an important ectoparasitoid of wood-boring insects, including various species of longhorn beetles, and is often used as biocontrol agent. When subjected to manipulated photoperiods, female *D. helophoroides* oviposited fewer eggs in longhorn beetle hosts, significantly reducing lifetime fecundity (Jiang et al. 2023) and reducing the effectiveness of pest control. Conversely, normally day-active parasitoid wasps *Venturia canescens* extend their period of activity into the nocturnal period when exposed to ALAN, and—in contrast to non-disrupted conspecifics—subsequently prioritize host location and ovipositing behaviours over foraging (Gomes et al. 2021).

The infective stages of parasites often exhibit sophisticated host-finding mechanisms that are frequently highly species-specific, often using host chemical gradients to locate and identify suitable host species (Haas 2003). For example, the parasitic lungworm *Rhabdias pseudosphaerocephala* uses chemical cues released in the skin secretions of cane toads *Rhinella marina* to facilitate location and invasion of hosts (Mayer et al. 2021). Infective larvae of human threadworms *Strongyloides stercoralis* and hookworms *Ancylostoma ceylanicum* have highly divergent olfactory preferences, suggesting the evolution of distinct strategies to infect human hosts (Gang et al. 2020). Such fine-tuned chemosensory mechanisms are susceptible to disruption by pollutants and other environmental perturbations that are known to impact the efficient functioning of sensory systems and/or the composition or degradation of chemical cues in other contexts (Candolin and Wong 2019).

Impacts on the behaviour, activity, and longevity of infectious agents

Environmental changes that alter the amount of time infectious agents spend in close association with prospective hosts also have the potential to alter patterns of exposure and infection. For the motile infective stages of parasites, which contain a finite energy resource to sustain life and fuel locomotion, increased activity levels under elevated temperatures (Koprivnikar et al. 2010) mean this energy is consumed more rapidly (Pechenik and Fried 1995), reducing the time they can remain motile and competent to infect susceptible hosts. Similarly, aquatic pollutants can negatively impact the longevity of trematode cercariae and their infectivity to amphibian hosts, though the precise effects depend on the trematode species involved (Koprivnikar et al. 2006a). Heavy metal pollutants, including cadmium and zinc, also affect the swimming behaviour (Cross et al. 2001) of trematode cercariae, as well as their phototaxis and geotaxis responses (Morley et al. 2003), and, as a consequence, reduce their infectivity in experimental studies (Pietrock and Goater 2005; Figure 17.2). In marine ecosystems, studies of the effects of ocean acidification have provided equivocal evidence for impact to date; end-of-century pCO$_2$ levels had no impact on a major economically damaging fish parasite, the salmon louse *Lepeophtheirus salmonis* (Thompson et al. 2019), whereas effects on trematode cercariae appear to vary between species and often interact with temperature (Franzova et al. 2019; Leiva et al. 2019).

17.2.3 Stage III: Impacts on the likelihood of infections establishing following exposure

Not all exposures lead to host organisms becoming infected; hosts have evolved a variety of defence mechanisms—including structural defences,

protective secretions, effective immune responses, and behaviours—that can prevent infectious agents from invading successfully and becoming established following initial contact (Klemme and Karvonen 2017). Anthropogenic environmental changes that impact the effectiveness of these structural defences, reduce the efficacy of immune responses, or impact normally protective behaviours, can therefore have a significant impact on the capability of hosts to respond effectively following exposure, with implications for infection levels.

Impacts on protective behaviours of hosts

Changes to normal patterns of behaviour, induced by anthropogenic stressors, may also play an important role in determining the outcome of parasite exposures. When exposed to infective trematode cercariae, tadpoles of wood frogs *Lithobates sylvaticus* normally elevate their activity levels, which, under normal conditions, is an effective response that reduces the ability of cercariae to establish and leads to reduced parasite loads (Koprivnikar et al. 2006b). However, when subject to ALAN, tadpoles significantly reduced their activity levels and were more susceptible to trematode infection; this effect occurred in the absence of any effect of ALAN on tadpole immune response, indicating that it was the change in their behaviour that enhanced susceptibility (May et al. 2019).

Impacts on immune systems

Stress responses activated through either the sympathetic adrenal-medullary axis or through the hypothalamic-pituitary-adrenal axis, in response to environmental or physiological stimuli, generate chronic production of glucocorticoid hormones and catecholamines, which can dysregulate immune function (Padgett and Glaser 2003). It is therefore not unexpected that a wide range of anthropogenic stressors have been demonstrated to negatively impact immunocompetence, with impacts on infection and disease, across a variety of taxa.

Anthropogenic noise, for example, has been demonstrated to induce an immunosuppressive effect in birds, fish, and amphibians (Wysocki et al.

2006; Troianowski et al. 2017; Brumm et al. 2021). In a controlled laboratory study, host guppies *Poecilia reticulata* experiencing acute noise developed significantly elevated ectoparasite (*Gyrodactylus*) burdens following a controlled experimental challenge, when compared with those in no-noise treatments; however, guppies exposed to chronic noise, over a longer time frame, exhibited increased mortality at lower parasite burdens, and so may have experienced reduced tolerance to infections (Masud et al. 2020).

Anthropogenic light can also impact the effectiveness of immune systems. Among dark-eyed juncos *Junco hyemalis*, exposure to ALAN during sensitive periods prior to their migration was associated with subsequent impaired immunocompetence (inflammation and elevated leucocyte counts) and elevated haemosporidian parasite loads (Becker et al. 2020; Figure 17.2).

17.2.4 Stage IV: Impacts on the growth, development, and proliferation of infections

Once established, the rate at which parasites grow and proliferate, and consequently, how infections impact the biology and health of hosts, can vary considerably. Hosts are capable of adopting a range of strategies—from mounting immune responses to adopting complex social behaviours that have evolved to reduce infection loads—to limit the impacts of parasite infections, but the capacity of hosts to use these strategies, or their efficacy, may depend on environmental conditions and, hence, be impacted in a changing world.

Immunometabolic impacts of climate warming

Immune responses are particularly susceptible to environmentally induced stress and so their ability to prevent disease proliferation following establishment may be compromised under anthropogenically altered conditions. Warming climates can favour the growth, development, or proliferation of parasites following establishment, either because they impact host immune systems, or because parasites are closer to their metabolic thermal optimum than hosts under altered regimes. In fish, elevated

water temperatures can be associated with suppressed innate immune responses (Dittmar et al. 2014), which can restrict the ability of fish to control the growth, proliferation, or development of infections.

Plerocercoid larvae of the cestode *Schistocephalus solidus* commonly parasitize three-spined sticklebacks *Gasterosteus aculeatus* in temperate freshwaters. Plerocercoids grow to a large size in the fish's body cavity (Figure 17.2) before achieving infectivity to definitive bird hosts, with transmission being facilitated by parasite-induced host behavioural change (Barber and Scharsack 2010). Experimental studies have demonstrated that plerocercoids grow more quickly at warmer temperatures (Macnab and Barber 2012) as a consequence of a shift in host immunometabolic processes that increase resource availability to parasites (Scharsack et al. 2021). Infected sticklebacks also exhibit behavioural preferences for warmer temperatures, generating positive feedback for parasite growth. This preference for temperatures that favour parasite growth contrasts with cases of 'behavioural fever' where infected animals select thermal regimes to counter the negative impacts of infection (e.g. desert locusts *Schistocerca gregaria*; Elliot et al. 2002) or reduce parasite loads (e.g. guppies *Poecilia reticulata*; Mohammed et al. 2016), and provides further evidence of the capacity for warming climates to impact the behavioural basis of host-parasite relationships in diverse ways.

Impacts on parasite removal behaviours

Many species are able to control parasite loads by adopting behaviours such as preening or grooming, which can either be performed individually or in social groups. Such behaviours are often time-constrained and, in urban environments, human disturbance can reduce the time that individuals invest in them. For example, bonnet macaques *Macaca radiata* that increased time spent monitoring people reduced their levels of allogrooming (Balasubramaniam et al. 2020; Figure 17.2). In marine systems, large fish often visit cleaning stations where specialist fish and invertebrates remove ectoparasites, and this behaviour is effective in reducing infection levels (Vaughan et al. 2017). Physiological studies of the tropical cleaner shrimp *Lysmata amboinensis* indicate that they are susceptible to warming ocean temperatures, with implications for parasite loads among client fish and the future health of reef ecosystems (Rosa et al. 2014).

17.2.5 Stage V: Impacts on onward transmission of infection

The various mechanisms operating at Stages I–IV not only have implications for the establishment and proliferation of infections and for the impact of disease on focal host individuals, but also have the potential to impact onward transmission to other individuals in the host population and beyond. However, there is one further behavioural mechanism that can be disrupted under altered environmental conditions and that is likely to have implications for transmission: the manipulated behaviour of infected individuals.

Parasite-altered host behaviour in the face of environmental change

Many parasites have evolved mechanisms to alter the behaviour of host organisms in ways that are expected to increase the efficiency of transmission to subsequent hosts (Poulin 2010; Hughes et al. 2012); for example, the anti-predator behaviour of intermediate hosts of trophically transmitted parasites can be disrupted in ways that enhance predation by susceptible predators (Berdoy et al. 2000). The mechanisms by which parasites induce these behavioural changes in host behaviour— often referred to as 'manipulation'—often involve complex neurophysiological manipulation of host systems (Klein 2003), suggesting fine-scale tuning over evolutionary timescales. Anthropogenic change might, therefore, impact the mechanistic basis of manipulation due to physiological or biochemical disruption of the host or parasite, and also its effectiveness, if it alters the likelihood that the behavioural change leads to increased transmission under changed ecological conditions (Labaude et al. 2015). For example, the neuromanipulatory mechanisms used by the infective larval stages of trophically transmitted helminth parasites to alter fish behaviour (Grecias et al. 2020) could be susceptible

to disruption by anthropogenic neuropharmaceuticals in urban waterways, whereas altered predator regimes resulting from species introductions might lead to ingestion by non-target predators. To date, few studies have investigated such phenomena, but given the critical importance of manipulative parasites in 'directing' ecological food web dynamics (Lafferty et al. 2008), it will be important to develop a better understanding of how changing environments and manipulation behaviour interact to determine the potential for disease transmission.

17.3 Harnessing behavioural knowledge to predict, prevent, and manage infectious disease outbreaks in a changing world

Anthropogenic activity is changing the environments in which hosts and parasites interact, with implications for the survival, spread, and impact of existing infections, and for the development of novel associations. There remain many challenges to developing a comprehensive understanding of the behavioural basis of the spread of infectious diseases under anthropogenic change that can be used to effectively predict and manage disease outbreaks. Nonetheless, a critical first step in achieving this goal is to understand the diversity of mechanisms through which behaviour plays a role in the dissemination, spread, and impact of disease in anthropogenically impacted ecosystems. A significant body of research, selectively and briefly reviewed in Section 17.2 above, is now emerging, and this is beginning to identify the potential for behaviour to play a role in determining the patterns of disease that are likely to arise in a changing world.

The findings of behavioural studies are providing insights that have allowed the development of some relatively simple, practical, and effective interventions that can be built into broader management plans for restricting the further spread of diseases that threaten the health and welfare of humans or livestock, or wildlife. Three examples are provided below, and in Box 17.1.

Figure 17.2 Examples of host-parasite interactions that are impacted in anthropogenically altered ecosystems (see text for more details). (a) Aggregations of raccoons *Procyon lotor* around anthropogenic food resources in urban environments can lead to increased opportunities for parasite transmission; (b) the ability of the parasitoid fly *Ormia ochracea* to locate cricket hosts is impaired by noise pollution; (c) bonnet macaques *Macaca radiata* spend less time engaged in allogrooming behaviour in the presence of people in urban environments; (d) *Schistocephalus solidus* plerocercoids infecting host three-spined sticklebacks *Gasterosteus aculeatus* grow more quickly and achieve infectivity sooner in warmer temperatures, and change host thermal preferences; (e) the survival, activity, and host-finding behaviour of trematode cercariae can be impacted by a range of aquatic pollutants; (f) exposure to chronic light pollution is associated with impaired immunocompetence and elevated parasite loads in dark-eyed juncos *Junco hyemalis*.

Photos: (a) iStock.com/dzphotovideo; (b) Alamy/Vinícius Souza; (c) Alamy/ Łukasz Szczepanski; (d) Iain Barber; (e) Shutterstock.com/F.Niedl; (f) Alamy/Ivan Kuzmin.

17.3.1 Guidelines for supplementary feeding of wildlife

Supplementary feeding, particularly in urban environments, can sustain wildlife populations during periods of food shortage, but behavioural and parasitological studies have shown that the aggregation of mammals and birds at feeding sites often leads to elevated parasite infections that threaten the health of individuals and populations (e.g. red kites *Milvus milvus*: Blanco et al. 2017; wild boar *Sus scrofa*: Oja et al. 2017; raccoons *Procyon lotor*: Wright and Gompper 2005). Supplementary feeding may also lead to reduced levels of immunity (Strandin et al. 2018). A meta-analysis undertaken by Murray et al. (2016) showed that 95% of studies undertaken identified close contact between individuals as a factor influencing the increased risk of transmission, with the accumulation of pathogens around feeders being influential in 77% of cases. A similar analysis by Lawson et al. (2018) focused on the disease risks to UK songbirds resulting from backyard bird feeding. Both sets of authors make clear recommendations for mitigation strategies, which have been adopted by leading wildlife charities and management organizations.

17.3.2 Design of LED lighting systems to minimize risk from vector-borne diseases

The attraction of biting insects, which serve as the vectors of infectious diseases of people and livestock, to artificial light sources in tropical ecosystems can lead to elevated spread of important infectious diseases, including Chagas disease, leishmaniasis, and malaria (Barghini and de Medeiros 2010). Behavioural research has identified the frequencies that are most and least attractive to different vector insect species, allowing specific recommendations to be made regarding the optimal specifications for the light spectrum of LED household lighting

systems, the design of plans to reduce overall nocturnal lighting, and the design of light traps (Deichmann et al. 2021; Wilson et al. 2023).

17.3.3 Biosecurity measures to combat the threat of bovine TB on farms (see also Box 17.1)

Land use changes, including the development of intensive agriculture, can lead to increased contact between species, which in turn can facilitate the transmission of disease between domestic animals and wildlife (Wiethoelter et al. 2015). Bovine tuberculosis (bovine TB, bTB) is a serious, globally distributed infectious disease of cattle that causes reduced productivity and premature death of infected animals, with significant consequences for animal health and for rural economies and well-being in farming communities, and has the potential to cause human infections (O'Reilly and Daborn 1995). The causative agent of the disease is the bacterium *Mycobacterium bovis*, which is spread between domestic cattle *Bos taurus*, but native mammalian wildlife—including various species of deer, wild pigs, possums, and buffalo (Fitzgerald and Kaneene 2013)—often serve as reservoir hosts for infection, creating challenges for effective control through routine stock management. While a range of approaches have been taken to manage levels of infection—including the culling and vaccination of native wildlife reservoir hosts—behavioural

---

Box 17.1 Bovine TB: behavioural challenges and opportunities for the management of an economically and socially devastating disease in the UK

Background

In the UK, bovine tuberculosis (bovine TB, bTB) transmission is often facilitated by native Eurasian badgers *Meles meles* (Figure 17.3a), which are often numerous in the lowland pasture environments that have been developed for cattle farming. In an attempt to manage infection levels in cattle, badger populations in infected regions of the UK were routinely culled from the 1970s through until the late 1990s, but the variable effectiveness of culling, coupled with increasingly vocal public concern over perceived persecution of a charismatic and iconic member of the native UK macrofauna,

led to a pause in the practice and the commissioning of a scientific report (Krebs et al. 1997), which recommended the implementation of a Randomised Badger Culling Trial (RBCT). The trial showed that that while proactive culling of badgers around farms led to a decrease in infection levels within cull areas, infection levels actually increased outside of them, whereas reactive culling—targeting badgers only after a bTB outbreak—also led to increased infections within cull zones (Donnelly et al. 2006). Consequently, reactive culling was suspended as a control strategy (Giesler and Ares 2018) and the trial was ended prematurely.

Box 17.1 *Continued*

The role of behaviour in transmission and spread of disease in culled populations

Behavioural studies examining badger-cattle interactions on farms have documented a range of contact points facilitating disease transmission, including badgers visiting cattle feed stores in farm buildings, exhibiting close 'nose-to-nose' contact with housed cattle, and frequent scent-marking and excretion in food, with a peak of activity during dry weather in the spring and summer (Tolhurst et al. 2009).

Behavioural research has also provided significant insight into the factors underlying the variable impacts of badger culling strategies on infection levels in cattle, identifying a key role of cull-induced changes in ranging behaviour. Badgers typically live in small groups within territories whose range is restricted by the territories of adjoining groups. An unforeseen consequence of the culling programme was the propensity of badgers that avoided the cull to expand their movements beyond the original territory (Woodroffe et al. 2006). The size of the monthly range explored by badgers in culled zones increased by 61% post-cull, with individual animals ranging 39% further each night from the sett. These range size increases occurred despite the animals being active for a shorter period, because they also moved more quickly, presumably because the reduction in conflict with adjacent territorial animals meant more direct routes could be taken. Consequently, badgers surviving the cull visited 45% more fields every month, increasing their opportunity to interact with a greater number of cattle (Ham et al. 2019).

How behavioural knowledge has influenced policy and practice in the control of bovine TB

The results of behavioural studies focusing on badger behaviour in and around farms, their interactions with cattle, and their responses to changes in local social structures arising from culling, have contributed significantly to the development of management approaches and policy for control of the disease in the UK, most notably in changing culling policy (see above) but also in promoting the importance of farm biosecurity. Insight from these studies has provided a far greater evidence-based focus on the importance of maintaining high levels of biosecurity on farms, aimed at restricting the potential for interactions. While on-farm biosecurity controls are unlikely to eliminate bTB, they can play a key role in reducing disease risk, and simple biosecurity measures—including the use of sheet metal gates, electrified fencing, and badger-resistant raised bucket feeders (Figure 17.3b)—are effective in restricting badger-cattle interactions in and around farm buildings (Judge et al. 2011). The UK Government's bTB biosecurity plan was launched in 2015, and farmers are now being provided with abundant practical biosecurity advice through government agencies (e.g. the Animal and Plant Health Agency's TBhub www.tbhub.co.uk and DEFRA's TB advisory service www.tbas.org.uk).

Figure 17.3 (a) A European badger *Meles meles*, the major reservoir host for bovine TB in the UK; (b) Hereford cattle *Bos taurus* using raised bucket feeders that prevent badgers reaching them and help prevent the spread of bovine TB.

Photos: (a) Alamy/Redmond Durrell; (b) Alamy/Wayne Hutchinson.

studies have been instrumental in identifying relatively simple, low-cost improvements to biosecurity in farm buildings and at pasture that can often provide effective solutions (Tolhurst et al. 2008; Judge et al. 2011; see Box 17.1).

17.4 Conclusions, and challenges to developing a more predictive understanding

Human activity is altering the environments in which host organisms and their many parasites, pathogens, and pests interact at an unprecedented rate, and this is having significant implications for the establishment of novel interactions and for the patterns of disease. As exemplified by the recent COVID-19 pandemic, behaviour can play a key role at all stages of the disease process, and the ways in which human and animal behaviours are altered in our changing world can have profound implications for the establishment and spread of both novel and existing infections. Behavioural research has begun to provide insights into the diversity of mechanisms that can contribute to the altered transmission of disease under anthropogenically altered conditions, and in some cases, this has facilitated the development of interventions and control strategies to manage infectious diseases.

As a mode of life, parasitism—in its broadest sense—is an extraordinarily successful strategy. Although estimates vary, perhaps 50% of known metazoan species are likely to be parasites (Dobson et al. 2008); parasitism has evolved independently over 200 times in animals alone (Weinstein and Kuris 2016), and parasitic organisms span the taxonomic spectrum. This taxonomic diversity is matched by the variation in the details of parasite life cycles, and in the modes of transmission. For example, while some parasites are mobile and capable of moving independently between individual hosts, others rely on vectors—frequently insects, such as biting flies or mosquitoes—for transmission between successive hosts. Similarly, while some parasites require only a single host individual or species to complete their life cycle, others have more complex, indirect life cycles that typically rely on ecological and behavioural interactions to allow transmission between a succession of separate hosts. Furthermore, while some parasites are capable of rapid multiplication on, or in, any particular host, others are not and must colonize each host independently, such that each individual parasite infecting a host represents the outcome of a successful transmission event.

An understanding of these details for any host-parasite system under investigation is critical before we can predict how parasites, and the host organisms they infect, are likely to be impacted by changing conditions. Yet, a detailed knowledge of hosts, life cycles, and transmission routes is often lacking for many parasites that are not of medical or veterinary importance, or have not been subject to intense research in an ecological context. This lack of basic knowledge places a limit on our ability to make general statements or predictions about how the behavioural impacts of global change will impact patterns of diseases.

Furthermore, impacted ecosystems are typically subject to multiple, simultaneous, and diverse anthropogenic stressors (Chapter 10), which can also act synergistically to create unexpected challenges for hosts and parasites separately. While our understanding of how environmental stressors interact to impact free-living organisms is improving, far less is known about how they impact parasites and disease. Furthermore, for infections that rely on multiple host species, the impact of environmental stressors on all hosts must be considered, and where these hosts represent diverse taxa, divergent responses to any environmental change might be expected (Barber et al. 2016).

We should therefore be cautious—and realistic about the significant challenges that exist—when considering our current capacity to identify general patterns and develop a more unified, comprehensive, and predictive understanding of the impact of global change on infectious diseases, and the role played by behaviour. There remains an urgent need for further research to address existing knowledge gaps.

References

Adamo, S.A., and Webster, J.P. (2013). Neural parasitology—how parasites manipulate host behaviour (Special Issue). *Journal of Experimental Biology*, 216, 1–160.

Alhashemi, S.H., Ahmadi, F., and Dehshahri, A. (2023). Lessons learned from COVID-19 pandemic: vaccine platform is a key player. *Process Biochemistry*, 124, 269–279.

Altizer, S., Ostfeld, R.S., Johnson, P.T.J., et al. (2013). Climate change and infectious diseases: from evidence to a predictive framework. *Science*, 341, 514–519.

Arneberg, P., Skorping, A., Grenfell, B., and Read, A.F. (1998). Host densities as determinants of abundance in parasite communities. *Proceedings of the Royal Society B: Biological Sciences*, 265, 1283–1289.

Baker, R.E., Mahmud, A.S., Miller, I.F., et al. (2022). Infectious disease in an era of global change. *Nature Reviews Microbiology*, 20, 193–205.

Balasubramaniam, K.N., Marty, P.R., Arlet, M.E., et al. (2020). Impact of anthropogenic factors on affiliative behaviors among bonnet macaques. *American Journal of Physical Anthropology*, 171, 704–717.

Balstad, L.J., Binning, S.A., Craft, M.E., et al. (2021). Parasite intensity and the evolution of migratory behavior. *Ecology*, 102, e03229.

Barber, I., Berkhout, B.W., and Ismail, Z. (2016). Thermal change and the dynamics of multi-host parasite life cycles in aquatic ecosystems. *Integrative and Comparative Biology*, 56, 561–572.

Barber, I., and Dingemanse, N.J. (2010). Parasitism and the evolutionary ecology of animal personality. *Philosophical Transactions of the Royal Society B: Biological Sciences*, 365, 4077–4088.

Barber, I., and Scharsack, J.P. (2010). The three-spined stickleback-*Schistocephalus solidus* system: an experimental model for investigating host-parasite interactions in fish. *Parasitology*, 137, 411–424.

Barghini, A., and de Medeiros, B.A.S. (2010). Artificial lighting as a vector attractant and cause of disease diffusion. *Environmental Health Perspectives*, 118, 1503–1506.

Barouki, R., Kogevinas, M., Audouze, K., et al. (2021). The COVID-19 pandemic and global environmental change: emerging research needs. *Environment International*, 146, 106272.

Becker, D.J., Singh, D., Pan, Q.Y., et al. (2020). Artificial light at night amplifies seasonal relapse of haemosporidian parasites in a widespread songbird. *Proceedings of the Royal Society B: Biological Sciences*, 287, 20201831.

Berdoy, M., Webster, J.P., and Macdonald, D.W. (2000). Fatal attraction in rats infected with *Toxoplasma gondii*. *Proceedings of the Royal Society B: Biological Sciences*, 267, 1591–1594.

Berkhout, B.W., Budria, A., Thieltges, D.W., and Slabbekoorn, H. (2023). Anthropogenic noise pollution and wildlife diseases. *Trends in Parasitology*, 39, 181–190.

Blanco, G., Cardells, J., and Garijo-Toledo, M.M. (2017). Supplementary feeding and endoparasites in threatened avian scavengers: coprologic evidence from red kites in their wintering stronghold. *Environmental Research*, 155, 22–30.

Bradley, C.A., and Altizer, S. (2007). Urbanization and the ecology of wildlife diseases. *Trends in Ecology & Evolution*, 22, 95–102.

Brumm, H., Goymann, W., Deregnaucourt, S., et al. (2021). Traffic noise disrupts vocal development and suppresses immune function. *Science Advances*, 7, eabe2405.

Byers, J.E. (2021). Marine parasites and disease in the era of global climate change. *Annual Review of Marine Science*, 13, 397–420.

Candolin, U., and Wong, B.B.M. (2012). *Behavioural Responses to a Changing World: Mechanisms and Consequences*. Oxford University Press, Oxford.

Candolin, U., and Wong, B.B.M. (2019). Mate choice in a polluted world: consequences for individuals, populations and communities. *Philosophical Transactions of the Royal Society B: Biological Sciences*, 374, 20180055.

Canning, E.U., and Wright, C.A. (1972). Behavioural aspects of parasite transmission. *Zoological Journal of the Linnean Society*, 51 (Supplement 1), 219.

Chantrey, J., Dale, T.D., Read, J.M., et al. (2014). European red squirrel population dynamics driven by squirrelpox at a gray squirrel invasion interface. *Ecology and Evolution*, 4, 3788–3799.

Choisy, M., and Rohani, P. (2006). Harvesting can increase severity of wildlife disease epidemics. *Proceedings of the Royal Society B: Biological Sciences*, 273, 2025–2034.

Coetzee, B.W.T., Gaston, K.J., Koekemoer, L.L., et al. (2022). Artificial light as a modulator of mosquito-borne disease risk. *Frontiers in Ecology and Evolution*, 9, 768090.

Combes, C. (2001). *Parasitism: The Ecology and Evolution of Intimate Interactions*. University of Chicago Press, Chicago, IL.

Cross, M.A., Irwin, S.W.B., and Fitzpatrick, S.M. (2001). Effects of heavy metal pollution on swimming and longevity in cercariae of *Cryptocotyle lingua* (Digenea: Heterophyidae). *Parasitology*, 123, 499–507.

Deichmann, J.L., Gatty, C.A., Navarro, J.M.A., et al. (2021). Reducing the blue spectrum of artificial light at night minimises insect attraction in a tropical lowland forest. *Insect Conservation and Diversity*, 14, 247–259.

Dittmar, J., Janssen, H., Kuske, A., et al. (2014). Heat and immunity: an experimental heat wave alters immune functions in three-spined sticklebacks (*Gasterosteus aculeatus*). *Journal of Animal Ecology*, 83, 744–757.

Dobson, A., Lafferty, K.D., Kuris, A.M., et al. (2008). Homage to Linnaeus: how many parasites? How many hosts? *Proceedings of the National Academy of Sciences*, 105, 11482–11489.

Donnelly, C.A., Woodroffe, R., Cox, D.R., et al. (2006). Positive and negative effects of widespread badger culling on tuberculosis in cattle. *Nature*, 439, 843–846.

Elliot, S.L., Blanford, S., and Thomas, M.B. (2002). Host-pathogen interactions in a varying environment: temperature, behavioural fever and fitness. *Proceedings of the Royal Society B: Biological Sciences*, 269, 1599–1607.

Fitzgerald, S.D., and Kaneene, J.B. (2013). Wildlife reservoirs of bovine tuberculosis worldwide: hosts, pathology, surveillance, and control. *Veterinary Pathology*, 50, 488–499.

Franzova, V.A., MacLeod, C.D., Wang, T.X., and Harley, C.D.G. (2019). Complex and interactive effects of ocean acidification and warming on the life span of a marine trematode parasite. *International Journal for Parasitology*, 49, 1015–1021.

Gang, S.S., Castelletto, M.L., Yang, E., et al. (2020). Chemosensory mechanisms of host seeking and infectivity in skin-penetrating nematodes. *Proceedings of the National Academy of Sciences*, 117, 17913–17923.

Giesler, R. and Ares, E. (2018). Badger culling in England. House of Commons Library Briefing Paper Number 6837.

Gomes, E., Rey, B., Debias, F., et al. (2021). Dealing with host and food searching in a diurnal parasitoid: consequences of light at night at intra- and trans-generational levels. *Insect Conservation and Diversity*, 14, 235–246.

Grecias, L., Hebert, F.O., Alves, V.A., et al. (2020). Host behaviour alteration by its parasite: from brain gene expression to functional test. *Proceedings of the Royal Society B: Biological Sciences*, 287, 20202252.

Gresh, T., Lichatowich, J., and Schoonmaker, P. (2000). An estimation of historic and current levels of salmon production in the Northeast Pacific ecosystem: evidence of a nutrient deficit in the freshwater systems of the Pacific Northwest. *Fisheries*, 25, 15–21.

Haas, W. (2003). Parasitic worms: strategies of host finding, recognition and invasion. *Zoology*, 106, 349–364.

Ham, C., Donnelly, C.A., Astley, K.L., et al. (2019). Effect of culling on individual badger *Meles meles* behaviour: potential implications for bovine tuberculosis transmission. *Journal of Applied Ecology*, 56, 2390–2399.

Hamede, R.K., McCallum, H., and Jones, M. (2013). Biting injuries and transmission of Tasmanian devil facial tumour disease. *Journal of Animal Ecology*, 82, 182–190.

Harrington, W.N., Kackos, C.M., and Webby, R.J. (2021). The evolution and future of influenza pandemic preparedness. *Experimental Molecular Medicine*, 53, 737–749.

Hart, B.L. (1990). Behavioral adaptations to pathogens and parasites—5 strategies. *Neuroscience and Biobehavioral Reviews*, 14, 273–294.

Harvell, C.D., Mitchell, C.E., Ward, J.R., et al. (2002). Ecology—Climate warming and disease risks for terrestrial and marine biota. *Science*, 296, 2158–2162.

Hughes, D.P., Brodeur, J., and Thomas, F. (2012). *Host Manipulation by Parasites*. Oxford University Press, Oxford.

Hulme, P.E. (2017). Climate change and biological invasions: evidence, expectations, and response options. *Biological Reviews*, 92, 1297–1313.

Jiang, X.L., Ren, Z., Hai, X.X., et al. (2023). Exposure to artificial light at night mediates the locomotion activity and oviposition capacity of *Dastarcus helophoroides* (Fairmaire). *Frontiers in Physiology*, 14, 1063601.

Jones, B.A., Grace, D., Kock, R., et al. (2013). Zoonosis emergence linked to agricultural intensification and environmental change. *Proceedings of the National Academy of Sciences*, 110, 8399–8404.

Jones, K.E., Patel, N.G., Levy, M.A., et al. (2008). Global trends in emerging infectious diseases. *Nature*, 451, 990–993.

Judge, J., McDonald, R.A., Walker, N. and Delahay, R.J. (2011). Effectiveness of biosecurity measures in preventing badger visits to farm buildings. *PLOS ONE*, 6, e28941.

Kernbach, M.E., Hall, R.J., Burkett-Cadena, N.D., et al. (2018). Dim light at night: physiological effects and ecological consequences for infectious disease. *Integrative and Comparative Biology*, 58, 995–1007.

Kernbach, M.E., Martin, L.B., Unnasch, T.R., et al. (2021). Light pollution affects West Nile virus exposure risk across Florida. *Proceedings of the Royal Society B: Biological Sciences*, 288, 20210253.

Klein, S.L. (2003). Parasite manipulation of the proximate mechanisms that mediate social behavior in vertebrates. *Physiology & Behavior*, 79, 441–449.

Klemme, I., and Karvonen, A. (2017). Vertebrate defense against parasites: interactions between avoidance, resistance, and tolerance. *Ecology and Evolution*, 7, 561–571.

Koprivnikar, J., Forbes, M.R., and Baker, R.L. (2006a). Effects of atrazine on cercarial longevity, activity, and infectivity. *Journal of Parasitology*, 92, 306–311.

Koprivnikar, J., Forbes, M.R., and Baker, R.L. (2006b). On the efficacy of anti-parasite behaviour: a case study of tadpole susceptibility to cercariae of *Echinostoma trivolvis*. *Canadian Journal of Zoology: Revue Canadienne de Zoologie*, 84, 1623–1629.

Koprivnikar, J., Lim, D., Fu, C., and Brack, S.H. (2010). Effects of temperature, salinity, and pH on the survival and activity of marine cercariae. *Parasitology Research*, 106, 1167–1177.

Krebs, J.R., Anderson, R., Clutton-Brock, T., et al. (1997). *Bovine Tuberculosis in Cattle and Badger: Report by the Independent Scientific Review Group*. Ministry for Agriculture, Fisheries and Food (UK Government), London.

Labaude, S., Rigaud, T., and Cezilly, F. (2015). Host manipulation in the face of environmental changes: ecological consequences. *International Journal for Parasitology: Parasites and Wildlife*, 4, 442–451.

Lafferty, K.D. (2012). Biodiversity loss decreases parasite diversity: theory and patterns. *Philosophical Transactions of the Royal Society B: Biological Sciences*, 367, 2814–2827.

Lafferty, K.D., Allesina, S., Arim, M., et al. (2008). Parasites in food webs: the ultimate missing links. *Ecology Letters*, 11, 533–546.

Lawson, B., Robinson, R.A., Toms, M.P., et al. (2018). Health hazards to wild birds and risk factors associated with anthropogenic food provisioning. *Philosophical Transactions of the Royal Society B: Biological Sciences*, 373, 20170091.

Lefèvre, P., Blancou, J., Chermette, R., and Uilenberg, G. (2010). *Infectious and Parasitic Diseases of Livestock*. CABI Publishing, Cambridge, MA.

Leiva, N.V, Manriquez, P.H., Aguilera, V.M., and Gonzalez, M.T. (2019). Temperature and pCO_2 jointly affect the emergence and survival of cercariae from a snail host: implications for future parasitic infections in the Humboldt Current system. *International Journal for Parasitology*, 49, 49–61.

Lewis, E.E., Campbell, J.F., and Sukhdeo, M.V.K. (eds). (2002). *The Behavioural Ecology of Parasites*. CABI Publishing, Cambridge, MA, USA.

MacDonald, D.W., Harmsen, B.J., Johnson, P.J., and Newman, C. (2004). Increasing frequency of bite wounds with increasing population density in Eurasian badgers, *Meles meles*. *Animal Behaviour*, 67, 745–751.

Macnab, V., and Barber, I. (2012). Some (worms) like it hot: fish parasites grow faster in warmer water, and alter host thermal preferences. *Global Change Biology*, 18, 1540–1548.

Masud, N., Hayes, L., Crivelli, D., et al. (2020). Noise pollution: acute noise exposure increases susceptibility to disease and chronic exposure reduces host survival. *Royal Society Open Science*, 7, 200172.

May, D., Shidemantle, G., Melnick-Kelley, Q., et al. (2019). The effect of intensified illuminance and artificial light at night on fitness and susceptibility to abiotic and biotic stressors. *Environmental Pollution*, 251, 600–608.

Mayer, M., Justicia, L.S., Shine, R., and Brown, G.P. (2021). Host defense or parasite cue: skin secretions mediate interactions between amphibians and their parasites. *Ecology Letters*, 24, 1955–1965.

Mohammed, R.S., Reynolds, M., James, J., et al. (2016). Getting into hot water: sick guppies frequent warmer thermal conditions. *Oecologia*, 181, 911–917.

Morley, N.J., Crane, M., and Lewis, J.W. (2003). Effects of cadmium and zinc toxicity on orientation behaviour of *Echinoparyphium recurvatum* (Digenea: Echinostomatidae) cercariae. *Diseases of Aquatic Organisms*, 56, 89–92.

Murray, M.H., Becker, D.J., Hall, R.J., and Hernandez, S.M. (2016). Wildlife health and supplemental feeding: a review and management recommendations. *Biological Conservation*, 204, 163–174.

Oja, R., Velstrom, K., Moks, E., et al. (2017). How does supplementary feeding affect endoparasite infection in wild boar? *Parasitology Research*, 116, 2131–2137.

O'Reilly, L.M., and Daborn, C.J. (1995). The epidemiology of *Mycobacterium bovis* infections in animals and man—a review. *Tubercle and Lung Disease*, 76, 1–46.

Padgett, D.A., and Glaser, R. (2003). How stress influences the immune response. *Trends in Immunology*, 24, 444–448.

Paladini, G., Shinn, A.P., Taylor, N.G.H., et al. (2021). Geographical distribution of *Gyrodactylus salaris* Malmberg, 1957 (Monogenea, Gyrodactylidae). *Parasites & Vectors*, 14, 34.

Pechenik, J.A. and Fried, B. (1995). Effect of temperature on survival and infectivity of *Echinostoma trivolvis* cercariae: a test of the energy limitation hypothesis. *Parasitology*, 111, 373–378.

Phillips, J.N., Ruef, S.K., Garvin, C.M., et al. (2019). Background noise disrupts host-parasitoid interactions. *Royal Society Open Science*, 6, 190867.

Pietrock, M., and Goater, C.P. (2005). Infectivity of *Ornithodiplostomum ptychocheilus* and *Posthodiplostomum minimum* (Trematoda: Diplostomidae) cercariae following exposure to cadmium. *Journal of Parasitology*, 91, 854–856.

Poulin, R. (2010). Parasite manipulation of host behavior: an update and frequently asked questions. *Advances in the Study of Behavior*, 41, 151–186.

Rosa, R., Lopes, A.R., Pimentel, M., et al. (2014). Ocean cleaning stations under a changing climate: biological responses of tropical and temperate fish-cleaner shrimp to global warming. *Global Change Biology*, 20, 3068–3079.

Scharsack, J.P., Wieczorek, B., Schmidt-Drewello, A., et al. (2021). Climate change facilitates a parasite's host exploitation via temperature-mediated immunometabolic processes. *Global Change Biology*, 27, 94–107.

Shaw, A.K., Craft, M.E., Zuk, M., and Binning, S.A. (2019). Host migration strategy is shaped by forms of parasite transmission and infection cost. *Journal of Animal Ecology*, 88, 1601–1612.

Short, E.E., Caminade, C., and Thomas, B.N. (2017). Climate change contribution to the emergence or re-emergence of parasitic diseases. *Infectious Diseases: Research and Treatment*, 10, 1178633617732296.

Smith, K.F., Sax, D.F., and Lafferty, K.D. (2006). Evidence for the role of infectious disease in species extinction and endangerment. *Conservation Biology*, 20, 1349–1357.

Strandin, T., Babayan, S.A., and Forbes, K.M. (2018). Reviewing the effects of food provisioning on wildlife immunity. *Philosophical Transactions of the Royal Society B: Biological Sciences*, 373, 20170088.

Sures, B., Nachev, M., Selbach, C., and Marcogliese, D.J. (2017). Parasite responses to pollution: what we know and where we go in 'Environmental Parasitology'. *Parasites & Vectors*, 10, 65.

Thompson, C.R.S., Fields, D.M., Bjelland, R.M., et al. (2019). The planktonic stages of the salmon louse (*Lepeophtheirus salmonis*) are tolerant of end-of-century pCO_2 concentrations. *PeerJ*, 7, e7810.

Tian, H.Y., Liu, Y.H., Li, Y.D., et al. (2020). An investigation of transmission control measures during the first 50 days of the COVID-19 epidemic in China. *Science*, 368, 638–642.

Tolhurst, B.A., Delahay, R.J., Walker, N.J., et al. (2009). Behaviour of badgers (*Meles meles*) in farm buildings: opportunities for the transmission of *Mycobacterium bovis* to cattle? *Applied Animal Behaviour Science*, 117, 103–113.

Tolhurst, B.A., Ward, A.I., Delahay, R.J., et al. (2008). The behavioural responses of badgers (*Meles meles*) to exclusion from farm buildings using an electric fence. *Applied Animal Behavioural Science*, 113, 224–235.

Tompkins, D.M., Dunn, A.M., Smith, M.J., and Telfer, S. (2011). Wildlife diseases: from individuals to ecosystems. *Journal of Animal Ecology*, 80, 19–38.

Torchin, M.E., Lafferty, K.D., Dobson, A.P., et al. (2003). Introduced species and their missing parasites. *Nature*, 421, 628–630.

Troianowski, M., Mondy, N., Dumet, A., et al. (2017). Effects of traffic noise on tree frog stress levels, immunity, and color signaling. *Conservation Biology*, 31, 1132–1140.

Vaughan, D.B., Grutter, A.S., Costello, M.J., and Hutson, K.S. (2017). Cleaner fishes and shrimp diversity and a re-evaluation of cleaning symbioses. *Fish and Fisheries*, 18, 698–716.

Weinstein, S.B., and Kuris, A.M. (2016). Independent origins of parasitism in Animalia. *Biology Letters*, 12, 20160324.

Wells, K., and Flynn, R. (2022). Managing host-parasite interactions in humans and wildlife in times of global change. *Parasitology Research*, 121, 3063–3071.

White, L.A., Forester, J.D., and Craft, M.E. (2018). Disease outbreak thresholds emerge from interactions between movement behavior, landscape structure, and epidemiology. *Proceedings of the National Academy of Sciences*, 115, 7374–7379.

White, R.J., and Razgour, O. (2020). Emerging zoonotic diseases originating in mammals: a systematic review of effects of anthropogenic land-use change. *Mammal Review*, 50, 336–352.

Wiethoelter, A.K., Beltran-Alcrudo, D., Kock, R., and Mor, S.M. (2015). Global trends in infectious diseases at the wildlife-livestock interface. *Proceedings of the National Academy of Sciences*, 112, 9662–9667.

Wilcove, D.S., and Wikelski, M. (2008). Going, going, gone: is animal migration disappearing? *PLOS Biology*, 6, 1361–1364.

Wilson, R., Cooper, C.E.C., Meah, R.J., et al. (2023). The spectral composition of a white light influences its attractiveness to *Culex pipiens* mosquitoes. *Ecology and Evolution*, 13, e9714.

Woodroffe, R., Donnelly, C.A., Cox, D.R., et al. (2006). Effects of culling on badger *Meles meles* spatial organization: implications for the control of bovine tuberculosis. *Journal of Applied Ecology*, 43, 1–10.

Worobey, M., Levy, J.I., Serrano, L.M., et al. (2022). The Huanan Seafood Wholesale Market in Wuhan was the early epicenter of the COVID-19 pandemic. *Science*, 377, 951–959.

Wright, A.N., and Gompper, M.E. (2005). Altered parasite assemblages in raccoons in response to manipulated resource availability. *Oecologia*, 144, 148–156.

Wysocki, L.E., Dittami, J.P., and Ladich, F. (2006). Ship noise and cortisol secretion in European freshwater fishes. *Biological Conservation*, 128, 501–508.

Yoon, S.-W., Webby, R.J., and Webster, R.G. (2014). Evolution and ecology of Influenza A viruses. In: R.A.W. Compans and M.B.A. Oldstone (eds), *Influenza Pathogenesis and Control*, Vol. I, pp. 359–375. Springer International Publishing, Heidelberg.

Young, H.S., Parker, I.M., Gilbert, G.S., et al. (2017). Introduced species, disease ecology, and biodiversity-disease relationships. *Trends in Ecology & Evolution*, 32, 41–54.

CHAPTER 18

How humans and their societies respond when the world changes

Robert C. Brooks

Overview

The most dramatic changes faced by contemporary organisms originate in human activities. These activities can, in large part, be traced to the Agricultural Revolution and the Scientific-Industrial Revolution, two periods of rapid and profound change in how humans make their living and organize their societies. I ask how the changes in human activities and ways of life brought about by these revolutions changed the world that people inhabit, and how our species has responded to those changes. Those responses include combinations of physiological, life-history, economic, cultural, and evolved genetic changes. Considering some of those many changes can provide insights that are peculiarly human, suggesting the value of an integrated biological and cultural approach to improving human lives in a changing world. More generally, an understanding of how two major revolutions altered human lives can provide tentative ideas about how non-human organisms might respond to rapid changes to the world they inhabit.

18.1 What kind of species is *Homo sapiens*?

By any comparative measure, humans are large, long-lived, slow-breeding, intelligent, highly cooperative, and social animals. Other members of the great ape family (the Hominidea) are also relatively large, long lived, slowbreeding, intelligent, and cooperative. But there exists a substantial gap, particularly as far as intelligence and cooperation go, between humans and the other great apes. As primatologist Thomas Suddendorf (2013: 2) put it,

our minds have spawned civilizations and technologies that have changed the face of the earth, while our closest living relatives sit unobtrusively in their remaining forests.

Indeed, those forests and the apes and other species that live in those forests are dwindling precisely because humans have proliferated so vastly, and our technologies and the consumption they enable have so changed entire ecosystems. For non-human apes, as for most other animal species, the world has changed rapidly as a direct result of the phenomenal growth in human population and related advances in human technologies.

At the same time, the world has changed for humans too. Whereas the loss of wilderness, increases in pollution, and anthropogenic climate change affect most animals in some way, population growth, technological progress, and associated cultural changes have also altered human social worlds in ways that bear few direct comparisons with other species. The bulk of this chapter will consider what is known about behavioural and evolutionary responses to two of the biggest changes humans have ever faced.

Robert C. Brooks, *How humans and their societies respond when the world changes*. In: *Behavioural Responses to a Changing World*. Edited by: Bob B. M. Wong and Ulrika Candolin, Oxford University Press. © Oxford University Press (2024). DOI: 10.1093/oso/9780192858979.003.0018

18.2 How the world changed for humans (as they changed the world)

Humanity's distinguishing traits evolved during the 5 million or so years since contemporary *Homo sapiens* last shared an ancestor with the two extant chimpanzee species, *Pan paniscus* and *Pan troglodytes*. Students of human behavioural evolution are particularly concerned with the period since the start of the Pleistocene epoch, approximately 2.6 million years ago, during which early *Homo* and, subsequently, *H. sapiens* arose in Africa. The traits that arose in East and Southern Africa during the Pleistocene enabled humanity's dramatic ecological success, including the successive spread of *Homo erectus*, *Homo neandertalis*, and then *H. sapiens* out of Africa (Stringer 2012).

There is healthy debate about the properties of the so-called 'environment of evolutionary adaptedness' (EEA)—the conditions to which humans are predominantly adapted (Symons 1992). Given the vast period during which all human ancestors lived in Pleistocene Africa, the EEA is heavily weighted toward the savannah conditions that prevailed across much of our ancestors' range at that time. Those conditions varied dynamically, but it is relatively uncontroversial that human psychology and physiology were extensively shaped by adaptation to a foraging (i.e. gathering and hunting) lifestyle, living in small nomadic groups (usually fewer than 200 people) with extensive parental and alloparental care of young (Symons 1992). That gathering and hunting way of life persisted, even as humans spread, in successive waves, throughout Africa and the world.

The spread of human foragers into such varied environments as the Arctic coast, inland arid Australia, the Andes Mountains, Mongolian steppe, and Ganges floodplain was made possible by a flexible, communicative, cooperative, and cultural intelligence that evolved on the African savannah but that worked in a variety of environments and contexts (Suddendorf 2013). The last great wave of gene flow throughout Africa and out into the rest of the world approximately 70–50,000 years ago was preceded by a quantum change in human social and cultural intelligence (Karmin et al. 2015; Posth et al. 2016). This change has been called 'The Cognitive Revolution' (Harari 2015), 'The Social Leap' (von Hippel 2018), and the 'Secret of our Success' (Henrich 2016).

The unique human psychology and capacity for cooperation toward long-term goals made possible all the subsequent achievements of human culture, technology, and society, including ones that created rapid and far-reaching changes to the world that humans and other organisms inhabit (see e.g. Part 1 of this book). Most of this chapter will consider the implications of two of those changes: agriculture and the post-Enlightenment spread of science and industry.

In Table 18.1, I compare—in highly simplified form—some important dimensions of the human environment across four simplified ways of life: (1) the foraging lifestyle that all humans occupied during the Pleistocene, (2) agriculture, (3) science and industrialization, and (4) the current day, which might be viewed as either late-stage Scientific-Industrial or part of a new incipient revolution driven by computerization, the Internet, and artificial intelligence. All four ways of life can be found among societies today, as well as a variety of societies that combine some features from two or more columns of the table.

The Agricultural and Scientific-Industrial Revolutions are two important inflection points toward the major anthropogenic changes to natural habitats, climate, and environmental processes that define our current changing world. Those changes have been amplified by technological progress, population spread, and population growth since the early Neolithic through to the present day. So central are farming and industrialization to the recent and ongoing anthropogenic changes in the living world that each has its supporters in the ongoing debate around when the Anthropocene epoch began (Crutzen 2002; Ruddiman 2003).

18.3 The Agricultural Revolution

Writing about human behavioural responses to a changing world is complicated by the fact that many of the most consequential ways the world has changed, or is changing, are due to human behaviour. Agriculture, industrialization, and the

Table 18.1 How human diet, fertility, mortality, population, and urbanization have changed across four major historic-ecological periods.

	Foraging	Agriculture	Scientific-Industrial	21st century
Diet: macronutrient composition	High protein, fat; carbohydrates hard to acquire	Carbohydrates easier; protein harder (compared to previous column)	Carbohydrates easier; protein harder	Carbohydrates and some fats easy; protein harder
Fertility	Modest	High	Dropping from high to low	Low
Mortality rates	Modest	Rising from modest to high	Dropping from high to low Improved child survival	Very low
World human population[1]	2 million	Growing to 600 million	Growing to > 6 billion	Currently 8 billion Slower growth, possible stabilization
Share of world population living in urban settings[1]	0	Growing from 0 to ~5% (regions vary, 0–13%)	Growing to ~47% (regions vary, 18–82%)	> 50%; growing

[1] Klein Goldewijk et al. (2010)

changes we are currently living through are, above all, changes in human behaviour made possible by new behavioural applications of culture and technology. As a result, it is often tortuous, and perhaps of dubious value, to attempt to partition the change from the response to the change.

The advent of crop cultivation and animal domestication in various parts of the world effected some of the most dramatic and consequential changes humans have ever experienced (Ruddiman 2003; Larsen 2006; Harari 2015). The earliest invention of farming began in Mesopotamia and the Levant around 9500 BCE, and spread outward into Europe, North Africa, and Asia over a period of millennia. Farming also appeared independently in Central America around 6500 BCE, in the Indus River Valley by 5000 BCE, in China and Southeast Asia around 3500 BCE (Larsen 2006; Head 2014), and likely in several other places over the last several millennia. The uptake of agriculture via both independent domestication events and the cultural spread of knowledge and technologies is a complex and contingent story (Head 2014). Put as simply as possible, before 12,000 years ago, all humans foraged for wild foods, and since that time farming and pastoralism have grown steadily more common.

The Agricultural Revolution revolutionized the environment that humans in Neolithic agricultural communities inhabited, compared with their Pleistocene foraging ancestors. Few aspects of human life were left untouched, notably their diets, microbiota, pathogen loads, population density,

family structures, gender relations, economic inequality, and the scale of inter-group conflict. A variety of sources of evidence provide insights into how humans and their societies responded to the fast-breaking changes of the Neolithic. Some of those insights may be useful in understanding how other species might respond to contemporary changes in their environments.

18.3.1 Foraging and diet

Perhaps the most immediate way in which humans experienced the changes of the transition to farming is also the most obvious: shifts in diet. Domestication of cereal grains altered the macronutrient ratio to higher carbohydrate and lower protein and fat (Simpson and Raubenheimer 2012). Reallocation of effort to farming, coupled with population growth (see below), meant fewer calories from wild-growing foods and hunted animals, and thus lower per-capita intake of protein, fat, and several micronutrients.

Analyses of human skeletons have shown certain consistent trends associated with the change in diet from foraging to farming. Intensive agriculture tends to reduce dietary diversity and to introduce the potential for periods of extreme food scarcity before a harvest or when a crop fails. Skeletons from Neolithic sites are typically smaller and show evidence of nutritional deficiencies like anaemia and osteoporosis (Ulijaszek 1991; Larsen 2006; Latham 2013), conditions that were rare in pre-Neolithic skeletons from similar locations. The decline in stature of early farmers, together with patterns of growth inferred from detailed analyses of bones, suggests a phenotypically plastic 'growth stunting' response commonly observed in humans in response to undernutrition (Goodman 1993). Teeth from early farmers also show patterns of growth arrest lines and enamel deposition associated with growth stunting (Goodman 1993; Larsen 2006).

Analyses of teeth from several Neolithic sites also show more dental caries and periodontal disease in early farmers than in late hunter-gatherers in the same locales (Latham 2013). These declines in dental health occurred due to the high starch content of cereal grains, and new processing and cooking methods that meant foods were often softer and stickier and remained stuck in dental spaces where cariogenic bacteria thrive (Meller et al. 2009). Softer foods that required less chewing led also to smaller skulls, with smaller jaws, and, even though tooth size also shrank, the net effect was more dental crowding (Latham 2013; Katz et al. 2017). Evidence from experiments on non-human animals and contemporary humans suggest considerable phenotypic plasticity in skull morphology associated with diet (Katz et al. 2017). So, it is likely that changes in skull morphology with agriculture were—at least initially—phenotypically plastic.

Transitions in diet also drove evolutionary change. Some of the simplest examples concern selection on digestive physiology, including increases in copy number of the salivary amylase gene *AMY1* in populations with histories of agriculture compared with those pastoralist and forager populations that depended less on starch (Perry et al. 2007). Population genetic evidence suggests that forms of the gene encoding lactase (*LCT*) that permit the ability to digest milk in adulthood have been driven to high frequencies over the past ~7000 years in European and African populations in which adults commonly consume milk from domesticated animals (Tishkoff et al. 2007).

These single-locus examples of evolution in response to farming and livestock domestication illustrate how new environments can lead to evolutionary responses. More precisely it is a *mismatch* between the phenotype and the changed conditions that drives selection (see also Chapter 14). Mismatch is an important concept in understanding both evolutionary and ecological responses to environmental change (Gluckman and Hanson 2006; Li et al. 2018; Change and Durante 2022). The transition to agriculture is thought to have introduced mismatches not only in diet, but in a variety of other aspects of human development, physiology, and behaviour (Gluckman and Hanson 2006; Latham 2013). Many of those changes were—or appear from our 21st century vantage point to have been—so detrimental to human well-being that the transition to farming has been popularly lamented as 'the worst mistake in the history of the human race' (Diamond 1999; for a more recent, scholarly take see Larsen 2006).

18.3.2 Reproduction, population growth, and urbanization

Farming greatly increased productivity, in terms of calories per hour of labour or unit of land. This productivity increase, together with the novel capacity to store cereal grain surpluses, and the more settled lifestyle associated with agriculture, led to much higher fertility rates in farmers than in foragers (Gage and DeWitte 2009; Bocquet-Appel 2011). Analyses of the age structure of skeletons in cemetery sites associated with multiple transitions to agriculture indicate a steady increase in human population by approximately two births per woman over the first 1000 years post-transition (Bocquet-Appel 2011).

The increased birth rates in the so-called 'Neolithic demographic transition' were likely due to shorter inter-birth intervals, as greater caloric availability allowed women to achieve energy balance sooner after parturition and even during lactation (Valeggia and Ellison 2004). Further, settled life eased the nomadic requirement of having to either carry children or have them walk when the family were on the move. Last, in agricultural societies children become important economic contributors early on via tasks like herding or care for younger children, creating an incentive for larger families (see also Section 18.3.3).

The rising birth rates in the Neolithic were later balanced out by higher mortality (Gage and DeWitte 2009). Some of that mortality occurred due to more effective pathogen transmission due to higher population densities and increasing connectivity between trading partners (see also Chapter 17), and due to contamination of water and other sanitation issues (Barrett et al. 1998; Latham 2013). Some of the pathogens involved had already been associated with humans, and new zoonoses infected humans as domesticated animals became important to human livelihoods (Barrett et al. 1998). Poor nutrition likely combined with more effective pathogen transmission to increase the burden of disease as population density rose (Harrison et al. 1990; Barrett et al. 1998; Latham 2013).

The population growth spurred by the Agricultural Revolution increased the global human population by two orders of magnitude, from less than 6 million to more than 600 million (Klein Goldewijk et al. 2010). Growth was rapid at first, and then slower as Malthusian mortality caught up. Aggregation of a small proportion of the overall population into cities led to new challenges for city-dwellers in terms of public health, cooperation, and economic diversification of roles.

18.3.3 Inequality and gender relations

Observations that contemporary foraging societies appear far more egalitarian—and averse to inequality—than contemporary farming or industrialized societies are often used to infer that within-society among-household inequality (hereafter 'economic inequality') was small or negligible in the Pleistocene. Modest capacities to accrue some material possessions and to rise in status within one's group were likely already in place in some Pleistocene foragers (e.g. Hayden 1997), but evidence generally supports a rise in economic inequality, in the most general sense of unevenness in access to and control of resources, during the Neolithic (Kohler et al. 2017; Bowles and Choi 2019; Fochesato et al. 2019).

Settling down on arable land created forms of property that can be owned and passed down to descendants. Livestock herds and crop surpluses, too, can be exchanged or sold, greatly increasing the capacity for individuals and groups to build wealth, defend it from others, and pass it on to offspring. The invention of money and trade, and the diversification of specialized roles (including elite leadership) that these made possible, furthered the capacity for individual property and wealth. All of these developments amplified inequalities among households and caused them to persist across generations in ways that were new to human experience (Borgerhoff Mulder et al. 2009; Bowles and Choi 2019).

At the same time as it was amplifying inequality among households, agriculture also changed the ways in which women and men work to obtain food for themselves and their families. In so doing, it altered gender relations and power dynamics, with far-reaching consequences. Contemporary societies are still feeling the effects of these shifts in both income inequality among households and gender inequality.

The capacity for economic inequality (i.e. among households or families) grew larger with the transition from small-scale hoe-based agriculture to the larger-scale intensive agricultures that were made possible by technologies like irrigation and animal-drawn ploughs. Contemporary hoe-based farming tends to rest more on women's labour than men's, and this appears to have been generally the case among past hoeing societies too. The great increase in productivity made possible by intensive agriculture led to greater men's involvement in farming and women being more likely to work in the home on processing and childcare-related tasks (Boserup 1970).

These shifts in gender roles, which began several thousand years ago, became culturally entrenched. The earlier in history that a society started intensive crop cultivation, the later women in that same geographic location gained the vote, the fewer women have been elected to parliament, the fewer women participate in wage labour, and the more harshly the societies judge pre-marital sex and female sexual autonomy to the present day (Boserup 1970; Alesina et al. 2013; Hansen et al. 2015).

This is not a pattern that can be explained by considering the household a harmonious and coordinated unit of production. Instead, it would appear that sexual conflict may play an important role. Intensive agriculture presents a pathway for men to accumulate wealth, rise in status, and outcompete other men for mates (Betzig 1994; Brooks 2021). Women's invisibility in public becomes an expression of male mate guarding. High fertility, too, leads to further building of wealth as children became important contributors to farming work. Agriculture, then, restricted women's autonomy and confined their contributions to reproductive and domestic work, eroding their agency, visibility, and thus their political power (Betzig 1994; Hansen et al. 2015). This gendered division had the further effect of stimulating human population growth as societies embraced high fertility and as improved agricultural productivity raised carrying capacity.

One might be tempted to view the responses of human societies and individuals to the seismic shifts in income and gender inequality as economic and cultural phenomena, beyond the realm of biology. Not only is this kind of distinction entirely spurious, it also is a major obstacle to progress in understanding human lives and societies (Pinker 2002). If behavioural ecologists' immense interest in sexual selection and sexual conflict over the last four decades has taught us anything, it is that mating competition imposes strong, ongoing selection on all sexually reproducing organisms all the time. The same is likely to have been true during and after the Agricultural Revolution (Betzig 1986; Betzig 1994).

18.3.4 What does the Agricultural Revolution teach about behavioural responses to a changing world?

In many ways human behavioural responses to the challenges of massive shifts in diet, growing population density, urbanization, and social stratification tell the story of civilization. Practices, rules, and norms regulating public sanitation, cooperation, peaceful coexistence, defence against invasion, markets, and dealing with inequality were all either invented or massively transformed in the Neolithic. What records exist—in skeletal remains, artefacts, and even traces of microbes—testify to the vast and often rapid changes in the world occupied by many, and eventually most, humans.

Genetic analyses also indicate strong evidence of selection as populations moved from foraging to agricultural economies. Some of those genetic changes, like those involving salivary amylase (Perry et al. 2007), lactase persistence (Tishkoff et al. 2007), and alcohol dehydrogenase function (Han

et al. 2007) involve well-characterized and simple responses to agriculture. Broader surveys indicate evidence of positive selection on large numbers of genetic loci associated with nutrition, particularly protein (Wang et al. 2006) and glucose (Haygood et al. 2007; Hancock et al. 2010) metabolism.

The genetic changes do not end with nutrition. There is also evidence of strong positive selection on the regulation of many genes involved in reproduction and host-pathogen interactions (Wang et al. 2006). This is consistent with the shifting challenges of sex, reproduction, and disease as populations became societies with higher-density living, new social challenges, and more extreme disease ecologies (Chapter 17). Signatures of strong selection on neural development and function (Haygood et al. 2007) may also indicate the increased cognitive demands that come with navigating more elaborate human societies, status networks, and economies.

18.4 The Scientific-Industrial Revolution

The second major suite of world changes to be considered are those that followed industrialization. Industrialization probably emerged where and when it did due to a climate of scientific discovery and exchange, and Enlightenment values, that proved every bit as important as the mechanization of industry in changing the world humans inhabit. For that reason, I favour the clunkier 'Scientific-Industrial Revolution'.

The invention of steam power, and later other forms of fossil fuel-driven power, unlocked new forms of productivity, transforming how people worked, travelled, and lived. The expensive machinery involved altered the ownership of the means of production, giving rise to capitalism and to new investments in infrastructure. Capitalism and industrialization grew alongside colonial conquest, which brought both plentiful raw materials and often ill-gotten colonial wealth to manufacturing economies. Later, it also depended on innovations in democratic government.

Like the Agricultural Revolution, the Scientific-Industrial Revolution constitutes a complex, contingent suite of changes that played out at different times and speeds in different places. And yet, over the last 300–400 years, the world most humans inhabit has moved from one of widespread subsistence farming to one of far greater urbanization, where a relatively few farmers produce most of the food. That has driven at least a tenfold increase in the global human population, and the percentage of people who live in urban areas has grown from about 5% to over 50% (Table 18.1). With these changes, there has again been an enormous transition in fertility and mortality, in inequality, and in diet.

18.4.1 Food and diet

The British textile industry is typically considered the first industry to be transformed by steam-powered machinery, beginning in the 1760s. Soon thereafter—as early as 1768—steam-powered engines were already being installed in some Jamaican sugar mills (Satchell 1995). The industrialization of food production, transport, and storage, together with improvements in communication, alleviated many local food shortages and thus both malnutrition and starvation. It also shifted the balance of macronutrients available to people.

It is not without relevance that sugar was the first food-related industry to be industrialized. Industrialization in food production and processing has generally made carbohydrates, in the form of both starch and sugars, a much more prominent part of human diets by reducing the price of those carbohydrates relative to protein (Drewnowski and Darmon 2005; Simpson and Raubenheimer 2012).

As with agriculture, the primary response to a changing dietary world has tended to create a mismatch with human well-being. Cheaper carbohydrates have, generally, turned the major global dietary challenge from undernutrition to overnutrition (WHO 2006). This change also saw obesity shift from a rare condition mostly experienced by those wealthy enough to eat to excess into a very common condition far more likely to afflict those who do not earn enough to get the protein, fibre, and micronutrients that assist healthy living (Drewnowski and Darmon 2005; Brooks et al. 2010; Simpson and Raubenheimer 2012).

The increase in lifespan as undernutrition and infectious disease dropped in the early part of the demographic transition (see Section 18.4.2), combined with the 20th century rise in obesity, shifted the majority of mortality from infectious to non-infectious causes. Diseases that manifest in middle and old age, such as heart disease, cancers, and dementia, became the most prominent causes of death in industrialized nations by the late 20th century (O'Keefe and Cordain 2004; McKeown 2009; Gersh et al. 2010).

This 'epidemiological transition' constitutes another mismatch between our evolved physiology and dietary preferences and the contemporary dietary environment. There exists considerable scope for behavioural responses, in the form of improved understanding of the problem and promising interventions, together with screening and education. Popular media attention on diet, obesity, cancer, and other non-infectious diseases can be viewed, at least in part, as a behavioural response to shifting mortality risks.

Obesity and the epidemiological transition are almost certainly exerting new forms of selection, particularly for those diseases that curtail people's reproductive lifespan or their ability to contribute as grandparents and alloparents to their kin. They may also be exerting selection in other ways due to effects on fertility. One intriguing example concerns the high incidence of polycystic ovary syndrome (PCOS), a condition that drastically reduces fertility and that is made more likely by being overweight. Corbett et al. (2009) suggested that extremely high rates of PCOS in recent (first and second generation) migrants from subsistence agriculture economies to fully modernized economies may belie a lack of selection against PCOS in their non-obesogenic environment of origin. The implication is that people whose ancestors have been through several generations in an obesogenic environment have already passed through a selection bottleneck, with their ancestors less likely to carry the genes that predispose a carrier to PCOS. There are likely to be all manner of interesting selective responses to this and other aspects of our changing world.

18.4.2 The demographic transition

The period since the start of the Industrial Revolution has been a time of such dramatic demographic change that it is referred to simply as 'the demographic transition'. The effects on individual lives, families, societies, and economies can hardly be overstated. As Lee (2003: 167) put it, '[t]his global demographic transition has brought momentous changes, reshaping the economic and demographic life cycles of individuals and restructuring populations'.

Towards the end of the 18th century, lives were short. Global life expectancy at birth was, by best estimates, 25–35 years (Lee 2003; Riley 2005). 'Malthusian' conditions of disease, starvation, and conflict prevailed, causing high mortality, especially among children. Fertility was high, with women giving birth to an average of six children in their lifetime (Lee 2003). This combination of high mortality and fertility meant that populations grew slowly, and the global population totalled around 1 billion people.

Then, mortality—especially child mortality—dropped. This began with decreases in deaths from infectious diseases due to improved understanding of infection, and thus improved sanitation, hygiene, quarantine, and vaccines. Improvements in food storage and distribution provided new buffering against local famine. Increases in both incomes and agricultural output improved nutrition. In Europe, life expectancy grew to 42.7 years by 1900, 62.0 by 1950, and 78.6 by 2019 (Riley 2005).

As with agriculture, the demographic transition took hold at different times in different places, as industrialization and economic modernization displaced subsistence economy. Places where the demographic transition began later saw more rapid improvements in lifespan as they benefitted from ever-growing medical knowledge and logistic expertise. Gains in life expectancy made over 200 years in Europe took just 100 years in India (1920–2019), where life expectancy rose from 24 to 70 (Lee 2003), and 80 years in China (1930–2010), where it improved from 32 to 75 years (Riley 2005; United Nations 2022; see Figure 18.1).

The second phase of the demographic transition is a decline in fertility. In Europe, between 1870 and 1930, fertility declined by a median of 40% (Lee 2003). In other parts of the world, the fertility decline, like the mortality decline, began later, and some countries are presently experiencing the early or middle stages of the drop in fertility. The lag of

several decades between the two phases made for rapid population growth, followed by population stabilization at or below the replacement fertility level of 2.1 children per woman as fertility began to match or even drop below mortality (Lee 2003; Galor 2012). As a result, world population grew by an order of magnitude (Table 18.1) in the first two centuries of the demographic transition, reaching 8 billion people toward the end of 2022. Slower growth due to more countries reaching the later phases of the transition is projected to result in the human population peaking in the second half of the 21st century at between 9.5 and 11 billion, before modestly declining (Vollset et al. 2020).

For wildlife species, such a demographic boom would be considered a highly successful response to a changing world, and below-replacement fertility as a worrying trend. The vast impact of burgeoning human population, growing consumption, and improving technologies that harvest resources, clear land, and generate planet-warming emissions makes the demographic legacy of the Scientific-Industrial Revolution far more ambiguous, especially for those who study and conserve wildlife. There is some hope, however, to be gained from understanding how and why human demographic changes took the trajectory that they did.

The human behavioural responses that triggered mortality decline in early phases of the demographic transition were reasonably straightforward. Improved understanding of, and interventions against, health and disease, coupled with improved food production and distribution, made for better child and adult survival. These behavioural responses depended on the unusual extent and transmission of human cultural knowledge, however, and bear few direct lessons regarding other species' likely responses to their changing worlds.

The fertility decline is far less straightforward and remains the subject of some debate. Improved contraceptive technology might seem an important ingredient in fertility declines, but the earliest demographic fertility declines, in Europe and North America, were well underway before reliable hormonal contraceptives became available (Lee 2003).

The idea that parents want to have a certain number of surviving children, and thus adjust their fertility to match falling childhood mortality, is

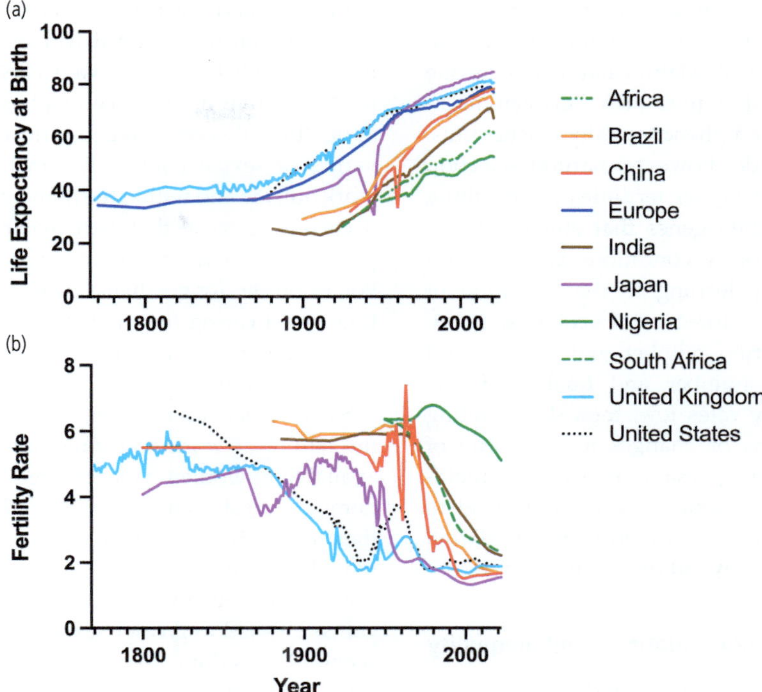

Figure 18.1 Changes in (a) life expectancy and (b) fertility rate over the last 250 years. The time period corresponds roughly with the time since before the Scientific-Industrial revolution started to affect the lives of most people, even in the places where that Revolution got going first. Panel (a) depicts 'period life expectancy', which is the average number of years a newborn would live if age-specific mortality rates in the year depicted remained the same throughout the individual's life. Panel (b) depicts total fertility rate, the number of children that would be born to a woman if she were to live to the end of her child-bearing years and give birth at the age-specific mortality rates for the given year.

Recreated from data made publicly available by OurWorldInData.org

likely overly simplistic, and at odds with the time-lag, often several generations, between mortality and fertility decline. Nonetheless, falling childhood mortality is likely to have played a role because parents invest more in each child, effecting a shift in the trade-off from offspring quantity toward greater quality (Galor 2012). Children became expensive, but the shift away from agriculture and toward high-value adult labour meant that children benefitted from parental investment in human capital via costly child-rearing and education (Lee 2003; Galor 2012). The declining importance of physical strength as economies modernized also increased women's capacity to earn money from their labour. Time women spent out of the workforce carrying and caring for children presented a growing opportunity cost (Galor and Weil 1993).

These economic arguments that fertility has declined due to greater investment in and higher opportunity costs of each child evoke evolutionary thinking about quality-number trade-offs, suggesting a possible evolutionary explanation for the demographic transition. In one recent and influential review, Lawson and Borgerhoff Mulder (2016) find that low fertility in the demographic transition does, indeed, increase socioeconomic success. It does not, however, improve reproductive success and, as far as can be told, fitness. This leads to the conclusion that even though evolved reaction norms may lead humans toward lower fertility when encountering low childhood mortality, higher costs per child, and greater resource abundance, the phenotypic plasticity is not currently adaptive (Lawson and Borgerhoff Mulder 2016).

Consider falling fertility and the attendant slower population growth that may yet limit the global population growth and forestall Malthusian misery as one positive example of a mismatch between modern environments and phenotypically plastic traits. On the negative side, however, natural selection waits for nobody. If higher fertilities are resulting in greater fitness, then genes that enhance fertility under contemporary conditions are likely on the rise, potentially leading to the evolution of faster growth rates. Moreover, if economic trade-offs between offspring number and quality, and between offspring number and quality of life, are keeping fertility rates low, then deteriorating economic conditions or changes to the value of female wage labour—possible as artificial intelligence remodels the labour market—could lead to higher fertilities and a resurgence of population growth that extends beyond the predicted peak.

18.4.3 Families, gender relations, and inequality

The demographic transition altered not only lifespan and fertility but also population age structure. The combination of fewer children and longer-lived adults made the average population much older. By the year 2100, people over 65 are expected to exceed the number of children (younger than 15) across the world. Some countries in the late-stage demographic transition (e.g. Japan and Germany) are already approaching this milestone.

One implication of an ageing population is that children tend to grow up with more living grandparents. Human parenting is supported by facultative contributions from parents, and kin and non-kin alloparents, of whom grandparents are especially important (Hawkes 2004). Sear and Coall (2011) showed that families in societies in the late stages of demographic transition are more likely to have living grandparents, and that if they live near to those grandparents, there is more alloparental help. But families in those societies are also more likely to move away from grandparents, in which case they lack that kind of alloparental help (Sear and Coall 2011).

I have already mentioned the fact that a shift in the nature of women's earning capacity that came with industrialization and the move away from

agriculture and toward manufacturing and services was likely an important driver of declining fertility. Although Galor and Weil (1993) tie this shift to the relative importance of physical strength in agriculture, theoretic and empirical progress on the nature of sexual conflict within that family over work outside and within the home, and over fertility decisions, suggests that there may be much more at play here than sex differences in physical strength. For example, longer lives, better health, and less time spent caring for small children freed women up to spend more of their adult lives working in the wage economy.

Since the Scientific-Industrial Revolution, shifts in the nature of work and, thus, the economic roles available to women have restored a degree of economic, domestic, and political gender equality. Further, in jurisdictions where women gained access to reliable contraception and safe means to terminate unwanted pregnancies, they have extended these gains. This led to greater sexual, reproductive, and economic freedom for women, declining rates of marriage, rising rates of divorce, and earlier ages of sexual debut (Stevenson and Wolfers 2007; Rissel et al. 2014; Myers 2017).

Sex roles and gendered behaviour, then, show considerable plasticity, especially due to shifts in the ways in which women and men make their livings, and in the extent of economic inequality. They are likely to be highly responsive to a changing world, and while the main trend since the start of the Scientific-Industrial Revolution has been toward greater male-female equity, there have been many local setbacks and the ongoing trend toward gender equity is by no means guaranteed (Hudson et al. 2020). When technologies of production or the reins of government can be commandeered by a small number of individuals, those individuals are likely to be predominantly male, just as was the case with intensive agriculture. Ross (2008) showed that oil production is more important than Islam as an explanation for contemporary gender inequity in North Africa and Asia Minor. As oil exports create unfavourable terms of trade for domestic manufacturing and food production, industries employing many women, women's visibility and political and economic power diminishes. Even oil-producing countries outside the Islamic world (e.g. Venezuela and Russia) retain stronger than expected patriarchal norms.

Recent history shows that when democratic regimes are replaced by despots, and when citizens' safety and the security of their property are no longer effectively provided by agents of third-party justice (e.g. police, independent courts), then security falls to powerful men and their coalitions, often composed of kin. In these situations, the rights of women, often hard-won over decades, regress with alarming speed (Hudson et al. 2020).

One might expect that advances toward social, political, and economic gender equity would erode sex differences in social roles and behaviours (e.g. see Eagly et al. 2004). Intuitive as it may sound, this prediction does not hold up well. Across 21 cross-cultural studies, sex differences in behavioural and personality traits diminished as gender equity rose in only 2 of 28 traits, but grew larger for 20 traits, with 6 traits remaining unchanged (Schmitt 2015). Further, economic experiments have shown that traits like patience, altruism, and risk-taking show a similar pattern (Falk and Hermle 2018). Across 80,000 people in 76 countries, sex differences in traits studied using economic games grew bigger with increasing gender equity.

Broadly, estimates of economic inequality in pre-industrial societies suggest that it was typically high. In Europe, after a drop due to the plague, inequality rose steadily from about 1450 until the start of the Industrial Revolution (Milanovic et al. 2011). It then declined, gradually from the onset of industrialization in the late 18th century, and then more rapidly from the early 20th century to the 1970s (Lindert and Williamson 1985). Since the 1970s, however, inequality has risen again in most countries (Piketty and Saez 2006).

Overall, present-day national estimates of income inequality show a similar distribution to the best pre-industrial estimates (Milanovic et al. 2011). We should, however, be cautious in dismissing inequality as one of the factors of our changing social world. Much of the post-1970s rise in inequality is driven by the increase in the very top incomes, rather than a fall in the poorest incomes (Piketty and Saez 2006). This, and burgeoning media access, leaves a large proportion of people with the opportunity to notice their stalled progress relative to

the elites. Social and other electronic media also grant ever-greater access to the lives of the richest people and the status-signalling behaviours of celebrities and 'influencers'. So even without shifts in economic inequality, that inequality may become more apparent, and more psychologically salient.

My collaborators, notably Khandis Blake, and I have shown that inequality at a variety of levels (town, commuting zone, state, country) predicts social media status-seeking behaviours like posting sexy selfies (Blake et al. 2018), as well as young men's online misogyny associated with being unable to find a mate (Brooks et al. 2022). Inequality is a powerful motivator of human behaviour (Daly et al. 2001; Bruce 2015; Wilkinson and Pickett 2017; Blake and Brooks 2019), and phenotypically plastic responses to inequality may become particularly important in coming decades as inequality becomes ever more apparent via social media.

18.4.4 What does the Scientific-Industrial Revolution teach about behavioural responses to a changing world?

As with the Agricultural Revolution, behavioural responses to the Scientific-Industrial Revolution have been of three broad kinds. First, phenotypically plastic changes in response to new, often evolutionarily novel environments. In most cases, as one might expect, these changes appear to be mismatched. That is not surprising given how rapidly the world changed.

Second, the immense human capacity for learning and cultural transmission has underpinned more deliberate responses. Examples include the invention and adoption of technologies like hormonal contraception to meet the newly pressing need to limit fertility. They also include large-scale attempts to understand how human environments and processes like economic development alter human behaviour and well-being, and then to intervene where suitable.

Third, the possibility of natural selection is often difficult to discern, but, as students of evolution will caution, it is never far away. Anything that changes lives, and especially anything that alters reproduction and lifespan, is going to alter selection.

Box 18.1 Ameliorating the problems that humans face in a changing world

This chapter has taken a necessarily cursory overview of the two most dramatic changes to the world that humans inhabit. In many ways the consequences for societies, institutions, laws, norms, and so on are evidence of the impact that agriculture, and then science and industrialization, wrought upon humans. But what good is this knowledge of how humans have responded to these changes? The following brief list numbers a few of the many impacts of understanding how humans have responded to these changes.

Developmental Origins of Health and Disease (DOHaD) is a scientific field that emerged to tackle the increased burden of non-communicable diseases that arose—directly and indirectly—from the Agricultural and Scientific-Industrial Revolutions. DOHaD is built on the foundation that non-communicable diseases such as obesity, arthritis, mood disorders, and vascular disease arise from mismatches between contemporary conditions (often during gestation and early childhood) and the environment in which humans evolved (Gluckman and Hanson 2006; Nesse and Stearns 2008).

Mismatch theory rests on the premise that humans have been unable to completely adapt, via natural selection, to the changes in their worlds over the past 12,000 years. In some cases that is because many people have only been exposed to agriculture in the last few generations, and some to industrialization only within the last generation or so. Further, phenotypic plasticity and behavioural/cultural responses may have buffered human populations from the full force of natural selection.

Apart from the clinical translation approach taken within the DOHaD discipline, other forms of impact are somewhat more nebulous. In part this is due to the fact that the adaptive explanations remain firmly contested within the broader 'culture wars' (Segerstråle 2000; Pinker 2002), and that people tend to pick and choose which aspects of this kind of evolutionary-historical analysis they adopt in order to advocate for their pre-existing preferred position. One needn't exercise the imagination too much to see how the historic overview in this chapter might be co-opted by those arguing for libertarian or socialist approaches to taxation and public spending, for or against interventions to foster ethnic and gender diversity in institutions, or for or against pro-natalist policy.

The history of evolutionary biology and human behavioural ecology suggests that an understanding of adaptation, reaction norms, and mismatches is seldom helpful when its main application is generalizing about what humans are and are not; what is and is not 'natural'. Instead, careful considerations of a given problem, such as the rise in mood disorders among 21st century youth, can benefit from a behavioural ecological view of how the world has changed, and how humans have reacted to those changes. But it must always be done in the context of the specific problem to be addressed (improving youth mental health) rather than forming blanket statements about history (blaming the victim or the entire system). The powerful insights that come from understanding how humans have responded to dramatic changes will best be achieved in partnership with specific disciplines with expertise in addressing those problems, such as psychology, education, technology policy, and public health.

It is still too early to tell, but the example of a possible history of selection on the fertility-limiting condition polycystic ovary syndrome (Corbett et al. 2009) presents one such example. Another is the extremely likely scenario that genes influencing fertility rates in low fertility societies (post-demographic transition) will favour the evolution of higher fertility rates, which is worth keeping a close eye on.

More subtle than selection acting directly on genes that influence fertility is the possibility of gene-culture coevolution. Already there are cogent arguments and accumulating evidence that evolutionary responses to the Agricultural Revolution are interdependent with cultural evolution (Laland et al. 2010; O'Brien and Laland 2012). The same is likely to be true for the Scientific-Industrial Revolution. That will be important, not only in understanding how genes and cultures are coevolving in response to the well-documented changes of the last two centuries, but also in tracking the likely responses to the coming dramatic changes as artificial intelligence, robotics, the so-called 'metaverse', and climate change take hold.

18.5 Conclusions

The origins of the recent changes that challenge most, if not all, of the living world emerge from two major historic transitions in human ways of living: the Agricultural Revolution and the Scientific-Industrial Revolution. Both of these revolutions created great upheaval in the lives of the humans that were subject to them. This chapter has considered some of the cultural, economic, physiological, and evolutionary responses to that upheaval.

It is not, however, as if the changes are entirely behind us. Many people are still going through the first generations of agriculture or industrialization, and some are undergoing both at more or less the same time. Moreover, the changes to climate and biodiversity in the Anthropocene will likely further shift human diet, demography, inequality, and social living in ways that reflect the earlier changes discussed in this chapter. Insights from history may yet provide useful strategies for managing, or at least coping with, the future (see Box 18.1).

While the historic context of this chapter is one of human history, the approach to understanding how organisms respond to change should be universal. Carefully considering how humans responded to the big changes of the Agricultural Revolution and the Scientific-Industrial Revolution could provide some instruction in how other species might respond to anthropogenic change to their worlds.

It is clear that rapid changes in human ways of living have generated mismatches between evolved phenotype and prevailing environmental conditions. Some of the mismatch has been buffered by behavioural responses, but it would appear that selection has played, and continues after many generations to play, an important part in the response to large-scale changes. Often the responses arise in the form of complex gene-environment interaction, including gene-culture interaction.

There is a temptation—when considering the responses of non-human organisms to rapid anthropogenic changes—to focus on demographic effects, including possible extinction, and effects within individual lifetimes. To these levels of analysis, we urge researchers to focus on natural selection, including the complex processes of genotype-environment interaction.

Acknowledgements

My research was supported by the Australian Research Council (DP220101023) during the writing of this chapter.

References

Alesina, A., Giuliano, P., and Nunn, N. (2013). On the origins of gender roles: women and the plough. *Quarterly Journal of Economics*, 128, 469–530.

Barrett, R., Kuzawa, C.W., McDade, T., and Armelagos, G.J. (1998). Emerging and re-emerging infectious diseases: the third epidemiologic transition. *Annual Review of Anthropology*, 27, 247–271.

Betzig, L. (1994). Sex, succession, and stratification in the first six civilizations. In: L. Ellis (ed.), *Social Stratification and Socioeconomic Inequality*, pp. 37–74. Praeger, Westport, CT.

Betzig, L.L. (1986). *Despotism and Differential Reproduction: A Darwinian View of History*s. Aldine Pub, New York, NY.

Blake, K.R., Bastian, B., Denson, T.F., et al. (2018). Income inequality not gender inequality positively covaries with female sexualization on social media. *Proceedings of the National Academy of Sciences*, 115, 8722–8727.

Blake, K.R., and Brooks, R. (2019). Income inequality and its implications for gendered conflict. In: J. Jetten and K. Peters (eds), *The Social Psychology of Income Inequality*, pp. 173–183. Springer, Cham.

Bocquet-Appel, J.-P. (2011). The agricultural demographic transition during and after the agriculture inventions. *Current Anthropology*, 52, S497–S510.

Borgerhoff Mulder, M., Bowles, S., Hertz, T., et al. (2009). Intergenerational wealth transmission and the dynamics of inequality in small-scale societies. *Science*, 326, 682–688.

Boserup, E. (1970). *Woman's Role in Economic Development*. Allen and Unwin, London.

Bowles, S., and Choi, J.-K. (2019). The Neolithic Agricultural Revolution and the origins of private property. *Journal of Political Economy*, 127, 2186–2228.

Brooks, R. (2021). *Artificial Intimacy: Virtual Friends, Digital Lovers and Algorithmic Matchmakers*. NewSouth, Sydney.

Brooks, R.C., Russo-Batterham, D., and Blake, K.R. (2022). Incel activity on social media linked to local mating ecology. *Psychological Science*, 33, 249–258.

Brooks, R.C., Simpson, S.J., and Raubenheimer, D. (2010). The price of protein: combining evolutionary and economic analysis to understand excessive energy consumption. *Obesity Reviews*, 11, 887–894.

Bruce, J.R. (2015). Power, economic inequality, and moral psychology. *Psychology and Society*, 7 (1), 12–28.

Chang, Y., and Durante, K.M. (2022). Why consumers have everything but happiness: an evolutionary mismatch perspective. *Current Opinion in Psychology*, 46, 101347.

Corbett, S.J., McMichael, A.J., and Prentice, A.M. (2009). Type 2 diabetes, cardiovascular disease, and the evolutionary paradox of the polycystic ovary syndrome: a fertility first hypothesis. *American Journal of Human Biology*, 21, 587–598.

Crutzen, P.J. (2002). Geology of mankind. *Nature*, 415, 23.

Daly, M., Wilson, M., and Vasdev, N. (2001). Income inequality and homicide rates in Canada and the United States. *Canadian Journal of Criminology*, 43, 219–236.

Diamond, J. (1999). The worst mistake in the history of the human race. *Discover*. 1 May 1999 ed.

Drewnowski, A., and Darmon, N. (2005). The economics of obesity: dietary energy density and energy cost. *The American Journal of Clinical Nutrition*, 82, 265S–273S.

Eagly, A.H., Wood, W., and Johannesen-Schmidt, M.C. (2004). Social role theory of sex differences and similarities: implications for the partner preferences of women and men. In: A.H. Eagly, A.E. Beall, and R.J. Sternberg (eds), *The Psychology of Gender*, pp. 269–295. Guilford Press, New York, NY.

Falk, A., and Hermle, J. (2018). Relationship of gender differences in preferences to economic development and gender equality. *Science*, 362(6412), eaas9899.

Fochesato, M., Bogaard, A., and Bowles, S. (2019). Comparing ancient inequalities: the challenges of comparability, bias and precision. *Antiquity*, 93, 853–869.

Gage, T.B., and Dewitte, S. (2009). What do we know about the agricultural demographic transition? *Current Anthropology*, 50, 649–655.

Galor, O. (2012). The demographic transition: causes and consequences. *Cliometrica*, 6, 1–28.

Galor, O., and Weil, D.N. (1993). *The Gender Gap, Fertility, and Growth*. National Bureau of Economic Research, Cambridge, MA.

Gersh, B.J., Sliwa, K., Mayosi, B.M., and Yusuf, S. (2010). Novel therapeutic concepts: the epidemic of cardiovascular disease in the developing world: global implications. *European Heart Journal*, 31, 642–648.

Gluckman, P., and Hanson, M. (2006). *Mismatch: Why Our World No Longer Fits Our Bodies*. Oxford University Press, New York, USA.

Goodman, A.H. (1993). *On the Interpretation of Health From Skeletal Remains*. University of Chicago Press, Chicago, IL.

Han, Y., Gu, S., Oota, H., et al. (2007). Evidence of positive selection on a Class I ADH locus. *The American Journal of Human Genetics*, 80, 441–456.

Hancock, A.M., Witonsky, D.B., Ehler, E., et al. (2010). Human adaptations to diet, subsistence, and ecoregion are due to subtle shifts in allele frequency. *Proceedings of the National Academy of Sciences*, 107, 8924–8930.

Hansen, C.W., Jensen, P.S., and Skovsgaard, C.V. (2015). Modern gender roles and agricultural history: the Neolithic inheritance. *Journal of Economic Growth*, 20, 365–404.

Harari, Y.N. (2015). *Sapiens: A Brief History of Humankind*. Harper, New York, NY.

Harrison, G.A., Waterlow, J.C., and International Commission on Anthropology of Food and Food Problems. (1990). *Diet and Disease: In Traditional and Developing Societies*. Cambridge University Press, Cambridge.

Hawkes, K. (2004). Human longevity: the grandmother effect. *Nature*, 428, 128–129.

Hayden, B. (1997). *The Pithouses of Keatley Creek: Complex Hunter-gatherers of the Northwest Plateau*. Harcourt Brace College Publishers, Fort Worth, TX.

Haygood, R., Fedrigo, O., Hanson, B., et al. (2007). Promoter regions of many neural- and nutrition-related genes have experienced positive selection during human evolution. *Nature Precedings*, 1, 1.

Head, L. (2014). Contingencies of the Anthropocene: lessons from the 'Neolithic'. *Anthropocene Review*, 1, 113–125.

Henrich, J. (2016). *The Secret of Our Success: How Culture Is Driving Human Evolution, Domesticating Our Species, and Making Us Smarter*. Princeton University Press, Princeton, NJ.

Hudson, V.M., Bowen, D.L., and Nielsen, P.L. (2020). *The First Political Order: How Sex Shapes Governance and National Security Worldwide*. Columbia University Press, New York, NY.

Karmin, M., Saag, L., Vicente, M., et al. (2015). A recent bottleneck of Y chromosome diversity coincides with a global change in culture. *Genome Research*, 25, 459–466.

Katz, D.C., Grote, M.N., and Weaver, T.D. (2017). Changes in human skull morphology across the agricultural transition are consistent with softer diets in preindustrial farming groups. *Proceedings of the National Academy of Sciences*, 114, 9050–9055.

Klein Goldewijk, K., Beusen, A., and Janssen, P. (2010). Long-term dynamic modeling of global population and built-up area in a spatially explicit way: HYDE 3.1. *The Holocene*, 20, 565–573.

Kohler, T.A., Smith, M.E., Bogaard, A., et al. (2017). Greater post-Neolithic wealth disparities in Eurasia than in North America and Mesoamerica. *Nature*, 551, 619–622.

Laland, K.N., Odling-Smee, J., and Myles, S. (2010). How culture shaped the human genome: bringing genetics and the human sciences together. *Nature Reviews Genetics*, 11, 137–148.

Larsen, C.S. (2006). The agricultural revolution as environmental catastrophe: implications for health and lifestyle in the Holocene. *Quaternary International*, 150, 12–20.

Latham, K.J. (2013). Human health and the Neolithic revolution: an overview of impacts of the agricultural transition on oral health, epidemiology, and the human body. Nebraska Anthropologist. https://digitalcommons.unl.edu/nebanthro/187/

Lawson, D.W., and Borgerhoff Mulder, M. (2016). The offspring quantity-quality trade-off and human fertility variation. *Philosophical Transactions of the Royal Society B: Biological Sciences*, 371, 20150145.

Lee, R. (2003). The Demographic Transition: three centuries of fundamental change. *Journal of Economic Perspectives*, 17, 167–190.

Li, N.P., Van Vugt, M., and Colarelli, S.M. (2018). The evolutionary mismatch hypothesis: implications for psychological science. *Current Directions in Psychological Science*, 27, 38–44.

Lindert, P.H., and Williamson, J.G. (1985). Growth, equality, and history. *Explorations in Economic History*, 22, 341–377.

McKeown, R.E. (2009). The epidemiologic transition: changing patterns of mortality and population dynamics. *American Journal of Lifestyle Medicine*, 3, 19s–26s.

Meller, C., Urzua, I., Moncada, G., and Von Ohle, C. (2009). Prevalence of oral pathologic findings in an ancient pre-Columbian archeologic site in the Atacama Desert. *Oral Diseases*, 15, 287–294.

Milanovic, B., Lindert, P.H., and Williamson, J.G. (2011). Pre-industrial inequality. *The Economic Journal*, 121, 255–272.

Myers, C.K. (2017). The power of abortion policy: reexamining the effects of young women's access to reproductive control. *Journal of Political Economy*, 125, 2178–2224.

Nesse, R.M., and Stearns, S.C. (2008). The great opportunity: evolutionary applications to medicine and public health. *Evolutionary Applications*, 1, 28–48.

O'Brien, M.J., and Laland, K.N. (2012). Genes, culture, and agriculture: an example of human niche construction. *Current Anthropology*, 53, 434–470.

O'Keefe, J.H., and Cordain, L. (2004). Cardiovascular disease resulting from a diet and lifestyle at odds with our paleolithic genome: how to become a 21st-century hunter-gatherer. *Mayo Clinic Proceedings*, 79, 101–108.

Perry, G.H., Dominy, N.J., Claw, K.G., et al. (2007). Diet and the evolution of human amylase gene copy number variation. *Nature Genetics*, 39, 1256–1260.

Piketty, T., and Saez, E. (2006). The evolution of top incomes: a historical and international perspective. *American Economic Review*, 96, 200–205.

Pinker, S. (2002). *The Blank Slate: The Modern Denial of Human Nature*. Viking, New York, NY.

Posth, C., Renaud, G., Mittnik, A., et al. (2016). Pleistocene mitochondrial genomes suggest a single major dispersal of non-Africans and a late glacial population turnover in Europe. *Current Biology*, 26, 827–833.

Riley, J.C. (2005). Estimates of regional and global life expectancy, 1800–2001. *Population and Development Review*, 31, 537–543.

Rissel, C., Heywood, W., De Visser, R.O., et al. (2014). First vaginal intercourse and oral sex among a representative sample of Australian adults: the Second Australian Study of Health and Relationships. *Sexual Health*, 11, 406–415.

Ross, M.L. (2008). Oil, Islam, and women. *American Political Science Review*, 102, 107–123.

Ruddiman, W.F. (2003). The Anthropogenic greenhouse era began thousands of years ago. *Climatic Change*, 61, 261–293.

Satchell, V.M. (1995). Early use of steam power in the Jamaican sugar industry, 1768–1810. *Transactions of the Newcomen Society*, 67, 221–231.

Schmitt, D.P. (2015). The evolution of culturally-variable sex differences: men and women are not always different, but when they are . . . it appears not to result from patriarchy or sex role socialisation. In: T.K. Shackelford and R.D. Hansen (eds), *The Evolution of Sexuality*, pp. 221–256. Springer, Cham.

Sear, R., and Coall, D. (2011). How much does family matter? Cooperative breeding and the demographic transition. *Population and Development Review*, 37, 81–112.

Segerstråle, U. (2000). *Defenders of the Truth: The Sociobiology Debate*. Oxford University Press, Oxford.

Simpson, S.J., and Raubenheimer, D. (2012). *The Nature of Nutrition: A Unifying Framework from Animal Adaptation to Human Obesity*. Princeton University Press, Princeton, NJ.

Stevenson, B., and Wolfers, J. (2007). Marriage and divorce: changes and their driving forces. *Journal of Economic Perspectives*, 21, 27–52.

Stringer, C. (2012). *Lone Survivors: How We Came to be the Only Humans on Earth*. Macmillan, London.

Suddendorf, T. (2013). *The Gap: The Science of What Separates Us from Other Animals*. Basic Books, New York.

Symons, D. (1992). On the use and misuse of Darwinism in the study of human behavior. In: J.H. Barkow, L. Cosmides, and J. Tooby (eds), *The Adapted Mind: Evolutionary Psychology and the Generation of Culture*, pp. 137–159. Oxford University Press, New York, NY.

Tishkoff, S.A., Reed, F.A., Ranciaro, A., et al. (2007). Convergent adaptation of human lactase persistence in Africa and Europe. *Nature Genetics*, 39, 31–40.

Ulijaszek, S.J. (1991). Human dietary change. *Philosophical Transactions of the Royal Society B: Biological Sciences*, 334, 271–279.

United Nations, Department of Economic and Social Affairs, Population Division. (2022). World Population Prospects, Online Edition. [online]. Available from: https://population.un.org/wpp/Download/Standard/Population/ [accessed 7 April 2023].

Valeggia, C., and Ellison, P.T. (2004). Lactational amenorrhoea in well-nourished Toba women of Formosa, Argentina. *Journal of Biosocial Science*, 36, 573–595.

Vollset, S.E., Goren, E., Yuan, C.-W., et al. (2020). Fertility, mortality, migration, and population scenarios for 195 countries and territories from 2017 to 2100: a forecasting analysis for the Global Burden of Disease Study. *The Lancet*, 396, 1285–1306.

Von Hippel, W. (2018). *The Social Leap: The New Evolutionary Science of Who We Are, Where We Come From, and What Makes Us Happy*. Harper Collins, New York, NY.

Wang, E.T., Kodama, G., Baldi, P., and Moyzis, R.K. (2006). Global landscape of recent inferred Darwinian selection for *Homo sapiens*. *Proceedings of the National Academy of Sciences*, 103, 135–140.

WHO. (2006). *World Health Organization Fact Sheet No 311: Obesity and Overweight*. World Health Organization, Geneva.

Wilkinson, R.G., and Pickett, K.E. (2017). The enemy between us: the psychological and social costs of inequality. *European Journal of Social Psychology*, 47, 11–24.

CHAPTER 19

Behavioural ecology for a changing world

Andrew Sih

Overview

Human-induced rapid environmental change (HIREC) is negatively impacting many organisms, but others are doing so well that we consider them to be 'pests'. For animals, behaviour provides a key first response to environmental change. A major issue is thus to explain variation in behavioural responses to environmental change. Here, my main goals are to: (1) outline major factors, using evolutionary match/mismatch as a conceptual framework, that can explain variation in how well animals respond effectively to HIREC, either immediately or via learning (including social learning); (2) provide an overview on an expanded role for behaviour in a multi-trait response to one or multiple stressors; (3) note the importance of consistent individual differences in behavioural responses to HIREC, focusing on the surprisingly understudied topic of fear generalization across multiple threats; and (4) offer suggestions on how to decide what one should study amidst the complexity of multiple responses to multiple stressors.

19.1 Introduction

With few exceptions, organisms on the Earth are facing multiple threats associated with human-induced rapid environmental change (HIREC). Of particular concern are the so-called 'Five Horsemen of the Apocalypse': habitat change (habitat loss, degradation, and fragmentation, including urbanization; see also Chapters 8 and 9), pollutants (e.g.

chemical: Chapters 3 and 5; noise: Chapter 2; and light: Chapter 4), exotic species (e.g. competitors, predators, parasites, and pathogens: Chapters 6 and 17), human harvesting of wildlife (Chapter 7), and climate change (Chapter 1; including both the effects of global warming per se, and the impacts of disturbances such as fires, storms, and droughts). While HIREC is having negative impacts on many organisms, others are faring well, and are becoming 'nuisance species' (Barrett et al. 2019b). Thus, a major issue is to explain this variation in species, population, and individual success in the modern world. For animals, their behavioural response is often a key first response that determines individual fitness and species persistence with HIREC (Sih et al. 2011; Sih 2013; Wong and Candolin 2015), which can then allow further evolution, including evolution of behaviour (Caspi et al. 2022; see Chapter 14).

Understanding behaviour in a changing world poses an important and intellectually fascinating challenge for behavioural ecologists. Behavioural ecology has traditionally used the optimality approach to predict and explain behaviour based on the assumption that natural selection has shaped behaviours to be reasonably adaptive (Westneat and Fox 2010). HIREC, however, means that many organisms are experiencing novel situations that differ in some way from the ecological or social conditions that shaped their traits, including behaviours. We are thus not surprised that while animals often still exhibit adaptive behaviours in the modern world, some (perhaps many) animals

Andrew Sih, *Behavioural ecology for a changing world*. In: *Behavioural Responses to a Changing World*. Edited by: Bob B. M. Wong and Ulrika Candolin, Oxford University Press. © Oxford University Press (2024). DOI: 10.1093/oso/9780192858979.003.0019

in some contexts do not exhibit fine-tuned adaptive behaviours (Saaristo et al. 2018; Bertram et al. 2022; Pollack et al. 2022a).

In this book, separate chapters have focused on each of various stressors or threats, discussing variation in behavioural responses and consequences for individuals, populations, communities, and even ecosystems. In this closing chapter, I first provide a brief overview on main, repeated themes in the single-stressor chapters. I then build on and extend these themes by: (1) outlining major factors, using evolutionary match/mismatch as a conceptual framework, that can explain variation in whether animals respond effectively to HIREC, either immediately or via learning or longer-term plasticity; (2) providing an overview on an expanded role for behaviour in a multi-trait response to one or multiple stressors; (3) discussing the importance of consistent individual differences in behavioural responses to HIREC, focusing on the surprisingly understudied topic of fear generalization across multiple threats; and (4) offering suggestions on the issue of deciding what one should study amidst the complexity of multiple responses to multiple stressors.

19.2 Behavioural responses to environmental change: a brief recap

Some of the threats of the modern world (e.g. exotic predators, human harvesting, novel toxic chemicals, automobiles, boats, and planes) can and often do directly kill animals if they do not exhibit appropriate behavioural responses. However, as we have seen throughout the chapters of this book, negative impacts can also be caused by a range of more subtle mechanisms. Many threats are physiological stressors (e.g. adverse temperature or humidity, chemical or acoustic pollutants) that induce a stress response that often results in energy-demanding—or otherwise costly—shifts in physiology, behaviour, and life history (Saaristo et al. 2018; Bertram et al. 2022; Chapter 10). If the stressor level is strong and/or persistent enough, it can cause significant reductions in condition and, ultimately, fitness (McEwen and Wingfield 2003; Wingfield 2013). Other novel 'stressors' have negative impacts by disrupting neural, sensory, or cognitive

systems. Chemical stressors, for example, have been shown to reduce anti-predator responses (e.g. Jellison et al. 2016; see also Chapters 3 and 5) or ability to choose habitats well (Gravinese et al. 2020) and also to disrupt social and mating behaviour, and thus sexual selection (Candolin and Wong 2019). It is not surprising, for example, that in the context of pharmaceutical pollution, drugs that are specifically designed to alter human behaviour (e.g. antidepressants) also affect the behaviour of many non-target wildlife (Brodin et al. 2014; McCallum et al. 2021). Similarly, some pesticides are endocrine disruptors and can therefore affect behaviour directly, or indirectly via effects on physiology or morphology (Bertram et al. 2022; Bertram et al. 2024).

Yet other environmental changes put animals in a bind where even the best behavioural response is still costly. For example, novel barriers (e.g. roads) can block adaptive movements (Berger-Tal et al. 2020; Chapter 8), and novel dangers (e.g. novel predators: Chapter 6; or harvesting pressure: Chapter 7) induce anti-predator responses that can be costly (i.e. the non-consumptive effects of predators; Peacor et al. 2020; Wirsing et al. 2021). Notably, with multiple stressors (see also Chapter 10), avoidance of one stressor often increases exposure to other stressors; for example, salamander larvae *Ambystoma barbouri* that darken and go deeper in the water column to reduce damage from ultraviolet light are then exposed to higher fish predation risk (Garcia and Sih 2003), and roe deer *Capreolus capreolus* that avoid open habitat during the hunting season then suffer increased predation by lynx *Lynx lynx* (Gehr et al. 2018).

Of course, environmental change can also be beneficial for some organisms. Notably, warmer temperatures have allowed many species to expand their latitudinal or altitudinal range (Walther et al. 2009; Chapter 1). Anthropogenic light can increase foraging rates of visual predators by causing prey to aggregate, by making prey easier to see, and by increasing the predator's daily foraging period (Chapter 4). And, in an example of the general 'enemy of my enemy' effect, species that are less susceptible to environmental stressors can gain a net benefit if the stressors have larger negative impacts on the focal species' enemies (e.g. predators or parasites). For example, although anthropogenic noise can reduce the efficacy of the mating calls of tungara frogs *Engystomops pustulosus* (see also Chapter 2), the frogs benefit from the fact that the noise significantly reduces the ability of parasitic frog-biting midges *Corethrella* spp. to detect them (McMahon et al. 2017).

A key point on all of the above is the fact that animals differ in their vulnerability to these novel or increased threats associated with HIREC. Variation in negative impacts can be due to variation in: (1) exposure, (2) conflicting demands if exposed, and (3) response to those conflicting demands. Those that respond poorly are often declining species of conservation concern, while those that respond well are sometimes 'pests' engaged in human-wildlife conflict (Barrett et al. 2019b; Chapter 13). An overarching goal is thus to develop a better explanation for this behavioural variation.

With regard to relative exposure, some animals simply live in habitats that are more disturbed or in situations where they are exposed to more novel predators or novel pollutants than others. For those that are exposed to novel/altered conditions, some had pre-existing traits that made the change a larger problem than for others. For example, urban noise (or boat noise) that is primarily low frequency poses a greater problem for species that communicate using low frequency wavelengths that overlap heavily with the anthropogenic noise (Slabbekoorn and Peet 2003; Patricelli and Blickley 2006; Senzaki et al. 2020; Chapter 2).

When HIREC puts animals in conditions that can potentially reduce fitness, how well they fare can depend on their relative morphological or physiological capacity to cope with the stressors that they encounter (Sokolova 2013; Todgham and Stillman 2013; Lopez et al. 2023). But beyond all that, relative efficacy of behavioural response is often critically important. Some animals avoid novel stressful situations—for example, by behavioural thermoregulation (Kearney et al. 2009; Huey et al. 2012), or avoidance of novel predators (Sih et al. 2010; Carthey and Blumstein 2018), or avoidance of pollutants, such as noise (Chapter 2), chemicals (Moreira-Santos et al. 2019; Chapter 3), or light (Gaynor et al. 2018; Chapter 4). Such avoidance can involve small-scale shifts in habitat use, or larger-scale dispersal. Again, however, while some respond well

behaviourally, others do not. Some animals have ignored novel threats that then prove deadly, and some have even preferred low-fitness situations (e.g. poor habitats: Robertson et al. 2013; Hale and Swearer 2016; Robertson and Blumstein 2019; Chapter 8). If animals survive their less-than-ideal initial responses to HIREC, then, over ontogeny, they can learn to respond better to HIREC. However, while some animals have learned to respond more adaptively to novel stressors, there are good reasons to actually expect some animals to not learn well (Barrett et al. 2019a; Greggor et al. 2019; Chapter 12).

To translate behavioural responses to HIREC up to population or community ecological effects (i.e. to effects on species success and species interactions), it is important to emphasize that animals typically face multiple stressors (Lopez et al. 2023; Chapter 10). Although the book remains open on which stressors are most important in any given system, and on how and when stressors might have synergistic negative impacts, it is clear that many species declines are caused by multiple stressors (Orr et al. 2022). Thus for both basic and applied ecology/behaviour—including for conservation and management—it is important to understand factors explaining variation in the animal's behaviour as part of its multi-trait response to multiple stressors (Lopez et al. 2023; Chapters 10, 11, and 12).

19.3 Evolutionary match/mismatch as an organizing framework

Evolutionary mismatch refers to the situation where a previously adaptive trait becomes maladaptive after a change in the environment, particularly when the change is rapid (see also Chapter 14). Figure 19.1 outlines the basic steps in the concept. We assume that past selection pressures have shaped a set of adaptive traits—morphology, physiology, behaviour, and/or life history. Immediately after environmental change, we further assume that there has not yet been enough time for substantial evolution; that is, organisms cope (or not) with the new conditions using traits that were adaptive in previous conditions, but may or may not be well

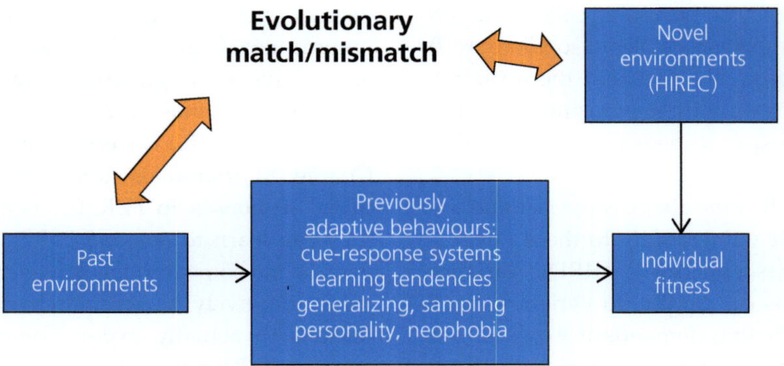

Figure 19.1 A conceptual framework for explaining variation in behavioural response to novel environments associated with human-induced rapid environmental change (HIREC). Selection pressures both biotic and abiotic in past environments have shaped a suite of previously adaptive traits, including behaviours, that have major effects on individual fitness. Literature on evolutionary traps has focused on cue-response systems; however, here I discuss several other aspects of behaviour—variation in personality and neophobia and in learning tendencies and tendencies to generalize or sample. Whether these previously adaptive behavioural tendencies produce appropriate versus maladaptive responses to novel environments depends on the degree of evolutionary match versus mismatch between the past and HIREC-altered environments.

adapted to the new, post-change conditions. When the previously adaptive traits still function well in the new conditions, there is an 'evolutionary match' and the species can do well despite the change. In contrast, if the previously adaptive traits are now maladaptive, the evolutionary mismatch results in reduced fitness.

This concept has long been invoked to explain maladaptive human behaviour in the modern world (e.g. Chapter 18). An often-cited but controversial example is the 'thrifty gene hypothesis', which suggests that because humans evolved with food uncertainty, we evolved metabolic efficiency when food is scarce, and a drive to consume large amounts of high calorie foods when food is plentiful. These traits then contribute to the high prevalence of obesity in the modern world. The application of the concept to humans has often been criticized for being 'just so stories' not backed by explicit theory or empirical tests of a priori predictions generated by explicit theory focused on specific traits and contexts (Chapter 18).

For non-humans, while the match/mismatch can involve previously adaptive morphological or physiological traits, our focus here is on behaviour; in particular, on ecological or evolutionary traps where the key previously adaptive trait is the behavioural cue-response system (Robertson et al. 2013; Hale and Swearer 2016; Robertson and

Blumstein 2019; Pollack et al. 2022a). The premise is that organisms use cues (e.g. visual, chemical, or auditory cues) to evaluate whether an encountered object is useful (e.g. good habitat, food, or a potential mate), dangerous (e.g. a predator or toxic 'food'), or neutral. Organisms can get 'trapped' if their previously adaptive cue-response system now results in bad decisions; for example, utilizing habitat or food that yields low fitness, or ignoring or even being attracted towards danger.

Many examples of evolutionary traps involve misguided habitat use. A classic, well-known example involves marine turtle hatchlings that evolved to follow light to navigate from the beach to the ocean being led landward by human lights (Kamrowski et al. 2012; see Chapters 4 and 11). Human lights have also attracted innumerable insects and birds to inappropriate sites where they can be temporarily blinded, disoriented, or fly until they are exhausted, and reflected light off human-produced surfaces has induced innumerable aquatic insects to attempt to oviposit not in water but on glass or wet asphalt (Horvath et al. 2014). In a perhaps more nuanced example, aquatic animals in urban environments are breeding in artificial stormwater wetlands that can be poor habitat due to high pollutant loads (Hale and Swearer 2016). Other examples of an evolutionary trap involve maladaptive consumption of toxic 'food'—for example, consumption of toxic cane toads *Bufo marinus* by snakes or mammals (Shine 2010, 2017; Chapter 6), consumption of plastics by seabirds (Savoca et al. 2016; Santos et al. 2021) or fish (Savoca et al. 2021; Pollack et al. 2022b), and possibly, consumption of unhealthy foods by humans (Chapter 18). Yet other examples of evolutionary mismatch focus on naïve prey exhibiting a lack of avoidance of novel dangers—exotic predators, parasites, or pathogens (Sih et al. 2010; Carthey and Blumstein 2018; Smith et al. 2021).

It is important to note that while the literature on evolutionary traps can give the impression that 'trapped' animals are exhibiting unavoidable, fixed, maladaptive behaviours, cue-response systems that guide their behaviour do not typically produce a fixed behaviour but, instead, usually produce a behavioural strategy—a pattern of behavioural plasticity. For example, whether an animal flexibly accepts or rejects potential food items depends on whether the item's cues (size, shape, movement, smell) meet some threshold criteria, and the threshold criteria can flexibly depend on the animal's state (e.g. its hunger level or energy reserves; Trimmer et al. 2017). An animal can get trapped if its previously adaptive, plastic cue-response system leads it to accept novel toxic 'foods' (see also Chapter 6).

The flipside of evolutionary traps involves what have been termed undervalued resources (Gilroy and Sutherland 2007); situations where animals do not utilize novel, valuable resources. For example, many insects that would thrive on nutritious crops do not feed on them, presumably because their cue-response systems do not induce them to 'take a chance' to feed on these novel foods (Pearse et al. 2013). Along parallel lines, many animals avoid humans or human structures even though they could potentially thrive with us. Given the rapid expansion of human presence, animals that over-avoid us have vanishingly little space to live.

An important point is that while some organisms get trapped, other organisms respond appropriately to the novel stimuli, and while some animals are not adopting valuable novel foods or habitats, other animals are utilizing useable novel resources or are moving in with us (see also Chapter 9). The evolutionary match/mismatch framework suggests that a key to explaining this variation lies in the match

or mismatch between the species' evolutionary history and modern conditions. To move beyond post hoc 'just so' stories, evolutionary ecologists have sought to be more rigorous about evolutionary match/mismatch by developing explicit, quantitative theory on past selection scenarios, and on the nature of the organism's decision-making system, to generate predictions on when we expect organisms to respond adaptively or not to HIREC (Sih et al. 2011; Sih 2013; Crowley et al. 2019; Ehlman et al. 2019; Pollack et al. 2022a).

For example, to address factors that influence decision-making using a cue-response system, we can use signal detection theory (SDT; Figure 19.2). In its simplest form, SDT distinguishes good versus bad objects (e.g. safe versus dangerous, nutritious versus toxic) where the adaptive response is to accept the good and avoid the bad. SDT acknowledges that animals often do not know for certain what is good versus bad, but instead use cues to assess situations; for example, they might judge that large animals approaching rapidly are dangerous, while smaller, slower animals are safe. Importantly, there can be overlap between the cues presented by safe versus dangerous situations. In that case, animals are uncertain about a range of intermediate situations. Is a medium-sized animal approaching at a medium speed dangerous or not? With uncertainty, animals are likely to make errors of two sorts: they can ignore dangerous situations or flee from safe ones. Balancing the probabilities of these errors and their costs produces an optimal response threshold for when the animal should flee versus ignore an approaching animal. To use SDT to address responses to novel foods, predators or habitat, we posit that, at least initially, animals use their previously adaptive response threshold (shaped by past cues, errors, and costs) to respond to novel situations. That is, when an animal detects a novel predator, it is not 'clueless' about what to do; it responds (flee or not) guided by its extant cue-response system. The key then is whether the previously adaptive response threshold causes animals to respond well or not to post-HIREC situations. Here, I summarize some of this field's main predictions on factors that affect when animals should behave adaptively or not when exposed to novel situations.

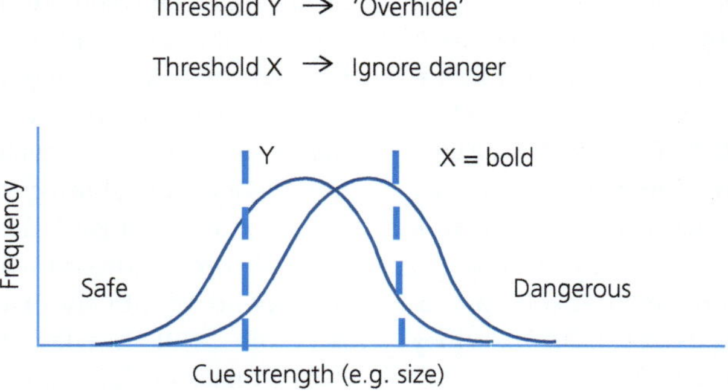

Figure 19.2 Illustrates the basic idea of signal detection theory (SDT) using an approaching animal as the stimulus. The focal animal must decide whether to respond (e.g. flee) or not. Based on the animal's past evolutionary or developmental history, it has an estimate of the cues put out by safe versus dangerous approaching animals. In basic SDT, multiple cues are combined in a single axis of 'cue strength'. For simplicity, imagine that the primary cue is size. Safe versus dangerous animals have a different frequency distribution of cue strengths, with dangerous ones tending to be larger than safe ones. Importantly, however, if the cue distributions for safe and dangerous stimuli overlap, the animal will make errors. SDT predicts an optimal response threshold balancing the estimated frequency and costs of over-avoiding safe stimuli (e.g. if it chose threshold Y) versus under-responding to dangerous ones (e.g. if it chose threshold X). See the text for more description.

An obvious prediction is the 'cue similarity hypothesis', which posits that even naïve animals are likely to respond adaptively to novel predators if they resemble familiar ones, but that naïve prey might often respond inappropriately to novel predators that do not resemble familiar predators (Sih et al. 2010; Carthey and Blumstein 2018; Guiden et al. 2019; Smith et al. 2021). A meta-analysis corroborated this prediction (Anton et al. 2020). Parallel predictions hold for novel toxic or inedible 'foods': animals are more likely to erroneously consume them if they resemble familiar edible foods. This idea can help explain why some (but not all) animals consume cane toads or plastics (Savoca et al. 2016; Shine 2017). Cue similarity also fits the observation that herbivores are more likely to consume high-quality exotic plants if they are phylogenetically similar (and thus often chemically similar) to native plants (Pearse et al. 2013).

Other predictions hinge on factors that affect an animal's boldness. In the SDT framework, bolder animals have a higher response threshold for fleeing (i.e. require a stronger cue to induce fleeing) and a lower response threshold for attraction (e.g. to potential food). Bolder animals should thus be more susceptible than cautious ones to accepting novel poor foods, and not avoiding novel dangers. An obvious prediction is that animals should be bolder if their familiar predators are rare or not very

dangerous (Ehlman et al. 2019). For example, naïve island prey, or prey from ephemeral aquatic habitats that lack dangerous predators, tend to respond poorly to exotic predators (Garvey et al. 2020). A parallel prediction for inedible/toxic foods (e.g. plastics) is that foragers should be more likely to mistakenly consume them if familiar foods that resemble the novel inedible foods are rarely toxic. Conversely, herbivores should be less likely to adopt novel, valuable plants (e.g. crops) if familiar native plants that resemble edible plants are highly toxic (Pearse et al. 2013).

An intriguing, less discussed SDT prediction is that the prevalence of safe objects that resemble dangerous ones should also affect an animal's boldness. Examples of organisms that are safe but can appear dangerous include large, active, herbivorous fish (that prey fish might view as predators) or human recreational hikers or ecotourists. If situations that are safe but appear dangerous are very common, cautious animals can waste a lot of time and energy avoiding them. This can favour animals being bolder (i.e. setting their response threshold high). Bolder animals, however, are more susceptible to actual dangers, including novel predators (Trimmer et al. 2017; Ehlman et al. 2019). We thus predict that, all else being the same, prey should be more susceptible to exotic predators if prey are in lakes where most large fish are herbivorous, or if prey are in areas heavily visited by ecotourists (Trimmer et al. 2017).

Finally, an intuitive idea is that smarter animals should be less likely to mis-assess novel options. For example, elephants *Loxodonta africana* (that are reasonably cognitively sophisticated) can distinguish more versus less dangerous humans by apparently determining the humans' ethnicity, age, and gender from just hearing their voices (McComb et al. 2014). Interestingly, however, SDT predicts that for a given level of cognitive ability, animals that evolved in situations where safe versus dangerous situations were easy to distinguish should be more 'confident' in their decision-making and, thus, more susceptible to novel mismatches. In contrast, animals that evolved in confusing circumstances where safe and dangerous situations were harder to distinguish (e.g. with ambush predators) should 'play it safe', be more cautious, and, thus, should be more likely to respond well to novel dangers (Ehlman et al. 2019).

19.4 Constraints on learning to adjust to environmental change

An intuitive expectation is that if animals initially respond poorly to HIREC, over time, they should learn to exhibit more appropriate behaviours. Interestingly, however, organisms sometimes do not appear to learn after responding poorly to novel situations despite having the ability to learn (Ward-Fear et al. 2016; Shine 2017). Why not? Greggor et al. (2019) turned the usual intuition on learning on its head by noting common scenarios where we would not expect animals to learn despite erring in their initial response to novel situations.

Consider the two main types of errors: not avoiding bad options (traps) and not utilizing good ones (undervalued resources). Perhaps the most obvious reason why animals might not learn to avoid bad options is that, if they respond poorly, they are dead or disabled beyond recovery. This might be a common unfortunate outcome of not avoiding a novel predator, pathogen, or toxic food (e.g. a cane toad; Chapter 6). Another type of reason why animals might not learn to avoid poor options arises if they do not get suitable feedback that informs them about their poor decision. This lack of feedback might be common when there is time lag between

the decision and the fitness outcome. For example, many adult animals lay eggs and then leave. If, as a result of HIREC, adults use previously adaptive—but now mismatched—cues to choose oviposition sites, adults will often oviposit in poor habitat (e.g. in heavily polluted waters, or habitats with novel predators). However, because they typically do not return (or perhaps do not survive) to see the fate of their offspring (e.g. tadpoles or insect larvae), they cannot easily learn to avoid those poor habitats. A within-generation example of delayed salient feedback involves animals not learning to avoid novel socially transmitted diseases. If an animal gets sick some time after interacting with a conspecific, it might not learn to avoid infected conspecifics if the illness emerged only after a time lag, particularly if the infected conspecific was asymptomatic (Bouwman and Hawley 2010).

The flipside involves animals not learning to utilize novel good resources. The problem here is if animals avoid a novel good option, they do not experience it, and thus might be unable to learn that it is, in fact, a novel valuable option. A parallel problem arises if animals avoid safe options. If their previously adaptive cue-response system tells them that the option is dangerous and they avoid encountering it, they might not gain the experience required to learn that it is, in fact, safe. This can help explain the observation that some animals have not habituated to non-dangerous humans. Their flight-initiation distance might be large enough that they flee before humans are close enough to initiate an attack, and thus do not learn that humans do not attack them.

A third type of error associated with environmental change that, in principle, animals should learn to adjust to involves changes not in the quality of a particular option, but in the quality of the overall environment. When animals are searching for a new home (or a mate or food item), a key prediction is that they should be less choosy if the background availability of high-quality options is reduced, as often happens with human-induced habitat change. If, however, animals do not know that the overall environment has become worse, they will retain their previously adaptive (but now too high) level of choosiness until they learn that general conditions have worsened. For example, an animal that is searching for a new territory in a degraded habitat could learn, as it encounters poor habitat, that the overall environment is not as good as expected. However, if it evolved in a world with substantial spatial variation in patch quality, but low temporal variation in overall average habitat quality, the animal should not assess that the world has become worse; it should instead chalk up its experience to just bad luck and continue searching. If, with habitat degradation, the cost of searching is high (e.g. high mortality while searching), the time taken to learn and adjust might be very costly (Crowley et al. 2019).

How can animals get around these constraints on learning? For lethal situations, one way that animals can learn to avoid them is by generalizing from a similar non-lethal situation (see also Chapters 11 and 12). If they survive an attack from a similar, but less lethal, novel predator, they might generalize to also avoid other, more lethal, novel predators. If an animal eats a nausea-inducing, but non-lethal, small cane toad, the hope is that they will generalize to avoid consuming a larger, lethally toxic one (Ward-Fear et al. 2016; Shine 2017; see also Chapter 6). Note that evolutionary history can play a role in shaping the tendency for animals to generalize. While generalizing can allow organisms to avoid novel dangers that are at least somewhat similar to familiar ones, or facilitate the adoption of novel resources that resemble familiar ones (the 'cue similarity hypothesis'), over-generalizing can be costly. It can cause animals to adopt maladaptive foods or habitats, or over-avoid safe situations. For example, animals with a history of human hunting can over-avoid safe humans and human presence long after hunting has ceased (see Chapter 7).

For learning to utilize novel resources that were initially avoided, animals need to sample and experience those resources. Again, the species' past history of costs and benefits of sampling can explain why some might sample, and thus be more likely to learn to adopt, novel resources. Such animals should be more likely to sample novel options if they were, indeed, often beneficial, particularly if the costs of sampling have been low. If, however, novel options have rarely been beneficial, and particularly if the costs of sampling have historically been very high (e.g. novel foods are often toxic,

or novel habitats are dangerous), then we expect animals to be less likely to learn to adopt novel, valuable resources.

Alternatively, social learning can be an important way for animals to safely learn to avoid novel dangers or to adopt novel resources. Although behavioural ecologists have often highlighted the role of social learning in facilitating the spread of behavioural innovations—for example, the spread of milk bottle-opening in blue tits *Cyanistes caeruleus* (Lefebvre 1995) or the spread of bin opening by sulphur-crested cockatoos *Cacatua galerita* (Klump et al. 2021)—it is important to note that social learning can either facilitate or hinder adaptive responses to environmental change. A survey of examples of social learning in response to environmental change (Barrett et al. 2019a) found examples of social learning of danger, but also social facilitation (e.g. via conspecific attraction) of the use of poor habitat (social traps) or of unhealthy foods (e.g. plastics; Pollack et al. 2022b). Theory suggests that whether social learning helps or hinders population response to HIREC depends on the group's social structure and dynamics (e.g. group size and social networks) that influence social learning strategies— who learns from whom (Barrett et al. 2019a). Some major forms of social learning, such as offspring learning from parents or conformity-based social learning where individuals adopt the group norm, tend to slow the spread of new (initially rare) adaptive behavioural responses to environmental change (Barrett et al. 2019a). In contrast, learning from successful individuals or from same-age individuals have better potential to spread new useful behavioural innovations.

Again, evolutionary history and past selection pressures have likely shaped both the group's social structure and the tendency of individuals to rely on individual versus social learning after environmental change. Individuals in a population are more likely to rely on social learning if the cost of individual learning has been high. If juveniles learning on their own has not just been time- and energy-consuming, but also highly risky, we would expect a greater population reliance on social learning (e.g. from parents or the group norm). Social learning is also favoured in a stable environment where the cost of being out-of-date after environmental change is less. In contrast, if a population has been in an unstable environment where social learning is often out-of-date—and behavioural innovations have frequently been required to thrive—the population is more likely to include individual learners who can, at least potentially, respond rapidly to environmental change.

19.5 Behaviour as a component of a multi-trait response to multiple stressors

Up to this point, the focus has largely been on behavioural responses to novel stressors one at a time. Environmental change, however, often requires organisms to exhibit multi-trait responses to multiple novel stressors (Lopez et al. 2023; Chapter 10). In this context, a key issue is to understand when exposure to multiple stressors produces negative synergistic impacts. While a large body of research has examined physiological responses to multiple stressors (e.g. Sokolova 2013; Todgham and Stillman 2013; Chapter 10), I focus here on four behavioural trade-offs that mediate the impact of multiple stressors on fitness (Figure 19.3; see Lopez et al. 2023). Importantly, these behavioural trade-offs generate three under-emphasized insights: (1) behavioural responses can produce strong indirect costs that are generally ignored in laboratory studies of physiological responses; (2) spatial and temporal correlations between stressors, resources, and other risks can play an important role in influencing how behaviour mediates the overall response to multiple stressors; and (3) behaviourally mediated multi-trait responses can produce 'fitness cliffs'—situations where a relatively small increase in stressor levels results in a large decrease in fitness (see Lopez et al. 2023).

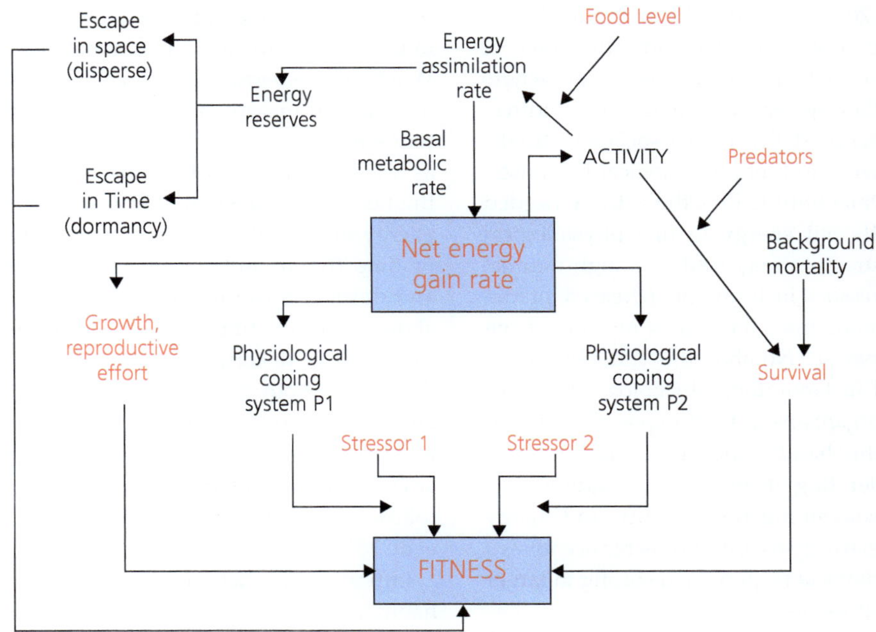

Figure 19.3 An overview on how four behavioural/life-history trade-offs mediate the impact of two stressors on fitness. The two stressors (1 and 2) have negative impacts on fitness. Animals can avoid those stressors; however, often with trade-offs where avoiding one increases exposure to the other. To reduce the negative impacts of stressors, animals can build and maintain physiological coping systems (P1 and P2); however, those require energy. To get energy, animals typically need to actively forage, which often exposes them to predation risk. Energy used for coping physiologically diverts energy away from growth and reproduction. Finally, animals can escape from stressors in space (dispersal) or time dormancy); however, doing so typically has its own costs, and requires advance investment in energy. Organisms have presumably evolved previously adaptive decision systems for balancing these multiple, behaviourally mediated trade-offs in responding to multiple stressors. A key question is: how well does this complex system respond to HIREC?

First, when animals attempt to avoid stressors in space or time, as discussed earlier, a trade-off arises if the stressors are negatively correlated, where avoiding one increases exposure to the other. A second trade-off emerges from the fact that to deal with stress, animals typically need more energy to mount their physiological responses (Romero et al. 2009). Importantly, to acquire the energy to cope with stressors can require organisms to adopt riskier behaviours (e.g. higher activity, longer foraging bouts, increased time in patches with high food but high risk). For instance, salamander larvae *Ambystoma barbouri* increased activity and reduced shelter use when exposed to higher pesticide concentrations (Rohr et al. 2004). Conversely, because animals often respond to high predation risk by exhibiting safer behaviours that reduce energy intake, this can constrain the energy available to cope with stressors physiologically. Predation risk per se can also induce physiological stress in prey (Clinchy et al. 2013; Zanette and Clinchy 2020).

Given these conflicting demands, the costs of stressors can involve a mix of direct costs, where the stressors themselves cause harm, and indirect costs (e.g. higher predation risk) associated with the need to get energy to fuel physiological responses. When food availability is low, the activity needed to acquire sufficient energy to fuel physiological responses to stressors may lead to a high indirect cost of the stressors in terms of increased predation risk instead of direct damage or mortality from the stressors per se. Notably, this indirect cost is not addressed in laboratory physiological studies where study organisms are not exposed to predators. On the other hand, if increased activity causes a sharp, accelerating increase in predation risk, then safety needs can constrain activity (and energy intake) to be relatively low, and thus reduce investment in physiological responses, resulting in larger direct costs of stressors.

A third behaviourally mediated trade-off emerges from the fact that physiological stressors divert energy that could otherwise be allocated to growth and reproduction (Rohr et al. 2004; Portner and Knust 2007). An important threshold effect exists when a small reduction in investment in a given demand results in a large decrease in fitness. Being near such a threshold could constrain organisms to allocate sufficient energy to a given demand to prevent falling over a 'fitness cliff'. For example, if offspring fitness is a strongly non-linear (e.g. sigmoidal) function of parental investment, this can cause parents to invest more into reproduction and less in coping with stressors, thus yielding larger direct costs of stressors. For males, mating success often depends heavily on possessing either large relative size or ornaments; in these cases, males can face a fitness cliff where reduced investment in sexually selected traits can result in little or no mating success. Strong sexual selection could then favour males diverting their limited energy into sexually selected traits, even at the cost of reduced investment in physiological responses to stressors. This scenario would result in a strong direct cost of exposure to stressors (e.g. mortality).

Finally, some organisms can escape environmental stressors in space or time through long-distance dispersal or some form of substantial, relatively long-term reduction in metabolic demands. Whether dispersal or dormancy is adaptive depends on the organism's fitness in both the current environment and in alternative environments, as well as on costs of escape (e.g. mortality and the need for a substantial front-end investment in energy stores that increases escape success) in space or time (Bonte and Dahirel 2017). Both dispersal and dormancy can involve substantial uncertainty about expected fitness. For long-distance dispersal, there is uncertainty about transit costs (that depend on both cost per unit of distance or time, and distance relative to mobility) and often great uncertainty over likely payoffs in prospective new environments. This might be especially true now, following habitat and climate change (Crowley et al. 2019).

Importantly, when organisms disperse to a new habitat, they might face a different set of stressors that require a different set of behavioural and physiological responses. Thus, the suitability of a new environment could hinge on the organism's plasticity in behaviour and physiology. In addition, because escaping to a new environment might also result in greater competition or predation risk, there is a game-theoretic aspect to this dynamic that can further complicate expectations.

This section has outlined how behaviour might often mediate a multi-trait response to multiple stressors. If organisms have had time to adapt to this complex scenario, they can, in principle, balance the multiple trade-offs reasonably adaptively. However, with novel circumstances associated with environmental change, it seems likely that there will be significantly more opportunities for mismatch between previously adaptive cue-response systems and post-HIREC conditions. Habitat change, climate change, and the introduction of novel risks and resources, including anthropogenic factors, has almost certainly made it more likely that animals will do a poor job of avoiding or escaping stressors in space or time. With greater complexity comes greater potential for evolutionary mismatch. However, even in the face of all this complexity, some animals are thriving. The relative roles of fortuitous pre-adaptation to HIREC (i.e. an evolutionary match) versus learning/social learning and evolution in explaining these 'success stories' needs more study (see Chapter 14).

19.6 Animal personalities, fear generalization, and neophobia

Numerous recent reviews have discussed the potential importance of consistent individual differences (CIDs) in behaviour (i.e. animal personalities, as observed, for example, through differences in boldness, aggressiveness, exploratory tendency, or sociability) in explaining variation in how animals respond to aspects of environmental change (e.g. Diaz Pauli and Sih 2017; Merrick and Koprowski 2017; Garvey et al. 2020; Goumas et al. 2020; Sadoul et al. 2021; Gunn et al. 2022). One repeated theme is that environmental change is often associated with animals being bolder or more exploratory, though that result clearly depends on the species and ecological context (Gunn et al. 2022). In some cases, HIREC is associated with a suite of correlated behaviours (i.e. a behavioural syndrome: Sih et al. 2004). For example, successful invaders into new habitats have been found to be bolder, more exploratory, and more aggressive (Chapple et al. 2012; Sih et al. 2012). Sadoul et al. (2021) added the intriguing idea that when animals living with humans experience reduced predation risk

(e.g. associated with urbanization, eco-tourism, and domestication), this favours a 'preactive' coping style where animals are bold, but unaggressive.

A relatively new twist on the concept of correlated CIDs in behaviour focuses on correlations in response to multiple threats that can be very different (e.g. predation risk versus chemical stressors; Sih et al. 2023). While the previous section on multiple stressors discussed behavioural avoidance of multiple types of threats, except for the large literature on avoidance of predation risk, the literature on behavioural avoidance of other threats is actually not well developed (Sih et al. 2023). In particular, while hundreds of studies have quantified consistent individual differences (CIDs) in boldness versus fearfulness relative to predation risk, very few studies on non-humans have examined CIDs in avoidance of other major threats, such as pathogens, human harvesting of wildlife, moving vehicles, chemical stressors, or fire. Interestingly, the only extensive literature on CIDs in avoidance of several of these stressors is on humans—for example, on disgust sensitivity as avoidance of pathogen risk (Tybur et al. 2018) and multiple chemical sensitivity (Zucco and Doty 2022). In the context of multiple threats, a key issue is whether fear generalizes. For example, is relative fear of predation risk correlated to relative fear of pathogens, moving vehicles, chemical stressors, or fire cues (smoke)? Here, I follow ecologists in using 'fear' (e.g. the landscape of fear) without implying deep emotional connotation, but instead as quantifiable avoidance of threats. Faced with conflicting demands, and often imperfect information, animals often either under-avoid or over-avoid risks. More fearful individuals tend to over-avoid, while less fearful ones under-avoid risks.

Why should fear generalization be particularly important in the context of HIREC? In parallel with the behavioural syndrome literature (Sih et al. 2004), if behaviours are correlated (either within or between individuals), then evolutionary or personal experience with one threat that affects an animal's fear of that threat can spill over to affect its response to multiple other threats. In humans, an example of generalized fear is post-traumatic stress disorder (PTSD), where a salient bad experience with one threat carries over to increase fear and anxiety of not just that threat, but also others (Cooper et al. 2022; Dunsmoor and Murphy 2015; Dunsmoor and Paz 2015). For animals facing novel threats in a human-altered world, fear generalization can be beneficial. If fear generalizes, then animals that have evolved with native predation risk that has made them relatively fearful could be more ready to respond appropriately to novel predators. Interestingly, however, to date, very few studies have examined whether CIDs in fear of native predators is actually correlated to CIDs in fear of exotic predators. This is an open question begging for further study. Along similar lines, although numerous studies have measured antipredator behaviour by quantifying flight-initiation distances (FIDs) in response to a harmless human (typically the investigator; Stankowich and Blumstein 2005; Samia et al. 2016; Samia et al. 2019), surprisingly few studies have quantified whether CIDs in fear of humans is actually correlated to fear of a real predator (Bar-Ziv et al. 2022). Focusing on the flipside, boldness, numerous studies have noted that urban animals tend to be bolder than their rural counterparts. Their lower fear of humans can allow them to live in urban environments, but does it also result in poor avoidance of urban predators, or cars, or pathogens? Along these lines, when animals habituate to humans (in urban environments, or in eco-tourism situations), do they also get less fearful of actual predators or other threats? This worry has been discussed (Geffroy et al. 2015; Trimmer et al. 2017), but not widely studied.

Finally, when faced with novel situations (threats or opportunities), relative neophobia, the fear of novelty per se, can obviously be important. For non-human animals, most of the work on neophobia has been on responses to novel objects (often placed near food) and exploration of novel space (Crane and Ferrari 2017). The adaptive value of neophobia depends on the type of novel situation, on whether novel situations of that type tend to be dangerous versus beneficial, and on the cost of over- versus under-avoiding dangerous versus beneficial novel situations (Crane et al. 2020). Thus, tendencies to be neophobic are thought to relate to major aspects of the species' evolutionary history, or the individual's personal history. Animals living in risky situations tend to be neophobic about unfamiliar objects and

stimuli (Brown et al. 2014) and those living in environments with toxic foods tend to avoid novel food (Ducatez et al. 2015). Neophilia is a related but distinct behavioural tendency where novelty is actively sought. Interestingly, a given individual can be neophobic about some aspects of novelty, but neophilic about others (Mettke-Hofmann 2014). For instance, migratory animals are often more likely to explore novel spaces, but may avoid unfamiliar local cues within that space (Mettke-Hofmann 2014). Additional study of neophobia/neophilia in the context of HIREC should clearly prove rewarding.

19.7 Concluding remarks

In this closing chapter, I offered conceptual frameworks for organizing and hopefully enhancing our understanding about variation in how animals respond behaviourally to novel or altered threats in the modern world. From an empirical perspective, several repeated themes on future directions emerged from other chapters. First, while many studies (and chapters in this book) have focused on one primary stressor, we clearly need more experimental work on how animals respond behaviourally to multiple stressors (see also Chapter 10). These studies would benefit from a stronger focus on ecologically relevant spatial and temporal variation in these stressors. Experiments contrasting outcomes when animals are exposed to different, but constant, temperatures, or light, sound, or chemical levels, is a start, but real organisms face combinations of complex, spatiotemporally varying temperature, light, sound, and chemical regimes. Second, more studies on abiotic stressors should be conducted in the context of common, natural biotic stressors (e.g. predators and other 'enemies', food limitation, and/or social stress). Third, given spatial variation in these stressors, a key behaviour that deserves more attention is habitat and microhabitat choice that influences exposure to these multiple stressors. Environment choice and associated plasticity can have major effects on individual fitness, which have potential impacts on longer-term, larger-scale ecological and evolutionary outcomes. Fourth, as noted in several chapters, there is a strong need

---

Box 19.1 A guide for choosing what to study

With so many species, so many stressors, and so many ways that behaviour can respond to those stressors, how should an animal behaviourist choose what to study? The complexity of it all may seem daunting. In this Box, I offer a simple, step-by-step framework for choosing a focus amidst the complexity (see Sih and Gleeson 1995). The over-arching theme is to focus on what is quantifiably important (which depends on the scale of your goal) and also on what is interesting in the sense of unsolved but solvable. To do this, draw on existing methods for assessing what is important. If you are connecting to community/ecosystem ecology, then why not study *keystone species* or *foundational ecosystem engineers*, species that are quantifiably important and threatened by environmental change? If you are connecting to population ecology, either for a species of community/ecosystem importance or for a species of conservation or other interest, then tap into ecology's long-standing focus on *limiting factors* (sometimes referred to as key factors). These are the factors that limit the population's performance. In addition, focus, in particular, on key life-history stages (identified, for example, by elasticity analyses; Caswell 2006) that have the largest effect on population success. Elasticity analyses have, for example, often found that for long-lived species, survival of young adults is far more important than survival of hatchlings (e.g. Crouse et al. 1987). Having identified the key factors that limit success

in a key life-history stage, we can then ask: are there key *'limiting behaviours'* that explain why individuals are limited by that key factor in that key life history stage? For example, if the species' ecology shows (or least strongly suggests) that predation on dispersing young adults, particularly those that were exposed to chemical stressors early in life, is the key limiting factor/stage, then study the anti-predator behaviour of those young adults and how and why early exposure to chemicals affects their anti-predator behaviour. The key limiting traits (here, behaviours) in any system are inherently intellectually interesting because if that trait limits the individual's fitness, there should likely be strong selection favouring an improvement in that behaviour. Why are these animals still exhibiting *the* behaviours that most importantly limit their fitness? What constraints (e.g. genetic, developmental, physiological, phylogenetic, or informational?) limit their ability to evolve or learn to exhibit more effective behaviour? While there are myriad motivations underlying why animal behaviourists choose the systems and questions that they study, this limiting traits framework, when applied well, should be useful for enhancing 'return on investment' with regard to increased understanding of how behavioural responses affect ecological and evolutionary outcomes in the modern world.

---

for more studies that actually bridge behaviour, fitness, and larger ecological and evolutionary outcomes. Ideally, these studies could address eco-evolutionary-plasticity feedbacks. All of the above would benefit from a better understanding of mechanisms (cognitive, sensory, physiological, neuroendocrine, developmental, and genetic) underlying key behaviours. Understanding mechanisms, in this regard, can be critically important for management. Finally, understanding human behaviour at individual and societal levels will, of course, be useful because it is our behaviour that ultimately determines the environmental changes that organisms experience (Chapter 18). An exciting, developing field examines the reciprocal feedbacks between human behaviour, environmental change, and animal behavioural responses to environmental change.

Of course, doing all of the above is a tall order. Fortunately, a bevy of relatively new technologies can help get better data than in the past, such as tracking individuals and their behaviour in nature, and ideally, simultaneously recording spatiotemporal variation in key environmental factors. Even as HIREC places more organisms under stressful conditions, with the recent technological progress, animal behaviourists can at least be excited and optimistic about the quality of the data we can collect in future decades. Of course, even with new tools, we clearly cannot and should not attempt to study all behaviours in all organisms in the modern, human-altered world. In Box 19.1, I offer my guide for choosing what to study. With ongoing feedback between conceptual frameworks and empirical study, the hope is that we will understand enough to provide useful guidance for managing nature in the face of human-induced environmental change.

Acknowledgements

Many thanks to Bob Wong and Ulrika Candolin for allowing me to read the other chapters in preparation for writing my chapter. The ideas in my chapter were shaped by numerous stimulating conversations with the Sih Lab's students and postdocs, with other collaborators, as well as with my wife, Caitlin McGaw. This work was supported in part by a grant from the National Science Foundation IOS 1456724.

References

Anton, A., Gerardi, N.R., Ricciardi, A., et al. (2020). Global determinants of prey naivete to exotic predators. *Proceedings of the Royal Society B: Biological Sciences*, 287, 198.

Barrett, L.P., Stanton, L.A., and Benson-Amram, S. (2019b). The cognition of 'nuisance' species. *Animal Behaviour*, 147, 167–177.

Barrett, B., Zepeda, E., Pollack, L., et al. (2019a). Counterculture: does social learning help or hinder adaptive response to human-induced rapid environmental change? *Frontiers in Ecology and Evolution*, 7.

Bar-Ziv, M., Sofer, A., Gorovoy, A., and Spiegel, O. (2022). Beyond simple habituation: anthropogenic habitats influence the escape behaviour of spur-winged lapwings in response to both human and non-human threats. *Journal of Animal Ecology*, 92, 417–429.

Berger-Tal, O., Blumstein, D., and Swaisgood, R. (2020). Conservation translocations: a review of common difficulties and promising directions. *Animal Conservation*, 23, 121–131.

Bertram, M., Martin, J., McCallum, E., et al. (2022). Frontiers in quantifying wildlife behavioural responses to chemical pollution. *Biological Reviews*, 97, 1346–1364.

Bonte, D., and Dahirel, M. (2017). Dispersal: a central and independent trait in life history. *Oikos*, 126, 472–479.

Bouwman, K., and Hawley, D. (2010). Sickness behaviour acting as an evolutionary trap? Male house finches preferentially feed near diseased conspecifics. *Biology Letters*, 6, 462–465.

Brodin, T., Piovano, S., Fick, J., et al. (2014). Ecological effects of pharmaceuticals in aquatic systems—impacts through behavioural alterations. *Philosophical Transactions of the Royal Society B: Biological Sciences*, 369.

Brown, G., Chivers, D., Elvidge, C., et al. (2014). Background level of risk determines the intensity of predator neophobia in juvenile convict cichlids. *Behavioral Ecology and Sociobiology*, 68, 127–133.

Candolin, U., and Wong, B. (2019). Mate choice in a polluted world: consequences for individuals, populations and communities. *Philosophical Transactions of the Royal Society B: Biological Sciences*, 374.

Carthey, A., and Blumstein, D. (2018). Predicting predator recognition in a changing world. *Trends in Ecology & Evolution*, 33, 106–115.

Caspi, T., Johnson, J., Lambert, M., et al. (2022). Behavioral plasticity can facilitate evolution in urban environments. *Trends in Ecology & Evolution*, 37, 1092–1103.

Caswell, H. (2006). *Matrix Population Models: Construction, Analysis and Interpretation*, 2nd edn. Oxford University Press, Oxford.

Chapple, D., Simmonds, S., and Wong, B. (2012). Can behavioral and personality traits influence the success of unintentional species introductions? *Trends in Ecology & Evolution*, 27, 57–64.

Clinchy, M., Sheriff, M., and Zanette, L. (2013). Predator-induced stress and the ecology of fear. *Functional Ecology*, 27, 56–65.

Cooper, S., van Dis, E., Hagenaars, M., et al. (2022). A meta-analysis of conditioned fear generalization in anxiety-related disorders. *Neuropsychopharmacology*, 47, 1652–1661.

Crane, A., Brown, G., Chivers, D., and Farrar, M. (2020). An ecological framework of neophobia: from cells to organisms to populations. *Biological Reviews*, 95, 218–231.

Crane, A., and Ferrari, M. (2017). Patterns of predator neophobia: a meta-analytic review. *Proceedings of the Royal Society B: Biological Sciences*, 284.

Crouse, D., Crowder, L., and Caswell, H. (1987). A stage-based population model for loggerhead sea turtles and implications for conservation. *Ecology*, 68, 1412–1423.

Crowley, P., Trimmer, P., Spiegel, O., et al. (2019). Predicting habitat choice after rapid environmental change. *The American Naturalist*, 193, 619–632.

Diaz Pauli, B., and Sih, A. (2017). Behavioural responses to human-induced change: why fishing should not be ignored. *Evolutionary Applications*, 10, 231–240.

Ducatez, S., Clavel, J., and Lefebvre, L. (2015). Ecological generalism and behavioural innovation in birds: technical intelligence or the simple incorporation of new foods? *Journal of Animal Ecology*, 84, 79–89.

Dunsmoor, J., and Murphy, G. (2015). Categories, concepts, and conditioning: how humans generalize fear. *Trends in Cognitive Sciences*, 19, 73–77.

Dunsmoor, J., and Paz, R. (2015). Fear generalization and anxiety: behavioral and neural mechanisms. *Biological Psychiatry*, 78, 336–343.

Ehlman, S., Trimmer, P., and Sih, A. (2019). Prey responses to exotic predators: effects of old risks and new cues. *The American Naturalist*, 193, 575–587.

Garcia, T., and Sih, A. (2003). Color change and color-dependent behavior in response to predation risk in the salamander sister species *Ambystoma barbouri* and *Ambystoma texanum*. *Oecologia*, 137, 131–139.

Garvey, P., Banks, P., Suraci, J., et al. (2020). Leveraging motivations, personality, and sensory cues for vertebrate pest management. *Trends in Ecology & Evolution*, 35, 990–1000.

Gaynor, K., Hojnowski, C., Carter, N., and Brashares, J. (2018). The influence of human disturbance on wildlife nocturnality. *Science*, 360, 1232.

Geffroy, B., Samia, D., Bessa, E., and Blumstein, D. (2015). How nature-based tourism might increase prey vulnerability to predators. *Trends in Ecology & Evolution*, 30, 755–765.

Gehr, B., Hofer, E., Pewsner, M., et al. (2018). Hunting-mediated predator facilitation and superadditive mortality in a European ungulate. *Ecology and Evolution*, 8, 109–119.

Gilroy, J., and Sutherland, W. (2007). Beyond ecological traps: perceptual errors and undervalued resources. *Trends in Ecology & Evolution*, 22, 351–356.

Goumas, M., Lee, V., Boogert, N., et al. (2020). The role of animal cognition in human-wildlife interactions. *Frontiers in Psychology*, 11, 589978.

Gravinese, P., Page, H., Butler, C., et al. (2020). Ocean acidification disrupts the orientation of postlarval Caribbean spiny lobsters. *Scientific Reports*, 10.

Greggor, A.L., Trimmer, P.C., Barrett, B.J., and Sih, A. (2019). Challenges of learning to escape evolutionary traps. *Frontiers in Ecology and Evolution*, 7, 408.

Guiden, P., Bartel, S., Byer, N., et al. (2019). Predator-prey interactions in the Anthropocene: reconciling multiple aspects of novelty. *Trends in Ecology & Evolution*, 34, 616–627.

Gunn, R., Hartley, I., Algar, A., et al. (2022). Understanding behavioural responses to human-induced rapid environmental change: a meta-analysis. *Oikos*, 2022, 08366.

Hale, R., and Swearer, S. (2016). Ecological traps: current evidence and future directions. *Proceedings of the Royal Society B: Biological Sciences*, 283, 20152647.

Horvath, G., Kriska, G., and Robertson, B. (2014). Anthropogenic polarization and polarized light pollution inducing polarized ecological traps. In: G. Horvath (ed.), *Polarized Light and Polarization Vision in Animal Sciences*, pp. 443–513. Springer Series in Vision Research, Springer Nature Switzerland, Cham.

Huey, R., Kearney, M., Krockenberger, A., et al. (2012). Predicting organismal vulnerability to climate warming: roles of behaviour, physiology and adaptation. *Philosophical Transactions of the Royal Society B: Biological Sciences*, 367, 1665–1679.

Jellison, B., Ninokawa, A., Hill, T., et al. (2016). Ocean acidification alters the response of intertidal snails to a key sea star predator. *Proceedings of the Royal Society B: Biological Sciences*, 283, 20160890.

Kamrowski, R., Limpus, C., Moloney, J., and Hamann, M. (2012). Coastal light pollution and marine turtles: assessing the magnitude of the problem. *Endangered Species Research*, 19, 85–98.

Kearney, M., Shine, R., and Porter, W. (2009). The potential for behavioral thermoregulation to buffer 'cold-blooded' animals against climate warming. *Proceedings of the National Academy of Sciences*, 106, 3835–3840.

Klump, B.C., Martin, J.M., Wild, S., et al. (2021). Innovation and geographic spread of a complex foraging culture in an urban parrot. *Science*, 373, 456–460.

Lefebvre, L. (1995). The opening of milk bottles by birds—evidence for accelerating learning rates, but against the wave-of-advance model of cultural transmission. *Behavioural Processes*, 34, 43–53.

Lopez, L., Gil, M., Crowley, P., et al. (2023). Integrating animal behaviour into research on multiple environmental stressors: a conceptual framework. *Biological Reviews*, 98, 1345–1364.

McCallum, E., Dey, C., Cerveny, D., et al. (2021). Social status modulates the behavioral and physiological consequences of a chemical pollutant in animal groups. *Ecological Applications*, 31, 2454.

McComb, K., Shannon, G., Sayialel, K., and Moss, C. (2014). Elephants can determine ethnicity, gender, and age from acoustic cues in human voices. *Proceedings of the National Academy of Sciences*, 111, 5433–5438.

McEwen, B., and Wingfield, J. (2003). The concept of allostasis in biology and biomedicine. *Hormones and Behavior*, 43, 2–15.

McMahon, T., Rohr, J., and Bernal, X. (2017). Light and noise pollution interact to disrupt interspecific interactions. *Ecology*, 98, 1290–1299.

Merrick, M., and Koprowski, J. (2017). Should we consider individual behavior differences in applied wildlife conservation studies? *Biological Conservation*, 209, 34–44.

Mettke-Hofmann, C. (2014). Cognitive ecology: ecological factors, life-styles, and cognition. *Wiley Interdisciplinary Reviews: Cognitive Science*, 5, 345–360.

Moreira-Santos, M., Ribeiro, R., and Araujo, C. (2019). What if aquatic animals move away from pesticide-contaminated habitats before suffering adverse physiological effects? A critical review. *Critical Reviews in Environmental Science and Technology*, 49, 989–1025.

Orr, J., Luijckx, P., Arnoldi, J., et al. (2022). Rapid evolution generates synergism between multiple stressors: linking theory and an evolution experiment. *Global Change Biology*, 28, 1740–1752.

Patricelli, G., and Blickley, J. (2006). Avian communication in urban noise: causes and consequences of vocal adjustment. *Auk*, 123, 639–649.

Peacor, S., Barton, B., Kimbro, D., et al. (2020). A framework and standardized terminology to facilitate the study of predation-risk effects. *Ecology*, 101, 3152.

Pearse, I., Harris, D., Karban, R., and Sih, A. (2013). Predicting novel herbivore-plant interactions. *Oikos*, 122, 1554–1564.

Pollack, L., Munson, A., Savoca, M., et al. (2022a). Enhancing the ecological realism of evolutionary mismatch theory. *Trends in Ecology & Evolution*, 37, 233–245.

Pollack, L., Munson, A., Zepeda, E., et al. (2022b). Variation in plastic consumption: social group size enhances individual susceptibility to an evolutionary trap. *Animal Behaviour*, 192, 171–188.

Portner, H., and Knust, R. (2007). Climate change affects marine fishes through the oxygen limitation of thermal tolerance. *Science*, 315, 95–97.

Robertson, B.A., and Blumstein, D.T. (2019). How to disarm an evolutionary trap. *Conservation Science and Practice*, 2019, e116.

Robertson, B., Rehage, J., and Sih, A. (2013). Ecological novelty and the emergence of evolutionary traps. *Trends in Ecology & Evolution*, 28, 552–560.

Rohr, J., Elskus, A., Shepherd, B., et al. (2004). Multiple stressors and salamanders: effects of an herbicide, food limitation, and hydroperiod. *Ecological Applications*, 14, 1028–1040.

Romero, L., Dickens, M., and Cyr, N. (2009). The reactive scope model—a new model integrating homeostasis, allostasis, and stress. *Hormones and Behavior*, 55, 375–389.

Saaristo, M., Brodin, T., Balshine, S., et al. (2018). Direct and indirect effects of chemical contaminants on the behaviour, ecology and evolution of wildlife. *Proceedings of the Royal Society B: Biological Sciences*, 285, 20181297.

Sadoul, B., Blumstein, D., Alfonso, S., and Geffroy, B. (2021). Human protection drives the emergence of a new coping style in animals. *PLOS Biology*, 19, 3001186.

Samia, D., Bessa, E., Blumstein, D., et al. (2019). A meta-analysis of fish behavioural reaction to underwater human presence. *Fish and Fisheries*, 20, 817–829.

Samia, D., Blumstein, D., Stankowich, T., and Cooper, W. (2016). Fifty years of chasing lizards: new insights advance optimal escape theory. *Biological Reviews*, 91, 349–366.

Santos, R., Machovsky-Capuska, G., and Andrades, R. (2021). Plastic ingestion as an evolutionary trap: toward a holistic understanding. *Science eLetters*, 373, 56.

Savoca, M., McInturf, A., and Hazen, E. (2021). Plastic ingestion by marine fish is widespread and increasing. *Global Change Biology*, 27, 2188–2199.

Savoca, M., Wohlfeil, M., Ebeler, S., and Nevitt, G. (2016). Marine plastic debris emits a keystone infochemical for olfactory foraging seabirds. *Science Advances*, 2, 1600395.

Senzaki, M., Kadoya, T., and Francis, C. (2020). Direct and indirect effects of noise pollution alter biological communities in and near noise-exposed environments. *Proceedings of the Royal Society B: Biological Sciences*, 287, 20200176.

Shine, R. (2010). The ecological impact of invasive cane toads (*Bufo marinus*) in Australia. *The Quarterly Review of Biology*, 85, 253–291.

Shine, R. (2017). New weapons in the toad toolkit. *The Quarterly Review of Biology*, 92, 125–149.

Sih, A. (2013). Understanding variation in behavioural responses to human-induced rapid environmental change: a conceptual overview. *Animal Behaviour*, 85, 1077–1088.

Sih, A., Bell, A., and Johnson, J. (2004). Behavioral syndromes: an ecological and evolutionary overview. *Trends in Ecology & Evolution*, 19, 372–378.

Sih, A., Bolnick, D., Luttbeg, B., et al. (2010). Predator-prey naivete, antipredator behavior, and the ecology of predator invasions. *Oikos*, 119, 610–621.

Sih, A., Chung, H.J., Neylan, I., et al. (2023). Fear generalization and behavioral responses to multiple dangers. *Trends in Ecology & Evolution*, 38, 269–380.

Sih, A., Cote, J., Evans, M., et al. (2012). Ecological implications of behavioural syndromes. *Ecology Letters*, 15, 278–289.

Sih, A., Ferrari, M., and Harris, D. (2011). Evolution and behavioural responses to human-induced rapid environmental change. *Evolutionary Applications*, 4, 367–387.

Sih, A., and Gleeson, S.K. (1995). A limits-oriented approach to evolutionary ecology. *Trends in Ecology & Evolution*, 10, 378–382.

Slabbekoorn, H., and Peet, M. (2003). Ecology: birds sing at a higher pitch in urban noise—great tits hit the high notes to ensure that their mating calls are heard above the city's din. *Nature*, 424, 267.

Smith, J., Gaynor, K., and Suraci, J. (2021). Mismatch between risk and response may amplify lethal and nonlethal effects of humans on wild animal populations. *Frontiers in Ecology and Evolution*, 9.

Sokolova, I. (2013). Energy-limited tolerance to stress as a conceptual framework to integrate the effects of multiple stressors. *Integrative and Comparative Biology*, 53, 597–608.

Stankowich, T., and Blumstein, D. (2005). Fear in animals: a meta-analysis and review of risk assessment. *Proceedings of the Royal Society B: Biological Sciences*, 272, 2627–2634.

Todgham, A., and Stillman, J. (2013). Physiological responses to shifts in multiple environmental stressors: relevance in a changing world. *Integrative and Comparative Biology*, 53, 539–544.

Trimmer, P., Ehlman, S., and Sih, A. (2017). Predicting behavioural responses to novel organisms: state-dependent detection theory. *Proceedings of the Royal Society B: Biological Sciences*, 284, 20162108.

Tybur, J., Cinar, C., Karinen, A., and Perone, P. (2018). Why do people vary in disgust? *Philosophical Transactions of the Royal Society B: Biological Sciences*, 373, 20170204.

Walther, G., Roques, A., Hulme, P., et al. (2009). Alien species in a warmer world: risks and opportunities. *Trends in Ecology & Evolution*, 24, 686–693.

Ward-Fear, G., Pearson, D., Brown, G., et al. (2016). Ecological immunization: in situ training of free-ranging predatory lizards reduces their vulnerability to invasive toxic prey. *Biology Letters*, 12, 20150863.

Westneat, D., and Fox, C., eds (2010). *Evolutionary Behavioral Ecology*. Oxford University Press, Oxford.

Wingfield, J. (2013). The comparative biology of environmental stress: behavioural endocrinology and variation in ability to cope with novel, changing environments. *Animal Behaviour*, 85, 1127–1133.

Wirsing, A., Heithaus, M., Brown, J., et al. (2021). The context dependence of non-consumptive predator effects. *Ecology Letters*, 24, 113–129.

Wong, B., and Candolin, U. (2015). Behavioral responses to changing environments. *Behavioral Ecology*, 26, 665–673.

Zanette, L., and Clinchy, M. (2020). Ecology and neurobiology of fear in free-living wildlife. *Annual Review of Ecology, Evolution, and Systematics*, 51, 297–318.

Zucco, G., and Doty, R. (2022). Multiple chemical sensitivity. *Brain Sciences*, 12, 12010046.

Index

Locators marked *t* refer to tables; *f* to figures and *b* to boxed content